P9-DXN-473

GENETICS AS A TOOL IN MICROBIOLOGY

Other Publications of the
*Society for General Microbiology**

THE JOURNAL OF GENERAL MICROBIOLOGY

THE JOURNAL OF GENERAL VIROLOGY

SYMPOSIA

1 THE NATURE OF THE BACTERIAL SURFACE
2 THE NATURE OF VIRUS MULTIPLICATION
3 ADAPTATION IN MICRO-ORGANISMS
4 AUTOTROPHIC MICRO-ORGANISMS
5 MECHANISMS OF MICROBIAL PATHOGENICITY
6 BACTERIAL ANATOMY
7 MICROBIAL ECOLOGY
8 THE STRATEGY OF CHEMOTHERAPY
9 VIRUS GROWTH AND VARIATION
10 MICROBIAL GENETICS
11 MICROBIAL REACTION TO ENVIRONMENT
12 MICROBIAL CLASSIFICATION
13 SYMBIOTIC ASSOCIATIONS
14 MICROBIAL BEHAVIOUR, 'IN VIVO' AND 'IN VITRO'
15 FUNCTION AND STRUCTURE IN MICRO-ORGANISMS
16 BIOCHEMICAL STUDIES OF ANTIMICROBIAL DRUGS
17 AIRBORNE MICROBES
18 THE MOLECULAR BIOLOGY OF VIRUSES
19 MICROBIAL GROWTH
20 ORGANIZATION AND CONTROL IN PROKARYOTIC AND EUKARYOTIC CELLS
21 MICROBES AND BIOLOGICAL PRODUCTIVITY
22 MICROBIAL PATHOGENICITY IN MAN AND ANIMALS
23 MICROBIAL DIFFERENTIATION
24 EVOLUTION IN THE MICROBIAL WORLD
25 CONTROL PROCESSES IN VIRUS MULTIPLICATION
26 THE SURVIVAL OF VEGETATIVE MICROBES
27 MICROBIAL ENERGETICS
28 RELATIONS BETWEEN STRUCTURE AND FUNCTION IN THE PROKARYOTIC CELL
29 MICROBIAL TECHNOLOGY: CURRENT STATE, FUTURE PROSPECTS
30 THE EUKARYOTIC MICROBIAL CELL

* Published by the Cambridge University Press, except for the first Symposium, which was published by Blackwell's Scientific Publications Limited.

GENETICS AS A TOOL IN MICROBIOLOGY

EDITED BY

S. W. GLOVER AND D. A. HOPWOOD

THIRTY-FIRST SYMPOSIUM OF THE
SOCIETY FOR GENERAL MICROBIOLOGY
HELD AT
THE UNIVERSITY OF CAMBRIDGE
APRIL 1981

Published for the Society for General Microbiology
CAMBRIDGE UNIVERSITY PRESS
CAMBRIDGE
LONDON NEW YORK NEW ROCHELLE
MELBOURNE SYDNEY

Published by the Press Syndicate of the University of Cambridge
The Pitt Building, Trumpington Street, Cambridge CB2 1RP
32 East 57th Street, New York, NY 10022, USA
296 Beaconsfield Parade, Middle Park, Melbourne 3206, Australia

First published 1981

Text set in 11/13 pt Linotron 202 Times, printed and bound
in Great Britain at The Pitman Press, Bath

British Library Cataloguing in Publication Data

Genetics as a tool in microbiology. – (Society
for General Microbiology. Symposia; 31).

1. Microbial genetics – Congresses
I. Glover, Stuart William
II. Hopwood, D A
576 QH434 80–41201
ISBN 0 521 23748 3

CONTRIBUTORS

ARST, H. N., Department of Genetics, Ridley Building, The University, Newcastle upon Tyne NE1 7RU, UK

BARTLETT, S. G., The Rockefeller University, New York, NY, USA

BOYNTON, J. E., Department of Botany, Duke University, Durham, NC, USA

DAY, P. R., Plant Breeding Institute, Maris Lane, Trumpington, Cambridge CB2 2LQ, UK

DIXON, R., ARC Unit of Nitrogen Fixation, University of Sussex, Brighton BN1 9RQ, UK

ELY, B., University of South Carolina, Columbia, South Carolina, USA

FERENCZY, L., Department of Microbiology, Attila József University, Szeged, PO Box 428, Hungary

GILLHAM, N. W., Department of Zoology, Duke University, Durham, NC, USA

GLOVER, S. W., Department of Genetics, Ridley Building, The University, Newcastle upon Tyne NE1 7RU, UK

HOPWOOD, D. A., John Innes Institute, Colney Lane, Norwich NR4 7UH, UK

KARI, C., MRC Laboratory of Molecular Biology, Hills Road, Cambridge CB2 2QH, UK

KENNEDY, C., ARC Unit of Nitrogen Fixation, University of Sussex, Brighton BN1 9RQ, UK

MAAS, W. K., Department of Microbiology, New York University Medical Center, New York, NY 10016, USA

MACE, H. A. F., MRC Laboratory of Molecular Biology, Hills Road, Cambridge CB2 2QH, UK

MÄKELÄ, P. H., Central Public Health Laboratory, SF-00280 Helsinki 28, Finland

MERRICK, M., ARC Unit of Nitrogen Fixation, University of Sussex, Brighton BN1 9RQ, UK

NISEN, P., Albert Einstein College of Medicine, Bronx, New York, USA

NURSE, P., School of Biological Sciences, The University of Sussex, Brighton, Sussex BN1 9QG, UK

PARKINSON, J. S., Biology Department, University of Utah, Salt Lake City, Utah 84112, USA

SHAPIRO, L., Albert Einstein College of Medicine, Bronx, New York, USA

SHERRATT, D., Department of Genetics, University of Glasgow, Glasgow, G11 5JS, UK

STOCKER, B. A. D., Department of Medical Microbiology, Stanford University School of Medicine, Stanford, California 94305, USA

TIMMIS, K. N., Max Planck Institute for Molecular Genetics, Berlin-Dahlem, West Germany

TRAVERS, A. A., MRC Laboratory of Molecular Biology, Hills Road, Cambridge CB2 2QH, UK

CONTENTS

EDITORS' PREFACE

The last Society symposium devoted exclusively to the genetics of micro-organisms was held twenty-one years ago (Hayes & Clowes, 1960). It is not surprising, therefore, that that symposium concentrated, in large measure, on the elucidation of genetic mechanisms in micro-organisms since the previous decade 1950–60 witnessed many important milestones in bacterial genetics including the discovery of genetic transduction and conjugation in bacteria and the description of a new class of genetic elements – the episomes (Jacob, Schaeffer & Wollman, 1960).

Twenty-one years later our understanding of these phenomena is, in many respects, almost complete. It is entirely appropriate, therefore, that this symposium concentrates on the application of genetics as a tool, an indispensable tool, in the analysis of many fundamental aspects of microbiology.

Those who edit books and organise symposia are faced with the formidable task of selection. Our aim has been to choose topics of wide interest and fundamental importance in which the application of genetic methods has led to rapid progress and new insights in recent years. The reader will judge how far this symposium achieves that aim. The limitations of time and of printed space have inevitably led to the exclusion of many interesting topics and it is our hope that these may be included in some future symposium of our Society.

We thank all our authors who laboured long to produce manuscripts in time to meet our stringent deadlines and we thank Cambridge University Press for their help in the production of this volume.

Department of Genetics
University of Newcastle upon Tyne S. W. Glover

John Innes Institute
Norwich D. A. Hopwood

MICROBIAL PROTOPLAST FUSION

L. FERENCZY

Department of Microbiology, Attila József University, Szeged, PO Box 428, Hungary

INTRODUCTION

The field of induced fusion of microbial protoplasts (i.e. cells completely deprived of the wall) is a rapidly expanding one, and many data have accumulated in both basic and applied areas since the first reports on complementation of auxotrophic mutants by controlled protoplast fusion (Ferenczy, Zsolt & Kevei, 1972; Ferenczy, Kevei & Zsolt, 1974). Results of intra- and interspecific fusion obtained in the past few years clearly indicate the possibilities and importance of this new method of genetic transfer. Different aspects of microbial protoplast fusion have been dealt with in various recent reviews (Ferenczy, Kevei & Szegedi, 1976; Peberdy, 1978, 1979a, 1980a, b; Alföldi, 1980; Fodor, Rostás & Alföldi, 1980; Cocking, 1980; Ferenczy, 1980). The aim of this paper is to give a short review of the whole area of induced protoplast fusion, from bacteria to algae, and to discuss the various consequences of genetic transfer.

The first observations on bacterial protoplasts and their fusion stem from as early as 1925 (Mellon, 1925). It might be supposed that artefacts were seen in the stained preparation; however, subsequent observations on preparations stained similarly and also claiming to demonstrate fusion events (Smith, 1944) were followed by direct examination of living material by time-lapse photography, and confirmed the validity of the finding of fusion of 'large bodies' (Dienes & Smith, 1944). This observation on *Bacteroides* strains was later corroborated, also by time-lapse microscopic photography, on *Proteus vulgaris* (Stempen & Hutchinson, 1951) and on *Bacillus anthracis* protoplasts (Stähelin, 1954).

The discovery of the spontaneous protoplast fusion of bacteria was followed by that of fungi, on *Saccharomyces* and *Candida* species (Müller, 1966, 1970), *Polystictus versicolor* (Strunk, 1967) and *Fusarium culmorum* (Lopez-Belmonte, Garcia Acha & Villanueva, 1966).

These observations in that period of protoplast research had the common features that: (i) the fusion took place mainly during protoplast formation; (ii) its frequency was rather low and incalculable; and (iii) both partners ('parents' of the fusion products) were wild-type, with identical genetic backgrounds.

It is interesting to note that Lederberg & St Clair (1958) gave an account of an unsuccessful experiment aiming at complementation by controlled protoplast fusion of genetically marked *Escherichia coli* strains. As they concluded: 'Attempts to detect the fusion of protoplasts of sexually incompatible (F^-) genotypes were unrewarding. The design of the experiments was similar to that for DNA-transduction, mixtures of protoplasts being evoked and grown together in penicillin agar or broth. We also tried graded osmotic shocks, and spinning a mixed protoplast suspension in 10% sucrose in an air turbine centrifuge of 80 000 *g* for 20 min. Whereas the pellet showed evidence of considerable lysis, there was no indication of fusion of protoplasts either from microscopy or tests for recombinants.' And: 'Further studies are needed to establish whether protoplasts stemming from different lines of cells can fuse with genetically interesting consequences.'

Fourteen years later the first controlled protoplast fusion and complementation had been achieved with the yeast-like filamentous fungus *Geotrichum candidum* (Ferenczy, Zsolt & Kevei, 1972). After another four years the first successful controlled intraspecific protoplast fusions were reported for bacteria (Fodor & Alföldi, 1976; Schaeffer, Cami & Hotchkiss, 1976) and yeast (Sipiczki & Ferenczy, 1976), as well as interspecific fusions of *Aspergillus* (Ferenczy, 1976) and *Penicillium* species (Anné, Eyssen & De Somer, 1976). The fusion of *Streptomyces* (Hopwood, Wright, Bibb & Cohen, 1977) and subsequently algal protoplasts (Matagne, Deltour & Ledoux, 1979) also proved successful.

BASIC METHODS OF PROTOPLAST FUSION

The principles and procedures whereby wall-deprived cells can be obtained by using lytic enzymes and/or inhibitors of cell wall synthesis in the presence of osmotic stabilizers, and whereby conditions can be created for the spherical protoplasts to revert to the normal, wall-bearing microbial form, are well established for both prokaryotic microbes (Weibull, 1953, 1958; Lederberg, 1956; McQuillen, 1960; Spizizen, 1962; Park, 1968; Kaback, 1971; Sagara *et al.*, 1971; Okanishi, Suzuki & Umezawa, 1974; Fodor, Hadlaczky & Alföldi, 1975; Hadlaczky, Fodor & Alföldi, 1976; Weiss, 1976; Marquis & Corner, 1976; Landman & De Castro-Costa, 1976; Peberdy, 1979b; Hopwood, Wright, Bibb & Ward, 1979) and eukaryotic ones (Nečas, 1956a, b, 1980; Eddy & Williamson, 1957; Villanueva, 1966; Villanueva & Garcia Acha, 1971; Ferenczy, Kevei & Szegedi, 1975b; Peberdy, 1976, 1979a; Ferenczy, Vallin & Maráz, 1977; Maráz & Ferenczy, 1979a).

In most cases, mutants with auxotrophic, antibiotic-resistance, temperature-sensitive, respiration-deficient, morphological and/or colour markers are used for fusion, and the resulting complementation is the indication of successful protoplast fusion. Of course, all the necessary controls have to be employed to detect reverse mutations or genetic transfer achieved in other ways than by protoplast fusion. The calculation of frequency of fusion is normally based upon the frequency of complementation, although these two events can never be exactly the same; for example fusion will also occur between non-complementary identical partners. The frequency of fusion is usually determined by comparing the number of complemented colonies (e.g. colonies growing on incomplete medium when auxotrophic partners are used) to that of the non-complemented ones (colonies growing only on complete medium). The presumed fusion products have to be characterized by using cytological, biochemical and genetic methods.

In early fusion experiments a centrifugal force (Ferenczy, Zsolt & Kevei, 1972; Ferenczy, Kevei & Zsolt, 1974; Binding & Weber, 1974) and/or intensive aggregation of protoplasts in cold KCl osmotic stabilizer (Ferenczy, Kevei & Szegedi, 1975a) were applied to induce their fusion. After the discovery that polyethylene glycol (PEG) acts as a fusogenic agent of plant protoplasts (Kao & Michayluk, 1974; Wallin, Glimelius & Eriksson, 1974), this compound was soon introduced into experiments with the aim of protoplast or cell fusion ranging from microbial protoplasts (Ferenczy, Kevei & Szegedi, 1975b, c; Anné & Peberdy, 1975a, b) to mammalian cells (Pontecorvo, 1975; Maggio, Ahkong & Lucy, 1976; Pontecorvo, Riddle & Hales, 1977; Gefter, Margulies & Scharff, 1977; Hales 1977; O'Malley & Davidson, 1977; Wacker & Kaul, 1977), and is nowadays used almost exclusively.

Interestingly, PEG has been utilized not only as a special dehydrating agent for enzyme precipitation (Foster, Dunnill & Lilly, 1973) or protein crystallization (McPherson, 1976), but at a lower concentration, also to stabilize protoplasts (Weibull, 1953; Wallin & Ericksson, 1973), and to stimulate nuclear division and wall formation (Wallin & Eriksson, 1973).

The methodological studies on PEG-induced fungal protoplast fusion (Ferenczy, Kevei & Szegedi, 1975b, 1976; Anné & Peberdy, 1975b, 1976) revealed that PEG preparations with molecular weights of 4000 or 6000 (PEG 4000 and PEG 6000, respectively) are optimum in fusion induction in the concentration range 25–40% in the presence of 10–100 mM $CaCl_2$. The addition of Ca^{2+} ions is critical for the attainment of high-frequency fusion. Ions or molecules as additional osmotic stabilizers in the PEG solution yield lower efficiencies, and this phenomenon is concentration-dependent. In experiments on bacterial protoplast fusion, high molecular weight PEG (PEG 6000) was employed (Fodor & Alföldi, 1976; Schaeffer, Cami & Hotchkiss, 1976; Kaneko & Sakaguchi, 1979; Tsenin, Karimov & Ribchin, 1978; Coetzee, Sirgel & Lacatsas, 1979), while in experiments with *Streptomyces* or *Micromonospora*, PEG preparations with different molecular weights were applied, such as PEG 1000 (Hopwood & Wright, 1978, 1979; Hopwood, Wright, Bibb & Ward, 1979), PEG 1540 (Hopwood, Wright, Bibb & Cohen, 1977), PEG 4000 (Ochi, Hitchcock & Katz, 1979), and PEG 6000 (Baltz, 1978).

Though the molecular mechanism of fusion by the PEG-Ca^{2+} fusogenic system is not exactly known, more and more of its details have recently become evident. Most of this newer knowledge derives from biophysical and freeze-fracture electron microscopic studies on the PEG-induced fusion of erythrocytes and erythrocyte membranes. Since the basic composition and structure of the cell membrane involved in the fusion are fairly similar throughout the whole of the living world, it is reasonable to assume that the fusion processes are also similar. This belief is strongly corroborated by the fact that fusion between erythrocyte cells and yeast protoplasts can be attained easily and with comparatively high yield (Ahkong *et al.*, 1975).

The sequence of fusion events starts with agglutination of the protoplasts caused by intensive dehydration, and the formation of aggregates of various extents. The protoplasts shrink and become highly distorted. Large areas of adjacent protoplasts come into very close contact (Ferenczy *et al.*, 1976). The next possible event, observed in erythrocyte membranes, is the translocation of intramembrane protein particles at the sites of close contact and their aggregation (Knutton, 1979; Knutton & Pasternak, 1979). The following step seems to be lipid–lipid interactions between the adjacent protein-denuded membranes. Perturbation and reorganization of the lipid molecules, strongly promoted by Ca^{2+} ions (Ito & Ohnishi, 1974; Papahadjopoulos, Poste, Schaeffer & Vail, 1974; Ahkong, Fischer, Tampson & Lucy, 1975; Lansman & Haynes, 1975; Papahadjopoulos, Vail, Pangborn & Poste, 1976; Cullis & Hope, 1978; Ingolia & Koshland, 1978; Sun, Hasang, Day & Ho, 1979), results in fusion in small regions of membranes in contact. Small cytoplasmic bridges are formed, which then enlarge, and the two protoplasts fuse.

Ever since the first observations on PEG-induced protoplast fusion (Kao & Michayluk, 1974), it has repeatedly been emphasized that at least partial removal of PEG from the suspension is needed to obtain high-frequency protoplast fusion. When microbial protoplasts are treated with PEG and the agglutinated protoplasts are then mixed with the osmotically-stabilized culture medium, the above-mentioned requirement for a high yield of fusion products is met. On the other hand, it has been found that a high frequency of nutritional complementation based upon protoplast fusion can be routinely obtained with amino acid-requiring auxotrophic mutants of *Aspergillus nidulans* in the fusogenic PEG-Ca^{2+} solution supplied with components of the minimal medium (L. Ferenczy, unpublished).

PROKARYOTIC PROTOPLAST FUSION

Protoplast fusion of Gram-positive bacteria

The protoplast fusion of prokaryotic organisms gives a unique opportunity for the bringing together of two (or more) complete genomes, instead of transferring fractions of DNA by transformation, transduction or conjugation.

Transfer of genetic information proved especially useful with *Bacillus megaterium*, the Gram-positive bacterium for which no genetic transfer mechanism had previously been known. Double auxotrophic mutants of *B. megaterium* were used by Fodor & Alföldi (1976) to obtain nutritionally complemented fusion products. Their progeny analysis showed the appearance of parental, stable recombinant and transitory segregating phenotypes. By fusing polyauxotrophic *B. megaterium* protoplasts it was clearly demonstrated that the physiological effects of cultivation and those of the different culture media employed during regeneration strongly influenced the yield of recombinants and distorted the expression of the genetic system (Fodor & Alföldi, 1979). A somewhat similar observation was reported for *Bacillus subtilis* (Gabor & Hotchkiss, 1979), with the conclusion that the regeneration of recombinant-forming fused protoplasts is different from the average regeneration for the population. The use of a fusion system for genetic analysis of these bacteria, therefore, seems to be complicated.

Thermal inactivation of one of the partners provides an opportunity for protoplast fusion and the selection of the recombinant fusion products, even when one of the partners is prototrophic. *B. megaterium* protoplasts lose the ability to revert to the bacillary form if incubated at 50 °C for 120 min. On the other hand, the heat-treated protoplasts can contribute to the formation of recombinants when fused with a viable partner (Fodor, Demiri & Alföldi, 1978).

From the genetic consequences of protoplast fusion of auxotrophic mutants of another Gram-positive bacterium, *B. subtilis*, similar conclusions can be drawn as for *B. megaterium*. Schaeffer, Cami & Hotchkiss (1976) reported the nutritional complementation of polyauxotrophic *B. subtilis* strains by protoplast fusion. It was concluded that transient diploids were obtained and that the only stable fusion products were haploid recombinants. Unselected markers segregated among the selected recombinants. No auxotrophic bacteria were found as segregants from prototrophic fusion products growing in a non-selective medium. The frequency of prototroph formation depended on the number and the chromosomal location of the auxotrophic markers used. It has further been demonstrated that not only viable, but also streptomycin-killed protoplasts of *B. subtilis* can be used as partners in fusion with living protoplasts of a streptomycin-resistant strain to obtain recombinants (Levi, Sanchez-Rivas & Schaeffer, 1977). When the protoplasts originated from sporulating cells of *B. subtilis*, protoplasts with two or more enclosed prespores could be observed in very high frequencies by electron microscopy (Frehel, Lheritier, Sanchez-Rivas & Schaeffer, 1979). High-frequency protoplast fusion of *B. subtilis* was also revealed by a prophage complementation test (Sanchez-Rivas & Garro, 1979) in which two strains, each lysogenic for a different

Sus mutant of the phage φ105, were induced by mitomycin-C, protoplasted, fused with PEG, and plated with φ105-sensitive indicator bacteria.

It was recently discovered that several per cent of the fusion products derived from polyauxotrophic *B. subtilis* strains are biparental, containing the unchanged genomes of both partners. Interestingly, a substantial proportion can be cloned as biparental cells for many generations. Their phenotype during the 'diploid' phase is that of one or the other auxotrophic parental type, and is not prototrophic. It is assumed that the particular chromosome can be replicated, but not expressed (Hotchkiss & Gabor, 1980).

Brevibacterium flavum protoplasts were induced by penicillin treatment from cells of this industrially-important bacterium, for which no transformation, transduction or conjugation mechanism has ever been found. Strains with auxotrophic and antibiotic (streptomycin and rifampicin) resistance markers were employed in the fusion process and selection was made for double antibiotic resistance. Stable haploid recombinants were obtained (Kaneko & Sakaguchi, 1979).

Protoplast (sphaeroplast) fusion of Gram-negative bacteria

Up till now, only two cases of fusion and complementation of protoplasts (sphaeroplasts) of Gram-negative bacteria have been reported (Tsenin, Karimov & Ribchin, 1978; Coetzee, Siergel & Lecatsas, 1979). With these microbes a part of the cell wall may not be completely dissolved by lysozyme and the additive compounds, so that the cell membrane, at least partially, will be covered by this. For this reason, the term 'sphaeroplast' may be more appropriate if the complete removal of the cell wall is not proved.

Two polyauxotrophic F^- mutants of *Escherichia coli* were selected, protoplasted (electron microscopic investigation revealed the formation of true protoplasts), fused, tested for nutritional complementation, and analysed genetically (Tsenin, Karimov & Ribchin, 1978). The overwhelming majority of the primary prototrophic colonies were not stable and the segregation of markers continued over many generations. With regard to segregation patterns and the requirements of the resulting clones, these colonies could be classified into the following groups: (i) mixed colonies containing stable recombinants and parental auxotrophs; (ii) mixed colonies with unstable prototrophs and a few unstable recombinants; and (iii) colonies uniformly composed of stable prototrophic cells.

The other Gram-negative bacterial species successfully used for sphaeroplast fusion was *Providencia alcalifaciens* (Coetzee, Siergel & Lecatsas, 1979). Though sphaeroplasts were produced, it was revealed by electron microscopy that about 15% of them showed breaks in the residual cell wall, these breaks exposing areas of underlying cytoplasmic membrane. The fusion-induced recombinants originating from the sphaeroplast mixture of the auxotrophic partners were haploids, as in the case of Gram-positive bacteria. Analysis of the prototrophic colonies revealed the presence of stable prototrophs, or mixtures of stable prototrophs and stable recombinants. Parental types were not found. Unselected markers segregated among the recombinants. The frequencies of recombination depended on the number of chromosomal loci used in selection.

Protoplast fusion in Actinomycetales

Streptomyces protoplast fusion achieved by Hopwood and coworkers (Hopwood, Wright, Bibb & Cohen, 1977; Hopwood & Wright, 1978, 1979; Hopwood, Wright, Bibb & Ward, 1979) opened up new possibilities of gaining more detailed knowledge on *Streptomyces* genetics, and also of making advances in the practical application of the protoplast fusion technique. At present, fusion of protoplasts seem to be far the most promising method for the strain improvement of *Streptomyces* and similar species known as producers of antibiotics and other important compounds.

With different *Streptomyces* species (*S. coelicolor, S. parvulus, S. lividans, S. griseus, S. acrymicini*), all the three major processes – formation, fusion and regeneration of protoplasts – can be carried out efficiently. Recombination frequencies are high: routinely above (sometimes much above) 1%. The frequency of recombinants in the progeny can be increased by adding dimethyl sulphoxide (14%, v/v) to the PEG solution and/or by UV irradiation of the protoplasts immediately before fusion (Hopwood & Wright, 1979). With *S. coelicolor*, recombinants can easily be obtained even under non-selective conditions, and these will constitute 10–20% of the total spore progeny of the regenerated cultures (Hopwood & Wright, 1978, 1979; Hopwood, Wright, Bibb & Ward, 1979). These extremely favourable conditions have the advantage that selectable markers need not be used to obtain recombinants in good yield.

Analysis of the recombinants of multiple crossover classes (Hopwood & Wright, 1979) revealed that, in contrast to conjugation or other means of gene transfer, genomes of both partners brought together by protoplast fusion are complete or nearly so, resulting in a transient diploid or quasi-diploid state of the fusion products. The similarity to events in eubacteria is obvious. The heterozygous diploid state was transient in *S. coelicolor* and fusion colonies frequently contained only recombinants without parentals; in other cases both recombinant and parental genomes could be found. In the same colony, the presence of different recombinant genotypes was also observed. It was postulated that the genomes of both partners became fragmented after fusion. Crossing-over between the fragments gave rise to stable haploid recombinants. Events of recombination were independent of the known sex factors SCP1 and SCP2. Not only two-partner, but successful simultaneous three- and four-partner fusions, generating crossing-over and production of recombinants, has been reported (Hopwood & Wright, 1978).

Despite marked technical dissimilarities, the above observations of high-frequency recombination were confirmed by Baltz (1978), who studied the genetic consequences of protoplast fusion of auxotrophic and antibiotic resistant mutants of *Streptomyces fradiae*.

Highly-efficient intraspecific gene transfer and nutritional complementation was achieved in auxotrophic strains of *Streptomyces parvulus* and *Streptomyces antibioticus* (Ochi, Hitchcock & Katz, 1979). For recombination, the efficiency of protoplast fusion over the mating technique was 10^4 times higher. In *S. parvulus* stable prototrophic recombinants were obtained, while in *S. antibioticus* both stable prototrophs and nutritionally complemented unstable heterokaryons could be isolated, which is the typical form of labile complementation predominant in filamentous fungi and mycelial (pseudomycelial) yeasts (Ferenczy, 1980).

Rifampicin-resistant and casamino acid-dependent double mutants were produced from *Micromonospora echinospora* and *M. inyoensis* and fused with the corresponding antibiotic-sensitive wild-type parental strains by Szvoboda *et al.* (1980). By both direct and indirect methods, prototrophic and antibiotic-resistant colonies were selected. A direct selection method could be used effectively if the selecting antibiotic was added only after a phenotypic lag period of two days. Over 90% of the resulting colonies proved to consist of stable recombinants.

A rewarding approach of direct selection of *Micromonospora* recombinants was to use heat-inactivated (2–3 hours at 55 °C) protoplasts of the wild-type partner, which are not able to revert into the normal microbial form, but are able to act as fusion partners to give rise to complemented colonies with an auxotrophic partner. One of the evident advantages of this system is that one can avoid the isolation of more than one auxotrophic mutant for the fusion experiment; another is the applicability of direct selection without waiting for the period of antibiotic sensitivity to pass. The idea had earlier proved successful in *B. megaterium* protoplast fusions (Fodor, Demiri & Alföldi, 1978). Similarly good results were achieved with *Micromonospora*; only recombinants could form colonies.

An indirect selection method was also applied successfully to obtain *Micromonospora* fusion products. PEG-treated mixed protoplasts of a cas^{-} rif^{R} mutant and the wild-type cas^{+} rif^{S} strain were plated on a nutritionally sufficient medium, and the colonies were then replica-plated for selection of hybrid colonies. Less than 1% of the prototrophic rifampicin-resistant colonies had the properties of stable recombinants.

EUKARYOTIC PROTOPLAST FUSION

Intraspecific protoplast fusion

Filamentous fungi

A series of basic discoveries on induced and controlled microbial protoplast fusion – e.g. the first successful fusion experiments and the subsequent characterization of the fusion products, the establishment of methods for obtaining high fusion frequency, the first interspecific protoplast fusion, and the recognition of the new possibilities of microbial gene transfer in both basic and applied fields – are all based upon research on protoplast fusion of filamentous fungi. Between 1972 and 1976, microbial protoplast fusion meant exclusively the protoplast fusion of filamentous fungi.

Stable auxotrophic mutants of *Geotrichum candidum* were selected, protoplasts were formed, and their fusion was induced by centrifugal force followed by long incubation. This was a lucky choice, since sexual or parasexual processes have never been observed in these mutants. The resulting nutritionally complemented protoplasts were able to regenerate a new cell wall and give rise to prototrophic colonies. The complemented hyphae could be transferred and maintained on minimal medium indefinitely. Complementation was at a low frequency, but reproducible. Most arthrospores developing in the complemented colonies proved to harbour only one of the parental genomes. A few arthrospores were always able to form colonies on minimal medium; however, these colonies again produced nutritionally deficient arthrospores. From complemented hyphae growing on minimal medium, protoplasts could be obtained, most of which were able to regenerate cell walls in nutritionally supplemented minimal medium to give rise to colonies of one of the parental types (Ferenczy, Zsolt & Kevei, 1972; Ferenczy, Kevei & Zsolt, 1974). These characteristics indicated heterokaryon formation as the reason for nutritional complementation.

In intraspecific fusion of filamentous fungi, heterokaryon formation is the basic and most frequent event leading to complementation, occasionally followed by the production of transient or stable diploids. Heterokaryosis, as the only known means of complementation after protoplast fusion, was reported with *Phycomyces blakesleeanus* (Binding & Weber, 1974) and *Mucor racemosus* (Genther & Borgia, 1978). Similarly, in *Cephalosporium acremonium*, protoplast fusion of complementing auxotrophic mutant strains resulted in heterokaryons (Anné & Peberdy, 1976). Diploidization also occurred, but this stage proved very unstable, yielding haploid recombinants (Hamlyn & Ball, 1979). In this way, strain improvement for higher cephalosporin production was also achieved.

Complementing heterokaryons were formed by the protoplast fusion of auxotrophic mutants of *Aspergillus nidulans* (Ferenczy, Kevei & Zsolt, 1974; Ferenczy, Kevei & Szegedi, 1975a, b, 1976; Ferenczy, Kevei, Szegedi, Frankó & Rojik, 1976; Anné & Peberdy, 1976; Ferenczy, 1976), including otherwise incompatible strains (Dales & Croft, 1977, 1980; Croft & Dales, 1979), *A. flavus* (Ferenczy, Kevei & Szegedi, 1975b, 1976), *A. niger* (Ferenczy, Kevei & Szegedi, 1975a, b, 1976; Anné & Peberdy, 1976), *A. fumigatus* (Ferenczy, 1976; Ferenczy, Szegedi & Kevei, 1977), *Penicillium frequentans* (Ferenczy, Kevei & Szegedi, 1975b, 1976), *P. ramigena* (Ferenczy, Kevei & Szegedi, 1975b, 1976), *P. chry-*

sogenum (Anné & Peberdy, 1975a, b), *P. patulum* (Anné & Peberdy, 1976), *P. roquefortii* (Anné & Peberdy, 1976) and *P. cyaneo-fulvum* (Peberdy, Eyssen & Anné, 1977). The heterokaryotic fusion products, especially those of *A. nidulans*, displayed somewhat irregular colony growth on minimal medium, rapid mycelial segregation to the parental types on complete medium, together with conidial segregation to the parental types on both media. Occasionally, normally growing sectors developed bearing large, prototrophic diploid conidia (Ferenczy, 1976; Anné & Peberdy, 1976; Croft & Dales, 1979; Dales & Croft, 1980). Haploids of both the parental and recombinant types could easily be induced from the diploid cells. Protoplast fusion may become a useful tool for genetic analysis or strain improvement in fungi, especially in imperfect fungi, and particularly in those cases where even anastomosis is impossible.

Yeasts

In the past four years an appreciable number of data have appeared on yeast protoplast fusion, including the genetic consequences, the diversity of complementation, and the possible application of the fusion technique in breeding for industrial purposes. It is essential to induce the process of fusion: Svoboda (1976) demonstrated that protoplasts would not fuse spontaneously even when derived from cells of opposite mating types.

After fusion induction, the first stage of complementation, if auxotrophic nuclear markers are employed, is heterokaryon formation. The heterokaryotic state can be either permanent or transient if the fusion products are kept under the selective pressure of a minimal medium, and transient in complete media. Characteristically, the heterokaryotic state is permanent in yeasts with frequent filament (pseudomycelium) formation (e.g. *Candida tropicalis*), resembling the complementation situation in filamentous fungi, and very transient in yeast species with the 'conventional' yeast form.

From auxotrophic mutants of *C. tropicalis*, heterokaryons were produced by protoplast fusion in minimal medium (Fournier, Provost, Bourquignon & Heslot, 1977; Ferenczy, Vallin & Maráz, 1977; Vallin & Ferenczy, 1978). The heterokaryons were unstable in nutritionally complete medium and readily dissociated into the parental mutants. Diploids arose from the haploid heterokaryons, and somatic segregants and recombinants could be obtained from the diploids. The existence of aneuploids was indicated by the fact that uninucleate prototrophs originating from diauxotrophs spontaneously yielded monoauxotrophs (Fournier *et al.*, 1977). If adenine-requiring (red) and cysteine-requiring (white) mutants were used in the fusion experiments, the multinucleate heterokaryotic prototrophs were different shades of pink. After occasional nuclear fusion the resulting diploids (and/or aneuploids) were white, rapidly growing and rather stable, giving rise to only very few parental haploids or recombinants even after induction (Vallin & Ferenczy, 1978). The technique of protoplast fusion provides an opportunity for genetic analysis in this and in similarly asexual species.

In many other yeast species the heterokaryotic stage is so transient that frequently it cannot be detected, and the first identified products are diploids. A general characteristic of these yeast species is that protoplasts of cells of identical mating type can be fused and complemented for the auxotrophic markers with frequencies similar to those in the case of opposite mating types. However, in diploids homozygous for mating type alleles, spore formation is drastically diminished or absent. The fusion cells are uninucleate, enlarged, and with a double amount of DNA per cell. By induced haploidization or by crosses with cells of the opposite mating types, both parental and recombinant haploids can be recovered. If protoplasts of opposite mating types are fused, the fusion products display the characteristics of normal crosses.

These rules were first established by fusion of protoplasts of complementing haploid auxotrophs from *Schizosaccharomyces pombe* (Sipiczki & Ferenczy, 1976, 1977a; Svoboda, 1977, 1978). On the basis of the above findings, protoplast fusion could recently be applied for genetic analysis of a sterile mutant of *S. pombe* (Thuriaux, Sipiczki & Fantes, 1980).

Understandably, fusion studies with *Saccharomyces cerevisiae* started simultaneously and independently in several laboratories, and papers giving a series of clear examples of the above-mentioned rules were sometimes received for publication by various journals within an interval of a few days (Ferenczy & Maráz, 1977; Solingen & Plaat, 1977; Svoboda, 1977, 1978; Yamamoto & Fukui, 1977; Maráz, Kiss & Ferenczy, 1978; Gunge & Tamaru, 1978; Kawakami, Mondo & Kawakami, 1978). The first publications were followed by several others (Christensen, 1979; Maráz & Ferenczy, 1979a, b; Ferenczy & Maráz, 1979; Arima & Takano, 1979a, b; Kozhina &

Chepurnaya, 1980; Böttcher *et al.*, 1980), including some reporting successful attempts to analyse or improve industrial strains (Stewart, 1978; Russell & Stewart, 1979; Snow, 1979; Hockney & Freeman, 1980; Spencer, Land & Spencer, 1980). Multiple fusion of haploids, leading to triploids and tetraploids in *Saccharomyces* yeasts, was also reported (Takano & Arima, 1979; Arima & Takano, 1979a). Triploid products were formed by protoplast fusion of two as well as three different haploid strains. Similarly, stable diploids could be obtained from the protoplast fusion of haploid strains of *Rhodosporidium toruloides* (Sipiczki & Ferenczy, 1977b; Böttcher *et al.*, 1980; Becher & Böttcher, 1980), *Kluyveromyces lactis* (Morgan, Heritage & Whittaker, 1977; Morgan, Hall, Brunner & Whittaker, 1980), *Pichia guillermondii* (Böttcher *et al.*, 1980; Klinner, Böttcher & Samsonova, 1980; Spata & Weber, 1980) and *Hansenula polymorpha* (Savchenko & Kapultsevich, 1980). On the other hand, several variations concerning polyploid or aneuploid formation have also been found after protoplast fusion.

Fusion of *Saccharomycopsis lipolytica* auxotrophs belonging to like and opposite mating types (Stahl, 1978; Spata & Weber, 1980) revealed transient heterokaryon formation coupled with unstable diploidy. Spore formation was induced in about 3% of the diploid fusion products originating even from auxotrophs of identical mating type (Stahl, 1978). The progeny bore the characteristics of the parental mating type. These results indicate that mating type alleles of this species, in contrast to those of many other yeasts, control only the initial steps in the mating process sequence.

Algae

Stable fusion products were obtained from haploid auxotrophs of *Chlamydomonas reinhardtii* having the same mating type and defective in cell wall synthesis. Electron microscopy revealed that these wall-defective mutants were actually true protoplasts without cell wall elements. Both partners were arginine auxotrophs, defective as regards the same enzyme, but able to complement, and the wild phenotype was restored in the fusion products. Diploids of mt^+ parental mating type resulted, with the ability to produce normal cell walls. The fusion products were very similar to sexual mating products in cellular and nuclear volume and nuclear DNA content, but had a significantly lower chloroplast DNA content and different mating type characteristics (Matagne, Deltour & Ledoux, 1979).

Interspecific protoplast fusion

Filamentous fungi

The attainment of gene transfer between two different species has been reported in two major genera of filamentous fungi, *Aspergillus* (Ferenczy, 1976; Ferenczy, Szegedi & Kevei, 1977; Kevei & Peberdy, 1977, 1979; Croft, Dales, Turner & Earl, 1980) and *Penicillium* (Anné, Eyssen & De Somer, 1976; Peberdy, Eyssen & Anné, 1977; Anné & Eyssen, 1978). Partners of these interspecific fusions cannot be 'parents' in sexual or parasexual processes, because of the lack of natural cell fusion between them. The technique of protoplast fusion seems to be the only means of obtaining interspecific heterokaryons (in other cases heteromitochondrial hybrids) and recombinants.

It is an acceptable general belief, supported by several facts, that protoplast fusion can be induced between two partners independently of their taxonomic relationship. On the other hand, very little is known about the genetics or biochemistry of compatibility–incompatibility events in these interspecific microbial fusion products. Fusions of protoplasts of taxonomically distant species may be fruitless in the sense of obtaining living products. Details of 'unsuccessful' experiments, not resulting in demonstrable genetic consequences, are normally not reported. It should be stressed that investigations on interspecific protoplast fusions, 'successful' or 'unsuccessful', will assist greatly in clarifying the genetics and biochemistry of species incompatibility, the second barrier after the cell wall–membrane system to protect species identity. The first few published experiments in this respect are decidedly promising.

In *Aspergillus* and in *Penicillium* species there are reported cases of protoplast fusion of both distant and closely related species. In *Aspergillus*, protoplast fusion could be induced between the two *distantly* related species *A. nidulans* and *A. fumigatus* (Ferenczy, 1976; Ferenczy, Szegedi & Kevei, 1977). From both species, stable auxotrophic mutants requiring adenine or lysine were produced, and reciprocal fusions were made between their protoplasts. Complemented colonies were obtained and their properties were compared with those of colonies resulting after intraspecific fusion of the auxotrophic mutants.

For clarity and brevity, it has been proposed (Ferenczy & Maráz, 1977) that bracketed names be used with the '+' sign between them to indicate products of protoplast fusion, at least until they have been characterized thoroughly. Accordingly, interspecific products [*A. nidulans lys* + *A. fumigatus ade*] and [*A. nidulans ade* + *A. fumigatus lys*] were obtained and their characteristics were compared with those of the intraspecific products [*A. nidulans ade* + *A. nidulans lys*] and [*A. fumigatus ade* + *A. fumigatus lys*].

In interspecific fusion, the complementation frequency proved to be much lower than in intraspecific fusion, the difference being more than five orders of magnitude. The morphological features of the colonies were also characteristically different: on minimal agar medium the interspecific colonies developed slowly and in a thick layer, with very distorted mycelia, in contrast to the rapidly growing, somewhat irregular, and thin intraspecific heterokaryotic colonies. In interspecific colonies, conidium formation was rare, and the conidia were colourless, uninucleate or nucleus-free, and of different sizes and forms; intraspecific colonies produced many conidiophores of normal shape, and a great number of uninucleate haploid conidia with colours characteristic of the parental strains. Conidia originating from interspecific fusion colonies, and especially from their central regions, were able to germinate and form slow-growing, thick colonies on minimal medium; these could be maintained indefinitely on minimal medium. Conidia produced by intraspecific fusion colonies were not able to germinate on minimal medium due to separation of nuclei of the parental strains during formation of the uninucleate conidia; only minimal medium supplemented with compounds required by the parental strains supported growth. If nutritionally complete medium was inoculated with conidia of interspecific colonies, the prototrophic characteristics were lost after germination, and only one of the partner species appeared. Conidia collected from intraspecific colonies always gave rise to both partner strains on the complete medium. Prototrophic mycelial growth of both interspecific and intraspecific colonies could be maintained under the selective pressure of the minimal medium. However, on a complete medium, only one of the partners, most often *A. nidulans*, segregated out from an interspecific mycelial inoculum, and both of them from an intraspecific one. The most surprising finding in connection with complementation in the [*A. nidulans* + *A. fumigatus*] fusion products was that it was not possible to obtain either heterokaryons or diploids among them, although these are the two main forms resulting from nutritional complementation in intraspecific fusion colonies. It seems reasonable to conclude that both the heterokaryotic and the diploid states are lethal, and that a precondition of interspecific complementation in otherwise incompatible fungi is the selective loss of certain chromosomes of one of the partners.

All the unusual properties of these interspecific colonies can be explained by assuming that one or a few chromosomes were retained from one of the partners in the complete chromosomal set of the other. The term 'interspecific aneuploid' was used (Ferenczy, 1976; Ferenczy, Szegedi & Kevei, 1977) in descriptions. Dales and Croft (1980) suggested the term 'heteroploid' instead. Unfortunately, the otherwise logical term 'heteroploid' has already been used in different meanings (Rieger, Michaelis & Green, 1976; Coetzee, Siergel & Lecatsas, 1979) and should also be coupled with the adjectives of 'interspecific' or 'allo'. The term 'alloploid' (with the adjectives of 'full' or 'partial') is also an acceptable possibility, though it has been used for sexual products, normally containing the whole chromosomal sets of the two parents. The term 'interploid', newly coined for interspecific, intergeneric, etc. fusion products, similarly seems appropriate to indicate unusual, artificial, interspecific to interkingdom ploidy situations.

Somewhat different results arose when the [*Penicillium roquefortii* + *P. chrysogenum*] fusion products were analysed. These two *Penicillium* species are also comparatively distantly related. From fusion of the auxotrophic partners, slow-growing hybrids were obtained, which could be classified into three groups according to the morphology and genetic characteristics. Prototrophic colonies of types 1 and 2 gave rise only to auxotrophic *P. roquefortii* conidia on rich medium, but produced penicillins of the same chemical composition as those produced by the *P. chrysogenum* strain. The type 3 colony released large prototrophic conidia resembling those of the *P. chrysogenum* partner, and likewise produced the same penicillins (Anné, Eyssen & De Somer, 1976).

Auxotrophic mutants of the taxonomically *closely* related *A. nidulans* and *A. rugulosus* gave high yields of prototrophic colonies (Kevei & Peberdy, 1977). From these colonies both parental strains segregated simultaneously on complete medium, indicating the heterokaryotic nature of the complemented colonies. These complemented colonies, growing irregularly and slowly on minimal medium, gave rise to vigorously developing hybrid diploid sectors producing diploid conidia, from which, in turn, parental and recombinant segregants were generated after haploidization with benomyl (Kevei & Peberdy, 1977; Peberdy & Kevei, 1979). From other segregation experiments, when the *A. nidulans* partner had a known genetic marker in each linkage group, it was concluded that the distribution of these groups (chromosomes) was random, and that a high degree of chromosomal homology might exist between the two species (Kevei & Peberdy, 1979).

Are they really different species? And what is the situation with *Penicillium* species – e.g. *P. chrysogenum* and *P. notatum* (Anné & Peberdy, 1975a, 1976), *P. chrysogenum* and *P. cyaneo-fulvum* (Peberdy, Eyssen & Anné, 1977), or *P. citrinum* and *P. cyaneo-fulvum* (Anné & Eyssen, 1978) – where similar observations were made but not analysed as thoroughly as in the case of [*A. nidulans* + *A. rugulosus*] (Kevei & Peberdy, 1979) in this respect? At present there is no clear answer to these questions. There can be no doubt that there are morphological differences between them, and the literature referred to also indicates genetic anomalies after protoplast fusion, i.e. some genetic differences do exist between the partners. At the same time, not much is known about whether the mutants used are typical representatives of their species or, because of cryptic mutations or other factors, differ significantly in genetic composition from one another. It may well be the time to reconsider taxonomic relationships and re-evaluate certain taxonomic boundaries of fungi, among others the species boundaries, in the light of results of protoplast fusions. If the chromosome homology is close enough for diploids to be obtained from the haploid fusion partners, then haploid recombinants can be induced, and especially if this procedure can be repeated with the new products, one can reasonably conclude that the two partners involved in the fusion belong to the same species. The method of protoplast fusion can help considerably in the determination of chromosome homology, even in imperfect fungi, and also in those cases when attempts to bring about anastomosis are unsuccessful.

Yeasts

Protoplasts of stable auxotrophic strains of *Kluyveromyces lactis* and *K. fragilis* were fused, and stable prototrophic colonies resulted (Whittaker & Leach, 1978). The fusion products were larger than cells of the parental strains, with an elevated DNA content. There were indications of both the loss of chromosomes after fusion, and also the occurrence of multiple fusion. At the same time, a selective retention of *K. fragilis* mitochondrial DNA and a loss of that of *K. lactis* was observed.

In the protoplast fusion of auxotrophic mutants of *Schizosaccharomyces pombe* and *S. octosporus*, the latter proved dominant in the fusion products, since only *S. octoporus* could be recovered from the hybrids. Protoplast fusion between the two species resulted in nutritionally complemented cells. As in the interspecific protoplast fusion of other taxonomically distant species, the physiological balance was disturbed, as reflected in the osmotic sensitivity of the cells due to incomplete wall formation and in their low viability (Sipiczki, 1979).

INTERGENERIC PROTOPLAST FUSION

No data were given about the genetic background of nutritional complementation of two diauxotrophic mutants of *Candida tropicalis* and *Saccharomycopsis fibuligera*, achieved by protoplast fusion (Provost *et al.*, 1978). The complemented cells proved uninucleate and exhibited the assimilation spectrum of *C. tropicalis* or *S. fibuligera* or both species. Similarly, the nutritionally complemented intergeneric fusion products [*Saccharomycopsis lipolytica* + *Pichia* (*Candida*) *guillermondii*] have not yet been genetically analysed. The hybrids of the auxotrophic strains were unstable uninucleate prototrophs, which rapidly segregated to the auxotrophic fusion partners (Spata & Weber, 1980).

Protoplast fusion of auxotrophs of the two best-known yeast species, *Saccharomyces cerevisiae* and *Schizosaccharomyces pombe*, was attempted with limited success by Svoboda (1980). After PEG-induced fusion, bi- tri- and tetranucleated protoplasts were obtained, and among them, with a frequency of 10^{-4} to 10^{-5}, cell wall formation and an increase in the volume of the cytoplasm of the cells were observed. In a combination of the strains, a very low frequency of colonies appeared, but with limited growing capacity. No genetic or biochemical analysis could be carried out.

INTERKINGDOM PROTOPLAST–CELL AND PROTOPLAST–PROTOPLAST FUSION

By PEG-induced fusion of yeast protoplasts with hen erythrocytes, fungal–animal heterokaryons were produced (Ahkong *et al.*, 1975). Cytological analysis revealed that, when the ratio of the two fusion partners was 1:1 or 1:10 (erythrocytes:protoplasts), a comparatively high proportion (about 0.1%) of the erythrocytes were involved in interkingdom fusion. On the other hand, with ratios of 1:50 or 1:250, giant yeast protoplasts appeared, presumably as products of protoplast–protoplast fusions. Of course, no products with immediately fruitful genetic consequences resulted from these experiments. Nevertheless, the results encourage basic studies on membrane structure and function, and point to the technical possibilities of fusing taxonomically very distant cells with the hope of transferring nuclear or other genetic elements (mitochondria, chloroplasts, plasmids, viruses).

Another interkingdom fusion experiment was reported recently (Kingsman, Clarke, Mortimer & Carbon, 1979), in which whole, 'partially lysed', and 'thoroughly lysed' bacterial protoplasts were fused with yeast protoplasts, by using PEG, to achieve yeast transformation. A plasmid, designated pLC544, was constructed from plasmid pBR313 and a 1.4 kb *Eco*RI fragment from the yeast *trp1* region, and used to transform yeast *trp1* mutants to Trp^+. The exact nature of the transfer is not clear since no transformants were obtained when whole bacterial protoplasts were employed; high-frequency transformation resulted with 'partially lysed' protoplasts; and only low-frequency transformation could be attained with 'thoroughly lysed' protoplasts. One can reasonably suppose that the transformation was the result of a fusion-mediated plasmid transfer, similarly to the first reported yeast transformations by a plasmid (Hicks, Hinnen & Fink, 1979; Hinnen, Hicks & Fink, 1978), rather than a direct bacterial-yeast protoplast fusion. Whatever the reason may be for the transformation, the attempt to transfer plasmid DNA from bacterial protoplasts to yeast protoplasts proved successful without it being necessary to purify the plasmid DNA.

ORGANELLE AND VIRUS TRANSMISSION BY PROTOPLAST FUSION

If the fusion partners are selected to harbour genetically different nuclei and mitochondria when two fungal protoplasts are fused, not merely two, but at least four genetic systems (the two kinds of nuclei and the two kinds of mitochondria) will interact within one cell. If an appropriate selection system is employed, the genetic and biochemical consequences of not only the nuclear, but also the mitochondrial transmission can be studied. The same holds true for algal chloroplasts. It is also possible to attain non-selective transfer of mitochondria or chloroplasts together with nuclei, or the selective transfer of organelles without the nuclear genetic system from the donor to the recipient. Similarly, smaller genetic elements, such as viruses or plasmids, can be transmitted by protoplast fusion. The term 'genetic transfusion' has been coined (Ferenczy *et al.*, 1976; Ferenczy & Maráz, 1977) and is being used for these cases to indicate the transfer of cell components via protoplast fusion.

Non-selective transfusion of mitochondria

As mitochondrial donor, a ϱ^+ haploid *Saccharomyces cerevisiae* strain of α mating type, with auxotrophic nuclear and antibiotic-resistance mitochondrial markers was selected, while the recipient was also a haploid strain of the same species, of identical mating type, but with another auxotrophic nuclear marker, and completely devoid of mitochondrial DNA (ϱ° *petite*). Protoplasts were formed and fused, the resulting diploid prototrophs with ϱ^+ characteristics were isolated and subjected to induced haploidization, and haploid products with the nuclear marker of the recipient but with the mitochondria of the donor strain were then selected (Ferenczy & Maráz, 1977, 1979). Similar results were reported in other strains too (Gunge & Tamaru, 1978).

The application of mitochondrial transfusion to gain new brewer's yeasts has also proved rewarding. With industrial strains it is not easy to produce auxotrophic mutants for breeding purposes, since they are mostly diploid, aneuploid or of higher ploidy; even the introduction of nuclear mutations can deleteriously alter the unique characteristics of the domesticated brewery strains. Production of mitochondrially marked strains and their fusion gives a good possibility of hybrid formation, identification and selection for breeding (Spencer, Land & Spencer, 1980).

Non-selective mitochondrial transfusion provided the means of genetic analysis in a respiratory-deficient mutant of the *petite*-negative yeast *Kluyveromyces lactis* (Morgan, Heritage & Whittaker, 1977; Allmark, Morgan & Whittaker, 1978; Morgan, Hall, Brunner & Whittaker, 1980).

In the fission yeast *Schizosaccharomyces pombe*, no significant differences were found in the average transmission of mitochondrial markers in the fusion process and in the 'traditional' zygote formation; however, the individual transmissions of the mitochondrial markers were very dissimilar in these processes, because of the time needed for cell wall regeneration (Lückemann, Sipiczki & Wolf, 1979).

Selective transfusion of mitochondria

During protoplast formation in budding yeasts, small protoplasts (1 μm in diameter or less) are frequently released from the buds. Although the great majority of these small protoplasts do not possess nuclei, some do contain mitochondria. Anucleate small protoplasts can easily be separated from those containing nuclei by low-speed centrifugation. An anucleate protoplast fraction was prepared from a double auxotrophic mutant of *S. cerevisiae* with mitochondrially-inherited resistance to erythromycin. The small protoplasts were fused with nucleus-containing protoplasts of an adenine-requiring strain of the same species lacking mitochondrial DNA. Under partially-selective conditions, the majority of the fusion products proved to be haploid, adenine-requiring with the mitochondrial genome of the donor cells (Maráz & Ferenczy, 1979b).

The isolation of mitochondria from an oligomycin-resistant strain of *S. cerevisiae* and their intraspecific fusion with protoplasts of a respiration-deficient mutant was reported recently (Gunge & Sakaguchi, 1979). On selective agar plates, fusion colonies appeared at a very low frequency. All the colonies carried the mitochondrial genotype of the donor and the nuclear genotype of the recipient strain. Similar results were reported not only intraspecifically, but interspecifically too (Yoshida, 1979).

Transfusion of chloroplasts

A fine model system was developed to induce uptake of higher plant chloroplasts into fungal protoplasts by Vasil & Giles (1975). Spinach chloroplasts were isolated and their uptake into cells of the slime strain of *Neurospora crassa* was induced by PEG. Protoplasts could accumulate as many as 40 chloroplasts each. For a period of time after the uptake, both systems remained intact and able to function.

The uptake of isolated photosynthetic organelles from the cyanobacterium *Anabena cylindrica* and from the green alga *Chlorella ellipsoidea* into protoplasts of *Saccharomyces cerevisiae* has also been reported (Kawakami *et al.*, 1980). These organelles were incorporated into the vacuoles of the yeast cell as thylakoid fragments.

Intraspecific chloroplast transfusion experiments were carried out recently to study the mode of chloroplast gene inheritance in the fusion product (Matagne & Hermesse, 1980). Wall-deficient and auxotrophic mutants of the same mating type, bearing different chloroplast markers, were used in PEG-induced fusion. About one-third of the fusion products showed the presence of chloroplast markers of both fusion partners, while another one-third had markers of one partner, and the remainder contained markers of the other partner. This experiment revealed that the preferential elimination of paternal chloroplast alleles depended upon heterozygosity at the mating type locus.

Transfusion of viruses

Fungal protoplast fusion can serve as the method of choice to transfer viruses from one cell to another. Earlier results indicated the applicability of protoplasts to enable fungal cells to take up viruses (Lhoas, 1971; Coutts & Cocking, 1972), while the possibility of cell-to-cell transfusion was demonstrated by Boissonnet-Menes & Lecoq (1976), who performed protoplast fusion between infected and healthy strains of *Pyricularia oryzae*. In these experiments the not very effective method of achieving protoplast fusion by using centrifugal force was employed. With PEG, the transfusion frequency of viruses might well be significantly elevated, since PEG has proved to be very effective in increasing the uptake of different viruses or viral nucleic acids into plant protoplasts (Cassels & Barlass, 1978; Dawson *et al.*, 1978; Maule, Boulton, Edmunds & Wood, 1980) or into animal cells (Rohde, Pauli, Henning & Friis, 1978).

CONCLUSIONS

The possibilities of the formation and fusion of protoplasts from a great variety of microorganisms, and the reversion of the fusion products into cell wall-bearing microbes, have opened up new prospects in different research directions.

The application of protoplasts in genetics started with higher plants, with the use of microbial enzymes to produce protoplasts from plant cells (Cocking, 1960) and to fuse them (Power, Cummings & Cocking, 1970; Power & Cocking, 1971; Potrykus, 1971), and in this way to utilize them in laboratory experiments almost as microorganisms (Carlson, 1975; Melchers, 1977; Cocking, 1977), with all the advantages of microbial cultures and procedures.

Because of the fast growth of microbes, the well-established microbial techniques, and the relatively advanced state of understanding in microbial genetics, the most rapid progress in protoplast fusion seems to have been achieved in microbiology. Significant advances have been made in the fundamental topics of fusion, as well as in those applied fields which are of great industrial and agricultural importance.

It has also turned out that PEG-induced prokaryotic or eukaryotic transformation processes (Bibb, Ward & Hopwood, 1978; Chang & Cohen, 1979; Hinnen, Hicks & Fink, 1978; Hicks, Hinnen & Fink, 1979; Beggs, 1978) cannot be separated from those of fusion, since the mechanism of introduction of transforming DNA molecules into protoplasts seems to involve the fusion of protoplasts (Hicks, Hinnen & Fink, 1979).

Fusion experiments have already given us a much better understanding of the basic regulatory processes in microbial cells. Nevertheless, it would appear that we are merely at the very beginning of the exploitation, for both theoretical and practical purposes, of the method of protoplast fusion. Protoplasts are no longer simply cytologic curiosities. Their ability to undergo fusion has led to their emergence as one of the important tools of modern biology.

REFERENCES

Ahkong, Q. F., Fischer, D., Tampson, W. & Lucy, J. A. (1975). Mechanism of cell fusion. *Nature, London*, **253**, 194–5.

Ahkong, Q. F., Howell, J. I., Lucy, J. A., Safwat, F., Davey, M. R. & Cocking, E. C. (1975). Fusion of hen erythrocytes with yeast protoplasts induced by polyethylene glycol. *Nature, London*, **255**, 66–7.

Alföldi, L. (1980). Fusion of microbial protoplasts. In *Fourteenth International Congress of Genetics*, August 21–30, 1978, Moscow, USSR (in press).

Allmark, B. M., Morgan, A. J. & Whittaker, P. A. (1978). The use of protoplast fusion in demonstrating chromosomal and mitochondrial inheritance of respiratory-deficiency in *Kluyveromyces lactis*, a petite-negative yeast. *Molecular and General Genetics*, **159**, 297–9.

Anné, J. & Eyssen, H. (1978). Isolation of inter-species hybrids of *Penicillium citrinum* and *Penicillium cyaneo-fulvum* following protoplast fusion. *Federation of European Microbiological Societies Microbiology Letters*, **4**, 87–90.

Anné, J., Eyssen, H. & De Somer, P. (1976). Somatic hybridization of *Penicillium roquefortii* and *P. chrysogenum* after protoplast fusion. *Nature, London*, **262**, 719–21.

Anné, J. & Peberdy, J. F. (1975a). Induced fusion of fungal protoplasts by polyethylene glycol. In *Fourth International Symposium on Yeast and Other Protoplasts*, p. 56, September 8–12, 1975. Nottingham, England: Abstracts.

Anné, J. & Peberdy, J. F. (1975b). Conditions for induced fusion of fungal protoplasts in polyethylene glycol solutions. *Archives of Microbiology*, **105**, 201–5.

Anné, J. & Peberdy, J. F. (1976). Induced fusion of fungal protoplasts following treatment with polyethylene glycol. *Journal of General Microbiology*, **92**, 413–17.

Arima, K. & Takano, I. (1979a). Multiple fusion of protoplasts in *Saccharomyces* yeasts. *Molecular and General Genetics*, **173**, 271–7.

Arima, K. & Takano, I. (1979b). Evidence for co-dominance of the homothallic genes. HMα/hmα and HMa/hm$a$, in *Saccharomyces* yeasts. *Genetics,* **93,** 1–12.

Baltz, R. H. (1978). Genetic recombination in *Streptomyces fradiae* by protoplast fusion and cell regeneration. *Journal of General Microbiology*, **107**, 93–102.

Becher, D. & Böttcher, F. (1980). Hybridization of *Rhodosporidium toruloides* by protoplast fusion. In *Advances in Protoplast Research*, ed. L. Ferenczy & G. L. Farkas, pp. 105–11. Budapest, Akadémiai Kiadó, Oxford: Pergamon Press.

Beggs, J. D. (1978). Transformation of yeast by a replicating hybrid plasmid. *Nature, London*, **275**, 104–9.

Bibb, M. J., Ward, J. M. & Hopwood, D. A. (1978). Transformation of plasmid DNA into *Streptomyces* at high frequency. *Nature, London*, **274**, 398–400.

Binding, H. & Weber, H. J. (1974). The isolation, regeneration and fusion of *Phycomyces* protoplasts. *Molecular and General Genetics*, **135**, 273–6.

Boissonnet-Menes, M. & Lecoq, H. (1976). Transmission de virus par fusion de protoplastes chez *Pyricularia oryzae* Briosi et Cav. *Physiologie Végétale*, **14,** 251–7.

Böttcher, F., Becher, D., Klinner, U., Samsonova, I. A. & Schilowa, B. (1980). Genetic structure of yeast hybrids constructed by protoplast fusion. In *Advances in Protoplast Research*, ed. L. Ferenczy & G. L. Farkas, pp. 99–104. Budapest: Akadémiai Kiadó, Oxford: Pergamon Press.

Carlson, P. S. (1975). The fungal-like genetics of higher plants. *Genetics*, **79**, 353–8.

Cassels, A. C. & Barlass, M. (1978). The initiation of TMV infection in isolated protoplasts by polyethylene glycol. *Virology*, **87**, 459–62.

Chang, S. & Cohen, S. N. (1979). High-frequency transformation of *Bacillus subtilis* protoplasts by plasmid DNA. *Molecular and General Genetics*, **168**, 111–15.

Christensen, B. E. (1979). Somatic hybridization in *Saccharomyces cerevisiae*: analysis of products of protoplast fusion. *Carlsberg Research Communication*, **44**, 225–33.

Cocking, E. C. (1960). A method for the isolation of plant protoplasts and vacuoles. *Nature, London*, **187**, 927–9.

Cocking, E. C. (1977). Uptake of foreign genetic material by plant protoplasts. *International Review of Cytology*, **48**, 323–43.

Cocking, E. C. (1980). Protoplasts: Past and present. In *Advances in Protoplast*

Research, ed. L. Ferenczy & G. L. Farkas, pp. 3–15. Budapest: Akadémiai Kiadó, Oxford: Pergamon Press.

Coetzee, J. N., Sirgel, F. A. & Lecatsas, G. (1979). Genetic recombination in fused spheroplasts of *Providence alcalifaciens*. *Journal of General Microbiology*, **114**, 313–22.

Coutts, R. H. A. & Cocking, E. C. (1972). Infection of protoplasts from yeast with tobacco mosaic virus. *Nature, London*, **240**, 466–7.

Croft, J. H. & Dales, R. B. G. (1979). Protoplast fusion and vegetative incompatibility in *Aspergillus nidulans*, in *Protoplasts – Applications in Microbial Genetics*, ed. J. F. Peberdy, pp. 27–34. Nottingham: University of Nottingham.

Croft, J. H., Dales, R. B. G., Turner, G. & Earl, A. (1980). The transfer of mitochondria between species of *Aspergillus*. In *Advances in Protoplast Research*, ed. L. Ferenczy & G. L. Farkas, pp. 85–92. Budapest: Akadémiai Kiadó, Oxford: Pergamon Press.

Cullis, P. R. & Hope, M. J. (1978). Effects of fusogenic agent on membrane structure of erythrocyte ghosts and the mechanism of membrane fusion. *Nature, London*, **271**, 672–4.

Dales, R. B. G. & Croft, J. H. (1977). Protoplast fusion and the isolation of heterokaryons and diploids from vegetatively incompatible strains of *Aspergillus nidulans*. *Federation of European Microbiological Societies Microbiology Letters*, **1**, 201–4.

Dales, R. B. G. & Croft, J. H. (1980). Protoplast fusion and the genetical analysis of vegetative incompatibility in *Aspergillus nidulans*. In *Advances in Protoplast Research*, ed. L. Ferenczy & G. L. Farkas, pp. 73–84. Budapest: Akadémiai Kiadó, Oxford: Pergamon Press.

Dawson, J. R. O., Dickerson, P. F., King, J. M., Sakai, F., Trim, A. R. H. & Watts, J. W. (1978). Improved methods for infection of plant protoplasts with viral ribonucleic acid. *Zeitschrift für Naturforschung*, **33c**, 548–51.

Dienes, L. & Smith, W. E. (1944). The significance of pleomorphism in *Bacteroides* strains. *Journal of Bacteriology*, **48**, 125–54.

Eddy, A. A. & Williamson, D. H. (1957). A method of isolating protoplasts from yeast. *Nature, London*, **179**, 1252–3.

Ferenczy, L. (1976). Some characteristics of intra- and interspecific protoplast fusion products of *Aspergillus nidulans* and *Aspergillus fumigatus*. In *Cell Genetics in Higher Plants*, ed. D. Dudits, G. L. Farkas & P. Maliga, pp. 171–82. Budapest: Akadémiai Kaidó.

Ferenczy, L. (1980). Fusion of protoplasts of auxotrophic fungal mutants: Diversity in the genetic background of nutritional complementation. In *Advances in Protoplast Research*, ed. L. Ferenczy & G. L. Farkas, pp. 55–62. Budapest: Akadémiai Kiadó, Oxford: Pergamon Press.

Ferenczy, L., Kevei, F. & Szegedi, M. (1975a). Increased fusion frequency of *Aspergillus nidulans* protoplasts. *Experientia*, **31**, 50–2.

Ferenczy, L., Kevei, F. & Szegedi, M. (1975b). High-frequency fusion of fungal protoplasts. *Experientia*, **31**, 1028–30.

Ferenczy, L., Kevei, F. & Szegedi, M. (1975c). Fusion of fungal protoplasts. In *Fourth International Symposium on Yeast and Other Protoplasts*, p. 29. September 8–12, 1975. Nottingham, England: Abstracts.

Ferenczy, L., Kevei, F. & Szegedi, M. (1976). Fusion of fungal protoplast induced by polyethylene glycol. In *Microbial and Plant Protoplasts*, ed. J. F. Peberdy, A. H. Rose, H. J. Rogers & E. C. Cocking, pp. 177–87. London, New York, San Francisco: Academic Press.

Ferenczy, L., Kevei, F., Szegedi, M., Frankó, A. & Rojik, I. (1976). Factors affecting high-frequency fungal protoplast fusion. *Experientia*, **32**, 1156–8.

Ferenczy, L., Kevei, F. & Zsolt, J. (1974). Fusion of fungal protoplasts. *Nature, London*, **248**, 793–4.

Ferenczy, L. & Maráz, A. (1977). Transfer of mitochondria by protoplast fusion in *Saccharomyces cerevisiae. Nature, London*, **268**, 524–5.

Ferenczy, L. & Maráz, A. (1979). Mitochondrial transfer in yeast by protoplast fusion. In *Protoplasts – Applications in Microbial Genetics*, ed. J. F. Peberdy, pp. 46–51. Nottingham: University of Nottingham.

Ferenczy, L., Szegedi, M. & Kevei, F. (1977). Interspecific protoplast fusion and complementation in *Aspergilli. Experientia*, **33**, 184–5.

Ferenczy, L., Vallin, C. & Maráz, A. (1977). A method of protoplast fusion for *Candida tropicalis* and other yeasts. In *Fifth International Specialized Symposium on Yeasts*, pp. 85–9. September 12–15, 1976, Keszthely, Hungary: Symposium Proceedings.

Ferenczy, L., Zsolt, J. & Kevei, F. (1972). Forced heterokaryon formation in auxotrophic *Geotrichum* strains by protoplast fusion. In *Third International Protoplast Symposium on Yeast Protoplasts*, p. 74. October 2–5, 1972, Salamanca, Spain: Abstracts.

Fodor, K. & Alföldi, L. (1976). Fusion of protoplasts of *Bacillus megaterium. Proceedings of the National Academy of Sciences, USA*, **73**, 2147–50.

Fodor, K. & Alföldi, L. (1979). Polyethylene-glycol induced fusion of bacterial protoplasts. Direct selection of recombinants. *Molecular and General Genetics*, **168**, 55–9.

Fodor, K., Demiri, E. & Alföldi, L. (1978). Polyethylene glycol-induced fusion of heat inactivated and living protoplasts of *Bacillus megaterium. Journal of Bacteriology*, **135**, 68–70.

Fodor, K., Hadlaczky, G. & Alföldi, L. (1975). Reversion of *Bacillus megaterium* protoplasts to the bacillary form. *Journal of Bacteriology*, **121**, 390–1.

Fodor, K., Rostás, K. & Alföldi, L. (1980). Bacterial protoplasts and their possible use in bacterial genetics. In *Advances in Protoplast Research*, ed. L. Ferenczy & G. L. Farkas, pp. 19–28. Budapest: Akadémiai Kiadó, Oxford: Pergamon Press.

Foster, P. R., Dunnill, P. & Lilly, M. D. (1973). The precipitation of enzymes from cell extracts of *Saccharomyces cerevisiae* by polyethylene glycol. *Biochimica et Biophysica Acta*, **317**, 505–16.

Fournier, P., Provost, A., Bourquignon, C. & Heslot, H. (1977). Recombination after protoplast fusion in the yeast *Candida tropicalis. Archives of Microbiology*, **115**, 143–9.

Frehel, C., Lheritier, A.-M., Sanchez-Rivas, C. & Schaeffer, P. (1979). Electron microscopic study of *Bacillus subtilis* protoplast fusion. *Journal of Bacteriology*, **137**, 1354–61.

Gabor, M. H. & Hotchkiss, R. D. (1979). Parameters governing bacterial regeneration and genetic recombination after fusion of *Bacillus subtilis* protoplasts. *Journal of Bacteriology*, **137**, 1346–53.

Gefter, M. L., Margulies, D. H. & Scharff, M. D. (1977). A simple method for polyethylene glycol-promoted hybridization of mouse myeloma cells. *Somatic Cell Genetics*, **3**, 231–6.

Genther, F. J. & Borgia, P. T. (1978). Spheroplast fusion and heterokaryon formation in *Mucor racemosus. Journal of Bacteriology*, **134**, 349–52.

Gunge, N. & Sakaguchi, K. (1979). Fusion of mitochondria with protoplasts in *Saccharomyces cerevisiae. Molecular and General Genetics*, **170**, 243–7.

Gunge, N. & Tamaru, A. (1978). Genetic analysis of products of protoplast fusion in *Saccharomyces cerevisiae. Japanese Journal of Genetics*, **53**, 41–9.

HADLACZKY, G., FODOR, K. & ALFÖLDI, L. (1976). Morphological study of the reversion to bacillary form of *Bacillus megaterium* protoplasts. *Journal of Bacteriology*, **125**, 1172–9.

HALES, A. (1977). A procedure for the fusion of cells in suspension by means of polyethylene glycol. *Somatic Cell Genetics*, **3**, 227–30.

HAMLYN, P. F. & BALL, C. (1979). Recombination studies with *Cephalosporium acremonium*. In *Genetics of Industrial Microorganisms*, ed. O. K. Sebek & A. I. Laskin, pp. 185–91. Washington: American Society for Microbiology.

HICKS, J. B., HINNEN, A. & FINK, G. R. (1979). Properties of yeast transformation. *Cold Spring Harbor Symposia on Quantitative Biology*, **43**, 1305–13.

HINNEN, A., HICKS, J. B. & FINK, G. R. (1978). Transformation of yeast. *Proceedings of the National Academy of Sciences, USA*, **75**, 1929–33.

HOCKNEY, R. C. & FREEMAN, R. F. (1980). Construction of polysaccharide-degrading brewing yeast by protoplast fusion. In *Advances in Protoplasts Research*, ed. L. Ferenczy & G. L. Farkas, pp. 139–44. Budapest: Akadémiai Kiadó, Oxford: Pergamon Press.

HOPWOOD, D. A. & WRIGHT, H. M. (1978). Bacterial protoplast fusion: Recombination in fused protoplast of *Streptomyces coelicolor*. *Molecular and General Genetics*, **162**, 307–17.

HOPWOOD, D. A. & WRIGHT, H. M. (1979). Factors affecting recombinant frequency in protoplast fusions of *Streptomyces coelicolor*. *Journal of General Microbiology*, **111**, 137–43.

HOPWOOD, D. A., WRIGHT, H. M., BIBB, M. J. & COHEN, S. N. (1977). Genetic recombination through protoplast fusion in *Streptomyces*. *Nature, London*, **268**, 171–4.

HOPWOOD, D. A., WRIGHT, H. M., BIBB, M. J. & WARD, J. M. (1979). Applications of protoplast in *Streptomyces* genetics. In *Protoplasts – Applications in Microbial Genetics*, ed. J. F. Peberdy, pp. 5–11. Nottingham: University of Nottingham.

HOTCHKISS, R. D. & GABOR, H. M. (1980). Recombination, segregation, and unequal chromosome expression in bacterial diploids created by protoplast fusion. In *Advances in Protoplast Research*, ed. L. Ferenczy & G. L. Farkas, pp. 29–36. Budapest: Akadémiai Kiadó, Oxford: Pergamon Press.

INGOLIA, T. D. & KOSHLAND, D. E. (1978). The role of calcium in fusion of artificial vesicles. *The Journal of Biological Chemistry*, **253**, 3821–9.

ITO, T. & OHNISHI, S.-I. (1974). Ca^{2+}-induced lateral phase separations in phosphatidic acid – phosphatidil-choline membranes. *Biochimica et Biophysica Acta*, **352**, 29–37.

KABACK, H. R. (1971). Biological membranes. In *Methods in Enzymology*, ed. W. B. Jakoby, vol. 22, pp. 99–120. New York & London: Academic Press.

KANEKO, H. & SAKAGUCHI, K. (1979). Fusion of protoplasts and genetic recombination of *Brevibacterium flavum*. *Agricultural and Biological Chemistry*, **43**, 1007–13.

KAO, K. N. & MICHAYLUK, M. R. (1974). A method for high-frequency intergeneric fusion of plant protoplasts. *Planta*, **115**, 355–67.

KAWAKAMI, N., MONDO, H. & KAWAKAMI, H. (1978). Protoplast fusion of *Saccharomyces cerevisiae* – optimal procedures and morphological evidences. *Transactions of Mycological Society of Japan*, **19**, 181–7.

KAWAKAMI, N., TANAKA, H., MONDO, H., KATAMINE, S. & KAWAKAMI, H. (1980). Incorporation of algal thylakoid membrane and DNA in yeast protoplasts. In *Advances in Protoplast Research*, ed. L. Ferenczy & G. L. Farkas, pp. 49–54. Budapest: Akadémiai Kiadó, Oxford: Pergamon Press.

KEVEI, F. & PEBERDY, J. F. (1977). Interspecific hybridization between *Aspergillus nidulans* and *Aspergillus rugulosus* by fusion of somatic protoplasts. *Journal of General Microbiology*, **102**, 255–62.

KEVEI, F. & PEBERDY, J. F. (1979). Induced segregation in interspecific hybrids of *Aspergillus nidulans* and *Aspergillus rugulosus* obtained by protoplast fusion. *Molecular and General Genetics*, **170**, 213–18.

KINGSMAN, A. J., CLARKE, L., MORTIMER, R. K. & CARBON, J. (1979). Replication in *Saccharomyces cerevisiae* of plasmid pBR313 carrying DNA from the yeast *trpl* region. *Gene*, **7**, 141–52.

KLINNER, U., BÖTTCHER, F. & SAMSONOVA, I. A. (1980). Hybridization of *Pichia guilliermondii* by protoplast fusion. In *Advances in Protoplast Research*, ed. L. Ferenczy & G. L. Farkas, pp. 113–18. Budapest: Akadémiai Kiadó, Oxford: Pergamon Press.

KNUTTON, S. (1979). Studies of membrane fusion. III. Fusion of erythrocytes with polyethylene glycol. *Journal of Cell Science*, **36**, 61–72.

KNUTTON, S. & PASTERNAK, C. A. (1979). The mechanism of cell-cell fusion. *Trends in Biochemical Sciences*, **4**, 220–3.

KOZHINA, T. N. & CHEPURNAYA, O. V. (1980). Sliyanie protoplastov kak metod polucheniya gibridov u drozzhej-Sakharomicetov. *Genetika*, **16**, 361–3.

LANDMAN, O. E. & DE CASTRO-COSTA, M. R. (1976). Reversion of protoplasts and L forms of bacilli. In *Microbial and Plant Protoplasts*, ed. J. F. Peberdy, A. H. Rose, H. J. Rogers & E. C. Cocking, pp. 201–17. London, New York, San Francisco: Academic Press.

LANSMAN, J. & HAYNES, D. H. (1975). Kinetics of a Ca^{2+}-triggered membrane aggregation reaction of phospholipid membranes. *Biochimica et Biophysica Acta*, **394**, 335–47.

LEDERBERG, J. (1956). Bacterial protoplasts induced by penicillin. *Proceedings of the National Academy of Sciences, USA*, **42**, 574–7.

LEDERBERG, J. & ST CLAIR, J. (1958). Protoplasts and L-type growth of *Escherichia coli*. *Journal of Bacteriology*, **75**, 143–60.

LEVI, C., SANCHEZ RIVAS, C. & SCHAEFFER, P. (1977). Further genetic studies on the fusion of bacterial protoplasts. *Federation of European Microbiological Societies Microbiology Letters*, **2**, 323–6.

LHOAS, P. (1971). Infection of protoplasts from *Penicillium stoloniferum* with double-stranded RNA virus. *Journal of General Virology*, **13**, 365–7.

LOPEZ-BELMONTE, F., GARCIA ACHA, I. & VILLANUEVA, J. R. (1966). Observations on the protoplasts of *Fusarium culmorum* and on their fusion. *Journal of General Microbiology*, **45**, 127–34.

LÜCKEMANN, G., SIPICZKI, M. & WOLF, K. (1979). Transmission, segregation, and recombination of mitochondrial genomes in zygote clones and protoplast fusion clones of yeast. *Molecular and General Genetics*, **177**, 185–7.

MCPHERSON, A. (1976). Crystallization of proteins from polyethylene glycol. *The Journal of Biological Chemistry*, **251**, 6300–3.

MCQUILLEN, K. (1960). Bacterial protoplasts. In *The Bacteria*, vol. 1, ed. I. C. Gunsalus & R. Y. Stanier, pp. 249–359. New York & London: Academic Press.

MAGGIO, B., AHKONG, Q. F. & LUCY, J. A. (1976). Polyethylene glycol, surface potential and cell fusion. *Biochemical Journal*, **158**, 647–50.

MARÁZ, A. & FERENCZY, L. (1979a). Mating-type independent protoplast fusion in *Saccharomyces cerevisiae*. In *Protoplasts – Applications in Microbial Genetics*, ed. J. F. Peberdy, pp. 34–45. Nottingham: University of Nottingham.

MARÁZ, A. & FERENCZY, L. (1979b). Transfer of mitochondria *via* anucleate protoplasts in *Saccharomyces cerevisiae*. In *Eighth Congress of the Hungarian Society of Microbiology*, p. 135, August 27–29, 1979. Budapest, Hungary: Abstracts.

MARÁZ, A., KISS, M. & FERENCZY, L. (1978). Protoplast fusion in *Saccharomyces cerevisiae* strains of identical and opposite mating types. *Federation of European Microbiological Societies Microbiology Letters*, **3**, 319–22.

MARQUIS, R. E. & CORNER, T. R. (1976). Isolation and properties of bacterial protoplasts. In *Microbial and Plant Protoplasts*, ed. J. F. Peberdy, A. H. Rose, H. J. Rogers & E. C. Cocking, pp. 1–22. London, New York, San Francisco: Academic Press.

MATAGNE, R. F., DELTOUR, R. & LEDOUX, L. (1979). Somatic fusion between cell wall mutants of *Chlamydomonas reinhardi*. *Nature, London*, **278**, 344–6.

MATAGNE, R. F. & HERMESSE, M.-P. (1980). Chloroplast gene inheritance studied by somatic fusion in *Chlamydomonas reinhardtii*. *Current Genetics*, **1**, 127–31.

MAULE, A. J., BOULTON, M. I., EDMUNDS, C. & WOOD, K. R. (1980). Polyethylene glycol-mediated infection of cucumber protoplasts by cucumber mosaic virus and virus RNA. *Journal of General Virology*, **47**, 199–203.

MELCHERS, G. (1977). Kombination somatischer und konventioneller Genetik für die Pflanzenzüchtung. *Naturwissenschaften*, **64**, 184–94.

MELLON, R. E. (1925). Studies on microbic heredity. I. Observations on a primitive form of sexuality (zygospore formation) in the colon-typhoid group. *Journal of Bacteriology*, **10**, 481–501.

MORGAN, A. J., HALL, J. L., BRUNNER, A. & WHITTAKER, P. A. (1980). Protoplast fusion in the study of mitochondrial genetics in the petite-negative yeast, *Kluyveromyces lactis*. In *Advances in Protoplast Research*, ed. L. Ferenczy & G. L. Farkas, pp. 93–8. Budapest: Akadémiai Kiadó, Oxford: Pergamon Press.

MORGAN, A. J., HERITAGE, H. & WHITTAKER, P. A. (1977). Protoplast fusion between petite and auxotrophic mutants of the petite-negative yeast. *Kluyveromyces lactis*. *Microbios Letters*, **4**, 103–7.

MÜLLER, R. (1966). Die Entstehung entwicklungsfähiger Protoplasten aus Hefezellen und ihre Reversion. *Wissenschaftliche Filme aus dem Zentralinstitut für Mikrobiologie und experimentelle Therapie der Wissenschaften der DDR*, No. T-HF 659.

MÜLLER, R. (1970). Contribution to the problem of protoplast fusion. *Acta Facultatis Medicae Universitatis Brunensis*, **37**, 39–41.

NEČAS, O. (1956a). Regeneration of yeast cells from naked protoplasts. *Nature, London*, **177**, 898–9.

NEČAS, O. (1956b). Die Regeneration von Zellfragmenten bei den Hefen. *Biologisches Zentralblatt*, **75**, 268–81.

NEČAS, O. (1980). Regeneration of protoplasts. In *Advances in Protoplast Research*, ed. L. Ferenczy & G. L. Farkas, pp. 151–61. Budapest: Akadémiai Kiadó, Oxford: Pergamon Press.

OCHI, K., HITCHCOCK, M. J. M. & KATZ, E. (1979). High-frequency fusion of *Streptomyces parvulus* or *Streptomyces antibioticus* protoplasts induced by polyethylene glycol. *Journal of Bacteriology*, **139**, 984–92.

OKANISHI, M., SUZUKI, K. & UMEZAWA, H. (1974). Formation and reversion of *Streptomycete* protoplasts: Cultural condition and morphological study. *Journal of General Microbiology*, **80**, 389–400.

O'MALLEY, K. A. & DAVIDSON, R. L. (1977). A new dimension in suspension fusion techniques with polyethylene glycol. *Somatic Cell Genetics*, **3**, 441–8.

PAPAHADJOPOULOS, D., POSTE, G., SCHAEFFER, B. E. & VAIL, W. J. (1974). Membrane fusion and molecular segregation in phospholipid vesicles. *Biochimica et Biophysica Acta*, **352**, 10–28.

PAPAHADJOPOULOS, D., VAIL, W. J., PANGBORN, W. A. & POSTE, G. (1976). Studies on membrane fusion. II. Induction of fusion in pure phospholipid membranes by

calcium ions and other divalent metals. *Biochimica et Biophysica Acta*, **448**, 265–83.
PARK, J. T. (1968). The mechanisms by which penicillin causes conversion of bacterial cells to spheroplasts. In *Microbial Protoplasts, Spheroplasts and L-Forms*, ed. L. B. Guze, pp. 52–4. Baltimore: The Williams & Wilkins Company.
PEBERDY, J. F. (1976). Isolation and properties of protoplasts from filamentous fungi. In *Microbial and Plant Protoplasts*, ed. J. F. Peberdy, A. H. Rose, H. J. Rogers & E. C. Cocking, pp. 39–50. London, New York, San Francisco: Academic Press.
PEBERDY, J. F. (1978). Protoplasts and their development. In *The Filamentous Fungi*, ed. J. E. Smith & D. R. Berry, vol. 3, pp. 119–31. London: Edward Arnold.
PEBERDY, J. F. (1979a). Fungal protoplasts: Isolation, reversion and fusion. *Annual Review of Microbiology*, **33**, 21–39.
PEBERDY, J. F. (1979b). Isolation and fusion of bacterial protoplasts. In *Protoplasts – Applications in Microbial Genetics*, ed. J. F. Peberdy, pp. 1–4. Nottingham: University of Nottingham.
PEBERDY, J. F. (1980a). Protoplast fusion – a new approach to interspecies genetic manipulation and breeding in fungi. In *Advances in Protoplast Research*, ed. L. Ferenczy & G. L. Farkas, pp. 63–72. Budapest: Adadémiai Kiadó, Oxford: Pergamon Press.
PEBERDY, J. F. (1980b). Protoplast fusion – a tool for genetic manipulation and breeding in industrial microorganism. *Enzyme and Microbial Technology*, **2**, 23–9.
PEBERDY, J. F., EYSSEN, H. & ANNÉ, J. (1977). Interspecific hybridization between *Penicillium chrysogenum* and *Penicillium cyaneo-fulvum* following protoplast fusion. *Molecular and General Genetics*, **157**, 281–4.
PEBERDY, J. F. & KEVEI, F. (1979). Interspecific hybridization in fungi by protoplast fusion. In *Protoplasts – Application in Microbial Genetics*, ed. J. F. Peberdy, pp. 23–6. Nottingham: University of Nottingham.
PONTECORVO, G. (1975). Production of mammalian somatic cell hybrids by means of polyethylene glycol treatment. *Somatic Cell Genetics*, **1**, 397–400.
PONTECORVO, G., RIDDLE, P. N. & HALES, A. (1977). Time and mode of fusion of human fibroblasts treated with polyethylene glycol (PEG). *Nature, London*, **265**, 257–8.
POTRYKUS, I. (1971). Intra and interspecific fusion of protoplasts from petals of *Torenia baillonii* and *Torenia fournieri*. *Nature, London*, **231**, 57–8.
POWER, J. B. & COCKING, E. C. (1971). Fusion of plant protoplasts. *Science Progress. Oxford*, **59**, 181–98.
POWER, J. B., CUMMINGS, S. E. & COCKING, E. C. (1970). Fusion of isolated plant protoplasts. *Nature, London*, **225**, 1016–18.
PROVOST, A., BOURGUIGNON, C., FOURNIER, P., RIBET, A. M. & HESLOT, H. (1978). Intergeneric hybridization in yeasts through protoplast fusion. *Federation of European Microbiological Societies Microbiology Letters*, **3**, 309–12.
RIEGER, R., MICHAELIS, A. & GREEN, M. M. (1976). *Glossary of Genetics and Cytogenetics*. Jena: VEB Gustav Fischer Verlag.
ROHDE, W., PAULI, G., HENNING, J. & FRIIS, R. R. (1978). Polyethylene glycol-mediated infection with avian sarcoma viruses. *Archives of Virology*, **58**, 55–9.
RUSSELL, I. & STEWART, G. G. (1979). Spheroplast fusion of brewer's yeast strain. *Journal of the Institute of Brewing*, **85**, 95–8.

SAGARA, Y., FUKUI, K., OTA, F., YOSHIDA, N., KASHIYAMA, T. & FUJIMOTO, M. (1971). Rapid formation of protoplasts of *Streptomyces griseoflavus* and their fine structure. *Japanese Journal of Microbiology*, **15**, 73–84.

SANCHEZ-RIVAS, C. & GARRO, A. J. (1979). Bacterial fusion assayed by a prophage complementation test. *Journal of Bacteriology*, **137**, 1340–5.

SAVCHENKO, G. V. & KAPULTSEVICH, YU. G. (1980). Hybridization of yeast from genus *Hansenula* by protoplast fusion. In *Advances in Protoplast Research*, ed. L. Ferenczy & G. L. Farkas, pp. 125–30. Budapest: Akadémiai Kiadó, Oxford: Pergamon Press.

SCHAEFFER, P., CAMI, B. & HOTCHKISS, R. D. (1976). Fusion of bacterial protoplasts. *Proceedings of the National Academy of Sciences, USA*, **73**, 2151–5.

SIPICZKI, M. (1979). Interspecific protoplast fusion in fission yeasts. *Current Microbiology*, **3**, 37–40.

SIPICZKI, M. & FERENCZY, L. (1976). Protoplast fusion of *Schizosaccharomyces pombe* auxotrophic mutants of identical mating-type. In *Eighth International Conference on Yeast Genetics and Molecular Biology*, August 29–September 4, 1976, Schliersee, W. Germany: Abstracts.

SIPICZKI, M. & FERENCZY, L. (1977a). Protoplast fusion of *Schizosaccharomyces pombe* auxotrophic mutants of identical mating-type. *Molecular and General Genetics*, **151**, 77–81.

SIPICZKI, M. & FERENCZY, L. (1977b). Fusion of *Rhodosporidium* (*Rhodotorula*) protoplasts. *Federation of European Microbiological Societies Microbiology Letters*, **2**, 203–5.

SMITH, W. E. (1944). Observations indicating a sexual mode of reproduction in a common bacterium (*Bacteroides funduliformis*). *Journal of Bacteriology*, **47**, 417–18.

SNOW, R. (1979). Toward genetic improvement of wine yeast. *American Journal of Enology and Viticulture*, **30**, 33–7.

SOLINGEN, P. VAN & PLAAT, J. B. VAN DER (1977). Fusion of yeast spheroplasts. *Journal of Bacteriology*, **130**, 946–7.

SPATA, L. & WEBER, H. (1980). A study on protoplast fusion and parasexual hybridization of alcane utilizing yeasts. In *Advances in Protoplast Research*, ed. L. Ferenczy & G. L. Farkas, pp. 131–7. Budapest: Akadémiai Kiadó, Oxford: Pergamon Press.

SPENCER, J. F. T., LAND, P. & SPENCER, D. M. (1980). The use of mitochondrial mutants in the isolation of hybrids obtained by fusion of protoplasts of brewing yeasts. In *Advances in Protoplast Research*, ed. L. Ferenczy & G. L. Farkas, pp. 145–50. Budapest: Akadémiai Kiadó, Oxford: Pergamon Press.

SPIZIZEN, J. (1962). Preparation and use of protoplasts. In *Methods of Enzymology*, vol. 5, ed. S. P. Colowick & N. Q. Kaplan, pp. 122–34. New York & London: Academic Press.

STAHL, U. (1978). Zygote formation and recombination between like mating types in the yeast *Saccharomycopsis lipolytica* by protoplast fusion. *Molecular and General Genetics*, **160**, 111–13.

STÄHELIN, H. (1954). Über osmotisches Verhalten und Fusion nackter Protoplasten von *Bac. anthracis*. *Schweizerische Zeitschrift für allgemeine Pathologie und Bakteriologie*, **17**, 296–310.

STEMPEN, H. & HUTCHINSON, W. G. (1951). The formation and development of large bodies in *Proteus vulgaris* Ox-19. I. Bright phase contrast observations of living bacteria. *Journal of Bacteriology*, **61**, 321–35.

STEWART, G. G. (1978). Application of yeast genetics within the brewing industry. *Journal of the American Society of Brewing Chemists*, **36**, 177–85.

STRUNK, Ch. (1967). Protoplastenenstehung und Reversion bei Basidiomyceten (*Polystictus versicolor*). In *Symposium über Hefe-Protoplasten*, ed. R. Müller, pp. 213–15, Berlin: Akademie Verlag.

SUN, S. T., HSANG, C. C., DAY, E. P. & HO, J. T. (1979). Fusion of phosphatidylserine and mixed phosphatidylserine-phosphatidylcholine vesicles. Dependence on calcium concentration and temperature. *Biochimica et Biophysica Acta*, **557**, 45–52.

SVOBODA, A. (1976). Mating reaction in yeast protoplasts. *Archives of Microbiology*, **110**, 313–18.

SVOBODA, A. (1977). Intraspecies fusion of yeast protoplasts. *Folia Microbiologica*, **22**, 441–2.

SVOBODA, A. (1978). Fusion of yeast protoplast induced by polyethylene glycol. *Journal of General Microbiology*, **109**, 167–75.

SVOBODA, A. (1980). Intergeneric fusion of yeast protoplast: *Saccharomyces cerevisiae* + *Schizosaccharomyces pombe*. In *Advances in Protoplast Research*, ed. L. Ferenczy & G. L. Farkas, pp. 119–24. Budapest: Akadémiai Kiadó, Oxford: Pergamon Press.

SZVOBODA, G., LÁNG, T., GADÓ, I., AMBRUS, G., KARI, C., FODOR, K. & ALFÖLDI, L. (1980). Fusion of *Micromonospora* protoplasts. In *Advances in Protoplast Research*, ed. L. Ferenczy & G. L. Farkas, pp. 235–40. Budapest: Akadémiai Kiadó, Oxford: Pergamon Press.

TAKANO, I. & ARIMA, K. (1979). Evidence of the insensitivity of the α-inc allele to the function on the homothallic genes in *Saccharomyces* yeasts. *Genetics*, **91**, 245–54.

THURIAUX, P., SIPICZKI, M. & FANTES, F. A. (1980). Genetical analysis of a sterile mutant by protoplast fusion in the fission yeast *Schizosaccharomyces pombe*. *Journal of General Microbiology*, **116**, 525–8.

TSENIN, A. N., KARIMOV, G. A. & RIBCHIN, V. N. (1978). Recombinaciya pri sliyanii protoplastov *Escherichia coli* K 12. *Dokladi Akademii Nauk SSSR*, **243,** 1066–8.

VALLIN, C. & FERENCZY, L. (1978). Diploid formation of *Candida tropicalis* via protoplast fusion. *Acta Microbiologica Academiae Scientiarum Hungaricae*, **25**, 209–12.

VASIL, I. K. & GILES, K. L. (1975). Induced transfer of higher plant chloroplasts into fungal protoplasts. *Science*, **190**, 680.

VILLANUEVA, J. R. (1966). Protoplast of fungi. In *The fungi*, vol. 2, ed. G. C. Ainsworth & A. S. Sussman, pp. 3–62. New York: Academic Press.

VILLANUEVA, J. R. & GARCIA ACHA, I. (1971). Production and use of fungal protoplasts. In *Methods in Microbiology*, vol. 4, ed. C. Booth, pp. 665–718. London & New York: Academic Press.

WACKER, A. & KAUL, S. (1977). Polyäthylenglycol-induzierte Fusion von Säugetierzellen mit Cytoplasten in Suspension. *Naturwissenschaften*, **64**, 146.

WALLIN, A. & ERIKSSON, T. (1973). Protoplast cultures from cell suspension of *Daucus carota*. *Physiologia Plantarum*, **28**, 33–9.

WALLIN, A., GLIMELIUS, K. & ERIKSSON, T. (1974). The induction and aggregation and fusion of *Daucus carota* protoplasts by polyethylene glycol. *Zeitschrift für Pflanzenphysiologie*, **74**, 64–80.

WEIBULL, C. (1953). The isolation of protoplasts from *Bacillus megaterium* by controlled treatment with lysozyme. *Journal of Bacteriology*, **66**, 688–95.

WEIBULL, C. (1958). Bacterial protoplasts. *Annual Review of Microbiology*, **12**, 1–26.

WEISS, R. L. (1976). Protoplast formation in *Escherichia coli*. *Journal of Bacteriology*, **128**, 668–70.

WHITTAKER, P. A. & LEACH, S. M. (1978). Interspecific hybrid production between the yeast *Kluyveromyces lactis* and *Kluyveromyces fragilis* by protoplast fusion. *Federation of European Microbiological Societies Microbiology Letters*, **4**, 31–4.
YAMAMOTO, M. & FUKUI, S. (1977). Fusion of yeast protoplasts. *Agricultural and Biological Chemistry*, **41**, 1829–30.
YOSHIDA, K. (1979). Interspecific and intraspecific mitochondria-induced cytoplasmic transformation in yeasts. *Plant & Cell Physiology*, **20,** 851–6.

IN VIVO GENETIC MANIPULATION IN BACTERIA

DAVID SHERRATT

Department of Genetics, University of Glasgow, Glasgow G11 5JS, UK

INTRODUCTION

One of the first observed consequences of *in vivo* genetic manipulation in bacteria was noted by Griffith in 1928, when he observed that injection of mice with a mixture of heat-killed virulent streptococci and live avirulent streptococci, resulted in their death from pneumonia. Because live virulent streptococci could be isolated from the dead mice, Griffith concluded that the presence of the heat-killed virulent cells *transformed* the live bacteria from the avirulent to virulent type. Later this transformation was repeated with cultures and extracts of bacteria in the test tube and eventually Avery and colleagues were able to identify the transforming substance as DNA.

These experiments illustrate the importance of being able to transfer genetic information between bacteria in order to study the continual interactions between DNA molecules and the expression of their gene products. Though genetic transformation is a powerful tool for introducing DNA into cells, most of the 'classical' experiments in microbial genetics with *Escherichia coli* used either plasmid-specified conjugation or virus-promoted specialised and generalised transduction (Hayes, 1968). Transformation is a relatively new genetic tool in *E. coli* and other Gram-negative bacteria. More recently protoplast fusion mediated by polyethylene glycol has been added to the armoury of the microbial geneticist (Fodor & Alföldi, 1976; Shaeffer, Cami & Hotchkiss, 1976; Hopwood, Wright, Bibb & Cohen, 1977; Hopwood & Wright, 1978; Coetzee, Sirgel & Lecatsas, 1979). This latter technique provides the only known means of bringing together two complete genomes within a bacterial cell. The uses of protoplast fusion will not be discussed any further in this article.

Some of the reasons for manipulating bacterial genes are outlined in Table 1. Though many of the examples that will be discussed use *Escherichia coli* and related Gram-negative bacteria, any bacterium

Table 1. *Genetic analysis in bacteria*

Genetic recombination	Genetic complementation
Requires physical interaction between DNA molecules. Recombination can occur between and within genes. Most recombination occurs between regions of extensive DNA homology, though *illegitimate* recombination between regions of little or no homology is important in the generation of genome rearrangements. *Homologous recombination*. Used to integrate homologous DNA after DNA transfer experiments. Continually occurs between homologous DNA sequences in a cell. Involved in some DNA repair processes. Important for gene mapping, strain construction and for investigating the mechanism of homologous recombination. *Illegitimate recombination*. Used extensively for *in vivo* genetic manipulation, for example to generate specialised transducing viruses and plasmid-primes; for gene enrichment and for gene mapping; to bring non-homologous genetic regions together and for genetic mutagenesis.	Used to define the unit of genetic function (*cistron*). No physical interaction between DNA molecules required, though two or more copies of the genetic region under test must be present within a cell. A cistron can generally be considered equivalent to a gene that encodes a diffusible function. If two mutations (each in genes specifying a product) on different chromosomes are introduced into a cell, the cell will produce both wild-type products if the mutations are in different genes (one product from each chromosome), but no wild-type product if both mutations are in the same gene, i.e. as a general rule, complementation occurs between genes, and does not result in inherited genetic change. Complementation is invaluable for assigning a series of mutations affecting a given phenotype into functional groups. Plasmid-primes, specialised transduction and virus infection provide the normal means of generating the necessary diploids in bacteria.

into which genetic material can be introduced is potentially amenable to genetic analysis.

METHODS FOR DNA TRANSFER BETWEEN CELLS

Genetic transformation

Genetic transformation involves the uptake, and usually the expression and inheritance, of DNA. The DNA can consist of random fragments or defined sequences and if it is to be inherited can establish itself as a replicon or integrate into the chromosome. In general, bacteria cannot distinguish the source of DNA, though *Haemophilus influenzae* has a requirement for DNA containing specific sequences (Deich & Smith, 1980). Whereas in *Bacillus subtilis* linear DNA transforms well and can integrate into the chromosome efficiently, in *E. coli* only circular DNA transforms wild-type strains well, though strains deficient in *exoV* (*rec*BC$^-$) can be transformed with linear DNA. The great advantage of genetic transformation is that the transforming DNA can be subject to mutagenesis or other *in vitro* treatments or manipulations prior to introduction into cells. The disadvantage of transformation is that with most bacteria only a small proportion of cells in a population can be transformed (about 0.1% with *E. coli*, 10–100 times higher with *B. subtilis*). In *E. coli* large plasmid DNA molecules (>20 kb) transform relatively inefficiently compared to smaller molecules.

Transduction

In specialised transduction, virus DNA covalently linked to non-viral DNA is packaged within a single virus particle. Because the virus heads are only big enough to accommodate the viral genome, specialised transducing particles must normally contain a deleted viral genome, which may or may not appear phenotypically defective. It also follows that for specialised transduction, the virus must become covalently attached to the sequences to be transduced, i.e. it needs to integrate into host chromosomal or extrachromosomal sequences. Bacteriophage λ is the classical *E. coli* specialised transducing phage. Because it normally integrates at a single locus in the *E. coli* chromosome, specialised transducing molecules contain chromosomal markers on either side of the integration site, but no others.

This restriction on the markers that bacteriophage λ can carry has been overcome by using integration of λ into secondary integration sites, by providing sequences in λ that can be used for homologous recombination with the same sequences elsewhere, or by using *in vitro* recombinant techniques which remove all restrictions on the nature of the DNA sequences that are carried on the virus genome. Specific *in vivo* genetic manipulations of λ are described later.

One of the major uses of specialised transduction has been to create cells diploid for the transducing DNA. This allows genetic complementation analysis to be carried out. Specialised transduction can also be used for genetic mapping, strain construction and for gene enrichment and purification. Expression of the purified sequences can then be studied. Because viral vectors normally have a limited host range, specialised transduction cannot be simply used for interspecies transfer of DNA.

In generalised transduction, random or near-random DNA fragments are occasionally packaged within viral particles during lytic growth. The virus can be temperate (e.g. P1) or virulent (e.g. T4; Wilson, Young, Edlin & Konigsberg, 1979), and the only restriction on the DNA carried is its size. Because such fragments will not normally carry replication origins, their inheritance depends on integration. In *E. coli* generalised transduction mediated by P1 has been widely used in genetic mapping and strain construction. The availability of P1 derivatives that are temperature-sensitive for lysogen maintenance and that carry drug-resistance determinants (Miller, 1972) has made generalised transduction in *E. coli* much simpler. Inter-species transfer is limited by the host-range of the viral vector. Fortunately P1 can grow on a number of different hosts. The *Salmonella typhimurium* virus P22 can be used both for specialised and generalised transduction.

Conjugation

Conjugal transfer is one of the most adaptable and has been one of the most useful methods for intra- and inter-species genetic transfer. It can result in the integration or replication of the transferred DNA and seems to be invariably plasmid mediated. For a recent review on the nature of conjugal transfer see Clark & Warren (1979). Though conjugation has been characterised in Gram-negative bacteria, conjugal transfer has been demonstrated in Gram-positive organisms. Though the classical conjugation experiments used the F plasmid, other conjugative plasmids carrying easily scorable phenotypic markers are often more convenient. Moreover the availability of broad host-range plasmids has made it possible to transfer genes amongst a wide range of Gram-negative bacteria (Datta *et al.*, 1971; Haas & Holloway, 1976; Beringer, Hoggan & Johnston, 1978). The ability of conjugative plasmids to integrate into the bacterial chromosome means that plasmid and/or chromosomal genes can be conjugally transferred. This integration into the chromosome can be mediated by *recA*-dependent homologous recombination or by *recA*-independent recombination (Cullum & Broda, 1979). Conjugal transfer of plasmid DNA is normally followed by its recircularisation and re-establishment as a replicon (repliconation) whilst transfer of chromosomal DNA is followed by its integration via homologous recombination into the recipient chromosome if it is to be inherited.

GENETIC RECOMBINATION

In *E. coli* and almost certainly in other organisms, recombination can be classified into two types. Homologous *recA*-dependent recombination requires fairly extensive homology to act efficiently and its action does not normally result in genome rearrangement. Much is now known about the mechanism of *recA*-mediate recombination. The *recA*-protein has been purified and its properties studied in detail: in part it promotes the 'aggression' of duplex DNA by single-stranded DNA (Shibata, DasGupta, Cunningham & Radding, 1979). At least within bacteriophage λ DNA, *recA*-mediated recombination requires specific sites known as CHI to act efficiently (Stahl, 1979). These sites are found every few kilobases on average in the *E. coli* chromosome, and they could be where *exoV* preferentially introduces single-strand breaks into DNA to produce recombinogenic aggressive single strands. Integration of chromosomal DNA after gene transfer experiments is normally by *recA*-mediated *exoV*-dependent homologous recombination.

Though homologous recombination is invaluable to the microbial geneticist, the use of *recA*-independent non-homologous recombination (*illegitimate* recombination) has revolutionised the way in which genomes can be manipulated *in vivo*. This recombination occurs continually in cells to generate deletions, insertions, duplications, translocations and inversions. Most spontaneous mutations in *E. coli* probably are the result of *illegitimate* recombination. If a strong selection is available, specific mutations or rearrangements can be selected directly without resort to deliberate mutagenesis or manipulations. However, since the realisation that some genetic elements can promote *illegitimate* recombination events very efficiently, the power of the microbial geneticist has increased enormously. Some applications are discussed in the next section.

RecA-independent non-homologous recombination can again be divided into two classes. The first includes so-called site-specific recombination systems. The prototype system is Int/Xis-mediated integration and excision of the bacterial virus λ in which recombination occurs between specific virus and host sequences that contain a 15 bp homology (Landy, Hoess, Bidwell & Ross, 1978; Hsu, Ross & Landy, 1980). Other temperate bacterial viruses use similar site-specific recombination systems for their integration and excision. Lambda Int/Xis-mediated recombination is most efficient between the viral site (*attP*) and the *primary attachment site* in the bacterial chromosome (*attB*), though other secondary chromosomal attachment sites can be used, albeit inefficiently (Shimada, Weisberg & Gottesman, 1972; Landy *et al.*, 1978). Site-specific recombination is used to invert segments of bacterial viruses Mu and P1, and to invert a region in *Salmonella typhimurium* that determines which of two flagella types are produced (Silverman & Simon, 1980). Interestingly, the same enzyme can mediate the recombination events in these three different systems (Kutsukake & Iino, 1980), indicating that they share a common origin. In Mu, and probably in P1, the ability to invert a DNA segment broadens the virus host range, apparently by allowing the synthesis of either of two types of tail fibre. These examples can also be considered as simple cases of 'antigenic variation'.

Another interesting class of site-specific recombination is conjugal transfer dependent (G. J. Warren & A. J. Clerk, personal communication). During conjugal transfer of ColE1-derived plasmids, intra-replicon and inter-replicon *recA*-independent recombination can occur between regions necessary in *cis* for transfer which apparently contain transfer origins.

The second class of illegitimate recombination events involves events in which there appears to be little specificity for unique sequences. For example, it appears that *recA*$^{-}$ *E. coli* can mediate recombination events across regions of homology as small as 5–8 base-pairs. This can result in deletions, duplications, etc., and is probably the mechanism for exact excision of transposable genetic elements and for the generation of many spontaneous deletions. The determinants for such recombinations appear to reside in the bacterial chromosome.

Transposable genetic elements (Starlinger, 1980) themselves mediate *illegitimate* recombination during transposition. Here, one of the components for the recombination events has little sequence specificity (the target sequence) whilst the other component (the element end) retains high specificity (Starlinger, 1980). At least some of the proteins required for genetic transposition are encoded by transposable elements. Current ideas for the mechanism of genetic transposition favour the use of an enzyme similar to the *cisA* protein of φX174 (Eisenberg, Griffith & Kornberg, 1977) for the breakage–reunion events.

SPECIFIC TOOLS FOR *IN VIVO* GENETIC MANIPULATIONS

Bacterial virus λ

Integration of λ, promoted by Int protein, occurs at *attB*, between *gal* and *bio* in the *E. coli* chromosome. Aberrant excision can generate specialised transducing viruses (Hayes, 1968) and selection for bacterial survivors of λ induction gives deletions extending out from the point of λ insertion. Generation of specialised transducing viruses and deletions can be extended to other regions of DNA if λ (or other temperate virus) can be integrated at sites other than *attB*. This has been achieved in two ways. In the first, λ is integrated at secondary attachment sites, of which there are many (Shimada *et al.*, 1972; McIntire & Willetts, 1978). In the second method, λ derivatives are constructed that contain DNA sequences which have homology with the desired integration site. This homologous recombination can be used for integration. A very powerful technique is to use hybrids between viruses Mu and λ. Such hybrids can be integrated using homologous recombination between Mu sequences in the virus and sites at which Mu has previously been integrated (Casadaban, 1975, 1976; MacNeil, Howe & Brill, 1980). Induction now yields λ specialised transducing viruses, containing the desired region. Casadaban (1976) has developed a modification of the general method in which the λ–Mu hybrid carries *lac* genes without a functional promoter. Integration of this phage through Mu homology, followed by selection for Lac$^+$ survivors of Mu induction (the Mu carries a thermoinducible repressor), gives derivatives in which *lac* expression is under the control of neighbouring promoters/operators. Since β-galactosidase is easily and sensitively assayed, the nature of given control systems can be probed. For example Chou, Casadaban, Lemaux & Cohen (1979) have examined the control of Tn3 transposition proteins and were able to isolate Tn3 mutants that were altered in normal control.

Transposable genetic elements

Transposable genetic elements are unique non-permuted segments of DNA that can insert at many loci within genomes (Starlinger, 1980). This integration occurs in the absence of a functional *recA*-dependent homologous recombination system. Many transposable elements encode proteins that are necessary for the transposition process. As tools for the bacterial geneticist, these elements are useful in a number of ways. By their insertion, they can mutate genomes; these mutations can be mapped both physically and genetically (Bukhari & Zipser, 1972; Dougan & Sherratt, 1977). Transposable elements that carry phenotypic markers can be used to 'mark' replicons that have no readily scorable phenotype, for example 'cryptic' plasmids (So, Heffron & Falkow, 1978). They can provide 'mobile sites of homology' upon which homologous recombination can act. The *recA*-dependent integration and excision of the plasmid F is one such example (Cullum & Broda, 1979). In addition to transposition, these elements can mediate a wide range of *illegitimate* recombination events, that include deletions, inversions, duplications and transversions. In general, the end points for such events are the boundaries of element and non-element DNA (Arthur & Sherratt, 1979; Shapiro, 1979; Starlinger, 1980).

Bacterial transposable elements can conveniently be divided into insertion sequences (IS elements), transposons and transposing viruses such as Mu. The last two classes are generally more useful to the bacterial geneticist; transposons because they carry selectable markers, and Mu because it can be induced to transpose at very high frequencies, with very little sequence specificity as regards its integration site. The availability of Mu derivatives that encode antibiotic resistance (Leach & Symonds, 1979) makes the use of this virus easier.

Bacterial virus Mu

Insertion mutants of Mu can be readily obtained after infection of cells with a Mu lysate. Selection of survivors from a temperature-induced thermoinducible lysogen gives derivatives deleted in the region of insertion. Since Mu can grow in many Gram-negative bacteria, insertion and deletion mutagenesis by Mu is widely applicable (see later). Moreover, Mu can efficiently promote the types of rearrangements alluded to above. For example, Mu-mediated transposition of chromosomal genes to a conjugative plasmid, to produce a plasmid-prime, can be accomplished in the following way (Faelen & Toussaint, 1976). A plasmid-containing donor, carrying a thermoinducible Mu prophage in the chromosome, is partially induced and mated with a suitable recipient. Recipients containing the required plasmid-prime are obtained after the appropriate selection. Chromosomal genes close to or distant from the prophage location can be transposed in this way. The transposed DNA is separated from the plasmid DNA by two directly repeated copies of Mu, as predicted by current models for transposition (Arthur & Sherratt, 1979; Shapiro, 1979).

Transposons

Transposons have been widely used as genetic mutagens, though it has now become clear that not all transposons are equally useful. For example, whereas Tn*1* and Tn*3* transpose readily into plasmids, chromosomal insertions can only be obtained with difficulty. Again, these transposons appear to transpose 'cleanly' into small plasmids, but transposition into large plasmids is often accompanied by adjacent deletions. The kanamycin resistant transposon Tn*5* is perhaps the most generally useful. It transposes readily with little target specificity, into both chromosomes and extrachromosomal elements. Tn*7* appears to insert at high frequency into a single locus in the *E. coli* chromosome (C. Lichtenstein, personal communication), but has many insertion sites in plasmids.

Transposon insertions can be readily mapped physically and sometimes genetically. Insertion of a transposon can provide usefully situated restriction targets for *in vitro* genetic manipulation and also homology for *recA*-dependent recombination. Insertion of a transposon into a region encoding a polycistronic message can result in blocked expression of genes distal to the insertion as well as the gene inserted into. This can be useful for identifying transcription units, but a nuisance when trying to map genes and do a complementation analysis. Though deletions and other rearrangements occur adjacent to transposon insertions, they usually cannot easily be selected.

It appears that at least some (and possibly all) transposable elements show preferred sites for integration. These appear to be A+T rich, and preferably contain sequences related to the element ends (Tu & Cohen, 1980).

A number of strategies are available for using transposon mutagenesis. One way is to take a cell population containing the target DNA and then to introduce the transposon on a DNA molecule that cannot replicate, either because it is on a non-replicon fragment, or because the replicon's replication is blocked in the recipient cells (e.g. using a temperature-sensitive replication mutant, a non-permissive host, incompatibility or superinfection immunity). A method for mutagenesis of either virus or plasmid

DNA is to introduce the target DNA into a transposon-containing cell. Transposon-containing virus or plasmid can then be recovered, for example in a virus lysate, in a plasmid DNA preparation by transformation, or after conjugal transfer.

Inter-replicon transposition of some transposons generates a cointegrate intermediate between the two replicons (Arthur & Sherratt, 1979; Shapiro, 1979). Tn*1* and Tn*3*, at least, then resolve this intermediate into the normal transposition products using a transposon-specified site-specific recombination system. This recombination system can also be used to enhance recombination between any two Tn*1*/*3*s. Moreover, using mutants or conditions that favour cointegrate formation, cointegrates between any two replicons can be obtained. Such molecules have many uses for the molecular biologist.

Broad host-range plasmids

The classical F plasmid can only be conjugally transferred to and maintained in close relatives of *E. coli*. It is therefore of limited use for the transfer of genes between bacterial species. Similarly, most viruses capable of propagation within *E. coli* do not infect other than a narrow range of hosts. The discovery of a set of conjugative plasmids, belonging to incompatibility group P1 and originally derived from *Pseudomonas*, that can transfer to and be maintained in almost any Gram-negative organism has revolutionised the molecular genetics of these organisms. The prototype of these plasmids is RP4 and its close (identical?) relatives RP1, RK2, and R68 (Burkhardt, Riess & Pühler, 1979). Though these plasmids can be efficiently transferred between species, they were found to mobilise *Pseudomonas aeruginosa* chromosomal genes poorly and spontaneous plasmid-primes were not easily obtained (Haas & Holloway, 1976; Jacob, Cresswell & Hedges, 1977). Haas & Holloway (1976) isolated derivatives of R68 that had increased ability to mobilise chromosomal genes. One of these derivatives, R68.45, has now been used extensively to mobilise chromosomal genes in a number of species, though its parent plasmid and its relatives have also been productively used (Towner, 1978; Beringer, Hoggan & Johnston, 1798; Johnston, Setchell & Beringer, 1978; Cees *et al.*, 1980).

Smaller, apparently related, non-conjugative plasmids conferring sulphonamide and streptomycin resistance and which have been isolated from a range of Gram-negative bacteria, also have a broad host-range (Barth & Grinter, 1974). R300B and RSF1010 are the prototypes of these plasmids. They have been transferred into cells by transformation and conjugation, and because of their size are useful vectors for *in vitro* genetic manipulation.

Insertion of transposable elements like Mu and transposons into broad host-range plasmids allows them to be transferred into species of choice, where they can be used to promote *in vivo* genetic manipulation in the ways described earlier. This is a particularly useful technique if the desired strain is refractory to genetic transformation and therefore is not readily amenable to *in vitro* genetic manipulation.

CONCLUDING REMARKS

Development of gene transfer mechanisms, particularly those that allow transfer between and within bacteria other than in *E. coli*, along with the exploitation of *illegitimate* recombination, particularly that specified by transposable elements, now allows detailed molecular analysis of genetic structure and function in a multitude of organisms. In combination with *in vitro* DNA technology, particularly that which generates defined mutations *in vitro, in vivo* genetic manipulation should allow a detailed structural and functional analysis to be made for the genetic regions of choice in bacteria amenable to the manipulations.

REFERENCES

Arthur, A. & Sherratt, D. (1979). Dissection of the transposition process: a transposon-encoded site-specific recombination system. *Molecular and General Genetics*, **175**, 267–74.

Barth, P. T. & Grinter, N. J. (1974). Comparison of the DNA molecular weights and homologies of plasmids conferring linked resistance to streptomycin and sulfonamides. *Journal of Bacteriology*, **120**, 618–30.

Beringer, J. E., Hoggan, S. A. & Johnston, W. B. (1978). Linkage mapping in *Rhizobium leguminosarum* by means of R-plasmid-mediated recombination. *Journal of General Microbiology*, **104**, 201–7.

Bukhari, A. I. & Zipser, D. (1972). Random insertion of Mu-1 DNA within a single gene. *Nature New Biology*, **236**, 240–3.

Burkhardt, H.-J., Riess, G. & Pühler, A. (1979). Relationship of group P1 plasmids revealed by heteroduplex experiments: RP1, RP4, R68 and RK2 are identical. *Journal of General Microbiology*, **114**, 341–8.

CASADABAN, M. J. (1975). Fusion of the *Escherichia coli lac* genes to the *ara* promoter: a general technique using bacteriophage Mu-l insertions. *Proceedings of the National Academy of Sciences, USA*, **72**, 809–13.
CASADABAN, M. J. (1976). Transposition and fusion of the *lac* genes to selected promoters in *Escherichia coli* using bacteriophage lambda and Mu. *Journal of Molecular Biology*, **104**, 541–55.
CEES, A. M., VAN DEN HONDEL, J. J., VERBEEK, S., VAN DER ENDE, A., WEISBEEK, P. J., BORRIAS, W. G., & VAN ARKEL, A. (1980). Introduction of transposon Tn*901* into a plasmid of *Anacystis nidulans. Proceedings of the National Academy of Sciences, USA*, **77**, 1570–4.
CHOU, J., CASADABAN, M. J., LEMAUX, P. & COHEN, S. N. (1979). Identification and characterization of a self-regulated repressor of translocation of the Tn*3* element. *Proceedings of the National Academy of Sciences, USA*, **76**, 4020.
CLARK, A. J. & WARREN, G. J. (1979). Conjugal transmission of plasmids. *Annual Review of Genetics 13*, 99.
COETZEE, J. N., SIRGEL, F. A. & LECATSAS, G. (1979). Genetic recombination in fused spheroplasts of *Providence alealifaciens. Journal of General Microbiology*, **114**, 313–22.
CULLUM, J. & BRODA, P. (1979). Chromosome transfer and Hfr formation by F in *rec*$^+$ and *recA* strains of *Escherichia coli* K12. *Plasmid*, **2**, 358.
DATTA, N., HEDGES, R. W., SHAW, E. J., SYKES, R. B. & RICHMOND, M. H. (1971). Properties of an R factor from *Pseudomonas aeruginosa. Journal of Bacteriology*, **108**, 1244–9.
DEICH, R. A. & SMITH, H. O. (1980). Mechanism of homospecific DNA uptake in *Haemophilus influenzae* transformation. *Molecular and General Genetics*, **177**, 369–74.
DOUGAN, G. & SHERRATT, D. (1977). The transposon Tn*1* as a probe for studying ColEl structure and function. *Molecular and General Genetics*, **151,** 151–60.
EISENBERG, S., GRIFFITH, J. & KORNBERG, A. (1977). φX174 *cistron* A protein is a multifunctional enzyme in DNA replication. *Proceedings of the National Academy of Sciences, USA*, **74**, 3198–202.
FAELEN, M. & TOUSSAINT, A. (1976). Bacteriophage Mu-l: A tool to transpose and to localize bacterial genes. *Journal of Molecular Biology*, **104**, 525–39.
FODOR, K. & ALFÖLDI, L. (1976). Fusion of protoplasts of *Bacillus megaterium. Proceedings of the National Academy of Sciences, USA*, **73**, 2147–350.
HAAS, D. & HOLLOWAY, B. W. (1976). R. factor variants with enhanced sex factor activity in *Pseudomonas aeruginosa. Molecular and General Genetics*, **144**, 243–51.
HAYES, W. (1968). *The Genetics of Bacteria and Their Viruses*. Oxford: Blackwell.
HOPWOOD, D. A., WRIGHT, H. M., BIBB, M. J. & COHEN, S. N. (1977). Genetic recombination through protoplast fusion in *Streptomyces. Nature, London*, **268**, 171–4.
HOPWOOD, D. A. & WRIGHT, H. M. (1978). Bacterial protoplast fusion. *Molecular and General Genetics*, **162**, 307–17.
HSU, P-L, ROSS, W. & LANDY, A. (1980). The λ phage *att* site: functional limits and interaction with Int protein. *Nature, London*, **285**, 85–91.
JACOB, A. E., CRESSWELL, J. M. & HEDGES, R. W. (1977). Molecular characterization of the P group plasmids R68 and variants with enhanced chromosome mobilizing ability. *Federation of European Microbiological Societies Letters*, **1**, 71–4.
JOHNSTON, A. W. B., SETCHELL, S. M. & BERINGER, J. G. (1978). Interspecific crosses between *Rhizobium leguminosarum* and *R. meliloti*: formation of haploid recombinants and of R.-primes. *Journal of General Microbiology*, **164**, 209–18.

KUTSUKAKE, K. & IINO, T. (1980). A *trans*-acting factor mediates inversion of a specific DNA segment in flagellar phase variation of *Salmonella*. *Nature, London*, **284**, 479–81.

LANDY, A., HOESS, R. H., BIDWELL, K. & ROSS, W. (1978). Site-specific recombination in bacteriophage λ: structural features of recombining sites. *Cold Spring Harbor Symposium on Quantitative Biology*, **43**, 1089–97.

LEACH, D. & SYMONDS, N. (1979). Isolation and characterisation of a plaque-forming derivative of bacteriophage Mu carrying a fragment of Tn*3* conferring ampicillin resistance. *Molecular and General Genetics*, **172**, 179–84.

MCINTIRE, S. & WILLETTS, N. (1978). Plasmid cointegrates of *Flac* and lambda prophage. *Journal of Bacteriology*, **134**, 184–92.

MACNEIL, D., HOWE, M. H. & BRILL, W. J. (1980). Isolation and characterization of lambda specialized transducing bacteriophages carrying *Klebsiella pneumoniae nif* genes. *Journal of Bacteriology*, **141**, 1264–71.

MILLER, J. (1972). *Experiments in Molecular Genetics*. Cold Spring Harbor Laboratory.

SHAEFFER, P., CAMI, B. & HOTCHKISS, R. D. (1976). Fusion of bacterial protoplasts. *Proceedings of the National Academy of Sciences, USA*, **73**, 2151–5.

SHAPIRO, J. A. (1979). Molecular model for the transposition and replication of bacteriophage Mu and other transposable sequences. *Proceedings of the National Academy of Sciences, USA*, **73**, 2151–5.

SHIBATA, T., DASGUPTA, C., CUNNINGHAM, R. P. & RADDING, C. M. (1979). Purified *Escherichia coli recA* protein catalyzes homologous pairing of superhelical DNA and single-stranded fragments. *Proceedings of the National Academy of Sciences, USA*, **76**, 1638–42.

SHIMADA, K., WEISBERG, R. A. & GOTTESMAN, M. E. (1972). Prophage λ at unusual locations. I location of the secondary attachment sites and the properties of the lysogens. *Journal of Molecular Biology*, **63**, 483–503.

SILVERMAN, M. & SIMON, M. (1980). Phase variation: genetic analysis of switching mutants. *Cell*, **19**, 845–54.

STAHL, F. W. (1979). Special sites in generalized recombination. *Annual Review of Genetics*, **13**, 7–24.

STARLINGER, P. (1980). IS elements and transposons. *Plasmid*, **3**, 241–59.

SO, M., HEFFRON, F. & FALKOW, S. (1978). Method for the genetic labeling of cryptic plasmids. *Journal of Bacteriology*, **133**, 1520–3.

TOWNER, K. J. (1978). Chromosome mapping in *Acinetobacter calcoaceticus*. *Journal of General Microbiology*, **104**, 175–80.

TU, C-P. D. & COHEN, S. N. (1980). Translocation specificity of the Tn*3* element: characterization of sites of multiple insertions. *Cell*, **19**, 151.

WILSON, G. G., YOUNG, K. K. Y., EDLIN, G. J. & KONIGSBERG, W. (1979). High-frequency generalised transduction by bacteriophage T4. *Nature, London*, **280**, 80–2.

GENE MANIPULATION *IN VITRO*

KENNETH N. TIMMIS

Max Planck Institute for Molecular Genetics
Berlin-Dahlem, West Germany

INTRODUCTION

The science of genetics has played a central role in the development of modern biology. Apart from providing important information on gene structure and organization, genetic methods have greatly facilitated studies of the nutrition, physiology and characteristic activities of living systems, and Man's exploitation of micro- and macro-organisms. The isolation of mutant organisms that exhibit altered behaviour in a particular activity allows the genes involved in that activity to be quantitated and mapped. Comparison of wild-type with mutant derivatives subsequently enables identification and characterization of specific gene products and provides information on their physiological roles and their regulation. In applied biology, genetic techniques have been used extensively to manipulate commercially important strains of organisms in order to achieve improved growth or cultivation characteristics, and/or higher yields of specific products.

In general, classical methods for the mutation and manipulation of organisms and the identification and analysis of mutant derivatives are inefficient and time consuming because they involve the random mutagenesis of total genomic DNA. The proportion of total mutants that are of the required phenotype is usually small therefore and such derivatives are obtained only after exhaustive screening of very large numbers of isolates. Moreover, such methods do not allow the isolation of mutant derivatives containing mutations (a) that are lethal, (b) for which no assay is available, or (c) which do not produce a marked change in phenotype. Classic methods therefore permit the analysis of only certain aspects of the function of any given gene and only certain genes in any given genome.

In the early 1970s the development of *in vitro* gene cloning techniques revolutionized the analysis and manipulation of genetic material by permitting the isolation, purification and selective amplification in convenient biological systems (such as *E. coli* cells) of almost any individual segment of DNA from practically any organism (Cohen, 1975). The deletion of DNA not containing relevant genes that is achieved by gene cloning enables the methods employed for analysis and manipulation to be focussed specifically on the gene region of interest, thus greatly increasing the efficiency of the method employed and simplifying the identification and characterization of new modified derivatives. More recently, several methods for the introduction of defined mutations at predetermined locations in cloned DNA have been developed. The examination of wild-type and manipulated cloned DNA segments in suitable *in vivo* and *in vitro* DNA expression systems permits the rapid identification of specific gene products, and the characterization of functional changes in these products which occur in response to the specific manipulation employed. The new techniques not only simplify classical genetic operations, such as the analysis of genetic modifications responsible for specific phenotypic changes, but also enable 'reversed genetics' experiments to be carried out, i.e. the specific alteration of a predetermined site within a gene or DNA sequence, followed by analysis of the functional change in the expression of that gene which occurs as a consequence (Weissmann, 1978; Weissmann *et al.*, 1979).

The gene cloning and associated *in vitro* procedures are extraordinarily powerful genetic tools because they (a) are rapid and universal, (b) permit the investigator to retain a high degree of control over the manipulation procedures and, most importantly, (c) permit the construction of new genetic combinations that hitherto could not have been selected in the laboratory (although they may arise at low frequencies in nature). In this review the principles and strategy of gene cloning will be discussed, as will several methods for the *in vitro* mutagenesis and manipulation of cloned DNA. Some examples of how these methods have been used to analyse gene structure and function, and to modify gene expression, will then be considered. The experimental procedures described are largely those employed to clone and manipulate genes in *E. coli* K-12 host–vector systems because these are currently the most advanced and widely used. It should be emphasized, however, that the procedures are in principle universal and are now being modified for use in a wide variety of different biological systems.

GENE CLONING

Cloning is the separation and individual propagation of a single element that is capable of reproduction from a mixture of similar elements. Cloning is, of course, widely used to maintain pure stocks of cellular organisms and their viruses but, more recently, has also been used to isolate and maintain individual autonomous genetic elements, such as plasmids and cryptic virus genomes, from heterogeneous mixtures (e.g. Timmis *et al.*, 1978b). The development of suitable gene cloning vectors or vehicles (virus and plasmid genomes) that provide functions for autonomous replication now enable DNA segments that do not have the ability to replicate autonomously to be cloned. Note that whereas the cloning of cellular organisms requires only a nutrient medium, the cloning of intracellular parasites or symbionts such as viruses and autonomous genetic elements requires a specific host cell system plus a nutrient medium, and the cloning of genes requires a vector replicon, a specific host cell system and a nutrient medium (Table 1).

Gene cloning *in vitro* involves the separation and individual propagation of a single DNA segment from a single DNA molecule in a suitable vector–host system. The gene mixture in any given DNA preparation consists of different genes covalently linked in linear arrays on individual DNA molecules and populations of the same or similar genes (alleles) on similar DNA molecules. Gene cloning therefore requires breakage of the DNA molecules to generate gene-sized DNA segments, followed by the covalent linkage of such segments to vector molecules in which they can be propagated, and introduction of the hybrid DNA molecules into suitable host cells. Note that gene separation in the *in vitro* gene cloning procedure is achieved at several distinct stages: (a) during DNA fragmentation both physical and spatial (dilution) separation of genes from one another occurs, (b) during linkage of DNA fragments to vector molecules enzymatic separation of the genes occurs, because DNA fragments which do not become linked to a vector molecule are not subsequently propagated in host cells, (c) during transformation biological separation of the different hybrid molecules occurs because the experimental conditions are usually so arranged that only one or at most two or three hybrid DNA molecules are taken up by each bacterial cell that is transformed, and (d) physical (dilution) and biological (selection) separation of transformed bacteria is achieved by plating diluted suspensions of

Table 1. *Cloning*

Mixture	Source (e.g.)	Separation	Growth medium	Amplification	Requirements
Bacteria	Soil	Physical, by dilution, plating on solid medium	Nutrient agar plate	Of individual bacteria to form colonies	Nutrient medium
Viruses	Sewage	Physical and biological, by dilution, infection of sensitive host cells by individual virus particles	Pure culture of host bacteria; nutrient agar plate	of individual viruses to form plaques in growing lawn of host cells	Host bacteria; nutrient medium
Plasmids	Bacterial lysate	Physical and biological, by dilution, transformation of competent cells by individual molecules	Pure culture of host bacteria; nutrient agar plate containing agent selective for a plasmid-encoded property	of individual plasmid molecules in individual bacteria that form colonies	Host bacteria; nutrient and selective medium
Genes[a]	Chromosome	Biochemical, physical and biological, by breakage of DNA into gene-sized fragments, linkage of single fragments to individual plasmid molecules, transformation of competent cells.	Pure culture of host bacteria; nutrient agar plate containing agent selective for a plasmid-encoded property	of individual plasmid molecules in individual bacteria that form colonies	Cloning vector; appropriate DNA fragmentation procedure; DNA ligase; host bacteria; nutrient and selective medium

[a] This entry refers to the *in vitro* cloning of genes; equivalent *in vivo* procedures are transduction and F-prime formation.

the bacteria on solid nutrient medium that is selective for expression of a vector function (e.g. antibiotic resistance).

The concept of separation and selective amplification of *individual* DNA molecules is of central importance, not only for gene cloning but also for the majority of other types of *in vitro* gene manipulation procedures, because these do not usually bring about a single, specific change (mutation) in the population of substrate DNA molecules, nor are 100% effective. After manipulation, therefore, the DNA preparation usually consists of a mixture of unmodified molecules and molecules containing a variety of different modifications, which must be separated from one another and individually propagated and analysed.

The following is an outline of the *in vitro* gene cloning procedure. For detailed discussion of the basic procedure and its variations, the reader is referred to recent reviews on the subject (Cohen, 1975; Nathans & Smith, 1975; Murray, 1976; Collins, 1977; Timmis, Cohen & Cabello, 1978d). Note that, because the principal objective in most gene cloning experiments is a reduction in the physical and genetic complexity of a genome, by elimination of all genetic information that does not pertain to the biological property under investigation, gene cloning is analogous to the older *in vivo* DNA deletion procedures, such as the isolation of transducing bacteriophages and F-prime factors. *In vitro* gene cloning does, however, represent a major advance on the *in vivo* procedures in two important respects, namely the fact that it can be used for essentially any DNA fragment from any organism, and the fact that the investigator can precisely specify the DNA segment to be cloned (i.e. the investigator has a degree of control over the DNA segment cloned that is not approachable in the earlier systems).

Gene cloning experiments are usually carried out in five stages: (a) the generation of DNA fragments that are suitable for cloning; (b) covalent linkage of the fragments to a vector plasmid or virus genome; (c) introduction of the vector:fragment hybrid molecules into an appropriate host cell system; (d) identification of cell clones that contain the desired hybrid molecule; and (e) amplification, isolation and characterization of the hybrid molecule.

Generation of DNA fragments

Most DNA segments that are to be cloned are not available as discrete fragments but must be generated by the cleavage of chromosomes or other long DNA molecules. Moreover, these DNA

Table 2. *Restriction endonucleases most commonly used for gene cloning*

Enzyme	Source	Recognition sequence and mode of cleavage	No. of cleavage sites in genome of:		
			φx174	SV40	λ
*Eco*RI	*Escherichia coli* RY13	5′-GAATTC-3′ 3′-CTTAAG-5′	0	1	5
*Hind*III	*Haemophilus influenzae* R_d	5′-AAGCTT-3′ 3′-TTCGAA-5′	0	6	6
*Bam*HI	*Bacillus amyloliquefasciens* H	5′-GGATCC-3′ 3′-CCTAGG-5′	0	1	5
*Pst*I	*Providencia stuatrii* 164	5′-CTGCAG-3′ 3′-GACGTC-5′	1	2	18
*Sal*I	*Streptomyces albus* G	5′-GTCGAC-3′ 3′-CAGCTG-5′	0	0	2

segments must be inserted at a precise location in the cloning vector molecule so as not to interfere with its ability to replicate and express its selection function, after introduction into a host cell system. Class II restriction endonucleases have two very important properties in this regard (see Roberts, 1976, for review). Firstly, they are sequence-specific and introduce double-strand breaks at precise locations in double-stranded DNA; they are therefore ideal for cleaving long DNA molecules into fragments of convenient length and for introducing a single break in a vector molecule at the site at which DNA fragments must be inserted. Secondly, many restriction endonucleases recognize symmetrical DNA sequences and make staggered breaks in the two strands at the recognition sequence, thereby generating cohesive or sticky ends, short 5′- or 3′-single-stranded termini, that are complementary to, and can anneal with, any other DNA fragment end that is generated by the same endonuclease. Cleavage of chromosomal DNA and the vector molecule with the same restriction endonuclease therefore permits subsequent annealing of the generated DNA fragments to vector molecules, and greatly facilitates their covalent linkage by DNA ligase. For these reasons, the majority of gene cloning experiments involve the linkage of DNA fragments generated by a specific restriction endonuclease to a vector molecule that is cleaved by the same endonuclease (e.g. see Fig. 1). The recognition sequences and modes of cleavage of the restriction nucleases most commonly used for gene cloning experiments are given in Table 2. These endonucleases all recognize a specific hexanucleotide sequence, and hence are 'infrequent cutters' (in principle, they should cleave DNA molecules of 50% average G+C composition once every 4096 base pairs), and generate DNA fragments having 3′- or 5′-tetranucleotide single-strand extensions. Other restriction endonucleases that recognize different sequences (and therefore generate different DNA fragments; for a recent summary of the endonucleases currently available, see Roberts, 1980) and/or produce longer, shorter, or no cohesive ends are, however, used quite frequently, especially where the more commonly used enzymes cut within the required fragment, or where they produce a much larger fragment than is required. In the latter case, a 'frequent cutter' which recognizes a tetranucleotide sequence may be employed. However, because most vector molecules have multiple cleavage sites for such endonucleases, additional steps must be carried out (e.g. tailing with linkers, see below) to join such fragments to a vector.

Although the isolation of individual genes is most conveniently accomplished by the cloning of DNA fragments generated by a single specific restriction endonuclease, the isolation of a group of genes, or all genes present on a chromosome, may only be accomplished by the cloning of randomly- or near randomly-cleaved DNA fragments, because of the presence of endonuclease cleavage sites within some of the genes. Near randomly-cleaved DNA fragments may be conveniently generated by partial digestion with a 'frequent cutter', such as *Sau*3A, whereas randomly-cleaved DNA fragments are generated physically by controlled shearing. The termini of randomly-cleaved DNA fragments must subsequently be modified before linkage to a vector molecule (see below).

The ease of cloning a DNA fragment that contains a specific gene depends upon its relative concentration in the total population of DNA fragments. Purification or selective enrichment of the required fragment greatly simplifies its cloning and may be accomplished if the fragment exhibits a physical property, such as size or buoyant density, that distinguishes it from other fragments, or contains a sequence that has affinity for an available DNA, RNA or protein probe (e.g. Timmis, Cabello & Cohen, 1975; Gautier,

Mayer & Goebel, 1976; Lovett & Helinski, 1976; Thomas, White & Davis, 1976; Weideli *et al.*, 1977). Recently, a high-resolution enrichment procedure involving RPC5 column chromatography and preparative electrophoresis has been developed that enables 100–1000-fold enrichment of any DNA segment in a complex mixture to be attained (e.g. Edgell, Weaver, Haigwood & Hutchison, 1979). The use of such an enrichment procedure is indicated when the initial DNA fragment mixture is known to contain $\geqslant 10^6$ genes.

In addition to DNA from natural sources, synthetic DNA may be cloned. Synthetic DNA is of two main types: cDNA that has been enzymatically synthesized by reverse transcriptase and DNA polymerase I from purified or partially purified messenger RNA (e.g. Maniatis, Efstratiadas, Kee & Kafatos, 1976; Efstratiadis & Villa-Komaroff, 1979) and DNA that has been chemically synthesized in the absence of a template (Itakura *et al.*, 1977; Crea, Kraszewski, Hirose & Itakura, 1978). In this case, the proportion of total DNA represented by the required fragment can approach 100%.

Joining of DNA fragments to vector molecules

In *in vitro* covalent linkage of DNA fragments to cloning vector molecules is ordinarily achieved by treatment of the DNA mixture with *E. coli* or bacteriophage T4 DNA ligase to form phosphodiester linkages between the fragment termini. DNA fragments and vector molecules that have identical cohesive termini (i.e. that were generated by the same endonuclease) are readily ligated because, under appropriate reaction conditions, their termini anneal and thereby provide a stable substrate for the ligase (Mertz & Davis, 1972). DNA fragments lacking single-stranded termini may also be joined by T4 (but not by *E. coli*) ligase but the reaction is inefficient because joining depends upon the random collision of fragment ends that are unable to form stable associations. Effective ligation therefore requires high concentrations of DNA ends and enzyme (Sgaramella, van de Sande & Khorana, 1970).

An important consideration in gene cloning strategy is the subsequent recovery of the cloned fragment from a hybrid molecule. Precise excision of the cloned fragment is always possible if the linear vector molecule and cloned fragment were both generated by the same endonuclease, because that endonuclease recognition sequence will have been reconstituted at the two vector:fragment junctions, and hence will again be susceptible to cleavage by that

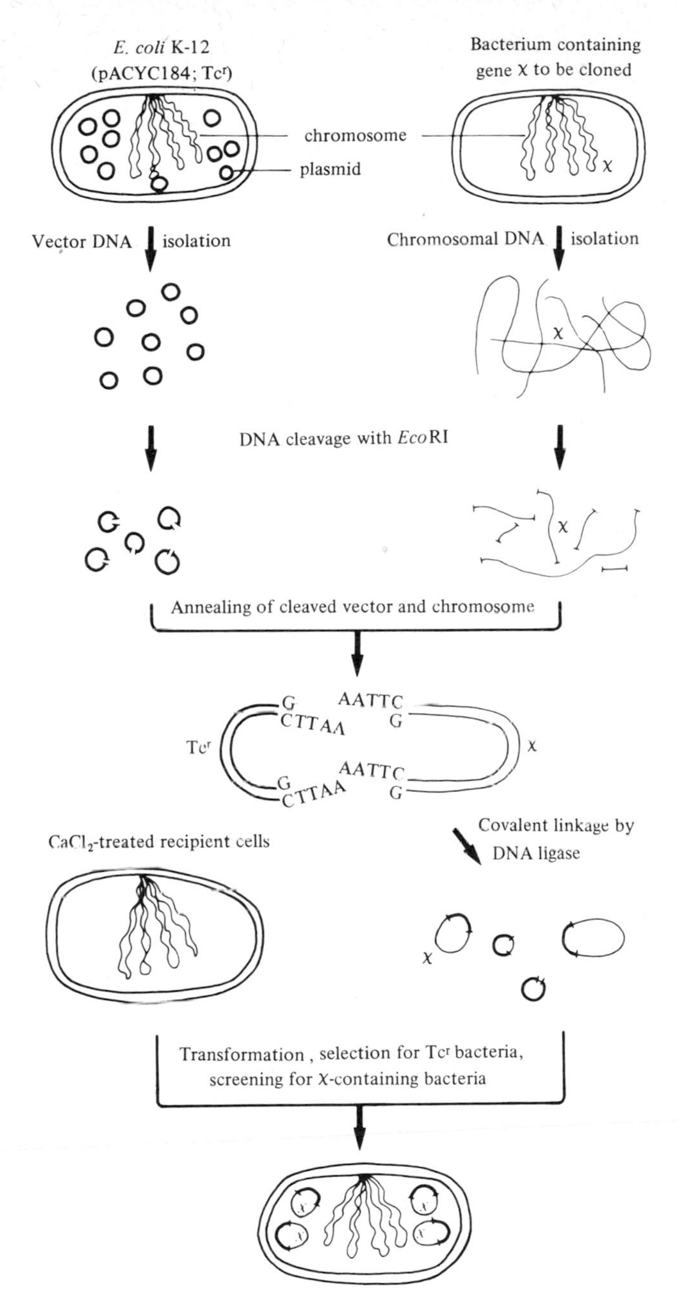

Fig. 1. The gene cloning procedure.

Table 3. *Some currently available linker molecules that contain restriction endonuclease recognition sequences*

Linker	Recognition[a] sequence
5′-GGAATTCC-3′ 3′-CCTTAAGG-5′	*Eco*RI
5′-GGTCGACC-3′ 3′-CCAGCTGG-5′	*Sal*I
5′-CCGGATCCGG-3′ 3′-GGCCTAGGCC-5′	*Bam*HI; *Hpa*II
5′-GCTGCAGC-3′ 3′-CGACGTCG-5′	*Pst*I
5′-CCAAGCTTGG-3′ 3′-GGTTCGAACC-5′	*Hin*dIII

[a] Indicated by the underlined nucleotides.

enzyme. An advantage of the flush-end ligation procedure is that it allows the linkage of two DNA fragments generated by different endonucleases, and hence permits the investigator much wider flexibility in his choice of enzyme for cleavage of the DNA under study. A disadvantage of joining two DNA fragments that were generated by different endonucleases is that a hybrid recognition sequence will be formed at the junctions of the two fragments. Ordinarily this hybrid recognition sequence cannot be cleaved by either of the two enzymes used in the construction of the hybrid.

This particular problem may be circumvented by the use of 'linkers', which are short synthetic polynucleotides containing one or more endonuclease recognition sequences (Greene *et al.*, 1975; Bahl *et al.*, 1976; Scheller *et al.*, 1977). The ligation of *Eco*RI linkers to a DNA fragment having flush ends thereby provides it with *Eco*RI termini, enables it to be cloned into the *Eco*RI site of a vector plasmid, and to be precisely excised subsequently by cleavage of the hybrid molecule with *Eco*RI. The availability of a large number of synthetic linkers (e.g. Table 3) enables the investigator to modify DNA fragment termini at will. It should be noted that linker usage is carried out in two stages: (a) blunt end ligation of the linker to the DNA fragment, and (b) cleavage of the linker with the appropriate endonuclease to generate the specific termini required (e.g. 5′-AATT, in the case of *Eco*RI linkers). The latter step will result in the internal cleavage of the DNA fragment if it contains one or more of the cleavage sites specified by the linker employed.

This problem can be avoided by the use of a linker containing a site for an endonuclease that does not cleave within the fragment to be cloned. One important use of linkers is to provide specific termini, that can be readily joined to a vector molecule, for randomly or near randomly generated fragments of genomic DNA (Maniatis *et al.*, 1978). In this case, some of the fragments must contain internal cleavage sites of the type present in the linker and these have to be protected from cleavage by treatment of the fragment mixture with the specific modification methylase (e.g. the *Eco*RI methylase) prior to addition of linkers (e.g. Maniatis *et al.*, 1978). The synthesis of asymmetric linkers that contain single, preformed, specific cohesive termini, and that do not require endonuclease cleavage subsequent to their ligation to DNA fragments, was recently described (Bahl *et al.*, 1978; Norris, Iserentant, Coutreras & Fiers, 1979). The use of such linkers, when they become generally available, will obviate the problem of internal cleavage of DNA fragments.

An alternative procedure for the modification of DNA fragment termini prior to cloning is the addition of homopolymer 'tails' of a nucleotide (often dC) to their 3′-ends by means of deoxynucleotidyl terminal transferase (Lobban & Kaiser, 1973; Bollum, 1974; Roychoudhury, Jay & Wu, 1976). DNA fragments that have been tailed in this way can stably anneal with linear vector molecules that have been tailed with the complementary nucleotide (e.g. dG). The non-covalently linked molecules may be used directly for transformation, after which repair of the junction region single-stranded stretches, and covalent linkage of the fragments to vector molecules, occurs *in vivo*. Judicious choice of the endonuclease used to cleave the vector molecule, and the nucleotides used in the tailing procedure (e.g. dG for *Pst*I-cleaved vector molecules), results in reconstruction of the original cleavage site in the vector molecule, thereby permitting subsequent excision of the cloned fragment. One particular advantage of this method is that circularization of individual fragments, i.e. *intra*molecular joining, is absolutely precluded because the tails of both ends of any particular fragment are identical. Regeneration of the original cloning vector with the resultant elevation of 'background' non-hybrid molecules is therefore avoided.

Introduction of hybrid molecules into host bacteria

The *in vitro* construction mixture used for infection of bacteria in cloning experiments usually consists of an extremely heterogeneous mixture of DNA molecules formed by the random joining of DNA fragments and vector molecules. Only a small proportion of the total molecules will be of the required type, namely those consisting of a single vector molecule linked to a single DNA fragment. The problem in subsequent stages in the gene cloning procedure is therefore to select and individually propagate the required hybrid molecules. This is usually achieved in two stages. Firstly, the crude DNA preparation is used to infect suitable host bacteria; any molecules not containing the cloning vector and hence not possessing the ability to replicate will be lost at this stage. Secondly, transformant bacteria that contain the desired hybrid molecules are identified by a specific screening procedure, purified, and their hybrid plasmids isolated and characterized.

The process of introduction of DNA into bacteria is termed transformation, when involving chromosomal or plasmid DNA, and transfection, when involving a bacteriophage genome. For the transformation of *E. coli*, DNA molecules are mixed with bacterial cells that have been made permeable ('competent') by treatment with a cold solution of calcium chloride. The cells subsequently receive a heat shock during which they take up exogenous DNA. After incubation for a period of time to allow expression of functions encoded by the newly acquired DNA, the bacteria are plated on a medium that selects those which have been transformed for a specific gene; in cloning experiments this is a selectable function (usually resistance to an antibiotic) encoded by the cloning vector. The procedure used for transfection is similar to that used for transformation, except that the transfected cells do not usually survive and must be detected by plating with indicator bacteria (Mandel & Higa, 1970; Cohen, Chang & Hsu, 1972).

Transformation and transfection efficiencies in *E. coli* are very low: typical values are 10^6 transformants per μg of purified plasmid or phage DNA (or 10^{-5} transformants per molecule) and 10^3 per μg cloning mixture DNA. Efforts to improve the efficiency of transformation have resulted in an increase of about two orders of magnitude (Kushner, 1978). Optimal transformation frequencies are, however, rarely achieved in gene cloning experiments due to the necessity to use saturating DNA concentrations to obtain a sufficiently high total number of transformants. When $>10^5$ independent clones are required from a cloning experiment (e.g. in the construction of a gene bank of a mammalian genome), these frequencies are inadequate. Recently, methods have been described for the *in vitro* packaging in λ bacteriophage particles of DNA from construction mixtures involving either λ genome vectors or plasmid vectors (cosmids) that contain the λ packaging recognition sequence, *cos* (Hohn & Murray, 1977; Sternberg, Tiemeier & Enquist, 1977; Collins & Hohn, 1978). Compared with naked DNA, the phage particles are very efficient in the infection of bacterial cells and the relative yield of clones containing recombinant DNA is increased by one or two orders of magnitude.

During the usual transformation procedure, little or no cell multiplication occurs until transformant bacteria are plated out on a selective medium for single colony isolation. Thus, rarely do two or more primary transformant clones contain identical recombinant DNA molecules that are the progeny of a single DNA molecule from the original cloning mixture. With some procedures, only hybrid DNA molecules transform, transfect, infect, or propagate in, host cells whereas in others, reconstituted cloning vector molecules are the majority of those that transform host cells. In this case, screening procedures must be used to identify the minority clones carrying hybrid DNA molecules. The specific procedure followed to obtain clones that carry hybrid molecules depends upon the properties of the cloning vector employed.

General purpose cloning vectors

General cloning vectors should be small, genetically and physically well-defined autonomous replicons that can be easily purified in large quantities. They must code for a property, such as resistance to an antibiotic, which can be used to select bacteria that have taken up a vector molecule, and neither this selectable property nor the vector's replication functions may be inactivated by the insertion of a DNA fragment. The vector should, moreover, contain single cleavage sites for as many of the commonly used cloning enzymes as possible, and should encode a property that permits the direct selection or rapid detection of bacterial clones containing hybrid DNA molecules. To obtain active transcription of cloned DNA the vector should contain one or more strong promoters located near to, and oriented towards, the cloning sites. Since the description of the first successful gene cloning experiment (Cohen, Chang, Boyer & Helling, 1973) a large number of plasmid and phage genome vectors have been described and the reader is referred to earlier reviews (Collins, 1977; Timmis *et al.*, 1978d; Brammar, 1979; Barnes, 1980;

Bernard & Helinski, 1980; Williams & Blattner, 1980) for more extensive listings.

Although for many gene cloning purposes, phage and plasmid vectors are equally appropriate, in this review I will focus on plasmid vectors because they are easier to use and somewhat more versatile. Plasmid vectors are especially suitable for the cloning of genes whose expression and regulation is to be studied in the host cell system used for cloning. On the other hand, phage vectors can be particularly useful for the cloning of poorly-expressed, autoregulated genes (infection of host cells with hybrid phage may result in a burst of synthesis of the product of the autoregulated gene, if no cellular regulating element exists in the uninfected cell) or genes that are lethal (e.g. if actively expressed, such as genes of ribosomal proteins and some membrane proteins) for the host cell.

Plasmids are autonomous covalently closed circular extrachromosomal genetic elements that have been found in a wide variety of microorganisms and in some higher organisms (for reviews, see Helinski, 1973; Falkow, 1975). They range in size from about 2 to 500 kb and are present in host bacteria at cellular levels between 1 and 100 copies per chromosome. Several of the smaller plasmids (e.g. ColE1 and its relatives) are not only maintained at high cellular copy numbers, and hence may induce the synthesis of large amounts of plasmid-coded products through gene dosage (e.g. Hershfield *et al.*, 1974) but also, unlike the host chromosome, are able to replicate extensively in cells that have been treated with a protein synthesis inhibitor. The selective amplification of plasmid vectors to ~1 000 copies/cell, by treatment of host cells with chloramphenicol, greatly facilitates the isolation of large quantities of hybrid DNA.

The currently most widely used cloning vector is the pBR322 plasmid (Bolivar *et al.*, 1977), which was itself generated by a number of *in vitro* cloning procedures, and which contains the replication determinants of the ColE1-related plasmid pMB1, the ampicillin resistance gene of plasmid R1 (Tn*3*), and the tetracycline resistance gene of plasmid pSC101 as selection markers (Table 4). pBR322 is small (4.4 kb), extremely well-characterized, and its entire polynucleotide sequence has been determined (Sutcliffe, 1979). It is maintained in host cells at a copy number of 40–60/chromosome, can be selectively amplified by treatment of host bacteria with chloramphenicol, and contains single cleavage sites for at least nine endonucleases. The pBR325 plasmid (Bolivar, 1978)

Table 4. *Some general purpose high copy number plasmid cloning vectors for* E. coli

Plasmid vector	Replicon	Cloning sites	Selection marker[a]	1. Insertional inactivation[b] or 2. Positive selection[c]	Size (kb)	Copy number	Amplification[d]	References[e]
pBR322	pMB1	*Pst*I	Tc^r	1, Ap^r	4.4	40–50	Yes, Cm	1
		*Cla*I, *Hind*III, *Bam*HI, *Sal*I	Ap^r	1, Tc^r				
		*Eco*RI, *Ava*I, *Pvu*II, *Bal*I	Ap^r, Tc^r	—				
pBR325	pMB1	*Pst*I	Tc^r, Cm^r	1, Ap^r	5.4	40–50	Yes, Sp	2
		*Hind*III, *Bam*HI, *Sal*I	Ap^r, Cm^r	1, Tc^r				
		*Eco*RI	Ap^r, Tc^r	1, Cm^r				
pACYC177	p15A	*Pst*I	Km^r	1, Ap^r	3.7	20	±, Cm	3
		*Hind*III, *Xho*I	Ap^r	1, Km^r				
		*Bam*HI, *Hinc*II	Ap^r, Km^r	—				
pACYC184	p15A	*Eco*RI	Tc^r	1, Cm^r	4.0	20	Yes, Sp	3
		*Hind*III, *Bam*HI, *Sal*I	Cm^r	1, Tc^r				
pKT235	ColD	*Eco*RI	Ap^r, Km^r	1, Cm^r	8.5	20	Yes, stationary phase	4
		*Hind*III, *Xho*I	Ap^r, Cm^r	1, Km^r				
		*Pst*I	Km^r, Cm^r	1, Ap^r				
		*Bst*EII, *Sst*I, *Bam*HI, *Kpn*I	Km^r, Cm^r, Ap^r	—				
pKY2289	ColE1	*Xma*I, *Eco*RI	Ap^r	2, MC	11.0	20	Yes, Cm	5
pKN80	ColE1	*Hinc*II, *Hpa*I	Ap^r	2	15.9	20	Yes, Cm	6

[a] Tc^r, tetracycline resistance; Ap^r, ampicillin resistance; Cm^r, chloramphenicol resistance; Km^r, kanamycin resistance.
[b] Insertional inactivation of the indicated property.
[c] Positive selection, in the case of pKY2289, on mitomycin C-containing medium, MC.
[d] By growth of host cells in chloramphenicol (Cm), spectinomycin (Sp) or stationary phase.
[e] References: 1, Bolivar *et al.*, 1977; 2, Bolivar, 1978; 3, Chang & Cohen, 1978; 4, Timmis *et al.*, 1980a; 5, Ozaki, Maeda, Shimada & Takagi, 1980; 6, Schumann, 1979.

additionally contains a chloramphenicol resistance gene and can be selectively amplified by treatment of host bacteria with spectinomycin. The pACYC177 and pACYC184 plasmids (Chang & Cohen, 1978), which contain replication determinants from the p15A plasmid, are also widely used vectors and have properties similar to those of pBR322 and pBR325. They do not, however, amplify extensively in cells inhibited for protein synthesis (pACYC177, 2×; pACYC184, 10×). The pKT235 plasmid (Timmis, Bagdasarian & Brady, 1980a) contains the replication determinants of the ColD plasmid, the ampicillin resistance gene of plasmid R1, and the chloramphenicol and kanamycin resistance genes of plasmid R6-5, rather than the tetracycline resistance gene of pSC101, which was found to exhibit sequence instability (K. N. Timmis & H. Danbara, unpublished data). pKT235 undergoes very extensive amplification as the host cells enter stationary phase and high yields of plasmid DNA are obtained without treatment of host cells with protein synthesis inhibitors.

The major part of the time and effort required for most cloning experiments is taken up with the identification of transformant clones that contain hybrid DNA molecules, or that contain a specific hybrid molecule, although occasionally a procedure may be available for the direct selection of transformant clones that carry a desired DNA fragment (e.g. Chang *et al.*, 1978). For this reason, much effort has been directed towards the development of cloning vectors that allow the rapid identification of clones carrying hybrid plasmids or, better still, their direct selection. Two principal methods have been developed to achieve these objectives. Firstly, the vector may encode a readily assayed property, such as resistance to an antibiotic, that is inactivated by insertion of a DNA fragment into the vector (*insertional inactivation*, Timmis, Cabello & Cohen, 1974, 1978c). The pBR322 vector, for example, contains genes for resistance to ampicillin (Ap^r) and tetracycline (Tc^r); transformant clones carrying hybrid pBR322 molecules containing *Pst*I fragments inserted in the unique *Pst*I cleavage site of the Ap^r gene are selected on Tc-containing medium and identified by their sensitivity to Ap. Similarly, clones carrying hybrid plasmids containing *Hin*dIII, *Bam*HI or *Sal*I fragments inserted in the unique *Hin*dIII, *Bam*HI or *Sal*I cleavage sites of the Tc^r gene are Ap^r Tc^s. Thus, a single replica plating of transformant clones from a medium containing one antibiotic onto plates containing a second rapidly identifies clones carrying hybrid molecules. The procedure of *insertional inactivation* may also be employed in the direct selection of hybrid molecules if the vector specifies a function that is conditionally lethal for host cells. Thus far, two such vector plasmids have been described (Table 4). pKN80 is a pBR322 derivative containing the bacteriophage Mu *kil* gene (Schumann, 1979). The vector is propagated in a Mu-lysogen in which expression of the *kil* gene is repressed. Introduction of the vector into a non-lysogenic host results in expression of the *kil* gene and death of the host cell, unless the *kil* gene has been inactivated by insertion of a DNA fragment into its unique *Hpa*I/*Hin*dII site. Thus, the pKN80 vector allows the direct selection of transformant clones carrying hybrid plasmids generated by the insertion of DNA fragments into its *Hpa*I/*Hin*dII site. The pKY2289 vector is a ColEl derivative containing a Tn*3* transposon in its colicin E1 immunity gene (Ozaki *et al.*, 1980). Induction of colicin E1 synthesis by treatment of host cells with mitomycin C results in the colicin E1-mediated killing of host cells, unless the colicin E1 gene has been inactivated by insertion of DNA fragments into its unique *Eco*RI or *Xma*I sites. The pKY2289 vector therefore allows the direct selection of transformant clones carrying hybrid plasmids generated by the insertion of a DNA fragment into its unique *Eco*RI or *Xma*I sites.

Secondly, the cloning procedure may include a biological size selection of DNA molecules that are larger than the vector itself and that hence are hybrid. λ DNA molecules are packaged into phage particles according to a 'headfull' mechanism, i.e. a genome must be about 47 kb in length in order to be packaged. Cloning into λ replacement vectors may involve removal of a non-essential segment of the phage genome prior to ligation to the DNA fragments; only those λ molecules that contain additional DNA, such that they are about 47 kb in length, are large enough to be packaged into phage heads after infection, and hence to produce viable phage particles. As indicated above, the packaging of λ DNA into phage heads may also be carried out *in vitro* with the result that the infection efficiency of gene cloning mixtures is greatly increased (Hohn & Murray, 1977; Sternberg *et al.*, 1977). The *in vitro* packaging procedure is therefore ideal for the cloning of a large number of *large* DNA fragments, as for example in the construction of gene banks or libraries: collections of bacterial clones or phages containing vector molecules plus randomly or near randomly generated DNA fragments that collectively cover the whole of a particular genome (Clarke & Carbon, 1976; Maniatis *et al.*, 1978). The probability that any particular sequence will be present in one or more of the clones in the gene bank depends on the average size of the fragments cloned, and upon the number of the hybrid molecules in the bank; for example an *E. coli* DNA gene bank of 940 clones containing hybrid plasmids with average-sized inserts of 13 kb will contain any given gene with a probability of 0.95 (Clarke & Carbon, 1976). The same probability is achieved with only 415 clones if the inserts are 30 kb. Thus, from the point of view of the number of clones that must be screened to find any given gene, it is important that the hybrid molecules which comprise the gene bank contain large DNA fragments. Cosmids, which are small plasmids that contain the λ *cos* site, are specifically designed for the cloning of very large DNA fragments in conjunction with the *in vitro* packaging procedure (Collins & Brüning, 1978; Collins & Hohn, 1978; Hohn & Hinnen, 1980). Cosmid pHC79 (Table 5), which selectively clones DNA fragments that are about 40 kb in size, is therefore very attractive for the generation of gene banks (B. Hohn & J. Collins, unpublished).

Special purpose cloning vectors

In addition to the general vectors and cosmids which can serve most cloning purposes, a number of vectors have recently been developed for specific needs. The cloning of certain types of DNA (e.g. from pathogenic organisms) may involve some degree of biological hazard and the investigator is obliged to carry out such experiments under conditions that provide a prescribed level of containment of the hybrid DNA molecules (e.g. NIH Guidelines, 1976, for the USA and some other countries). Containment procedures may involve the use of physical and/or biological barriers. One element of biological containment is the inability of host bacteria used for cloning to survive without the particular *in vitro* cultivation conditions provided by the investigator (Curtiss *et al.*, 1977). Another element of biological containment is the inability of vector molecules to transfer from their laboratory host cells to other bacterial strains. The *E. coli* plasmid pHSG415 is a vector that cannot be mobilised to other bacteria by conjugative plasmids, and which is unable to replicate at 37 °C (Table 5), i.e. the temperature of the normal environment of *E. coli* (Hashimoto-Gotoh, Nordheim & Timmis, 1980). Plasmid pRK646 is an incomplete replicon which lacks the *pir* gene whose product is absolutely essential for replication; this vector can therefore replicate only in cells that contain the *pir* gene (Kahn *et al.*, 1980). Vectors pHSG415 and pRK646 appear to provide a high degree of biological containment for cloned DNA fragments.

It is known that certain DNA segments are difficult or impossible to clone on high copy number vectors. These include certain segments of virulent bacteriophage genomes and genes of structural proteins of cell organelles (membrane, ribosome) whose synthesis appears to be very carefully regulated; over-production of such proteins presumably causes major disturbances in cell physiology. λ vectors are often ideal for the cloning of these genes because the brief disturbance in cell physiology caused by expression of these genes does not normally prevent the production of λ phage particles. On the other hand, it is often desirable to establish the cloned gene in viable cells in order to investigate its activity and, here, plasmid vectors are more appropriate. Several low copy number vector plasmids have therefore been developed for this purpose, e.g. pHSG415 (copy number 5; Hashimoto-Gotoh *et al.*, 1980), pRK2501 (copy number 3; Kahn *et al.*, 1980), and pMF3 (copy number 1; Manis & Kline, 1977) (Table 5). In order to obtain large quantities of a DNA segment (e.g. of a virulent phage) that is lethal for *E. coli*, and that may interfere with the normal growth of a λ vector, a host–vector system may be utilized in which the cloned genes are not expressed in, and/or are not detrimental to, the host cell. This strategy has been successfully used to clone in *Staphylococcus aureus* a DNA segment that carries the origin of replication of coliphage T7 (Scherzinger, Lauppe, Voll & Wanke, 1980).

Although most cloning experiments are carried out in *E. coli*, investigations of the physiological activities of organisms that are unrelated to *E. coli* require the cloning to be carried out in the organism under investigation. Cloning vector systems are now being developed for many classes of organisms, including *Bacillus subtilis* and *Staphylococcus aureus*, *Streptomyces*, yeast, fungi, and plant and animal cell lines, as well as intact plants. One group of organisms that exhibits a wealth of metabolic activities (photosynthesis, nitrogen fixation, catabolism of exotic organic compounds, pathogenicity, symbiotic associations, etc.) are soil bacteria. Many of these are gram-negative and can propagate the so-called broad host range plasmids such as RP4/RK2, RSF1010, and S-a, although not the ColE1-related vectors, like pBR322. Considerable effort therefore has been invested in the development of cloning vectors

Table 5. *Special purpose plasmid cloning vectors for* E. coli

Plasmid	Replicon	Cloning sites	Selection	Size	Copy numbers[a]	Remarks	References[b]
1. *Biological containment vector*							
pHSG415	pSC101	*Eco*RI	Km^r, Ap^r	7.1	5	Temperature-sensitive replication; poorly mobilizable	1
		*Xho*I, *Hin*dIII	Cm^r, Ap^r				
		*Pst*I	Km^r, Cm^r				
		*Bam*HI	Km^r. Cm^r, Ap^r				2
pRK646	R6K	*Bam*HI, *Bgl*II, *Pst*I	Ap^r	3.4	(20)	Can only replicate in bacteria that contain the R6K pir gene	
2. *Low copy number vector*							
pHSG415	See above						
pMF3	F	*Bam*HI, *Hin*dIII, *Eco*RI	Ap^r	11.0	1		3
pRK2501	RK2	*Hin*dIII, *Xho*I	Tc^r	11.1	2–4		2
		*Sal*I	Km^r				
		*Bgl*II, *Eco*RI	Km^r, Tc^r				
3. *Broad host range vector*							
pRK2501	see above					Can be propagated in many gram-negative bacteria	
pKT231	RSF1010	*Eco*RI	Km^r			Can be propagated in many gram-negative bacteria	4
		*Hin*dIII, *Xho*I	Sm^r	12.5	15–20		
		*Xma*I	Km^r, Sm^r				
4. *Cosmid for packaging in λ heads*							
pHC79	pMB1	*Pst*I	Tc^r	6.4	(20)	Can be packaged in λ heads when linked to DNA fragments 30–40 kb in size	5
		*Hin*dIII, *Bam*HI, *Sal*I	Ap^r				
		*Eca*I	Ap^r, Tc^r				
5. *Bifunctional vector*							
pVH14	pBR322/pC194	*Eco*RI, *Bam*HI	Cm^r, Ap^r	6.9	(40–60) *E. coli*)	Cloning in *E. coli*, *Bacillus subtilis* and *Staphylococcus aureus*	6
YRp7	pBR322/trp1	*Bam*HI, *Sal*I	Ap^r, Tc^r, *trp*1	5.7	(40–60 *E. coli*)	Cloning in *E. coli* and yeast	7

[a] Copy numbers in brackets are assumed values.
[b] References: 1, Hashimoto-Gotoh *et al.*, 1980; 2, Kahn *et al.*, 1980; 3, Manis & Kline, 1977; 4, Bagdasarian, Nordheim & Timmis, 1980; 5, B. Hohn & J. Collins, unpublished; 6, Ehrlich, 1978; 7, Struhl, Stinchcomb, Scherer & Davis, 1979.

Table 6. *Detection of transformant clones carrying required hybrid molecules*

Method	Detects	Requirement	Reference[a]
Colony hybridization	DNA segment	Radioactive DNA or RNA probe homologous to required sequence	1, 2, 3
Immunodetection	Polypeptide gene product	Specific antibody probe against required polypeptide	4, 5, 6, 7
Phenotypic change	Gene product	Specific assay or host cell system, for change in phenotype	8, 9
Microscale hybrid DNA analysis	Specific size insert	—	10, 11, 12, 13, 14

[a] References: 1, Grunstein & Hogness, 1975; 2, Benton & Davis, 1977; 3, Hanahan & Meselson, 1980; 4, Sanzey *et al.*, 1976; 5, Skalka & Shapiro, 1976; 6, Broome & Gilbert, 1978; 7. Erlich *et al.*, 1978; 8, Chang *et al.*, 1978; 9, Moll, Manning & Timmis, 1980; 10, Barnes, 1977; 11, Telford, Boseley, Schaffner & Birnstiel, 1976; 12, Eckhardt, 1978; 13, Birnhoim & Doly, 1979; 14, Klein, Selsing & Wells, 1980.

from the broad host range plasmids and so far two vectors of this type have been developed, namely the RK2 derivative pRK2501 (Kahn *et al.*, 1980), and the RSF1010 derivative pKT231 (Bagdasarian *et al.*, 1979, 1980) (Table 5).

Even when a cloning system for a non-*E. coli* organism is available, the great convenience of *E. coli* usually makes it worthwhile to carry out an initial cloning experiment in this host and thereafter to transfer hybrid molecules into the non-*E. coli* system. For this, bi-functional cloning vectors that can be propagated in both types of host cell system are required; bi-functional plasmid vectors for cloning in *E. coli/Bacillus subtilis* (Ehrlich, 1978) and *E. coli*/yeast (Struhl *et al.*, 1979) have already been developed (Table 5) and others are now being constructed.

Identification of clones carrying specific hybrid DNA molecules

Although it is now a straightforward matter to identify transformant clones carrying hybrid DNA molecules, the identification of clones carrying a particular hybrid depends upon the availability of an assay for the specific DNA sequence required or its gene product (Table 6). Several colony and plaque hybridization tests have been developed for the rapid detection of clones or plaques that contain a specific DNA sequence for which a homologous radioactive DNA or RNA probe is available (Grunstein & Hogness, 1975; Benton & Davis, 1977; Hanahan & Meselson, 1980), as have immunological methods for the identification of colonies or plaques containing a gene product against which an antibody is available (Sanzey, Mercereau, Ternynck & Kourilsky, 1976; Skalka & Shapiro, 1976; Broome & Gilbert, 1978; Erlich, Cohen & McDevitt, 1978). These tests are highly sensitive and very effective in the cloning of double-stranded cDNAs, genes from genomic DNA whose cDNAs have already been cloned, and genes whose products have been immunologically characterized.

In many instances, however, no DNA/RNA or antibody detection probes are available and hybrid identification must rely on the known size or restriction endonuclease cleavage pattern of the desired DNA fragment. To facilitate the characterization of hybrid plasmids from many clones, microscale procedures for the isolation of plasmid DNA have been developed. One example of these is the so-called 'toothpick assay' in which single bacterial colonies from nutrient agar plates are transferred to the wells of an agarose gel, lysed, and their plasmid DNA sized by electrophoresis through the gel (Barnes, 1977; Telford *et al.*, 1976; Eckhardt, 1978). Another example is the microscale preparation of plasmid DNA for restriction endonuclease cleavage analysis. Isolation procedures of this type are so designed as to enable all steps to be performed in Eppendorf tubes, thus greatly reducing the time taken for the various centrifugation steps (Birnhoim & Doly, 1979; Klein *et al.*, 1980).

Once a transformant clone containing the required hybrid DNA molecule has been identified, a small amount of plasmid DNA is prepared and used for a preliminary analysis. If this analysis confirms the identification made by the screening procedure, a small (nanogram) amount of the DNA is used to re-transform (re-clone) a plasmid-free host cell, to obtain a clone that is certain to carry only one DNA species. This step is of great importance because the initial transformation, involving saturating amounts of ligation mixture DNA, sometimes produces primary transformant clones that contain two or three distinct hybrid molecules. Plasmid DNA is subsequently prepared in quantity from the secondary transformant clone and used for a detailed characterization of the hybrid molecule.

Structural analysis of a cloned DNA fragment

The isolation of a cloned DNA fragment is readily achieved by excision from the hybrid molecule, by cleavage with the restriction endonuclease used in the construction, and separation of the cloned fragment from its vector molecule by electrophoresis through an agarose gel. UV irradiation of the ethidium bromide-stained gel visualizes the DNA bands and enables band-containing segments of the gel to be cut out. The gel pieces may then be dissolved in a saturated solution of KI, followed by absorption of the DNA to, and subsequent elution from, glass beads (Vogelstein & Gillespie, 1979). DNA fragments prepared in quantity in this way are of high purity and suitable for most forms of biochemical analysis.

A detailed restriction endonuclease cleavage map normally forms the basis of structural and genetic studies of a cloned DNA fragment and is important for the design of a DNA sequencing strategy. It is constructed by careful determination of the sizes of subfragments generated by digestion of the fragment with single and pairs of 'infrequent cutter' endonucleases. The cleavage sites of 'frequent cutters' may be mapped by measurement of the sizes of partial digestion products of a fragment that is radioactively labelled at one end (Smith & Birnstiel, 1976). Further structural analysis of a cloned fragment may involve comparison with an alternative DNA form (e.g. if it is cDNA, it may be compared with genomic DNA, and *vice versa*; Jeffreys & Flavell, 1977) or with related fragments or purified RNA molecules (Thomas *et al.*, 1976; Garapin *et al.*, 1978), by hybridization methods such as heteroduplex formation (Davis, Simon & Davidson, 1971) or 'Southern' blotting (Southern, 1975). Electron microscopic methods may also be employed to localize regions of sequence symmetry (inverted repeats) or RNA polymerase binding sites (potential transcription promoters). Alternatively, filter binding techniques can be used to detect specific binding sites for RNA polymerase and other proteins. Ultimately, the polynucleotide sequence of the cloned fragment, or a part thereof which has been identified during preliminary characterization of the fragment, may be determined by the rapid DNA sequencing methods that have been developed recently (Sanger & Coulson, 1975; Maxam & Gilbert, 1977; Sanger, Nicklen & Coulson, 1977). Where determination of the polynucleotide sequence of a number of cloned fragments is the primary objective of the cloning experiment, the use of single-stranded phage genome cloning vectors (Messing, Gronenborn, Muller-Hill & Hofschneider, 1977; Herrmann, Neugebauer, Schaller & Zentgraf, 1978; Barnes, 1980) may simplify the sequencing procedure. Recently, a rapid sequencing strategy for cloned fragments was developed by H. Lehrach (in preparation) which consists of (a) random cleavage of pBR322 hybrid DNA molecules containing an insert at the *Pst*I cleavage site by DNase I in the presence of Mn^{2+}, (b) addition of *Eco*RI linkers to the fragment termini, (c) treatment with *Eco*RI, which cleaves both the linkers and the unique *Eco*RI cleavage site within pBR322, (d) separation of the random length DNA fragments by electrophoresis through an agarose gel, (e) isolation of fragments in size classes that differ by approximately 0.1 kb, and that are shorter than that of the complete hybrid molecule by more than 0.7 kb (this procedure selects for deletions of the hybrid that begin at the *Eco*RI site of pBR322 and terminate either within the cloned fragment, or within the Tc^r gene), (f) circularization of these selected fragments by treatment with DNA ligase and (g) transformation and selection for Tc^r transformants. This yields plasmid derivatives with random deletions extending into the cloned fragment. Cleavage of these with *Eco*RI, end-labelling with polynucleotide kinase, and cleavage with *Hin*dIII provides end-labelled fragments for sequencing that will provide a series of overlapping sequences of the original cloned fragment.

Functional analysis of a cloned DNA fragment

Functional analysis of a cloned DNA fragment normally involves characterization of its gene products and the regulation of their synthesis, in appropriate *in vivo* and *in vitro* systems. RNA species transcribed from the fragment may be characterized and their operator–promoter sequences identified. Gene products for which assays are available can be readily detected and their regulation studied. Assays may depend on the functioning of the gene product (e.g. enzymatic activity), or on its structure (e.g. immunodetection). If the host chromosome ordinarily codes for a product that is similar or identical to that specified by the cloned fragment, it is necessary to utilize mutant host cells that no longer synthesize this product or that synthesize a distinguishable version of it.

If no specific assay exists for a product encoded by the cloned fragment, its detection in *viable* cells is rendered difficult, if not impossible, due to high levels of synthesis of chromosomal (i.e. non-plasmid) gene products. In such cases, it is necessary to severely repress the expression of chromosomal genes. For this purpose the *minicell* mutants of bacteria provide excellent *in vivo* systems for the identification of plasmid-encoded products. Minicell mutants are defective in cell division functions such that the septum, instead of forming in the middle of a growing cell, thereby dividing the cell into two equal halves and ensuring the equipartition of the chromosomes, forms close to one of the cell poles producing a large, but otherwise normal cell, and a 'minicell'. Minicells do not receive a copy of the host chromosome and are therefore inviable, although they do possess most of the metabolic potential of normal cells (Adler, Fisher, Cohen & Hardigree, 1967; Frazer & Curtiss, 1975). If the parental cells contain a plasmid, some copies are transmitted into the minicells, which then contain only plasmid DNA as genetic information, and which actively express this information. Minicells are therefore highly useful *in vivo* systems for the identification of plasmid-encoded transcription and translation products (Roozen, Fenwick & Curtiss, 1971).

Minicells are separated from viable cells by sucrose density gradient centrifugation. After incubation in nutrient medium for a period of time, to allow *in vivo* degradation of long-lived mRNA molecules derived from the parental cells, purified plasmid-containing minicells are labelled with ^{32}P-phosphate or radioactive RNA or protein precursors. They are then lysed and the radioactive products analysed by electrophoresis through polyacrylamide gels, which are subsequently dried and autoradiographed. The employment for gene cloning experiments of small vector plasmids that specify few gene products greatly simplifies analysis of the gene products encoded by cloned fragments. Fig. 2 shows a polyacrylamide gel electrophoretic analysis of the proteins present in minicells carrying plasmid pKT107, a pACYC184 hybrid plasmid containing a DNA segment from the *tra* region of plasmid R6-5 (tracks a), and pKT107 *tra*T$^-$ insertion mutant derivatives containing the ampicillin resistance transposon Tn*3* (tracks b, c, d, e, and f). Part A shows the gel stained with Coomassie Blue, which reveals all proteins present in the minicells, and part B shows the autoradiogram of the gel, which reveals just the radioactive proteins that were newly synthesized in the purified minicells. It is clear that, with one exception (the *tra*T protein, which is produced by pKT107 but not the pKT107 Tn*3* derivatives) the plasmid-encoded proteins cannot be detected

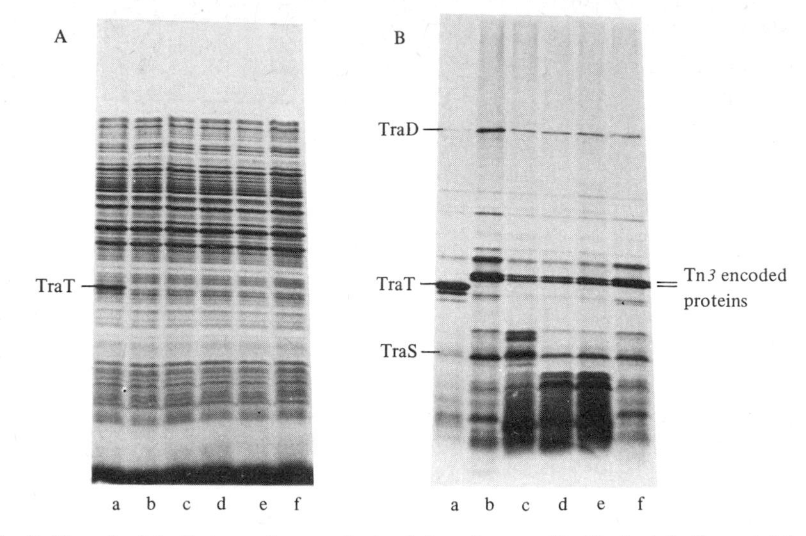

Fig. 2. Use of minicells to analyse products of cloned genes. Purified minicells containing the pKT107 plasmid (pACYC184 plus the *Eco*RI fragment E-7 of plasmid R6-5) which encodes a function (serum resistance) that elevates the resistance of host bacteria to the lethal activities of serum (track a), or different pKT107::Tn*3* insertion mutant derivatives that no longer confer serum resistance upon host bacteria (tracks b–f), were labelled with ^{35}S methionine and the proteins analysed by electrophoresis through a polyacrylamide gel (figure courtesy of P. A. Manning). Part A, gel stained with Coomassie Blue; Part B, gel autoradiograph. The autoradiograph readily reveals the plasmid-encoded proteins which, with the exception of the *tra*T protein, are completely obscured by cellular proteins in the stained gel. It also reveals that the *tra*T protein is responsible for R6-5 mediated serum resistance (Moll *et al.*, 1980).

in the stained gel but are clearly visible in the autoradiogram. Minicells are also useful for the identification of gene products encoded by DNA segments cloned in λ vectors. In this case, minicells carrying a plasmid that encodes the λ repressor are infected with λ phage that contain hybrid genomes; the λ repressor serves to depress the synthesis of λ proteins which otherwise are actively produced in the minicells (Reeve, 1977). More recently, a 'maxicell' system for the analysis of plasmid gene products was described (Sancar, Hack & Rupp, 1979). Maxicells are cells of a mutant bacterial strain that is highly sensitive to UV irradiation and that rapidly degrades UV-damaged DNA. Because the chromosome is such a large target compared with a plasmid molecule, a UV dose can be selected that causes almost complete chromosome degradation but that produces no significant number of lesions in the plasmid molecules, which therefore continue to express their genes. The ease of preparation of maxicells compared with minicells makes the former system rather attractive for the identification of the products of actively expressed genes.

In vitro procedures for the analysis of gene products are more time-consuming, at least to set up the experimental systems, than the corresponding *in vivo* procedures. They are, however, generally more sensitive and informative because they enable the investigator to modify the synthesis conditions and to add or subtract components. The analysis of the transcriptional units of a cloned DNA fragment may be made using RNA polymerase holoenzyme and radioactive ribonucleotides, and either supercoiled plasmid DNA or isolated fragment as template (it has been reported that supercoiled DNA is required for the efficient transcription from some promoters: Levine & Rupp, 1978). The transcripts made *in vitro* (or for that matter *in vivo*) may be sized by electrophoresis through polyacrylamide gels, and their template regions located by hybridization to electrophoretically separated restriction endonuclease-generated subfragments of the cloned DNA segment, which have been transferred to a nitro-cellulose filter (Southern, 1975). Cell-free coupled transcription–translation systems (Zubay, Chambers & Cheong, 1970; Gold & Schweiger, 1971) have been used successfully to synthesize and analyse the regulation of plasmid-encoded proteins *in vitro* (Yang, Zubay & Levy, 1976; Kennedy *et al.*, 1977). Moreover, comparison of the proteins made *in vitro* and *in vivo* has been used to investigate the possible processing of plasmid-encoded proteins that are localized in the bacterial membrane (Achtman, Manning, Edelbluth & Herrlich, 1979).

Applications of gene cloning

The objective of most gene cloning experiments is the isolation and selective amplification of a DNA segment either for detailed physical and/or functional analysis (see above) or for manipulation of the expression of a particular gene (see below). Gene cloning experiments can, however, in some instances directly provide information on gene structure and function and/or provide products of direct utility. One simple question that may be answered by gene cloning is whether the gene of a particular enzyme in a biochemical pathway has its own promoter, or whether it is expressed from a promoter that serves genes of other enzymes in the pathway. This may be tested, for instance, by cloning *Pst*I-generated fragments that contain only one or a few genes of a pathway into the *Pst*I cleavage site of pBR322: if the gene assayed is transcribed from its own promoter, or at least from a promoter contained within the

Table 7. *Selection probe fragments for generating mini-replicons or specific insertion mutant derivatives*

Probe fragment	Size (kb)	Origin	Available in	Vector	References[a]
*Eco*RI-Ap^r	6.8	pI258	pSC122	pSC101	1, 2
*Eco*RI-Km^r	6.3	R6-5	pSC105	pSC101	3, 4
*Pst*I-Cm^r	3.6	S-a	pKT205	pBR322	5
*Pst*I-Km^r	4.3	R6-5	pKT231	RSF1010	6
*Sal*I-Km^r	4.5	R6-5	pKT086	pBR322	7
*Kpn*I-Km^r	4.5	R6-5	pKT029	pBR322	7
*Bam*HI-Sm^r Su^r	4.2	R6-5	pKT218	pBR322	8
*Bam*HI/*Bgl*II-Sm^r	3.8	R6-5	pKT218	pBR322	9
*Hin*dIII-Tc^r	4.8	R6	pKT007	pML21	10
*Hae*II-Cm^r	1.5	R6-5	pACYC184	pACYC184	11
*Hae*II-Km^r	1.5	R6-5	pACYC177	pACYC177	11
*Hha*I-Cm^r	1.2	R6-5	pACYC184	pACYC184	11

[a] References: 1, Chang & Cohen, 1974; 2, Timmis *et al.*, 1975; 3, Cohen *et al.*, 1973; 4, Lovett & Helinski, 1976; 5, Bagdasarian, Bagdasarian, Coleman & Timmis, 1979; 6, M. Bagdasarian & K. N. Timmis, in preparation; 7, Andres *et al.*, 1979; 8, 9, M. Bagdasarian & K. N. Timmis, in preparation; 10, Timmis *et al.*, 1978c; 11, Chang & Cohen, 1978.

cloned fragment, it will be expressed irrespective of the orientation of the fragment within the vector molecule; if not, it will be expressed only when the fragment is cloned in the orientation that brings the assayed gene under the control of the β-lactamase promoter of the vector. Cloning may also be used to investigate the cellular location of gene products. *In vitro* or *in vivo* fusion of a gene under investigation, in the correct orientation and reading frame, to the gene of a readily identified protein such as β-galactosidase, enables standard cell fractionation techniques to be employed to determine the cellular location of the test protein, and hence the gene products of interest (Silhavy, Shuman, Beckwith & Schwartz, 1977).

Gene cloning has been particularly fruitful in the analysis of replication functions of, and the organization of their determinants in, replicons of interest. Specific probes, restriction endonuclease-generated DNA fragments that specify a selectable function (e.g. resistance to an antibiotic) but lack replication ability, have been isolated for the selective cloning of DNA segments which encode functions for autonomous replication (Timmis *et al.*, 1975; Lovett & Helinski, 1976; see Table 7 for examples). These probes have been used to determine whether essential plasmid replication genes are clustered on plasmid genomes, and to generate minimum-sized replicons for the analysis of replication determinants. Several studies have revealed that, in the majority of plasmids, essential replication determinants are clustered together whereas, in broad host range plasmids, they are distributed around the plasmid genome (Timmis *et al.*, 1975; Lovett & Helinski, 1976; Thomas & Helinski, 1979). The minimum amount of contiguous DNA required to form an autonomous *regulated* replicon has been shown to be about 1.8–2.0 kb for the F and R6-5 plasmids (Timmis, Andrés & Slocombe, 1978a; Thomas & Helinski, 1979). On the other hand, the minimum amount of contiguous DNA to form a non- or poorly-regulated ColE1 replicon which, unlike F and R6-5, requires no positive-acting translation product for its replication, is only 0.4–0.5 kb (Backman, Betlach, Boyer & Yanofsky, 1979). Minireplicon formation has also been successfully employed to obtain autonomously replicating origins of replication in cells that provide all essential *trans*-acting replication functions and, thus far, DNA fragments carrying the replication origins of plasmids RK2 (fragment 435 bp) and R6K (fragment 420 bp) (Thomas & Helinski, 1979) and the *E. coli* chromosome (416 bp) (Yasuda & Hirota, 1977; Messer *et al.*, 1978) have been obtained. Gene cloning techniques have also been used to fuse two different replicons in order to investigate replication control mechanisms and to generate replication mutants of one of the replicons (Timmis *et al.*, 1974; Cabello, Timmis & Cohen, 1976; Lusky & Hobom, 1979a, b).

One major application of gene cloning is in the dissection of complex, multifactorial phenomena, such as bacterial pathogenicity, symbiosis, etc. The cloning of individual pathogenicity determinants (e.g. genes for toxins, adhesion antigens, serum resistance: So, Boyer, Betlach & Falkow, 1976; Cabello & Timmis, 1979; Moll *et al.*, 1980), will allow manipulation of the cellular levels of individual pathogenicity factors, and the construction of pathogenic derivatives from non-pathogenic bacteria, for evaluation of the roles and relative importance of each pathogenicity component.

The other main direct application of gene cloning is the generation of DNA hybridization probes for identifying, mapping and characterizing homologous or partly homologous sequences in or from biological material. Cloned cDNAs are widely used for analysis of the structure of genomic DNA (Jeffreys & Flavell, 1977) and cloned determinants of importance (antibiotic resistance genes, toxin genes) in pathogenic bacteria are being increasingly used for epidemiological studies of bacterial isolates from clinical material using colony hybridization methods (Timmis *et al.*, 1978c, d; Dallas, Moseley & Falkow, 1979).

MUTAGENESIS *IN VITRO*

Mutagenesis of DNA may involve minor or major polynucleotide sequence alterations at the site of mutation. Gross sequence rearrangements do not usually provide useful information about the genetic structure of the DNA, whereas deletion, insertion and point mutations do. Deletion mutants, which suffer the greatest sequence alteration of these three types, are particularly useful where large DNA segments must be analysed and approximate locations of specific genes and signals are to be determined. Deletion formation usually produces an all or none response. Moreover, the deletion of small DNA segments containing promoter sequences or positive regulatory genes of polycistronic operons can completely inactivate expression of large DNA segments, and may often result in the production of fusion proteins or frame shifts at the site of deletion. In other words, although deletion formation can be very useful for genetic analysis, it sometimes results in phenotypic changes that are out of proportion to the size of the DNA segment involved in the deletion event. Insertion mutations are less pronounced in their effects because they involve physical change of the target DNA only at the position of the two nucleotide pairs between which the DNA insert is introduced. Of course, if this site is located between a promoter and its gene or within a positive acting regulatory element, then polar effects will also be observed. Nevertheless, insertion formation is a widely used form of mutagenesis because it is more sensitive than deletion mutagenesis and the mutations obtained are ordinarily simpler to analyse than point mutations. Point mutations are, of course, more subtle in their effects than mutations involving large nucleotide blocks, and can therefore provide more detailed information about the structure and function of small DNA segments. Point mutants are obligatory for the analysis of individual genes and provide information on the functional parts of their products, i.e. the domains (Miller, 1978), regulatory segments, and sequences involved in the maturation, transport etc., of proteins. Point mutations are therefore the most powerful of all mutations for the generation of detailed information; their analysis may be, however, correspondingly difficult or time consuming.

One of the most powerful gene technologies to be developed during the last few years is *in vitro* mutagenesis which, unlike *in vivo* procedures, permits the mutagenesis method to focus on a small segment of DNA (and, in some instances, on predetermined individual nucleotides) under investigation (Table 8). Coupled with gene cloning, it is an immensely powerful technique for the analysis and manipulation of gene structure and function and permits not only classical genetic, physiological and biochemical studies to be carried out on mutant DNA segments, but also several procedures that are ordinarily difficult if not impossible. For example, the cloning of wild-type and mutant DNA segments, or two different mutant DNA segments, on compatible vector plasmids (e.g. pBR322 and pACYC184) enables the introduction of both hybrid molecules into recombination deficient (RecA) bacteria, and establishment of the diploid state for the cloned fragment. Thus complementation analysis and the *cis/trans* test may be carried out with genes from one cellular system (e.g. *Pseudomonas*) that are cloned in another (e.g. *E. coli*). Moreover, it is possible to clone a gene on a high copy number vector plasmid, mutate it *in vitro*, and reintroduce it into its original host where, after a period of time, recombination events exchange the wild-type allele with the mutant allele. This process of homogenotization, the introduction of a mutant allele by recombination into another replicon, such as the host chromosome, carrying the wild-type allele, is an important component of 'reversed genetics' (Weissmann, 1978), the generation of specific mutations *in vitro* followed by their phenotypic analysis. In some host systems (e.g. multicellular organisms), effective homogenotization may not be possible and the phenotypic change caused by the mutation must be investigated in simple *in vivo* (e.g. frog oocytes: de Robertis & Gurdon, 1979) or *in vitro* expression systems.

Deletion mutations

In vitro methods for the generation of deletion derivatives of genomes fall into three classes: those that are carried out entirely *in vitro*, those that are primarily dependent upon *in vitro* reactions, and those that largely depend upon *in vivo* events. The formation of deletion derivatives entirely by *in vitro* manipulation involves standard gene cloning procedures. For example, it was possible to localize a serum resistance gene of plasmid R6-5 by the generation of deletion mutants of a pACYC184 hybrid plasmid containing the serum resistance-encoding *Eco*RI fragment E-7 of R6-5. pACYC184 contains no cleavage sites for the *Pst*I endonuclease, whereas the E-7 fragment contains six. Partial cleavage of the hybrid plasmid with *Pst*I, followed by recircularization of the linear fragments and their transformation into *E. coli*, generated a large series of hybrid plasmids deleted for one or several contiguous *Pst*I DNA subfragments (Moll *et al.*, 1980). Examination of the serum resistance phenotype of host bacteria carrying the deletion derivatives accurately localized the serum resistance gene. The *in vitro* generation of miniplasmids (Timmis *et al.*, 1975; Lovett & Helinski, 1976) and origin of replication replicons (Yasuda & Hirota, 1977; Messer *et al.*, 1978) are further examples of the formation of deletion derivatives entirely *in vitro*. In both types of example, the deletion derivatives have lost DNA segments having precisely defined end points (i.e. restriction endonuclease cleavage sites).

Deletion derivatives that are obtained primarily by *in vitro* manipulations are generated by nuclease digestion of genomes that have been linearized (a) at a unique site, by digestion with a restriction endonuclease that cleaves the genome once, (b) at one of several specific sites, by partial digestion with a restriction endonuclease that cuts the genome at multiple locations, or (c) at random locations, by cleavage with pancreatic DNase I in the presence of $MnCl_2$ (Shenk *et al.*, 1976). Because uncleaved molecules transform much more effectively than cleaved molecules, linear monomeric molecules are usually separated from uncleaved and more extensively degraded molecules by electrophoresis through an agarose gel. The purified full-length linear molecules are then degraded from their ends by a nuclease and subsequently introduced by transformation into host cells. The nuclease may be S1, if small deletions corresponding to loss of restriction endonuclease-generated cohesive ends are required (Backman *et al.*, 1979), or can be λ exonuclease (Covey *et al.*, 1976; Shenk, *et al.*, 1976) or exonuclease III (Heffron *et al.*, 1977), if longer deletions are required. In the latter case, the degree of exonuclease digestion is usually monitored by following the appearance of acid soluble radioactivity from radioactive DNA, to obtain derivative genomes exhibiting deletions of the desired size. The $3'$ (λ exonuclease) or $5'$ (exonuclease III) single-stranded termini that remain after exonuclease digestion, may be removed by treatment with S1 nuclease (Heffron *et al.*, 1977), but this is not essential for deletion forma-

Table 8. *Mutagenesis* in vitro

Mutant type	Mutagenesis procedure[a]	Site or regional specificity	Predictable change[b] in cleavage pattern	References[c]
Deletion	*Transformation* of randomly cleaved linear monomers	None	–	1
	Transformation of specifically cleaved linear monomers	Regional	+	2, 3
	S1 removal of cohesive ends of rest. endonuclease-generated linear monomers	Site	+	4
	Exonuclease treatment of randomly cleaved linear monomers	None	–	1, 5
	Exonuclease treatment of specifically cleaved linear monomers	Regional	(+)	6, 7
	Deletion of specific rest. endonuclease-generated fragments	Site	+	2, 8, 9
Insertion	Random or specific linearization; *insertion of synthetic linkers*	None or site	–	4, 10
	'Filling in' of cohesive termini	Site	+	4
	Insertion of fragment encoding selectable function at specific cleavage site	Site	+	4, 11
	Fusion of replicons at specific cleavage site	Site	+	4, 12
Point	*Hydroxylamine treatment*	None	–	13, 14, 15
	Sodium bisulphite treatment	Regional	(–)	16
	Incorporation of nucleotide analogues	Site	(–)	17
	Incorporation of synthetic oligonucleotides	Site	(–)	18, 19

[a] Step that produces the mutation is italicized.
[b] Enables recognition of mutant genomes.
[c] References: 1, Shenk, Carbon & Berg, 1976; 2, Lai & Nathans, 1974; 3, Murray & Murray, 1974; 4, Backman *et al.*, 1979; 5, Heffron, Bedinger, Champoux & Falkow, 1977; 6, Carbon, Shenk & Berg, 1975; 7, Covey, Richardson & Carbon, 1976; 8, Timmis *et al.*, 1975; 9, Lovett & Helinski, 1976; 10, Heffron, So & McCarthy, 1978; 11, M. Bagdasarian & K. N. Timmis, unpublished experiments (see Table 7 for list of some selectable fragments); 12, Timmis *et al.*, 1974; 13, Borrias *et al.*, 1976; 14, Hashimoto-Gotoh & Sekiguchi, 1976; 15, Humphreys, Willshaw, Smith & Anderson, 1976; 16, Shortle & Nathans, 1978; 17, Flavell, Sabo, Bandle & Weissmann, 1974; 18, Schott & Kössel, 1973; 19, Hutchinson *et al.*, 1978.

tion. After transformation, *in vivo* processing of the termini of the fragments occurs and a small proportion of the fragments become circularized and thereby survive as viable genomes. The remainder are presumably degraded. Deletion formation in the genomes that do survive therefore is predominantly due to the *in vitro* exonuclease digestion, but in some cases is also due to *in vivo* processing of the fragment termini, prior to ring formation. Because neither *in vitro* exonuclease digestion nor *in vivo* processing results in the removal of a precise number of nucleotides, the deletions generated by this method have only *regional* specificity, if initiated at a specific endonuclease cleavage site, and no specificity, if initiated at a random location in the genome. S1 removal of cohesive ends does, however, result in the generation of site-specific deletions in the majority of cases, particularly if recircularization of the cleaved molecule is carried out *in vitro* by blunt-end ligation, prior to transformation of the DNA into a host cell system.

Because deletion of the terminal regions of linear DNA occurs readily *in vivo* after transformation (Lai & Nathans, 1974; Murray & Murray, 1974; Shenk *et al.*, 1976), *in vitro* exonuclease digestion may be dispensed with entirely if a lower yield (1–20%, compared with close to 100% of all viable genomes) of deletion derivatives is acceptable (i.e. if a rapid assay or screening procedure is available for the detection of deletion derivatives). Again, deletions from specific or random locations may be obtained according to whether the genomes are linearized specifically or randomly. In all cases where deletions originating from a unique restriction endonuclease cleavage site are required, a simple assay for deletion derivative genomes exists, namely the loss of that cleavage site and loss of susceptibility to the corresponding endonuclease.

The location and extent of deletion of a genome is readily mapped by heteroduplex formation with the parent genome, followed by visualization in the electron microscope (Davis *et al.*, 1971) or digestion of the single-stranded region with S1 nuclease (Shenk, Rhodes, Rigby & Berg, 1975) and restriction endonuclease cleavage analysis of the remaining duplex DNA.

These *in vitro* deletion procedures have been widely used to examine the functions of small genomes, such as that of SV40 (Lai & Nathans, 1974; Covey *et al.*, 1976; Shenk *et al.*, 1976), and have identified essential elements for transposition of the ampicillin resistance transposon Tn*3*, namely a Tn*3*-encoded ‘transposase’ and the terminal inverted repeat sequences of the Tn*3* element (Heffron *et al.*, 1977).

Insertion mutations

The usual method for generating insertion mutants of hybrid replicons in bacteria is transposition mutagenesis, the isolation of hybrid replicons carrying an antibiotic resistance transposon (see Sherratt, this volume). While transposition mutagenesis is relatively simple and can provide a large number of mutant derivatives of any given molecule, recent evidence indicates that there is some specificity in the site of insertion of transposons (Tu & Cohen, 1980), and hence the method may not be suitable if random mutations are required.

In vitro insertion of DNA segments into target molecules may be site-specific or random. A simple method for the introduction of insertions at cleavage sites of restriction endonucleases that generate fragments with 5′ cohesive single-stranded termini involves linearization of the molecules with the appropriate endonuclease, 'filling in' of the single-stranded termini with DNA polymerase I, followed by recircularization of the molecules by blunt-end ligation and re-introduction of the molecules into the host cell system (Backman *et al.*, 1979). If the fragment single-stranded termini are four nucleotides long, there will be a net insertion of 4 bp at the cleavage site, if two nucleotides long, the insertion will be two bp, etc. Note that the specific recognition sequence at the site of insertion is ordinarily lost by this procedure, thus providing a specific assay for the insertion derivative molecules.

Analysis by insertion mutagenesis of the replication origin region of the pMB1 replicon (Backman *et al.*, 1979) demonstrated that small insertions of this type at certain sites within the 600 nucleotide region upstream of the origin inactivated the replication ability of the plasmid whereas at other sites they did not. However, larger insertions at these latter sites did inactivate the replication ability of the plasmid. Larger DNA segments may be inserted at specific restriction endonuclease cleavage sites in a genome by standard gene cloning procedures, preferably employing DNA segments that encode a selectable function, such as resistance to an antibiotic. A selection of such fragments is given in Table 7. If insertional inactivation of replication functions of the replicon is anticipated, the DNA fragment used for insertion may itself be a replicon that can provide an alternate replication system for the target DNA (Timmis *et al.*, 1974; Cabello *et al.*, 1976; Backman *et al.*, 1979; Lusky & Hobom, 1979a, b).

All of the above insertion procedures are site-specific. They can, however, be used to mutagenize a genome at random locations if the target DNA is linearized not with a restriction endonuclease but by DNAse I in the presence of Mn^{2+} (Shenk *et al.*, 1976). The blunt-end ligation of 'linkers' to randomly linearized hybrid molecules is a highly useful method for introducing short insertions at random locations in small genomes. Generally, the addition of linkers to the molecule termini is followed by cleavage of the linker with the specific endonuclease whose recognition sequence it carries, and recircularization of the molecule by treatment with DNA ligase. Internal cleavage of the molecule is avoided by appropriate choice of the linker, or by protection of internal cleavage sites by prior treatment with the specific methylase. Linker mutagenesis has the virtue that it permits rapid localization of each random insertion, because a new restriction endonuclease cleavage site is created at the insertion site. This powerful method has been used by Heffron *et al.* (1978) to dissect the functions involved in transposition of the ampicillin resistance transposon Tn*3*. This study confirmed the previous conclusions, derived from *in vitro* deletion analysis, that Tn*3* encodes a transposase, a function absolutely required for transposition, and also identified the gene of a second element which appears to regulate negatively the frequency of transposition. The large number of insertions obtained in this study, which could be readily localized, further provided a fine-structure map of the Tn*3* element. It is clear that the use of linkers to generate readily mapped random insertions in cloned DNA fragments is a very powerful method for the functional analysis of any cloned gene or operon and is expected to find widespread use in the future.

Point mutations

A widely used general method for the introduction of point mutations at random locations in a DNA segment or small replicon is through treatment of the DNA *in vitro* with hydroxylamine (Borrias *et al.*, 1976; Hashimoto-Gotoh & Sekiguchi, 1976; Humphreys *et al.*, 1976), which induces $C \rightarrow T$ transitions. Treatment of hybrid DNA molecules that encode resistance to two antibiotics (or one resistance plus a second readily scored marker) for increasing periods of time, prior to transformation, provides not only the kinetics of loss of molecule viability (i.e. mutational loss of replication ability, or the selectable function), but also the frequency of mutations in the surviving molecules. This permits the investigator to choose the optimal mutagenesis conditions for the particular requirements of the experiment, i.e. weak mutagenesis in most cases, to avoid the occurrence of multiple mutations in single molecules, and heavy mutagenesis to obtain rare mutations in short DNA segments.

Sodium bisulphite is a single-strand specific mutagen which deaminates cytosine to uracil. The procedure developed by Shortle & Nathans (1978) for the *in vitro* local mutagenesis of DNA replicons, involves (a) the introduction of a random- or site-specific nick in a circular genome, by treatment of purified DNA with either DNase I or a 'frequent cutter' restriction endonuclease, both in the presence of ethidium bromide, (b) extension of the nick into a small gap, by limited exonucleolytic digestion with DNA polymerase I or exonuclease III, and (c) treatment of the short exposed single-stranded region with bisulphite. The mutagenized DNA may be used directly for infection of host cells, or may be repaired with DNA polymerase and polynucleotide ligase. As pointed out by Shortle & Nathans (1979b), this mutagenesis procedure is very precise because bisulphite produces the specific change C→U and gap filling *in vivo* or *in vitro* results in the substitution of a U/A pair for the original C/G pair, thus preventing reversion of the mutation by cellular repair enzymes *in vivo*. Moreover, the size of the gap and the extent of deamination of exposed C residues within the gap can be precisely regulated by the investigator by the appropriate choice of experimental conditions. Note that in this procedure, only one strand in any given molecule is gapped, and that different nucleotides are exposed in each of the two DNA strands.

In a very elegant study of the function of the SV40 origin of replication, Shortle & Nathans (1979a) introduced single mutations at the *Bgl*I cleavage site located at the SV40 origin. From the *Bgl*I recognition sequence

5′– G C C N N N N N G G C – 3′
3′– C G G N′N′N′N′N′C C G – 5′

six distinct mutant genomes, resulting from single C→T transitions, that would be resistant to *Bgl*I cleavage were expected. In fact, three such mutants, plus a C→A transversion mutant, were isolated and all exhibited distinct phenotypes (e.g. virus plaque morphology, temperature dependence, etc.). None of the mutants could be complemented by the wild-type virus genome, indicating that the mutations were located in a *cis*-dominant element. Most of the mutant genomes replicated at a higher or lower rate than the wild-type genome. This indicated that the mutations were located within a replication origin sequence whose structure regulates the frequency of initiation of SV40 replication. The SV40 T-antigen is known to interact with the origin region and to be involved in the initiation of replication. The question therefore arose as to whether mutations in the T-antigen could compensate for the origin mutations. The isolation of pseudo-revertants of the SV40 origin mutants, by local mutagenesis of gaps generated from random nicks in the mutant SV40 genome, confirmed that restoration of the normal functioning of the SV40 origin could be accomplished by secondary mutations in the T-antigen (Shortle, Margolskee & Nathans, 1979), thus demonstrating that the *cis*-dominant element containing the unique SV40 *Bgl*I cleavage site is involved in interactions with T-antigen that determine the frequency of initiation of replication.

Two other *in vitro* techniques for the introduction of point mutations into small DNA segments, namely the enzymatic incorporation of a nucleotide analogue at a predetermined location in a genome, and the enzymatic incorporation of a synthetic oligonucleotide containing a predetermined mutant sequence, are currently less widely used than the former methods, because they are more time-consuming and require a prior knowledge of the sequence to be mutated. They are, however, in principle more powerful than the simpler methods because they involve the introduction of a single predetermined mutational change in a large proportion of the population of molecules treated, and hence do not depend upon the availability of a specific phenotypic assay for the detection of rare mutant molecules. For this reason the sites of lethal mutations, which cannot be identified by other procedures, are readily detected (see below).

One method of site-specific mutagenesis that was developed by the group of Weissmann involves the insertion of a nucleotide analogue, e.g. N^4hydroxyCMP instead of CMP or UMP/TMP, at a predetermined location in a DNA molecule, using DNA polymerase under conditions of substrate-limited synthesis (Flavell *et al.*, 1974). The synthesis is initiated near the site in the genome to be mutated, either at a specific nick produced by a restriction endonuclease in the presence of ethidium bromide, or more generally, at the 3′ terminus of a specific single-stranded DNA fragment or oligonucleotide primer hybridized to an intact complementary DNA strand. Knowledge of the polynucleotide sequence of the DNA segment containing the site to be mutated permits the conditions of the DNA synthesis/nick translation to be chosen such that polymerization *in vitro* proceeds only as far as the nucleotide to be mutated; this may be carried out in one step or several consecutive steps in the presence of one, two or three substrate nucleotide triphosphates. The triphosphates are then removed by chromatography and the nucleotide analogue added to the replication complex. After incorporation of the analogue, all four normal triphosphates are added (that corresponding to the analogue in 200–fold excess) and the synthesis reaction completed. Introduction of the mutated genomes into a suitable host cell results in the production of 20–30% mutant progeny molecules. The identification and recloning of pure mutant genomes is then readily accomplished.

Weissmann and his group have used this procedure to analyse the functions of the 3′ extracistronic region of the Qβ RNA bacteriophage genome (Weissmann, 1978; Weissman *et al.*, 1979). *In vitro* incorporation of N⁴hydroxyCMP instead of UMP at position 39 of the Qβ − strand permitted the synthesis of + strands that were either wild-type, or that contained a G at position –40 (corresponds to position 39 of − strand), and enabled the isolation of mutant phage genomes with an A→G transition at position −40. Phage containing the mutant genome grew at only 25% the rate of wild-type phage (Domingo, Flavell & Weissmann, 1976), presumably due to the fact that the mutant position −40 is contained within the sequence −63 to −38, which binds both host factor (an essential replication component) and ribosomal protein S1, a subunit of Qβ replicase. It was subsequently shown that the mutant genome binds S1 protein less efficiently than the wild-type genome. A similar procedure was used to obtain mutant Qβ + strands with a G→A transition at position −16 (Flavell *et al.*, 1974), a G→A transition at position −25 (Weissmann *et al.*, 1979) and an A→G transition at position −29 (Weissmann *et al.*, 1979). No infectious virus could be obtained from the −16 mutant although the mutant RNA replicated normally *in vitro* (Flavell, Sabo, Bandle & Weissman, 1975; Sabo *et al.*,

1977). Mutants −25 and −29 were infectious although they exhibited a diminished growth rate when compared with wild type Qβ. Furthermore, the −25 mutant, but not the −29 mutant, produced a high proportion of non-infectious phage particles. These experiments suggested that the 3′ extracistronic region of the Qβ genome exercises at least two distinct functions, one that is involved in RNA replication, and another that is involved in virus morphogenesis (Weissmann *et al.*, 1979).

In order to investigate the possible role of the polypeptide initiation codon AUG in the formation of 70S ribosome/messenger RNA initiation complexes, Qβ RNA molecules were prepared containing G → A transitions at the third and fourth nucleotides of the phage coat cistron: sequence AUGG (Taniguchi & Weissmann, 1978). Mutant virus genomes AUAG, AUGA and AUAA bound ribosomes with efficiencies <0.1, 3.2 and 0.3, respectively, relative to that of the wild-type genome. This suggested that the AUG codon, and hence the interaction of the ribosome binding site with fMet-tRNA, is important for the formation of an initiation complex.

Another method for site-specific mutagenesis of small genomes is the use of chemically or enzymatically synthesized short primer oligodeoxyribonucleotides that are homologous with a selected segment of the genome to be mutated, and that contain a predetermined mutation (Schott & Kössel, 1973; Hutchinson *et al.*, 1978; Gillam & Smith, 1979a; Gillam *et al.*, 1979). Hybridization of the oligodeoxyribonucleotide to the complementary strand of the genome, followed by elongation with DNA polymerase I (Klenow fragment), ligation, and transformation results in the production of mutant and wild-type genomes. The use of a mixture of the two types of genome as a template, with the same mutant primer, in a DNA synthesis reaction carried out at an elevated temperature, which favours hybridization of the mutant primer to mutant template, greatly increases the proportion of mutant genomes synthesized, resulting in close to 100% mutants after one or two cycles of *in vitro* synthesis (Gillam & Smith, 1979b). This method has thus far only been utilized to mutate simple genomes, such as that of φX174, but is now expected to find more general use in the mutation of cloned DNA segments, with the recent development of rapid methods for the chemical synthesis of oligonucleotides.

MANIPULATION OF GENE EXPRESSION AND FUNCTION *IN VITRO*

The manipulation of gene expression and function is carried out to accomplish a variety of objectives. In homologous systems (e.g. *E. coli* genes cloned in an *E. coli* host/vector system) such manipulations may be carried out to provide information on (a) the regulation of expression of a cloned gene, (b) the influence of the cellular concentration of its product on the activity of related genes and products, (c) the influence of the cellular concentration of its product on cell physiology, if the product is a structural element or is involved in one of the centrally regulated cellular pathways, such as DNA replication, cell division, ribosome or membrane biosynthesis, etc., or (d) the cellular localization of the gene product. Cloned genes may also be manipulated to elevate the level of gene expression or to promote the excretion of an ordinarily intracellular gene product, for the purpose of isolating quantities of the product. In heterologous systems (e.g. mammalian genes cloned in *E. coli*) manipulations may be carried out to achieve (a) active transcription of the cloned gene, (b) correct translation and processing of the gene product, and (c) an increase in the resistance of the gene product to endogenous proteolysis.

Vectors for the manipulation of cloned genes

A significant increase in the level of gene expression is frequently obtained, via gene dosage, by cloning the gene into a high copy number vector plasmid (Hershfield *et al.*, 1974; Uhlin, Molin, Gustafsson & Nordström, 1979). A further increase may be achieved by insertion of the gene in a cloning vector carrying a strong promoter located close to, and oriented towards, the cloning site. The constitutive β-lactamase promoter of pBR322 and pACYC177 is reasonably strong and promotes the expression of many genes on DNA fragments cloned in the *Pst*I cleavage site of these vectors. The *Pst*I cleavage site is located 181–182 codons downstream of the N-terminal end of the β-lactamase gene. Fusion proteins, consisting of the N-terminal end of β-lactamase and the C-terminal end of a protein partly coded by the cloned fragment, may therefore be created by this type of construction, if the cloned gene is present in the same orientation and reading frame as the β-lactamase gene. Although the β-lactamase promoter may be active enough for some purposes, it is inadequate for others, and inappropriate where conditional (inducible) expression of a cloned gene is required.

Three different, strong promoters whose activities can be manipulated by the investigator, namely the promoters of the tryptophan operon (P_{trp}) and the lactose operon (P_{lac}), and the leftward promoter (P_L) of bacteriophage λ, have been incorporated into vector molecules close to one or more cloning sites to produce vectors that, under appropriate experimental conditions, provide high level expression of genes on cloned DNA fragments. These expression vectors have been comprehensively reviewed by Bernard & Helinski (1980), and only a few examples will therefore be described here (Table 9). The pOMPO vector is a pBR322 derivative plasmid with P_{lac} inserted next to the *Eco*RI cloning site (Mercereau-Puijalon *et al.*, 1978). Incorporation into this vector of a DNA fragment coding for most of the ovalbumin protein, and introduction of the hybrid molecule into *E. coli*, resulted in the synthesis of 30 000–90 000 molecules per cell (or 1% of the total cellular protein) of an ovalbumin-like fusion protein containing the first eight amino acids of β-galactosidase. Introduction of a compatible plasmid that overproduces *lac* repressor into a pOMPO hybrid plasmid-carrying cell results in repression of the activity of P_{lac} and enables it to be specifically induced by the addition of IPTG. Plasmid pHUB4 carries the P_L promoter of λ upstream of *Hpa*I, *Bam*HI and *Sal*I cloning sites (Bernard *et al.*, 1979). Promoter activity is negatively regulated by a temperature-sensitive cI gene that is located either on the host chromosome, or on a co-existing compatible plasmid. Growth of pHUB4-*trpA* gene-containing cells at 42 °C resulted in active transcription of the cloned gene for several hours, such that *trpA* gene product finally constituted 1–7% of the total cellular protein. Plasmid p*trp*ED5-1 is a pBR322 derivative containing P_{trp}, the *trpE* gene and 15% of the *trpD* gene, upstream of the Tc^r gene (Hallewell & Emtage, 1980). Insertion of DNA fragments at the *Hin*dIII, *Bam*HI, or *Sal*I sites in the Tc^r gene brings them under control of P_{trp}, which is completely switched off by *trp* repressor in normal *E. coli* cells. Addition of 3-β-indolylacrylic acid, a specific inducer of the *trp* operon, to cells carrying p*trp*ED5-1 switched on the promoter and caused high-level expression of *trpE* for several hours until *trpE* gene product constituted 25% of the total cellular protein. The p*trp*ED5-1 vector can be used to obtain fusion proteins consisting of the N-terminal

Table 9. *Plasmid vectors for the manipulation and analysis of cloned genes and genetic signals*

Vector	Replicon	Cloning sites	Selection	Size	Copy number	Remarks	References[a]
1. *High-level gene expression*							
pOMPO	pMB1	*Eco*RI	Ap^r	4.0	40–60	Fragments cloned into *Eco*RI site under P_{lac} control	1
pHUB4	pMB1	*Hpa*I, *Bam*HI, *Sal*I	Tc^r	6.5	40–60	Fragments cloned into *Hpa*I, *Bam*HI or *Sal*I site under λP_L control; expression induced by thermoinactivation of cI	2
p*trp*ED5-1	pMB1	*Hind*III, *Bam*HI, *Sal*I	Ap^r	6.7	40–60	Fragments cloned into *Hind*III, *Bam*HI or *Sal*I sites under P_{trp} control; expression induced with 3β-indolylacrylic acid	3
pKN410	R1	*Bam*HI, *Eco*RI	Ap^r	15.0	2–3 at 30 °C	Shift of bacteria to 42 °C results in selective amplification of plasmid and its gene products	4
2. *Production of fusion proteins*							
pPCΦ1, pPCΦ2, pPCΦ3	pMB1	*Eco*RI	Ap^r	4	40–60	Fusion with the β-galactosidase gene; expression under control of P_{lac}	5
pWT111, pWT121, pWT131	pMB1	*Hind*III	Ap^r	4.8	40–50	Fusion with the *trp*E gene; expression under control of P_{trp} and induced by 3β-indolylacrylic acid	6
3. *Transcription promoter and terminator detection*							
pMC81	ColE1	*Hind*III, *Kpn*I	Ap^r	25.6	20	Cloning sites between *lac* genes and P_{ara}. Expression of *lac* in absence of arabinose indicates presence of active promoter on cloned fragment; lack of expression of *lac* in presence of arabinose indicates presence of transcription terminator	7
pBRH1	pMB1	*Eco*RI	Ap^r	5.2	40–60	*Eco*RI fragments that contain a promoter result in expression of Tc^r when cloned in correct orientation	8
pGA39	p15A	*Hind*III, *Xma*I, *Pst*I	Cm^r	4.6	<10	*Hind*III, *Xma*I and *Pst*I fragments that contain a promoter result in expression of Tc^r when cloned in correct orientation	9
pGA46	p15A	*Hind*III, *Bgl*II, *Pst*I	Cm^r	4.4	<10	*Hind*III, *Bam*HI, *Bgl*II, *Bcl*I, *Sau*3A, *Mbo*I, *Pst*I fragments that contain a promoter result in expression of Tc^r when cloned in correct orientation	9

[a] References: 1, Mercereau-Puijalon *et al.*, 1978; 2, Bernard *et al.*, 1979; 3, Hallewell & Emtage, 1980; 4, Uhlin *et al.*, 1979; 5, Charnay, Perricaudet, Galibert & Tiollais, 1978; 6, Tacon, Carey & Emtage, 1980; 7, Casadaban & Cohen, 1980; 8, West, Nere & Rodriguez, 1979; 9, An & Friesen, 1979.

end of the *trpD* protein and the C-terminal end of polypeptides or proteins (e.g. a *trpD*-human growth hormone fusion protein: Martial, Hallewell, Baxter & Goodman, 1979) encoded by DNA sequences inserted at the single *Hin*dIII site of the vector.

It should be noted that, although these and similar expression vectors that contain highly active promoters have been shown to increase greatly the expression of some cloned genes, they do not substantially increase the expression of others, such as some ribosomal protein genes, whose products regulate their own synthesis at a post-transcriptional stage (see below).

Expression vectors have also been successfully employed to obtain high transcription rates of cloned genes in heterologous systems. High transcription rates are, however, only one aspect of improving the expression of DNA present in a foreign cellular environment (e.g. Gilbert & Villa-Komaroff, 1980). The expression of eukaryotic genes in prokaryotes, which do not specify mRNA splicing systems, requires that the gene does not contain intervening sequences, i.e. a DNA copy of mature (spliced) mRNA (cDNA) rather than genomic DNA must be cloned in order to express split genes (Mercereau-Puijalon *et al.*, 1978). Secondly, unless a fused protein is required, translational start and stop signals that are recognized by the translation apparatus of the host cell must bracket the cloned structural gene. This may be accomplished by the use of vectors containing such signals, or by *in vitro* fusion of synthetic oligonucleotides containing these signals to the fragments before cloning (Goeddel *et al.*, 1979a). Alternatively, it was shown that mouse dihydrofolate reductase cDNA was functionally expressed in *E. coli* when trailed with dC residues (Change *et al.*, 1978); here it appears that a Shine-Dalgarno-like sequence (see Argetsinger-Steitz, 1979, for review) was created next to the ATG protein start codon, which promoted binding of the dihydrofolate reductase mRNA to ribosomes via homology of its 5′-end with the 3′-end of the 16S ribosomal RNA. Thirdly, some proteins, particularly secreted mammalian proteins, are synthesized as precursors which are subject to post-translational modification, e.g. by specific cleavage to form an active gene product. If this is not accomplished in the cloning system employed, the protein must either be synthesized in a form that does not require post-translational modification, or must be modified *in vitro* by the investigator after purification. Human growth hormone, like many secreted polypeptides, is synthesized as a precursor molecule containing an N-terminal signal polypeptide that is cleaved off during secretion from the producing cells. In order to obtain the synthesis of this hormone in *E. coli* cells, Goeddel *et al.* (1979a) linked a cDNA segment coding for the major C-terminal portion of the hormone (amino acids 24–191) to a synthetic oligonucleotide coding for the N-terminal end amino acids 1–23, plus an ATG translational start codon. This synthetic gene, which does not code for the signal peptide, was then cloned into pBR322 using dC/dG homopolymer tailing. A series of subsequent *in vitro* manipulations, in which the cloned fragment was trimmed and inserted next to the *lac* promoter and ribosome binding site of a specifically-engineered vector, resulted in the formation of a hybrid plasmid that directed the synthesis in *E. coli* of almost 200 000 copies/cell of a polypeptide that is similar in size and immunological properties to human growth hormone.

Foreign mRNAs and proteins are often rapidly degraded in *E. coli* cells. Protein degradation has been shown to be greatly reduced if the foreign protein is 'buried' within (i.e. fused to) an *E. coli* protein. Gene fusion may thereby obviate problems of gene product degradation and a lack of correct translational start and stop signals. Fusion proteins may, however, lack biological activity, although in some instances it may be regained by purification and specific cleavage of the protein *in vitro* (Itakura *et al.*, 1977; Goeddel *et al.*, 1979b). Gene fusion requires that the vector and cloned genes are spliced in the same orientation and reading frame and, to facilitate this exercise, two series of cloning vectors that allow fusion of genes to a bacterial gene in all three reading frames have been developed (Table 9). The pPCφ1, 2 and 3 vectors (Charnay *et al.*, 1978) permit the cloning of *Eco*RI fragments and the fusion of genes on these fragments to the *E. coli* β-galactosidase gene; expression of a fusion protein is then under control of P_{lac}. The pWT111, 121 and 131 vectors (Tacon *et al.*, 1980) permit the cloning of *Hin*dIII fragments and the fusion of genes to the *trpE* gene product; expression of a fusion protein is then under control of P_{trp}.

The purification of extracellular or periplasmic proteins is generally much simpler than the purification of intracellular proteins, because the total number of proteins in the primary extract is small. Fusion of the foreign gene to a secreted protein, such as β-lactamase or β-galactosidase, may result in the secretion of the fusion protein from the producing cells, and hence a simplification of its purification (Villa-Komaroff *et al.*, 1978).

Just as the *manipulation* of gene expression may be achieved by removal of the natural promoter of a gene and its replacement by a highly active substitute promoter, so the *study* of gene expression may involve an analysis of promoter activity by removal of the natural gene and its replacement with a 'test' gene, whose product is readily assayed (e.g. β-galactosidase). The *in vitro* fusion of promoters of interest to test genes has been greatly simplified by the construction of vectors for the selective cloning and subsequent analysis of transcription promoters. These vectors usually contain a test gene that lacks a promoter, and that is situated immediately downstream of one or more cloning sites. Some vectors that are suitable for the selective cloning and analysis of properties of transcription promoters are given in Table 9. Three such cloning vectors rely on the expression of tetracycline resistance gene of pSC101 for the detection of promoters on cloned fragments: pBRH1 (West *et al.*, 1979) is suitable for cloning *Eco*RI fragments, pGA39 (An & Friesen, 1979) for *Hin*dIII, *Xma*I and *Pst*I fragments and pGA46 (An & Friesen, 1979) for *Hind*III, *Bam*HI, *Bcl*I, *Sau*3A (*Mbo*I) and *Pst*I fragments. The pMC81 plasmid (Casadaban & Cohen, 1980) contains *Hind*III and *Kpn*I cloning sites upstream of a β-galactosidase gene and downstream of the arabinose operon promoter P_{ara}. In this plasmid, the synthesis of β-galactosidase is dependent upon the presence of the specific inducer, arabinose. Insertion of *Hin*dIII or *Kpn*I fragments that contain a promoter results in the synthesis of β-galactosidase in the absence of arabinose. This vector can also be used for selective cloning of DNA fragments containing transcription termination signals; in this case, insertion of such fragments prevents arabinose-induced β-galactosidase synthesis.

Analysis of gene regulation by in vitro *manipulation procedures*

In a very elegant study, Ptashne and his colleagues have studied the interplay of the two bacteriophage λ control proteins, λ repressor and cro, with the two overlapping promoters P_R and P_{RM}, that are involved in regulating the bacteriophage lytic and lysogenic cycles (see Ptashne *et al.*, 1980, for review). The λ repressor, which is required for lysogeny, and cro, which is required for lytic growth of the phage, both bind to three adjacent operators in the P_R–P_{RM} sequence to regulate lambda development. By fusing P_R or P_{RM} to the β-galactosidase gene in a transducing phage it was possible to examine the activity of these promoters by measuring β-

galactosidase synthesis. By fusing the λ repressor or the cro gene to the *lac*OP sequence of a cloning vector, and introducing this hybrid plasmid into a cell that overproduces *lac* repressor, the synthesis of these regulatory proteins could be manipulated by growth of host cells in the presence of different concentrations of IPTG. By bringing together all three components, i.e. the λ promoter linked to the β-galactosidase gene, the overproducing *lac* repressor gene, and the λ repressor or cro gene linked to the *lac*OP sequence, it was possible to observe how the activity of P_R and P_{RM} respond to different cellular levels of repressor and cro products and to elucidate how the control proteins regulate positively and negatively the activities of P_R and P_{RM}.

In vitro gene fusion techniques have also been employed to investigate the control of gene expression by the regulation of translation events, in particular those involved in the synthesis of ribosomal proteins. Ribosomes from *E. coli* contain approximately 50 proteins. All of these proteins, with the exception of L7/L12, exist in a single copy per ribosome. In exponentially growing cells, there appears to be neither a significant pool of free ribosomal proteins (r-proteins) nor significant degradation of newly synthesized proteins. Therefore, all the r-proteins are apparently synthesized co-ordinately and stoichiometrically. It is known that the genes for r-proteins are organized in at least 11 different operons. The co-ordinate regulation of these operons, and the responsiveness of the regulation system to small changes in bacterial growth rates, is remarkable. It was known from earlier studies with r-protein operon transducing phages that the transcription rate, but not the translation rate, of r-proteins is gene dosage dependent (Fallon, Jinks, Strycherz & Nomura, 1978a; Fallon, Jinks, Yamamoto & Nomura, 1979b). *In vitro* fusion of the genes for r-proteins L2, L4, and L23 to *lacOP* in a vector plasmid enabled Lindahl & Zengel (1979) to investigate the influence of overproduction of these proteins (by induction with IPTG) on the synthesis of other r-proteins. Within 10 minutes of the induction of L2, L4, and L23, the synthesis of r-proteins S3, S19, L3, L16, L22, and L29, whose genes are located in the same operon as those of L2, L4, and L23, had almost ceased. In contrast, the synthesis of other r-proteins was unaffected. Subsequent studies on the *in vitro* synthesis of r-proteins demonstrated that, for several r-protein operons, the addition of one specific r-protein of the operon to the DNA-driven *in vitro* protein synthesis system abolished the synthesis of all other r-proteins of the operon (Yates *et al.*, 1980). This suggests that each r-protein operon is autoregulated at the level of translation of mRNA by one of the products of translation, a promoter-proximal r-protein, which presumably binds to the mRNA and blocks ribosome binding or translocation. A similar control mechanism for the synthesis of Qβ proteins has been described (e.g. Argetsinger-Steitz, 1979).

High copy number vector plasmids carrying r-protein genes and promoters are detrimental to the growth of host bacteria, because they depress the synthesis of other proteins of the operon and thereby destroy the fine co-ordinate regulation of r-protein synthesis. This fact was used to isolate point mutations in such plasmids which overcome the detrimental effect. Nucleotide sequence analysis of the mutants of a plasmid that carries the L10 gene and its promoter revealed that the mutations all occur within the leader region of the L10 operon and are well removed from its promoter. The mutant plasmids exhibit normal transcription of L10 but no translation of the messenger RNA. This suggests that the mutations define a regulatory region within the leader sequence of the RNA transcript which regulates the translational efficiency of the messenger RNA (Fiil, Friesen, Downing & Dennis, 1980).

Gene cloning techniques have also been instrumental in elucidating the mechanism of phase variation in *Salmonella. Salmonella* strains have two genes that are responsible for the synthesis of two antigenically distinct and alternatively expressed flagella. Phase 1 flagella are composed of the flagella filament subunit protein, flagellin, encoded by the H1 gene whereas phase 2 flagella are composed of the flagellin encoded by the H2 gene. Strains can switch from one phase to the other at frequencies of about 10^{-4} per cell per generation. Switching is known to be controlled by a genetic element associated with the H2 gene which is located in a region of the *Salmonella* genome that is different from that of the H1 gene. Evidence has been presented that H1 → H2 switching results from the co-ordinate synthesis of H2 protein and an H1 gene repressor, whereas H2 → H1 switching results from a failure to synthesize either protein. In order to elucidate the molecular basis of switching, Simon and his colleagues separately cloned segments of the *Salmonella* genome carrying the H1 gene, and carrying the H2 gene plus the element required for switching (Zieg, Silverman, Hilman & Simon, 1978; Silverman, Zieg & Simon, 1979). Heteroduplex studies with these hybrid molecules identified a 900 bp DNA segment located adjacent to the H2 gene which could invert and thereby switch on or off the H2 gene and the H1 gene repressor determinant. This finding suggested that the invertible segment carries a promoter that, in one orientation of the segment, is responsible for expression of H2. *In vitro* fusion of the H2 gene and invertible DNA sequence to the *trpB* gene of a vector plasmid resulted in unstable *trpB* gene expression, which indicates that promotion of *trpB* gene synthesis resulted from a promoter in the invertible DNA sequence. Phase variation in *Salmonella* therefore appears to result from the RecA-independent inversion of a DNA sequence that carries a promoter for the H2 gene and a gene that encodes a repressor of the H1 gene.

Another example of antigenic variation, which probably serves the same function as phase variation in *Salmonella*, namely evasion of the host immune system, is variation in the surface glycoprotein antigens of trypanosomes. The number of immunologically different surface antigen types that one clone of *T. brucei* can successively make is not known, but exceeds 100 and may be much greater than this. Successive surface glycoproteins are completely different: they differ in amino acid composition, conformational features, and N-terminal amino acid sequences. Three types of mechanism have been proposed to account for antigenic variation in trypanosomes: (1) alteration of a limited number of genes by mutation, sequence alteration, etc., (2) reassortment of sequences from a limited number of genes at the mRNA level, by differential splicing of long precursors, and (3) differential expression of a large number of different genes. The cloning of cDNAs of mRNAs of four different variant glycoproteins from *T. brucei* enabled Hoeijmakers *et al.* (1980) to show that the genes for these four glycoproteins exhibit no major DNA sequence homology, thus ruling out mechanism 1. The use of the cloned cDNAs as hybridization probes to localize the variant surface glycoprotein genes on the trypanosome chromosomes should readily distinguish mechanisms 2 and 3.

CONCLUDING REMARKS

Gene manipulation *in vitro* is a very recently-developed branch of molecular genetics. Despite its newness, however, this technology has already found wide use in many areas of the biological sciences and its potential in applied fields appears to be enormous (Gilbert & Villa-Komaroff, 1980).

The combination of gene cloning, site-specific mutagenesis, and the manipulation of gene expression already enables the investigator to probe the roles of individual nucleotides in a genetic element under study and thereby gain a profound understanding of the functioning of genetic information and its products. The sensitivity and high degree of investigator control inherent in these methods are unprecedented in the field of genetics and will undoubtedly lead to a renaissance in several fields in the biomedical sciences in the near future. In applied biology, gene manipulation is being aggressively exploited to improve existing commercial processes and to create products such as hormones, pharmaceuticals (interferon) and vaccines that can for the first time be produced by bacterial fermentations. Moreover, the creation of modified gene products (hormones, enzymes, vaccines, antisera) with altered specificities, by local or site specific mutagenesis, is expected to find widespread use as an alternative to screening for new pharmaceutical and industrial products (Weissman, 1978). The technology of immobilized enzymes and cells is also expected to benefit from gene cloning methods with the development of vectors in which an enzyme gene can be fused to the gene of a cell surface protein and so enable the production of bacteria that are coated with an active gene product (e.g. enzyme) (e.g. Manning, Echarti & Timmis, 1980). Such vectors will also find use in the production of vaccines, particularly live oral vaccines against intestinal pathogens.

Considering the rapidly growing world population, and the equally rapidly decreasing supplies of natural resources, one of the most exciting prospects for *in vitro* gene manipulation is in the area of conservation of raw materials, particularly the conservation and recycling of reduced carbon compounds, and in the improvement of food animal and plant yields. Many industrial concerns dispose of tons of organic compounds each day. These wastes not only constitute a major environment pollution problem but in principle could be converted to useful products, such as single cell protein or other forms of biomass (which, after appropriate treatment to remove toxic substances such as heavy metals, can be fed directly to animals) and methane (a valuable energy source), if bacteria are developed that can rapidly degrade the major components of the wastes. Bacteria that are able to degrade various noxious organic compounds, including heavily substituted xenobiotics, are readily found in nature and the manipulation of these to increase the spectrum of compounds degraded, and the rapidity of degradation is an important, immediate task for gene technologists (e.g. Franklin, Bagdasarian & Timmis, 1981). Food yields could also be improved by manipulation of N_2-fixing symbiotic bacteria, like *Rhizobium*, to increase their N_2-fixation activities and to broaden their plant host specificities. This will increase the yields of some food plants and decrease fertilizer costs.

It can be seen then that gene technology is unique in the field of biology in that it is at the same time a very fundamental and a very applied science. The long-term consequences of this are difficult to foresee but the developing interactions between the basic and applied areas of molecular genetics will probably, on balance, be invigorating for both disciplines.

I am grateful to members of the Plasmid Group at the Max-Planck-Institute for Molecular Genetics for stimulating discussions and to H. Markert, I. Schallehn and J. Timmis for secretarial assistance. Work carried out in my laboratory was supported by grants from the Deutsche Forschungsgemeinschaft and the Bundesministerium für Forschung und Technologie.

REFERENCES

Achtman, M., Manning, P. A., Edelbluth, C. & Herrlich, P. (1979). Export without proteolytic processing of inner and outer membrane proteins encoded by F sex factor *tra* cistrons in *Escherichia coli* minicells. *Proceedings of the National Academy of Sciences, USA*, **76**, 4837–41.

Adler, H. I., Fisher, W. D., Cohen, A. & Hardigree, A. A. (1967). Miniature *Escherichia coli* cells deficient in DNA. *Proceedings of the National Academy of Sciences, USA*, **57**, 321–6.

An, G. & Friesen, J. D. (1979). Plasmid vehicles for direct cloning of *Escherichia coli* promoters. *Journal of Bacteriology*, **140**, 400–7.

Andres, I., Slocombe, P. M., Cabello, F., Timmis, J. K., Lurz, R., Burkardt, H.-J. & Timmis, K. N. (1979). Plasmid replication functions. II. Cloning analysis of the *RepA* replication region of antibiotic resistance plasmid R6-5. *Molecular and General Genetics*, **168**, 1–25.

Argetsinger-Steitz, J. (1979). Genetic signals and nucleotide sequences in messenger RNA. In *Biological Regulation and Development, Vol. I. Gene Expression*, ed. R. F. Goldberger, pp. 349–99. New York: Plenum Press.

Backman, K., Betlach, M., Boyer, H. W. & Yanofsky, S. (1979). Genetic and physical studies on the replication of ColE1-type plasmids. *Cold Spring Harbor Symposia on Quantitative Biology*, **43**, 69–76.

Bagdasarian, M., Bagdasarian, M. M., Coleman, S. & Timmis, K. N. (1979).

New vector plasmids for gene cloning in *Pseudomonas*. In *Plasmids of Medical, Environmental and Commercial Importance*, ed. K. N. Timmis & A. Pühler, pp. 411–22. Amsterdam: Elsevier/North Holland.

BAGDASARIAN, M., NORDHEIM, A. & TIMMIS, K. N. (1980). Specific purpose plasmid cloning vectors. II. Broad host range RSF1010-derived vectors. Submitted for publication.

BAHL, C. P., MARIANS, K. J., WU, R., STAWINSKY, J. & NARANG, S. A. (1976). A general method for inserting specific DNA sequences into cloning vehicles. *Gene*, **1**, 81–92.

BAHL, C. P., WU, R., BROUSSEAU, R., GOOD, A. K., HSIUNG, H. M. & NARANG, S. A. (1978). Chemical synthesis of versatile adapters for molecular cloning. *Biochemical and Biophysical Research Communications*, **81**, 695–703.

BARNES, W. M. (1977). Plasmid detection and sizing in single colony lysates. *Science*, **195**, 393–4.

BARNES, W. M. (1980). DNA cloning with single-stranded phage vectors. In *Genetic Engineering, Principles and Methods*, vol. 2, ed. J. K. Setlow & A. Hollaender, pp. 185–200. New York: Plenum Press.

BENTON, W. D. & DAVIS, R. W. (1977). Screening λgt recombinant clones by hybridization to single plaques *in situ*. *Science*, **196**, 180–2.

BERNARD, H. U. & HELINSKI, D. R. (1980). Bacterial plasmid cloning vectors. In *Genetic Engineering, Principles and Methods*, vol. 2, ed. J. K. Setlow & Hollaender, pp. 133–67. New York: Plenum Press.

BERNARD, H. U., REMAUT, E., HERSHFIELD, M. V., DAS, H. K. & HELINSKI, D. R. (1979). Construction of plasmid cloning vehicles that promote gene expression from the bacteriophage lambda P_L promoter. *Gene*, **5**, 59–76.

BIRNHOIM, H. C. & DOLY, J. (1979). A rapid alkaline extraction procedure for screening recombinant plasmid DNA. *Nucleic Acids Research*, **7**, 1513–23.

BOLIVAR, F. (1978). Construction and characterization of new cloning vehicles. III. Derivatives of plasmid pBR322 carrying unique *Eco*RI site for selection of *Eco*RI generated recombinant DNA molecules. *Gene*, **4**, 121–36.

BOLIVAR, F., RODRIGUEZ, R. L., GREENE, P. J., BETLACH, M. C., HEYNEKER, H. L., BOYER, H. B., CROSA, J. H. & FALKOW, S. (1977). Construction and characterization of new cloning vehicles. II. A multipurpose cloning system. *Gene*, **2**, 95–113.

BOLLUM, F. J. (1974). Terminal deoxynucleotidyl transferase. In *The Enzymes*, vol. 10, ed. P. D. Boyer, pp. 145–71. New York: Academic Press.

BORRIAS, W. E., WILSCHUT, I. J. C., VEREIKEN, J. M., WEISBEK, P. J. & VAN ARKEL, G. A. (1976). Induction and isolation of mutants in a specific region of gene A of φX 174. *Virology*, **70**, 195–7.

BRAMMAR, W. J. (1979). Safe and useful vector systems. In *Biochemistry of Genetic Engineering*, ed. P. B. Garland & R. Williamson, pp. 13–27. London: The Biochemical Society.

BROOME, S. & GILBERT, W. (1978). Immunological screening method to detect specific translation products. *Proceedings of the National Academy of Sciences, USA*, **75**, 2746–9.

CABELLO, F. & TIMMIS, K. N. (1979). Plasmids of medical importance. In *Plasmids of Medical, Environmental and Commercial Importance*, ed. K. N. Timmis & A. Pühler, pp. 55–69. Amsterdam: Elsevier/North Holland.

CABELLO, F., TIMMIS, K. N. & COHEN, S. N. (1976). Replication control in a composite plasmid constructed by *in vitro* linkage of two distinct replicons. *Nature, London*, **259**, 285–90.

CARBON, J., SHENK, T. E. & BERG, P. (1975). Biochemical procedure for production of small deletions in simian virus DNA. *Proceedings of the National Academy of Sciences, USA*, **72**, 1392–6.

CASADABAN, M. J. & COHEN, S. N. (1980). Analysis of gene control signals by DNA fusion and cloning in *Escherichia coli*. *Journal of Molecular Biology*, **138**, 179–207.

CHANG, A. C. Y. & COHEN, S. N. (1974). Genome construction between bacterial species *in vitro*: replication and expression of *Staphylococcus* plasmid genes in *Escherichia coli*. *Proceedings of the National Academy of Sciences, USA*, **71**, 1030–4.

CHANG, A. C. Y. & COHEN, S. N. (1978). Construction and characterization of amplifiable DNA cloning vehicles derived from the p15A cryptic miniplasmid. *Journal of Bacteriology*, **134**, 1141–56.

CHANG, A. C. Y., NUNBERG, J. H., KAUFMAN, R. J., ERLICH, H. A., SCHIMKE, R. T. & COHEN, S. N. (1978). Phenotypic expression in *E. coli* of a DNA sequence coding for mouse dihydrofolate reductase. *Nature, London*, **275**, 617–24.

CHARNAY, P., PERRICAUDET, M., GALIBERT, F. & TIOLLAIS, P. (1978). Bacteriophage lambda and plasmid vectors, allowing fusion of cloned genes in each of three translational phases. *Nucleic Acids Research*, **5**, 4479–94.

CLARKE, L. & CARBON, J. (1976). A colony bank containing synthetic ColE1 hybrid plasmids representative of the entire *E. coli* genome. *Cell*, **9**, 91–9.

COHEN, S. N. (1975). The manipulation of genes. *Scientific American*, **233**, 24–33.

COHEN, S. N., CHANG, A. C. Y., BOYER, H. W., & HELLING, R. B. (1973). Construction of biologically functional bacterial plasmids *in vitro*. *Proceedings of the National Academy of Sciences, USA*, **70**, 3240–4.

COHEN, S. N., CHANG, A. C. Y. & HSU, L. (1972). Non chromosomal antibiotic resistance in bacteria: genetic transformation of *Escherichia coli* by R-factor DNA. *Proceedings of the National Academy of Sciences, USA*, **69**, 2110–14.

COLLINS, J. (1979). Gene cloning with small plasmids. *Current Topics in Microbiology and Immunology*, **78**, 121–70.

COLLINS, J. & BRÜNING, H. J. (1978). Plasmids useable as gene-cloning vectors in an *in vitro* packaging by coliphage λ: 'cosmids'. *Gene*, **4**, 85–107.

COLLINS, J. & HOHN, B. (1978). Cosmids: a type of plasmid gene-cloning vector that is packageable *in vitro* in bacteriophage λ heads. *Proceedings of the National Academy of Sciences, USA*, **75**, 4242–6.

COVEY, C., RICHARDSON, D. & CARBON, J. (1976). A method for the deletion of restriction sites in bacterial plasmid deoxyribonucleic acid. *Molecular and General Genetics*, **145**, 155–8.

CREA, R., KRASZEWSKI, A., HIROSE, T. & ITAKURA, K. (1978). Chemical synthesis of genes for human insulin. *Proceedings of the National Academy of Sciences, USA*, **75**, 5765–9.

CURTISS, R., PEREIRA, D. A., HSU, J. C., HULL, S. C., CLARK, J. E., MATURIN, SR., L. J., GOLDSCHMIDT, R., MOODY, R., INOUE, M. & ALEXANDER, L. (1977). Biological containment: the subordination of *Escherichia coli* K-12. In *Recombinant Molecules: Impact on Science and Society*, ed. R. F. Beers, Jr. & E. G. Bassett, pp. 45–56, New York: Raven Press.

DALLAS, W. S., MOSELEY, S. & FALKOW, S. (1979). The characterization of an *Escherichia coli* plasmid determinant that encodes for the production of a heat-labile enterotoxin. In *Plasmids of Medical, Environmental and Commercial Importance*, ed. K. N. Timmis & A. Pühler, pp. 113–22. Amsterdam: Elsevier/ North-Holland.

DAVIS, R. W., SIMON, M. & DAVIDSON, N. (1971). Electron microscope heteroduplex methods for mapping regions of base sequence homology in nucleic acids. In *Methods in Enzymology*, vol. 21, ed. L. Grossman & K. Moldave, pp. 413–28. New York: Academic Press.

DE ROBERTIS, E. M. & GURDON, J. B. (1979). Gene transplantation and the analysis of development. *Scientific American*, **241**, 60–8.

DOMINGO, E., FLAVELL, R. A. & WEISSMANN, C. (1976). *In vitro* site directed mutagenesis: generation and properties of an infectious extracistronic mutant of bacteriophage Qβ. *Gene*, **1**, 3–25.

ECKHARDT, T. (1978). A rapid method for the identification of plasmid desoxyribonucleic acid in bacteria. *Plasmid*, **1**, 584–8.

EDGELL, M. H., WEAVER, S., HAIGWOOD, N. & HUTCHISON, C. A. (1979). Gene Enrichment. In *Genetic Engineering, Principles and Methods*, vol. 1, ed. J. K. Setlow & A. Hollaender, pp. 37–49. New York: Plenum Press.

EFSTRATIADIS, A. & VILLA-KOMAROFF, L. (1979). Cloning of double-stranded cDNA. In *Genetic Engineering, Principles and Methods*, vol. 1, ed. J. K. Setlow & A. Hollaender, pp. 15–36. New York: Plenum Press.

EHRLICH, S. D. (1978). DNA cloning in *Bacillus subtilis. Proceedings of the National Academy of Sciences, USA*, **75**, 1433–6.

ERLICH, H. A., COHEN, S. N. & MCDEVITT, H. O. (1978). A sensitive radioimmune assay for detecting products translated from cloned DNA fragments. *Cell*, **13**, 681–9.

FALKOW, S. (1975). *Infectious Multiple Drug Resistance*. London: Pion.

FALLON, A. M., JINKS, C. S., STRYCHERZ, G. D. & NOMURA, M. (1979a). Regulation of ribosomal protein synthesis in *Escherichia coli* by selective mRNA inactivation. *Proceedings of the National Academy of Sciences, USA*, **76**, 3411–15.

FALLON, A. M., JINKS, C. S., YAMAMOTO, M. & NOMURA, M. (1979b). Expression of ribosomal protein genes cloned in a hybrid plasmid in *Escherichia coli*: gene dosage effects on synthesis of ribosomal proteins and ribosomal protein messenger ribonucleic acid. *Journal of Bacteriology*, **138**, 383–96.

FIIL, N. P., FRIESEN, J. D., DOWNING, W. L. & DENNIS, P. P. (1980). Post-transcriptional regulatory mutants in a ribosomal protein-RNA polymerase operon of *E. coli. Cell*, **19**, 837–44.

FLAVELL, R. A., SABO, D. L., BANDLE, E. F. & WEISSMANN, C. (1974). Site directed mutagenesis: generation of an extracistronic mutation in bacteriophage Qβ RNA. *Journal of Molecular Biology*, **89**, 255–72.

FLAVELL, R. A., SABO, D. L. O., BANDLE, E. F. & WEISSMANN, C. (1974). Site directed mutagenesis: effect of an extracistronic mutation on the *in vitro* propagation of bacteriophage Qβ RNA. *Proceedings of the National Academy of Sciences, USA*, **72**, 367–71.

FRANKLIN, F. C. H., BAGDASARIAN, M. & TIMMIS, K. N. (1981). Manipulation of degradative genes of soil bacteria. In *Microbial Degradation of Xenobiotics and Recalcitrant Molecules*, ed. R. Hütter & T. Leisinger, in press. New York: Academic Press.

FRAZER, A. C. & CURTISS, R. (1975). Production, properties and utility of bacterial minicells. *Current Topics in Microbiology and Immunology*, **69**, 1–84.

GARAPIN, A. C., CAMI, B., ROSKAM, W., KOURILSKY, P., LE PENNEC, J. P., PERRIN, F., GERLINGER, P., COCHET, M. & CHAMBON, P. (1978). Electron microscopy and restriction enzyme mapping reveal additional intervening sequences in the chicken ovalbumin split gene. *Cell*, **14**, 629–39.

GAUTIER, F., MAYER, H. & GOEBEL, W. (1976). Cloning of calf thymus satellite I DNA in *Escherichia coli. Molecular and General Genetics*, **149**, 23–31.

GILBERT, W. & VILLA-KOMAROFF, L. (1980). Useful proteins from recombinant bacteria. *Scientific American*, **242**, 68–82.

GILLAM, S., JAHNKE, P., ASTELL, C., PHILIPS, S., HUTCHISON, C. A. & SMITH, M. (1979). Defined transversion mutations at a specific position in DNA using synthetic oligodeoxyribonucleotides as mutagens. *Nucleic Acids Research*, **6**, 2973–85.

GILLAM, S. & SMITH, M. (1979a). Site specific mutagenesis using synthetic oligodeoxyribonucleotide primers. I. Optimum conditions and minimum oligodeoxyribonucleotide length. *Gene*, **8**, 81–97.

GILLAM, S. & SMITH, M. (1979b). Site specific mutagenesis using synthetic oligodeoxyribonucleotide primers. II. *in vitro* selection of mutant DNA. *Gene*, **8**, 99–106.

GOEDDEL, D. V., HEYNEKER, H. L., HOZUMI, T., ARENTZEN, R., ITAKURA, K., YANSURA, D. G., ROSS, M. J., MIOZZARI, G., CREA, R. & SEEBURG, P. H. (1979a). Direct expression in *Escherichia coli* of a sequence coding for human growth hormone. *Nature, London*, **281**, 544–8.

GOEDDEL, D. V., KLEID, D. G., BOLIVAR, F., HEYNEKER, H. L., YANSURA, D. G., CREA, R., HIROSE, T., KRASZEWSKI, A., ITAKURA, K. & RIGGS, A. D. (1979b). Expression in *Escherichia coli* of chemically synthesized genes for human insulin. *Proceedings of the National Academy of Sciences, USA*, **76**, 106–10.

GOLD, L. M. & SCHWEIGER, M. (1971). Synthesis of bacteriophage-specific enzyme directed by DNA *in vitro*. In *Methods in Enzymology*, vol. 20, ed. K. Moldave & L. Grossman, pp. 537–42. New York: Academic Press.

GREENE, P. J., POONIAN, M. S., NUSSBAUM, A. L., TOBIAS, L., GARFIN, D. E., BOYER, H. W. & GOODMAN, H. M. (1975). Restriction and modification of a self-complementary octanucleotide containing the *Eco*RI substrate. *Journal of Molecular Biology*, **99**, 237–61.

GRUNSTEIN, M. & HOGNESS, D. S. (1975). Colony hybridization: a method for the isolation of cloned DNAs that contain a specific gene. *Proceedings of the National Academy of Sciences, USA*, **72**, 3961–5.

HALLEWELL, R. A. & EMTAGE, S. (1980). Plasmid vectors containing the tryptophan operon promoters suitable for efficient regulated expression of foreign genes. *Gene*, **9**, 27–47.

HANAHAN, D. & MESELSON, M. (1980). Plasmid screening at high colony density. *Gene*, **10**, 63–7.

HASHIMOTO-GOTOH, T., NORDHEIM, A. & TIMMIS, K. N. (1980). Specific purpose plasmid cloning vectors. I. A low copy number, pSC101-derived biological containment vector. Submitted for publication.

HASHIMOTO-GOTOH, T. & SEKIGUCHI, M. (1976). Isolation of temperature-sensitive mutants of R plasmid by *in vitro* mutagenesis with hydroxylamine. *Journal of Bacteriology*, **127**, 1561–3.

HEFFRON, F., BEDINGER, P., CHAMPOUX, J. J. & FALKOW, S. (1977). Deletions affecting the transposition of an antibiotic resistance gene. *Proceedings of the National Academy of Sciences, USA*, **74**, 702–6.

HEFFRON, F., SO, M. & MCCARTHY, B. J. (1978). *In vitro* mutagenesis of a circular DNA molecule by using synthetic restriction sites. *Proceedings of the National Academy of Sciences, USA*, **75**, 6012–16.

HELINSKI, D. R. (1973). Plasmid determined resistance to antibiotics: molecular properties of R factors. *Annual Review of Microbiology*, **27**, 437–70.

HERRMANN, R., NEUGEBAUER, K., SCHALLER, H. & ZENTGRAF, H. (1978). Integration of DNA fragments coding for antibiotic resistances into the genome of phage fd *in vivo* and *in vitro*. In *The Single-Stranded DNA Phages*, ed. D. T. Denhardt, D. Dressler & D. S. Ray, pp. 473–6. Cold Spring Harbor, New York: Cold Spring Harbor Laboratory.

HERSHFIELD, V., BOYER, H. W., YANOFSKY, C., LOVETT, M. A. & HELINSKI, D. R. (1974). Plasmid ColE1 as a molecular vehicle for cloning and amplification of DNA. *Proceedings of the National Academy of Sciences, USA*, **71**, 3455–9.

HOEIJMAKERS, J. H. J., BORST, P., VAN DEN BURG, J., WEISSMAN, C. & CROSS, G. A. M. (1980). The isolation of plasmids containing DNA complementary to messenger RNA for variant surface glycoproteins of *Trypanosoma brucei. Gene*, **8**, 391–417.

HOHN, B. & HINNEN, A. (1980). Cloning with cosmids in *E. coli* and yeast. In *Genetic Engineering, Principles and Methods*, vol. 2, ed. J. K. Setlow & A. Hollaender, pp. 169–83. New York: Plenum Press.

HOHN, B. & MURRAY, K. (1977). Packaging recombinant DNA molecules into bacteriophage particles *in vitro. Proceedings of the National Academy of Sciences, USA*, **74**, 3259–63.

HUMPHREYS, G. O., WILLSHAW, C. A., SMITH, H. R. & ANDERSON, E. S. (1976). Mutagenesis of plasmid DNA with hydroxylamine: isolation of mutants of multi-copy plasmids. *Molecular and General Genetics*, **145**, 101–8.

HUTCHISON, C. A., PHILIPS, S., EDGELL, M. H., GILLAM, S., JAHNKE, P. & SMITH, M. (1978). Mutagenesis at a specific position in a DNA sequence. *Journal of Biological Chemistry*, **253**, 6551–60.

ITAKURA, K., HIROSE, T., CREA, R., RIGGS, A. D., HEYNEKER, H. L., BOLIVAR, F. & BOYER, H. W. (1977). Expression in *Escherichia coli* of a chemically synthesized gene for the hormone somatostatin. *Science*, **198**, 1056–63.

JEFFREYS, A. J. & FLAVELL, R. A. (1977). A physical map of the RNA regions flanking the rabbit β-globin gene. *Cell*, **12**, 429–39.

KAHN, M., KOLTER, R., THOMAS, C., FIGURSKI, D., MEYER, R., REMAUT, E. & HELINSKI, D. R. (1980). Plasmid cloning vehicles derived from plasmids ColE1, F, R6K, and RK2. *Methods in Enzymology*, **68**, 268–80.

KENNEDY, N., BEUTIN, L., ACHTMAN, M., SKURRAY, R., RAHMSDORF, U. & HERRLICH, P. (1977). Conjugation proteins encoded by the F sex factor. *Nature, London*, **270**, 580–5.

KLEIN, R. D., SELSING, E. & WELLS, R. D. (1980). A rapid microscale technique for isolation of recombinant plasmid DNA suitable for restriction enzyme analysis. *Plasmid*, **3**, 88–91.

KUSHNER, S. R. (1978). An improved method for transformation of *Escherichia coli* with ColE1 derived plasmids. In *Genetic Engineering*, ed. H. W. Boyer & S. Nicosia, pp. 17–23. Amsterdam: Elsevier/North Holland.

LAI, C.-J. & NATHANS, D. (1974). Deletion mutants of simian virus 40 generated by the enzymatic excision of DNA segments from the viral genome. *Journal of Molecular Biology*, **89**, 179–93.

LEVINE, A. D. & RUPP, W. D. (1978). Small RNA product from the *in vitro* transcription of ColE1 DNA. In *Microbiology-1978*, ed. D. Schlessinger, pp. 163–6. Washington DC: American Society for Microbiology.

LINDAHL, L. & ZENGEL, J. M. (1979). Operon-specific regulation of ribosomal protein synthesis in *Escherichia coli. Proceedings of the National Academy of Sciences, USA*, **76**, 6542–6.

LOBBAN, P. E. & KAISER, A. D. (1973). Enzymatic end-to-end joining of DNA molecules. *Journal of Molecular Biology*, **78**, 453–71.

LOVETT, M. A. & HELINSKI, D. R. (1976). Method for the isolation of the replication region of a bacterial replicon: construction of a mini-F′Km plasmid. *Journal of Bacteriology*, **127**, 982–7.

LUSKY, M. & HOBOM, G. (1979a). Inceptor and origin of DNA replication in lambdoid coliphages. I. The λ DNA minimal replication system. *Gene*, **6**, 137–72.

LUSKY, M. HOBOM, G. (1979b). Inceptor and origin of DNA replication in lambdoid coliphages. II. The λ DNA maximal replication system. *Gene*, **6**, 173–97.

MANDEL, M. & HIGA, A. (1970). Calcium-dependent bacteriophage DNA infection. *Journal of Molecular Biology*, **53,** 159–62.

MANIATIS, T., EFSTRATIADIS, A., KEE, S. G. & KAFATOS, F. C. (1976). *In vitro* synthesis and molecular cloning of eukaryotic structural genes. In *Molecular Mechanisms in the Control of Gene Expression*, ed. D. P. Nierlich, W. J. Rutter & C. F. Fox, pp. 513–33, New York: Academic Press.

MANIATIS, T., HARDISON, R. C., LACY, E., LAUER, J., O'CONNELL, C., QUON, D., SIM, G. K. & EFSTRATIADIS, A. (1978). The isolation of structural genes from libraries of eukaryotic DNA. *Cell*, **15**, 687–701.

MANIS, J. J. & KLINE, B. C. (1977). Restriction endonuclease mapping and mutagenesis of the F sex factor replication region. *Molecular and General Genetics*, **152**, 175–82.

MANNING, P. A., ECHARTI, C. & TIMMIS, K. N. (1980). A plasmid-coded major outer membrane protein of *E. coli* that mediates both plasmid surface exclusion and resistance to bactericidal activities of serum regulates its own synthesis. Submitted for publication.

MARTIAL, J. A., HALLEWELL, R. A., BAXTER, J. D. & GOODMAN, H. M. (1979). Human growth hormone: complementary DNA cloning and expression in bacteria. *Science*, **205**, 602–7.

MAXAM, A. M. & GILBERT, W. (1977). A new method for sequencing DNA. *Proceedings of the National Academy of Sciences, USA*, **74**, 560–4.

MERCEREAU-PUIJALON, O., ROYAL, A., CAMI, B., GARAPIN, A., KRUST, A., GANNON, F. & KOURILSKY, P. (1978). Synthesis of an ovalbumin-like protein by *Escherichia coli* K12 harbouring a recombinant plasmid. *Nature, London*, **275**, 505–10.

MERTZ, J. E. & DAVIS, R. W. (1972). Cleavage of DNA by R_I restriction endonuclease generates cohesive ends. *Proceedings of the National Academy of Sciences, USA*, **69**, 3370–4.

MESSER, M., BERGMANS, H. E. N., MEIJER, M., WOMACK, J. E., HANSEN, F. G. & VON MEYENBERG, K. (1978). Minichromosomes: plasmids which carry the *E. coli* replication origin. *Molecular and General Genetics*, **162**, 269–75.

MESSING, J., GRONENBORN, B., MULLER-HILL, B. & HOFSCHNEIDER, P. H. (1977). Filamentous coliphage M13 as a cloning vehicle: insertion of a *Hin*dII fragment of the *lac* regulatory region in M13 replicative form *in vitro*. *Proceedings of the National Academy of Sciences, USA,* **74,** 3642–6.

MILLER, J. H. (1978). The *lac*I Gene: Its role in *lac* Operon control and its use as a genetic system. In *The Operon*, ed. J. H. Miller & W. S. Reznikoff, pp. 31–88. Cold Spring Harbor, New York: Cold Spring Harbor Laboratory.

MOLL, A., MANNING, P. A. & TIMMIS, K. N. (1980). Plasmid-determined resistance to serum bactericidal activity: a major outer membrance protein, the *tra*T gene product, is responsible for plasmid-specified serum resistance in *Escherichia coli. Infection and Immunity*, **28**, 359–67.

MURRAY, K. (1976). Biochemical manipulation of genes. *Endeavour*, **126**, 129–33.

MURRAY, N. E. & MURRAY, K. (1974). Manipulation of restriction targets in phage λ to form receptor chromosomes for DNA fragments. *Nature, London*, **251**, 476–81.

NATHANS, D. & SMITH, H. O. (1975). Restriction endonucleases in the analysis and restructuring of DNA molecules. *Annual Review of Biochemistry*, **44,** 273–93.

NIH GUIDELINES (1976). National Institutes of Health Guidelines for Research Involving Recombinant DNA molecules. NIH, Bethesda.

NORRIS, K. E., ISERENTANT, D., COUTRERAS, R. & FIERS, W. (1979). Asymmetric linker molecules for recombinant DNA constructions. *Gene*, **7**, 355–62.

OZAKI, L. S., MAEDA, S., SHIMADA, K. & TAKAGI, Y. (1980). A novel ColE1::Tn3 plasmid vector that allows direct selection of hybrid clones in *E. coli*. *Gene*, **8**, 301–14.

Ptashne, M., Jeffrey, A., Johnson, A. D., Maurer, R., Meyer, B. J., Pabo, C. O., Roberts, T. M. & Sauer, R. T. (1980). How the λ repressor and Cro work. *Cell*, **19**, 1–11.

Reeve, J. N. (1977). Bacteriophage infection of minicells. A general method for identification of 'in vivo' bacteriophage directed polypeptide biosynthesis. *Molecular and General Genetics*, **158**, 73–9.

Roberts, R. J. (1976). Restriction endonucleases. *Critical Reviews in Biochemistry*, **4**, 123–64.

Roberts, R. J. (1980). Restriction and modification enzymes and their recognition sequences. *Nucleic Acids Research*, **8**, r63–r80.

Roozen, K. J., Fenwick, Jr., R. G., Curtiss, R. (1971). Synthesis of ribonucleic acid and protein in plasmid-containing minicells of *Escherichia coli* K-12. *Journal of Bacteriology*, **107**, 21–33.

Roychoudhury, R., Jay, E. & Wu, R. (1976). Terminal labeling and addition of homopolymer tracts to duplex DNA fragments by terminal deoxynucleotidyl transferase. *Nucleic Acids Research*, **3**, 101–16.

Sabo, D. L., Domingo, E., Bandle, E. F., Flavell, R. A. & Weissmann, C. (1977). A guanosine to adenine transition in the 3′ terminal extracistronic region of bacteriophage Qβ RNA leading to loss of infectivity. *Journal of Molecular Biology*, **112**, 235–52.

Sancar, A., Hack, A. M. & Rupp, W. D. (1979). Simple method for identification of plasmid-coded proteins. *Journal of Bacteriology*, **137**, 692–3.

Sanger, F. & Coulson, A. R. (1975). A rapid method for determining sequences of DNA by primed synthesis with DNA polymerase. *Journal of Molecular Biology*, **94**, 441–8.

Sanger, F., Nicklen, S. & Coulson, A. R. (1977). DNA sequencing with chain terminating inhibitors. *Proceedings of the National Academy of Sciences, USA*, **74**, 5463–7.

Sanzey, B., Mercereau, O., Ternynck, T. & Kourilsky, P. (1976). Methods for identification of recombinants of phage λ. *Proceedings of the National Academy of Sciences, USA*, **73**, 3394–7.

Scheller, R. H., Dickerson, R. E., Boyer, H. W., Riggs, A. D. & Itakura, K. (1977). Chemical synthesis of restriction enzyme recognition sites useful for cloning. *Science*, **196**, 177–80.

Scherzinger, E., Lauppe, H.-F., Voll, N. & Wanke, M. (1980). Recombinant plasmids carrying promoters, genes and the origin of DNA replication of the early region of bacteriophage T7. *Nucleic Acids Research*, **8**, 1287–305.

Schott, H. & Kössel, H. (1973). Synthesis of phage specific deoxyribonucleic acid fragments. I. Synthesis of four undecanucleotides complementary to a mutated region of the coat protein cistron of fd phage deoxyribonucleic acid. *Journal of the American Chemical Society*, **95**, 3778–85.

Schumann, W. (1979). Construction of an *Hpa*I and *Hin*dII plasmid vector allowing direct selection of transformants harbouring recombinant plasmids. *Molecular and General Genetics*, **174**, 221–4.

Sgaramella, V., van de Sande, J. H. & Khorana, H. G. (1970). Studies on polynucleotides. C. A novel joining reaction catalyzed by the T4-polynucleotide ligase. *Proceedings of the National Academy of Sciences, USA*, **67**, 1468–75.

Shenk, T. E., Carbon, J. & Berg, P. (1976). Construction and analysis of viable deletion mutants of simian virus 40. *Journal of Virology*, **18**, 664–71.

Shenk, T. E., Rhodes, C., Rigby, P. & Berg, P. (1975). Biochemical method for mapping mutational alterations in DNA with S1 nuclease: the location of deletions and temperature sensitive mutations in SV40. *Proceedings of the National Academy of Sciences, USA*, **72**, 989–93.

Shortle, D. R., Margolskee, R. F. & Nathans, D. (1979). Mutational analysis of the simian virus 40 replicon: pseudorevertants of mutants with a defective replication origin. *Proceedings of the National Academy of Sciences, USA*, **76**, 6128–31.

Shortle, D. & Nathans, D. (1978). Local mutagenesis: a method for generating viral mutants with base substitutions in preselected regions of the viral genome. *Proceedings of the National Academy of Sciences, USA*, **75**, 2170–4.

Shortle, D. & Nathans, D. (1979a). Regulatory mutants of simian virus 40: constructed mutants with base substitutions at the origin of DNA replication. *Journal of Molecular Biology*, **131**, 801–17.

Shortle, D. & Nathans, D. (1979b). Mutants of simian virus 40 with base substitutions at the origin of DNA replication. *Cold Spring Harbor Symposia on Quantitative Biology*, **43**, 663–8.

Silhavy, T. J., Shuman, H. A., Beckwith, J. & Schwartz, M. (1977). Use of gene fusions to study outer membrane protein localization in *Escherichia coli*. *Proceedings of the National Academy of Sciences, USA*, **74**, 5411–15.

Silverman, M., Zieg, J. & Simon, M. (1979). Flagellar-phase variation: isolation of the rh1 gene. *Journal of Bacteriology*, **137**, 517–23.

Skalka, A. & Shapiro, L. (1976). *In situ* immunoassays for gene translation products in phage plaques and bacterial colonies. *Gene*, **1**, 65–79.

Smith, H. O. & Birnstiel, M. L. (1976). A simple method for DNA restriction site mapping. *Nucleic Acids Research*, **3**, 2387–98.

So, M., Boyer, H. W., Betlach, M. & Falkow, S. (1976). Molecular cloning of an *Escherichia coli* plasmid determinant that codes for the production of heat-stable enterotoxin. *Journal of Bacteriology*, **128**, 463–72.

Southern, E. M. (1975). Detection of specific sequences among DNA fragments separated by gel electrophoresis. *Journal of Molecular Biology*, **98**, 503–17.

Sternberg, N., Tiemeier, D. & Enquist, L. (1977). *In vitro* packaging of a λ *Dam* vector containing endo R. *Eco*RI DNA fragments of *Escherichia coli* and phage P1. *Gene*, **1**, 255–80.

Struhl, K., Stinchcomb, D. T., Scherer, S. & Davis, R. W. (1979). High-frequency transformation of yeast: autonomous replication of hybrid DNA molecules. *Proceedings of the National Academy of Sciences, USA*, **76**, 1035–9.

Sutcliffe, J. G. (1979). Complete nucleotide sequence of the *Escherichia coli* plasmid pBR322. *Cold Spring Harbor Symposia on Quantitative Biology*, **43**, 77–90.

Tacon, W., Carey, N. & Emtage, S. (1980). The construction and characterization of plasmid vectors suitable for the expression of all DNA phases under the control of the *E. coli* tryptophan promoter. *Molecular and General Genetics*, **177**, 427–38.

Taniguchi, T. & Weissmann, C. (1978). Site-directed mutations in the initiator region of the bacteriophage Qβ coat cistron and their effect on ribosome binding. *Journal of Molecular Biology*, **118**, 533–65.

Telford, J., Boseley, P., Schaffner, W. & Birnstiel, M. (1976). Novel screening procedure for recombinant plasmids. *Science*, **195**, 391–3.

Thomas, C. M. & Helinski, D. R. (1979). Plasmid DNA replication. In *Plasmids of Medical, Environmental and Commercial Importance*, ed. K. N. Timmis & A. Pühler, pp. 29–46. Amsterdam: Elsevier/North-Holland.

Thomas, M., White, R. L. & Davis, R. W. (1976). Hybridization of RNA to double-stranded DNA: formation of R-loops. *Proceedings of the National Academy of Sciences, USA*, **73**, 2294–8.

TIMMIS, K. N., ANDRÉS, I. & SLOCOMBE, P. M. (1978a). Plasmid incompatibility: cloning analysis of an *Inc*FII determinant of R6-5. *Nature, London*, **273**, 27–32.

TIMMIS, K. N., BAGDASARIAN, M. & BRADY, G. (1980a). Specific purpose plasmid cloning vectors. II. A multicopy self-amplifying ColD-derived vector containing multiple cloning sites. Submitted for publication.

TIMMIS, K. N., CABELLO, F., ANDRÉS, I., NORDHEIM, A., BURKARDT, H. J. & COHEN, S. N. (1978b). Instability of plasmid DNA sequences: macro and micro evolution of the antibiotic resistance plasmid R6-5. *Molecular and General Genetics*, **167**, 11–19.

TIMMIS, K. N., CABELLO, F. & COHEN, S. N. (1974). Utilization of two distinct modes of replication by a hybrid plasmid constructed *in vitro* from separate replicons. *Proceedings of the National Academy of Sciences, USA*, **71**, 4556–60.

TIMMIS, K. N., CABELLO, F. & COHEN, S. N. (1975). Cloning, isolation, and characterization of replication regions of complex plasmid genomes. *Proceedings of the National Academy of Sciences, USA*, **72**, 2242–6.

TIMMIS, K. N., CABELLO, F. & COHEN, S. N. (1978c). Cloning and characterization of *Eco*RI and *Hin*dIII restriction endonuclease-generated fragments of antibiotic resistance plasmids R6-5 and R6. *Molecular and General Genetics*, **162**, 121–37.

TIMMIS, K. N., COHEN, S. N. & CABELLO, F. C. (1978d). DNA cloning and the analysis of plasmid structure and function. In *Progress in Molecular and Subcellular Biology*, vol. 6, ed. F. E. Hahn, pp. 1–58. Berlin-Heidelberg-New York: Springer Verlag.

TIMMIS, K. N., DANBARA, H., BRADY, G. & LURZ, R. (1980b). Inheritance functions of group IncFII transmissible antibiotic resistance plasmids. *Plasmid*, in press.

TU, C.-P. D. & COHEN, S. N. (1980). Translocation specificity of the Tn*3* element: characterization of sites of multiple insertions. *Cell*, **19**, 151–60.

UHLIN, B. E., MOLIN, S., GUSTAFSSON, P. & NORDSTRÖM, K. (1979). Plasmids with temperature-dependent copy number for amplification of cloned genes and their products. *Gene*, **6**, 91–106.

VILLA-KOMAROFF, L., EFSTRATIADIS, A., BROOME, S., LOMEDICO, P., TIZARD, R., NABER, S. P., CHICK, W. L. & GILBERT, W. (1978). A bacterial clone synthesizing proinsulin. *Proceedings of the National Academy of Sciences, USA*, **75**, 3727–31.

VOGELSTEIN, B. & GILLESPIE, D. (1979). Preparative and analytical purification of DNA from agarose. *Proceedings of the National Academy of Sciences, USA*, **76**, 615–19.

WEIDELI, H., SCHEDL, P., ARTAVANIS-TSAKONAS, S., STEWARD, R., UAN, R. & GERING, W. J. (1977). Purification of a protein from unfertilized eggs of *Drosophila* with specific affinity for a defined DNA sequence and the cloning of this DNA sequence in bacterial plasmids. *Cold Spring Harbor Symposia on Quantitative Biology*, **42**, 693–700.

WEISSMANN, C. (1978). Reversed genetics. *Trends in Biochemical Sciences*, **3**, N109–N111.

WEISSMANN, C., NAGATA, S., TANIGUCHI, T., WEBER, H. & MEYER, F. (1979). The use of site-directed mutagenesis in reversed genetics. In *Genetic Engineering, Principles and Methods*, vol. 1, ed. J. K. Setlow & A. Hollaender, pp. 133–50. New York: Plenum Press.

WEST, R. W., NERE, R. L. & RODRIGUEZ, R. L. (1979). Construction and characterization of *E. coli* promoter probe plasmid vectors. I. cloning of promoter-containing DNA fragments. *Gene*, **7**, 271–88.

WILLIAMS, B. G. & BLATTNER, F. R. (1980). Bacteriophage lambda vectors for DNA cloning. In *Genetic Engineering, Principles and Methods,* vol. 2, ed. J. K. Setlow & A. Hollaender, pp. 201–81. New York: Plenum Press.

YANG, H.-L., ZUBAY, G. & LEVY, S. B. (1976). Synthesis of an R plasmid protein associated with tetracycline resistance is negatively regulated. *Proceedings of the National Academy of Sciences, USA*, **73**, 1509–12.

YASUDA, S. & HIROTA, Y. (1977). Cloning and mapping of the replication origin of *Escherichia coli*. *Proceedings of the National Academy of Sciences, USA*, **74**, 5458–62.

YATES, J. L., ARFSTEN, A. E. & NOMURA, M. (1980). *In vitro* expression of *Escherichia coli* ribosomal protein genes: autogenous inhibition of translation. *Proceedings of the National Academy of Sciences, USA*, **77**, 1437–41.

ZIEG, J., SILVERMAN, M., HILMAN, M. & SIMON, M. I. (1978). The mechanism of phase variation. In *The Operon*, ed. J. H. Miller & W. S. Reznikoff, pp. 411–23. Cold Spring Harbor, New York: Cold Spring Harbor Laboratory.

ZUBAY, G., CHAMBERS, D. A. & CHEONG, L. C. (1970). Cell-free studies on the regulation of the *lac* operon. In *The Lactose Operon*, ed. J. R. Beckwith & D. Zipser, pp. 375–91. Cold Spring Harbor, New York: Cold Spring Harbor Laboratory.

TRANSCRIPTIONAL REGULATION BY BACTERIAL RNA POLYMERASE

ANDREW A. TRAVERS, CSABA KARI AND HILARY A. F. MACE

MRC Laboratory of Molecular Biology,
Hills Road, Cambridge
CB2 2QH, UK

INTRODUCTION

During exponential growth the bacterium *Escherichia coli* balances the production of ribosomes and other necessary components of the translation machinery of the cell so that no more ribosomes are produced than are required for the maintenance of growth (Maaløe, 1978; Nierlich, 1978). This homeostatic mode of gene expression contrasts with the substantial qualitative changes in the pattern of cellular protein synthesis observed when the environment of the organism is grossly perturbed.

One of the most studied examples of such a perturbation is the stringent response manifested when certain strains of *E. coli* are starved of a required amino acid (Gallant, 1979). Upon such deprivation the initiation of transcription of stable RNA species is reduced by 10–20 fold (Stomato & Pettijohn, 1971) while the synthesis of certain mRNA species is scarcely affected (Primakoff & Berg, 1970). Concomitant with this switch in transcriptional selectivity the cells accumulate the nucleotide, guanosine 3′ diphosphate 5′ diphosphate to millimolar levels (Cashel & Gallant, 1969). *In vitro* this nucleotide effects a qualitatively identical switch in the pattern of RNA chain initiation (Yang *et al.*, 1974; van Ooyen, Gruber & Jorgensen, 1976; Travers, 1976a) by altering the promoter selectivity of the transcribing enzyme, RNA polymerase (Travers, Buckland & Debenham, 1980). Although the accumulation of ppGpp can thus explain the reduction of stable RNA transcription during the stringent response, similar changes in the pattern of gene expression are observed following other shock treatments which do not result in ppGpp accumulation. Such treatments perturb the energy balance of the cell suggesting that transcriptional selectivity is directly coupled to energy metabolism. We shall argue that this response is mediated by adenine nucleotides directly regulating the promoter selectivity of RNA polymerase in an analogous manner to ppGpp.

RNA POLYMERASE–PROMOTER INTERACTIONS

The regulation of promoter selectivity of RNA polymerase implies that the enzyme must both recognise and discriminate between a wide variety of promoter sites. Such regulation must require the alteration of the interaction of the enzyme with its binding site.

Genetic and structural analysis of bacterial promoters has shown that they all have two regions of homologous sequences which are required for promoter function. These regions are located at about 10 and 35 base pairs upstream from the startpoint for transcription (Pribnow, 1975; Schaller, Gray & Herrman, 1975; Takanami, Sugimoto, Sugisaki & Okamoto, 1976; Seeburg, Nüsslein & Schaller, 1977) and have the most probable sequences TATAAT and TTGACA respectively (Siebenlist, Simpson & Gilbert, 1980). The separation between the −10 and −35 regions is however variable and can differ by up to four base pairs. In addition to these ubiquitous conserved sequences there exists in promoters known or believed to be under stringent control an additional conserved sequence spanning positions −5 to +2 (Travers, 1980a) defining the −10 region as spanning base pairs −12 to −7 inclusive upstream from the average startpoint for transcription (Rosenberg & Court, 1979). This additional conserved region, termed the discriminator region, has the consensus sequence $\mathrm{C}^{\mathrm{cc}}_{\mathrm{gg}}\mathrm{C{-}CC}$ in the anti-sense strand. This sequence is in general absent from other promoters.

How does RNA polymerase interact with these sequences to form a productive polymerase–promoter initiation complex? Current models suggest that the polymerase binds first to the −35 region and then locates the −10 region to form the 'closed' complex (Fig. 1). The simultaneous binding to these two recognition sites results in the disruption of a short segment of the DNA double helix of the order of ten base pairs (Hsieh & Wang, 1978) to form the 'transition' polymerase–promoter complex. This disruption is then stabilised, accompanied by melting of a short section of the DNA between positions −9 and +2 (Siebenlist, 1979). The transition to this 'open' complex is highly temperature dependent and is thought to require the interaction of RNA polymerase with the single-stranded DNA downstream from the −10 region (Travers, 1980a).

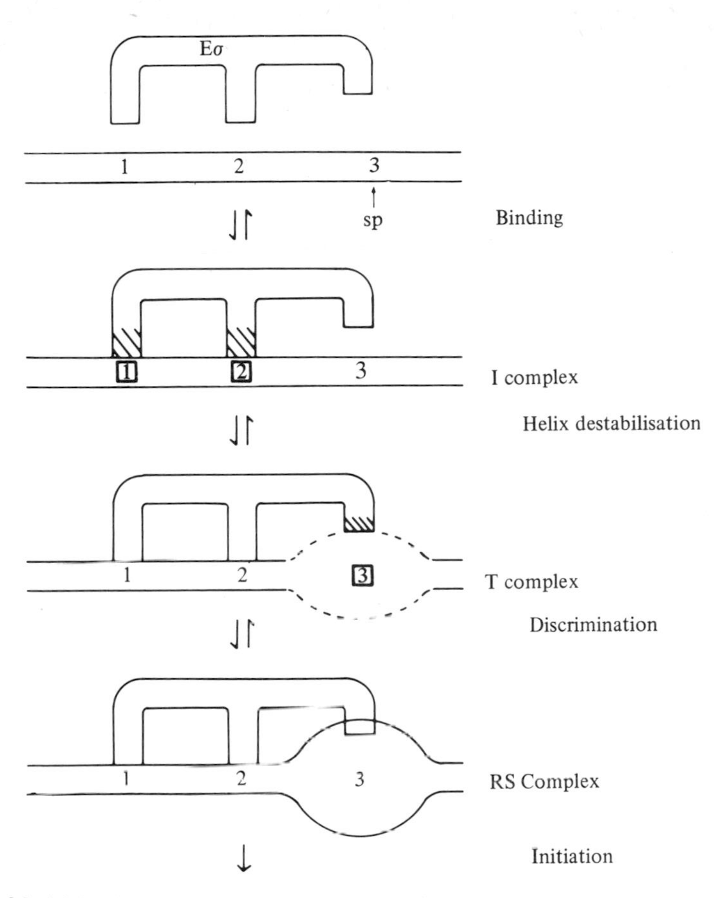

Fig. 1. Model for the initiation of transcription by *E. coli* RNA polymerase. 1,2,3 indicate the −35, −10 and discriminator regions respectively of the promoter. The I, T and RS complexes correspond to initial (or closed), transition, and rapid starting (or open) complexes.

Using an assay in which polymerase is restricted to the synthesis of the first internucleotide bond of a promoter-specific transcript McClure (1980) has functionally defined two steps in the initiation process, a primary association of the enzyme with the promoter followed by an isomerisation step. These steps probably correspond to the formation of the closed and open complexes respectively. Mutations in the –35 region strongly affect this inital association. One such mutation, *x*3 in the λP_R promoter, results in a twenty fold decrease in the association constant for the initial binding and a five fold reduction in the rate of isomerisation (Hawley & McClure, 1980). Mutations in the −10 region also influence the extent of formation of the open complex (Majors, 1978) but their effect, if any, on the rate of association is unknown. No natural mutations are known to exist in the discriminator region but the synthetic mutation of four base pairs in the promoter sequence of the tRNA^{Tyr} gene (Ryan *et al.*, 1979) completely alters the *in vitro* regulation of the promoter by ppGpp (Travers, 1980b) such that RNA synthesis from the mutant promoter responds to increasing concentrations of ppGpp in a reciprocal manner to that from the wild-type promoter. This result demonstrates directly that the discriminator sequence is a functional determinant of promoter selection.

THE ROLE OF ppGpp IN PROMOTER SELECTION

How does an alteration in the discriminator sequence of a promoter change the regulation of that promoter by ppGpp? The target of ppGpp has been shown to be RNA polymerase itself (Cashel, 1970; Travers, 1976a). One model for the regulation of the enzyme by the nucleotide proposes that the enzyme can exist in distinct structural states, each of which possesses a different promoter preference. We will argue that RNA polymerase can assume either of two major structural states, termed the A and G states. These states differ functionally both in their response to regulatory molecules and in their ability to utilise different DNA conformers as templates. Each of these major structural states of the enzyme can be further sub-divided into promoter specific forms such that a given promoter is used efficiently by a sub-population of both the A and G states of the enzyme. The A and G states can be partially separated as two peaks of enzyme activity on zone sedimentation, the A state sedimenting faster than the G state. The promoter-specific forms of each state of the enzyme can also be distinguished on the basis of their sedimentation coefficient. Thus a particular promoter will be utilised optimally by both a fast sedimenting and a slow sedimenting form of the enzyme. Experimentally this is manifested as a bimodal pattern of promoter utilisation across a gradient profile of polymerase activity. However, sedimentation of the enzyme at low temperatures (e.g. $\sim$ 4 °C) results in a unimodal pattern of promoter utilisation corresponding to the promoter specific forms of the G state. On this model regulators such as ppGpp alter the pattern of promoter selection by changing the number and proportions of forms present (Travers, 1976b). It follows that the different forms of the enzyme should distinguish between promoters on the basis of the sequence in the discriminator region.

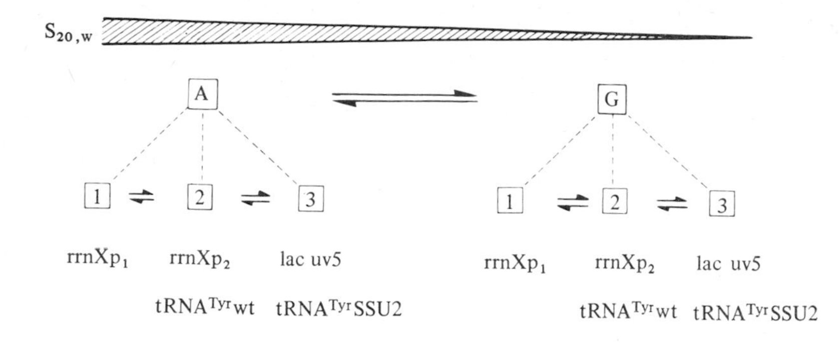

Fig. 2. Promoter specificity of structural forms of RNA polymerase. Cartoon of the separation of the structural forms of RNA polymerase on a glycerol gradient. A and G represent the major states of the enzyme and 1,2, and 3 the promoter specific sub-forms. The promoters used as examples include the proximal and distal promoters of the *rrnX* cistron ($rrnXp_1$ and $rrnXp_2$ respectively) and the wild-type and mutant (SSU2) $tRNA^{Tyr}$ promoters.

If RNA polymerase has the potential to assume different structural states the enzyme should normally exist as an equilibrium mixture of the different forms. Resolution of functionally distinct sub-populations of RNA polymerase can be achieved by sedimenting polymerase holoenzyme through a 15–30% glycerol gradient at 4 °C (Travers, Buckland & Debenham, 1980). Under these conditions the enzyme sediments as a broad peak with an average *s* value of 14 S (Richardson, 1966). To test the promoter selectivity of polymerase in different fractions equal weights of enzyme from such a gradient peak are assayed with DNA restriction fragments each containing a single promoter. In such an experiment the wild-type $tRNA^{Tyr}$ promoter and the tandem promoters of the *rrnX* cistron, whose activity is strongly inhibited by ppGpp, are utilised optimally by fractions from the leading shoulder of the polymerase sedimentation profile (Fig. 2). By contrast the *lac* UV5 promoter, whose activity is increased by ppGpp, is utilised optimally by enzyme molecules sedimenting at the trailing edge. This is also true for the mutant $tRNA^{Tyr}$ promoter with a four base alteration in the discriminator region resulting in an activation by ppGpp (Travers, 1980b).

These observations show that RNA polymerase can exhibit heterogeneity with respect to promoter preference, this heterogeneity being apparent as a correlation between the sedimentation position of the enzyme and its template selectivity. In particular the faster sedimenting polymerase molecules preferentially initiate at promoters whose activity is reduced by ppGpp while the slower sedimenting molecules preferentially initiate at promoters whose activity is enhanced by ppGpp. Thus if ppGpp alters promoter selectivity by switching the enzyme from one promoter-specific form to another, rather than by preferentially altering the activity of a particular form, the nucleotide should increase the proportion of slower sedimenting molecules and correspondingly decrease the proportion of faster sedimenting molecules. In other words the polymerase should sediment slower in the presence of this nucleotide. This is indeed the case, 10 μM ppGpp reducing the $S_{20,w}$ of polymerase holoenzyme by ~ 0.7 S while the adenine analogue of ppGpp, ppApp, increases the $S_{20,w}$ by ~ 0.4 S at the same concentration (Travers *et al.*, 1979).

Such a correlation between the effects of ppGpp on promoter selectivity and polymerase structure could be fortuitous. If it were not, the ability of different sub-populations of enzyme molecules to respond to ppGpp should depend on the apparent *s* value of the enzyme. This prediction was tested by determining the effect of increasing concentrations of ppGpp on rRNA synthesis by individual fractions of a glycerol gradient profile of the enzyme. In agreement with the prediction it was found that ppGpp inhibited rRNA synthesis by polymerase molecules sedimenting on the leading edge to a substantially greater extent than those sedimenting on the trailing edge (Travers *et al.*, 1980).

Another *in vitro* regulation of polymerase selectivity, fMet-$tRNA_f^{Met}$, has a functional effect very similar to that elicited by high concentrations of ppGpp; i.e. the charged tRNA preferentially inhibits the initiation of stable RNA synthesis and stimulates *lac* RNA synthesis (Debenham, Pongs & Travers, 1980). When different fractions from a polymerase sedimentation profile are assayed for their capacity to bind this tRNA by using limiting concentrations of ligand, the maximal binding activity is exhibited by polymerase molecules which sediment in the same position as those which optimally utilise the *lac* UV5 promoter. Thus, as with ppGpp, there is a correlation between the change in promoter selectivity effected by the tRNA and its capacity to interact with a particular class of polymerase molecules. Were fMet-tRNA to act like ppGpp by stabilising a particular structural form of the enzyme, the form which was stabilised should possess the highest affinity for the ligand. The observation that those polymerase molecules with a preference for the *lac* UV5 promoter and those with the highest capacity for fMet-tRNA binding co-sediment is consistent with this view but does not prove that these two functional characteristics are shared by the *same* polymerase molecules.

The differences in promoter preference between different sub-populations of polymerase molecules are not absolute, the extent of quantitative variation being only three to five fold. However, the ratio of transcription of different RNA species, for example *lac* and $tRNA^{Tyr}$, between different sub-populations can vary by up to ten fold. The data show that there is no correlation between the sequence of the −10 region of a promoter and the sub-population of polymerase molecules which utilise that promoter maximally. For example the *lac* UV5 and proximal *rrnX* promoters have the same −10 region sequence, TATAATA, yet are utilised optimally by different populations of polymerase molecules. Similarly the distal *rrnX* promoter and the $tRNA^{Tyr}$ promoters have TAATATA and TATGATG, respectively, in the −10 region but are utilised optimally by similar polymerase populations. In a like manner the presence of the sequence CTTTACA in the −35 region of the *lac* UV5 and $tRNA^{Tyr}$ promoters again does not correlate with maximum utilisation by a particular class of polymerase molecules. However, a change in the discriminator region of the $tRNA^{Tyr}$ promoter results in a transfer of preferential utilisation to a different class of polymerase molecules.

These observations show that RNA polymerase can exist in different structural forms which distinguish between different promoters on the basis of the sequence of the discriminator region. ppGpp acts by switching the enzyme from a form which recognises the particular discriminator sequence in promoters under stringent control to a form which fails to interact productively with this sequence. However, the stimulation of transcription from some promoters by ppGpp suggests that alternative discriminator sequences specific for other structural forms may exist.

REGULATION BY A GUANINE NUCLEOTIDE COUPLE

Although the effect of ppGpp on stable RNA transcription *in vitro* qualitatively parallels the stringent response *in vivo* the *in vitro* response does not quantitatively mimic the *in vivo* effects (Gallant, 1979). In particular the extent to which stable RNA transcription is curtailed *in vitro* by physiological concentrations of ppGpp is significantly less than that observed *in vivo*. Further the K_i observed *in vitro* for the selective inhibition of $tRNA^{Tyr}$ synthesis by ppGpp is 5 μM (Debenham & Travers, 1977), a concentration considerably below the normal *in vivo* basal concentration of 30 μM (Gallant, 1979). These anomalies suggest that the *in vitro* systems used lack a factor(s) necessary to reproduce accurately the pattern of *in vivo* regulation.

The selective curtailment of stable RNA synthesis *in vivo* can occur in the absence of ppGpp accumulation. For example guanine starvation reduces the rate of such transcription three fold (Gallant, Erlich, Hall & Laffler, 1970). This reduction of stable RNA production is accompanied by a drop in the GTP pool from 1 000 to ~250 μM suggesting that high GTP concentrations may be necessary for the efficient production of stable RNA species. A concentration of 250 μM GTP is substantially greater than the K_m of ~15 μM required for RNA elongation (Goldthwait, Anthony & Wu, 1970) yet the majority of rRNA transcripts, comprising the bulk of stable RNA synthesis, initiate with either ATP or CTP (Young & Steitz, 1979; Gilbert, de Boer & Nomura, 1979; Glaser & Cashel, 1979). This argues against the possibility that a requirement for high GTP concentrations reflects a high K_m for the initiating triphosphate.

We therefore investigated the effect of high guanosine triphosphate concentrations on stable RNA synthesis *in vitro*. Transcription of the *rrnX* cistron was measured in the presence of both a physiological concentration of GTP, i.e. ~1 000 μM and a concentration, 125 μM, typical of those used *in vitro*. The nucleotide stimulated transcription of rRNA sequences by up to three fold, this enhancement being particularly marked at low ionic strength. The higher GTP concentration had the additional consequence of altering the response of rRNA synthesis to ppGpp. At 125 μM GTP the K_i for the inhibition of rRNA transcription by ppGpp was ~5 μM. By contrast at 1 000 μM GTP, ppGpp at concentrations $<$ 100 μM had no inhibitory effects on rRNA synthesis. At higher ppGpp concentrations, corresponding to the normal *in vivo* levels a strong inhibition was observed, rRNA synthesis being diminished by 50% at ~300 μM ppGpp. Since the rRNA precursors from the *rrnX* cistron initiate with ATP and CTP but not with GTP (Young & Steitz, 1979) the effect of GTP on ppGpp regulation suggests that GTP may act as a regulatory antagonist to ppGpp, stabilising the forms of the polymerase with a preference for stable RNA promoters.

The *in vitro* data suggest that the effect of ppGpp on stable RNA synthesis is dependent on the GTP concentration, a conclusion which differs from that drawn from the *in vivo* data (Fiil, von Meyenburg & Friesen, 1972). One possible reason for this discrepancy is that the binding constants for the two nucleotides individually to RNA polymerase differ substantially. Thus the observed constant for GTP is 150 μM (Goldthwait *et al.*, 1970) whereas that for ppGpp is ~5 μM (Cashel, Hamel, Shapshak & Bouquet, 1976). With such a difference the small fluctuations in GTP concentrations *in vivo* compared with the much greater variations in ppGpp concentrations might not be sufficient for a competitive effect between the two nucleotides to be detected. If the *in vitro* result does reflect the situation *in vivo* then the inhibition of stable RNA synthesis observed during ppGpp accumulation may be a consequence of at least two effects, a rise in the ppGpp pool and a fall in the GTP pool (Gallant *et al.*, 1970) particularly since the fall in GTP concentration occurs in a range over which substantial effects are observed *in vitro*.

BIMODALITY OF TEMPLATE SELECTIVITY AND AUTOREGULATION

The pattern of promoter preference and fMet–tRNA binding observed across a polymerase sedimentation profile depends crucially on the conditions of sedimentation. When the enzyme is sedimented at a higher temperature or at a higher protein concentration, a different although related pattern of template utilisation to that previously described is apparent. Instead of a single subpopulation of enzyme molecules with a preference for the *lac* UV5 promoter *two* distinct populations with this property can be distinguished. Similarly there are two other populations of polymerase molecules which exhibit a preference for the tRNA^{Tyr} promoter. The pattern of fMet–tRNA binding is also duplicated, the greatest extent of binding again correlating with the maximal utilisation of the *lac* UV5 promoter, while those polymerase molecules which bind fMet–tRNA least sediment in the same position as those which optimally utilise the tRNA^{Tyr} promoter.

The difference in *s* value between the different populations of the enzyme with the same promoter preference is ~1 S. Thus the duplication of the pattern of promoter preference is unlikely to be a trivial consequence of the existence of different aggregational states of the enzyme. What then is the functional significance of this bimodality? One possibility is that the duplication could allow the efficient recognition of different DNA conformations at a particular promoter site. To test this hypothesis the different fractions of a polymerase sedimentation profile were challenged with the relaxed and supercoiled forms of pER 24 DNA, a circular chimaera containing both the promoters and the proximal portion of the 16 S rRNA sequence of the *rrnB* cistron (Brosius, Palmer, Kennedy & Noller, 1978). The slower sedimenting form of the enzyme specific for rRNA promoters transcribed rRNA sequences from the supercoiled DNA five times more efficiently than from the relaxed DNA (Wharton & Travers, unpublished observations). By contrast the faster sedimenting form of the enzyme showed no preference between the different conformers of DNA. A similar pattern of rRNA synthesis was observed with CpC as a primer. This dinucleotide specifically primes synthesis at the P_2 promoter of the *rrnB* cistron (Glaser & Cashel, 1979). Neither fraction of polymerase molecules nicked the supercoiled DNA (Wharton & Travers, unpublished experiments). These observations thus show that RNA polymerase can discriminate between promoter sites on the basis of the tertiary structure of the DNA, and that the basis for this discrimination is again the ability of the enzyme to exist in different structural states.

RNA polymerase can thus recognise both the tertiary structure and the sequence of the DNA at a promoter site. We suggest that the bimodal pattern of promoter preference represents the separation of the enzyme into two major populations, a faster sedimenting form, the A state, and a slower sedimenting form, the G state. Each of these functional forms of the enzyme can then be further divided into promoter-specific forms such that the discriminator sequence of a given promoter is recognised by a sub-population of both the A and G states of the enzyme (Fig. 2). The relative efficiency of transcription of this promoter by the A and the G states will then depend in part on the tertiary structure of that promoter.

The A and the G states of the enzyme are interconvertible. In particular the A state is favoured by high enzyme concentration as well as by low ionic strength or high temperature. Conversely the G state is favoured by low enzyme concentration, high ionic strength and low temperature (Travers *et al.*, 1979). The effect of polymerase concentration suggests that the enzyme may regulate its own specificity, such that the transition between the A and G states of

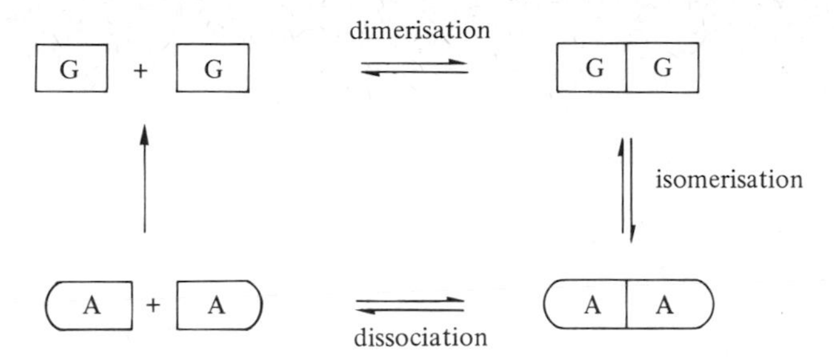

Fig. 3. Model for the autoregulation of RNA polymerase involving the interconversion of the A and G states.

the enzyme requires an interaction between polymerase molecules. Such a model would be fully consistent with the effect of low ionic strength which promotes the dimerisation of polymerase molecules (Richardson, 1966). We propose that two holoenzyme molecules in the G state dimerise, inducing a structural change to the A state in one or both of the monomers (Fig. 3). On dissociation of the dimer the monomers in the A state would be metastable and slowly revert to the G state. We have argued previously that the enzyme molecules isolated on a glycerol gradient in A state do not sediment as dimers but predominantly as monomers, which on the hypothesis above should be metastable. Their apparent stability on the glycerol gradient is dependent on the presence of high glycerol concentrations and on low temperatures. When these constraints are removed the enzyme rapidly re-equilibrates (Travers *et al.*, 1980).

COUPLING OF TRANSCRIPTION AND ENERGY METABOLISM

Although the accumulation of ppGpp can explain the reduction of rRNA transcription observed following amino acid starvation, changes in the pattern of transcription very similar to those observed during the stringent response can occur in the absence of amino acid starvation. In particular temperature-sensitive mutations affecting fructose 1:6 diphosphate aldolase (Bock & Neidhardt, 1966; Frey, Newlin & Atherley, 1975) and adenylate kinase (Kari, unpublished results) both result in a cessation of stable RNA accumulation without concomitant ppGpp accumulation. Similarly, in the presence of α methyl glucoside, a competitive inhibitor of glucose uptake, there is no correlation between stable RNA accumulation and ppGpp levels (Hansen *et al.,* 1975). All these treatments perturb the energy balance of the cell and the adenylate energy charge (Andersen & von Meyenberg, 1977) defined as $(ATP + \frac{1}{2}ADP)/(ATP + ADP + AMP)$ (Atkinson & Walton, 1967).

Although there is no correlation between the intracellular concentration of ATP alone and the extent of stable RNA synthesis (Friesen, Fiil & von Meyenburg, 1975) there is some evidence that adenine nucleotides can regulate RNA polymerase. In particular 5′ ADP alters the sedimentation characteristics of crude RNA polymerase preparations (Travers & Buckland, 1973) while ppApp affects both the sedimentation coefficient (Travers *et al.,* 1979) and the salt sensitivity of transcription by the enzyme. A further example of the interaction of adenine nucleotides with RNA polymerase is the stoichiometric modification of the α sub-units of the enzyme by the covalent addition of ADP-ribose residues following phage T4 infection (Walter, Seifert & Zillig, 1968; Goff, 1974).

The effect of adenine nucleotides on the sedimentation characteristics of the enzyme suggested that they might also influence transcriptional selectivity in an analogous manner to ppGpp. We observed *in vitro* that high concentrations of ATP, or its analogue $\alpha\beta$ methylene ATP, stimulated transcription of rRNA and $tRNA^{Tyr}$ by three to four fold. Conversely the transcription of *lac* mRNA from the *lac* UV5 promoter was inhibited to a similar extent. In both cases the effect of the nucleoside triphosphate was reversed by the addition of 5′ ADP, ppApp or 5′ AMP at five to ten fold lower concentrations than that of the triphosphate. 3′ AMP and 3′5′ cyclic AMP also reversed the effect of ATP but were required at higher concentrations. The concentration of ATP necessary to stimulate stable RNA transcription *in vitro*, 500–2 000 μM, is considerably above the K_s of 15 μM for nucleoside triphosphates for chain elongation, and similar to physiological concentrations of the nucleotide. High concentrations of ATP do not alter transcriptional selectivity when added after the formation of the polymerase–promoter complex and thus act prior to the formation of the first internucleotide bond. The ability of $\alpha\beta$ methylene ATP, which cannot be cleaved by the polymerase, to substitute for ATP shows that ATP does not act by altering the extent of elongation. Nevertheless both ATP and $\alpha\beta$ methylene ATP could influence selectivity by acting as primers at the 5′ terminus of the RNA transcript. However, the ATP-mediated stimulation of the $tRNA^{Tyr}$

transcript with a 5′ terminal pppG residue and the inhibition of the *lac* mRNA transcript with a 5′ terminal pppA residue argue against such a mechanism.

Since the effects of the adenine nucleotides on promoter selection are not apparently a consequence of the nucleotides acting as a substrate for the enzyme we propose that these nucleotides act as regulators of polymerase selectivity in a manner analogous to the guanine nucleotide regulatory couple. Thus ATP activates transcription of the same RNA species which are stimulated by GTP while 5′ ADP and 5′ AMP reverse this stimulation in a similar manner to ppGpp.

If the adenine nucleotides do act as regulatory effectors of the polymerase, do they stabilise particular structural forms of the enzyme? In contrast to ppGpp both 5′ ADP and ppApp increase the $S_{20,w}$. The effect of ATP on the sedimentation characteristics of the enzyme is unknown. However, ATP increases the salt optimum of transcription *in vitro* in a manner resembling that of EF-Ts, a protein which increases the sedimentation coefficient of RNA polymerase (Debenham, 1978). We therefore propose that the adenine nucleotides stabilise the different promoter-specific forms of the A state of polymerase while the guanine nucleotides stabilise the promoter-specific forms of the G state.

TRANSCRIPTION–TRANSLATION COUPLING

We have argued that the gross changes in the pattern of gene expression observed on amino acid starvation or energy deprivation are mediated in part by nucleotide induced changes in the template selectivity of RNA polymerase. One hypothesis to explain the homeostatic pattern of gene expression observed during balanced growth (Travers, 1976b) proposes that a facet of this control is the interaction of the cycling factors of the translation machinery with RNA polymerase. These would maintain a balanced distribution between the major structural forms of the enzyme and could adjust the rate of synthesis of rRNA, tRNA or r-protein mRNA depending on which component of the translation machinery is momentarily limiting.

A number of macromolecules which can affect the function and structure of RNA polymerase holoenzyme have so far been identified. They included EF-Tu (Travers, Kamen & Cashel, 1970), EF-Ts

(Biebricher & Druminski, 1980), fMet-tRNA (Pongs & Ulbrich, 1976; Debenham *et al.*, 1980), IF-2 (Travers, Debenham & Pongs, 1980) and EF-G (Travers *et al.*, manuscript in preparation). This list is unlikely to be exhaustive. How would these factors act? *In vitro* IF-2 acts in a similar manner to GTP, stimulating the synthesis of rRNA and reducing the sedimentation coefficient of polymerase. *In vivo* this factor is released following the initiation of a polypeptide chain. This release would constitute a signal that a ribosome had been engaged in protein synthesis. In the presence of an excess of ribosomes the factor would once more be engaged in the initiation of protein synthesis. However, if ribosomes were limiting the factor would be available to direct the polymerase to synthesise rRNA. Similarly the availability of free EF-Tu, which *in vitro* alters polymerase selectivity in a very similar manner to IF-2, would signal a shortage of tRNA and ribosomes. Again the modulation of polymerase specificity by EF-Tu would act to redress that imbalance. By contrast to EF-Tu and IF-2 the effect of fMet-tRNA on transcription resembles that of ppGpp. The initiator tRNA is removed from the cycling pool by the act of initiation and so will only accumulate when initiation is blocked. Such a blockage would result in a functional excess of ribosomes. Thus the action of fMet-tRNA in inhibiting stable RNA synthesis would prevent further accumulation of ribosomes.

In general an *in vivo* role for macromolecular effectors of transcription is poorly documented. One exception is the factor EF-Ts which *in vitro* resembles ATP in its effects on transcriptional selectivity. This factor increases the salt optimum for rRNA synthesis (Debenham & Travers, manuscript in preparation) resulting in an inhibition of such synthesis at low ionic strength (Biebricher & Druminski, 1980) and a stimulation at high ionic strength. *In vivo* the addition of kirromycin, a specific inhibitor of EF-Tu, to growing *E. coli* preferentially increases the rate of rRNA transcription (Young & Neidhardt, 1978). A major consequence of kirromycin addition would be a rise in the intracellular concentration of EF-Ts, which would then act directly to stimulate rRNA production. By contrast the addition of L-1-tosylamido-2-phenylethyl chloromethyl ketone (TPCK), a compound which specifically inactivates EF-Tu, to growing bacteria induces a stringent response in the absence of significant ppGpp accumulation (Young & Neidhardt, 1978). In this case EF-Ts would not be released from the EF-TuTs complex and consequently both EF-Tu and EF-Ts would no longer be available to interact with RNA polymerase. Since both these factors enhance stable RNA synthesis *in vitro* the *in vivo* result is thus fully consistent with the *in vitro* data.

ROLE OF POLYMERASE REGULATION IN TRANSCRIPTIONAL CONTROL

In vitro, three systems regulating the promoter selectivity of RNA polymerase have been identified; a guanine nucleotide couple, an adenine nucleotide couple and a system of macromolecular effectors coupling translation and transcription (Fig. 4). For all these systems the *in vitro* response to the presumptive regulators parallels the *in vivo* changes in transcription pattern observed when the metabolism of the regulators is perturbed. There is however very little evidence to prove that the *in vitro* response is mediated by the same mechanism as the *in vivo* control.

The best evidence that the mechanisms are the same is provided by a mutant of the sigma subunit of RNA polymerase. This mutant, *rpoD2* (previously known as *alt-1*), was originally selected on the basis of its ability to express the arabinose operon in the absence of the normally required cAMP (Silverstone, Goman & Scaife, 1972). The mutant is temperature sensitive for growth and its transcription patterns are grossly altered in comparison with the parent strain. At the permissive temperature both the arabinose and lactose operons are expressed in the mutant. However, at the restrictive temperature the mutant no longer expresses the *lac* genes but synthesises only stable RNA (as opposed to unstable RNA). *In vitro* the polymerase isolated from *rpoD2* behaves as though the equilibrium between the A and G states of the enzyme is shifted strongly in favour of the G state. In particular the sedimentation coefficient of the mutant enzyme is less than that of the wild-type enzyme. In addition at low but not at high temperatures the mutant enzyme transcribes up to five times more *lac* RNA in the absence of cAMP and CAP protein than does the wild-type enzyme. This property of the mutant enzyme thus parallels the *in vivo* phenotype.

We can explain the phenotype of the *rpoD2* mutant on the basis of the inability of the enzyme to switch from the G to the A states. At low temperatures the mutant enzyme will exist predominantly as the *lac* promoter-specific form of the G state while the wild-type

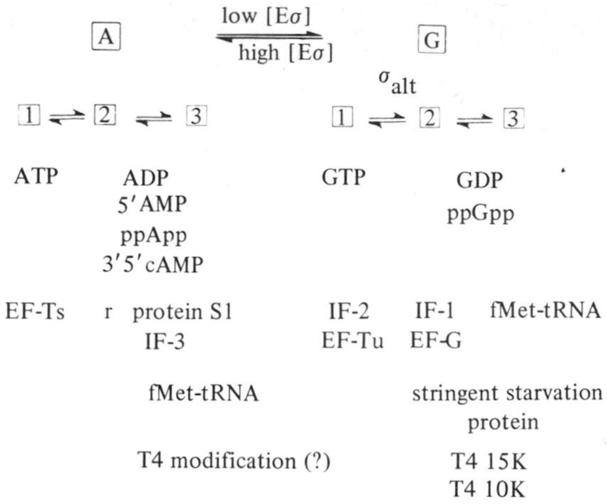

A $\underset{\text{high [E}\sigma\text{]}}{\overset{\text{low [E}\sigma\text{]}}{\rightleftharpoons}}$ G

σ_{alt}

1 ⇌ 2 ⇌ 3 1 ⇌ 2 ⇌ 3

ATP ADP GTP GDP
5′AMP ppGpp
ppApp
3′5′cAMP

EF-Ts r protein S1 IF-2 IF-1 fMet-tRNA
IF-3 EF-Tu EF-G

fMet-tRNA stringent starvation protein

T4 modification (?) T4 15K
T4 10K

Fig. 4. Regulation of the structural forms of RNA polymerase. The effectors of polymerase selectivity are arranged so that a given effector is listed beneath the enzyme form or state it is presumed to stabilise.

enzyme will be distributed between the *lac* and stable RNA forms. Thus in the mutant the effective concentration of polymerase molecules able to recognise the *lac* promoter will be higher and initiation at that promoter will be forced by a mass action effect even in the absence of cAMP. As the temperature is raised so the wild-type enzyme will switch to the A state while the mutant enzyme can only switch to stable RNA forms of the G state. Thus in the mutant there will no longer be a population of enzyme molecules able to recognise the discriminator sequence of the *lac* promoter. Instead stable RNA will be transcribed to the exclusion of many other RNA species. The switch between the A and G states in the wild-type strain would ensure that at all temperatures there is a sub-population of polymerase molecules able to specifically recognise the *lac* promoter. In this case, however, efficient initiation always requires cAMP and CAP protein.

The responses of RNA polymerase to both nucleotides and macromolecular translation components is rapid and results in a switch in the capacity of the transcription machinery of the cell to initiate at particular promoters. Characteristically this type of change is readily reversible. Nevertheless such changes in the pattern of gene expression also affect the production of the components of the transcription system itself and hence in the long term, may re-programme the cell. For example the accumulation of ppGpp *in vivo* is accompanied by the preferential synthesis of two proteins, B56.5 and stringent starvation protein (Reeh, Pedersen & Friesen, 1976). The latter protein binds to RNA polymerase (Ishihama & Saitoh, 1979) and alters the salt sensitivity of the enzyme in a similar manner to ppGpp itself. Should this change in enzyme function be accompanied by a parallel change in promoter selectivity, ppGpp *in vivo* would induce the synthesis of a protein which would affect gene expression in the same way as the low molecular weight effector. In other words the accumulation of the nucleotide, which need only be transient, could result in a re-programming of polymerase selectivity mediated by the stringent starvation protein. Such a switch might thus be regarded as a simple form of differentiation.

C. K. thanks UNESCO for a fellowship.

REFERENCES

ANDERSEN, K. B. & VON MEYENBURG, K. (1977). Charges of nicotinamide adenine nucleotides and adenylate energy charge as regulatory parameters of the metabolism in *Escherichia coli*. *Journal of Biological Chemistry*, **252**, 4151–6.

ATKINSON, D. E. & WALTON, G. M. (1967). Adenosine triphosphate conservation in metabolic regulation. Rat liver citrate cleavage enzyme. *Journal of Biological Chemistry*, **242**, 3239–41.

BOCK, A. & NIEDHARDT, F. C. (1966). Properties of a mutant of *Escherichia coli* with a temperature sensitive fructose-1, 6-diphosphate aldolase. *Journal of Bacteriology*, **92**, 470–6.

BIEBRICHER, C. K. & DRUMINSKI, M. (1980). Inhibition of RNA polymerase activity by the *Escherichia coli* protein biosynthesis elongation factor Ts. *Proceedings of the National Academy of Sciences, USA*, **77**, 866–9.

BROSIUS, J., PALMER, M. L., KENNEDY, P. J. & NOLLER, H. F. (1978). Complete nucleotide sequence of a 16 S ribosomal RNA gene from *Escherichia coli*. *Proceedings of the National Academy of Sciences, USA*, **75**, 4801–5.

CASHEL, M. (1970). Inhibition of RNA polymerase by ppGpp, a nucleotide accumulated during the stringent response to amino acid starvation in *E. coli*. *Cold Spring Harbor Symposium on Quantitative Biology*, **35,** 407–13.

CASHEL, M. & GALLANT, J. (1969). Two compounds implicated in the function of the RC gene of *Escherichia coli*. *Nature, London*, **221**, 838–41.

CASHEL, M., HAMEL, E., SHAPSHAK, P. BOUQUET, M. (1976). Interaction of ppGpp structural analogues with RNA polymerase. In *Control of Ribosome Synthesis*, ed. N. O. Kjeldgaard & O. Maaløe, pp. 279–90. Munksgaard, Copenhagen.

DEBENHAM, P. G. (1978). *In vitro* transcription of su_3^+ $tRNA^{Tyr}$. Ph.D. Thesis, University of Cambridge.

DEBENHAM, P. G., PONGS, O. & TRAVERS, A. A. (1980). Formylmethionyl-tRNA alters RNA polymerase specificity. *Proceedings of the National Academy of Sciences, USA*, **77**, 870–4.

DEBENHAM, P. G. & TRAVERS, A. (1977). Selective inhibition of $tRNA^{Tyr}$ transcription by guanosine 3′ diphosphate 5′ diphosphate. *European Journal of Biochemistry*, **72**, 515–23.

FIIL, N. P., VON MEYENBURG, K. & FRIESEN, J. D. (1972). Accumulation and turnover of guanosine tetraphosphate in *Escherichia coli. Journal of Molecular Biology*, **71**, 769–83.

FREY, T., NEWLIN, L. L. & ATHERLEY, A. G. (1975). Strain of *Escherichia coli* with a temperature sensitive mutation affecting ribosomal ribonucleic acid accumulation. *Journal of Biological Chemistry*, **121**, 923–32.

FRIESEN, J. D., FIIL, N. P. & VON MEYENBURG, K. (1975). Synthesis and turnover of basal level guanosine tetraphosphate in *Escherichia coli. Journal of Biological Chemistry*, **250**, 304–9.

GALLANT, J. (1979). Stringent control in *E. coli. Annual Review of Genetics*, **13**, 393–415.

GALLANT, J., ERLICH, H., HALL, B. & LAFFLER, T. (1970). Analysis of the RC function. *Cold Spring Harbor Symposium on Quantitative Biology*, **35**, 397–405.

GILBERT, S. F., DE BOER, H. A. & NOMURA, M. (1979). Identification of initiation sites for the *in vitro* transcription of rRNA operons *rrnE* and *rrnA* in *Escherichia coli. Cell*, **17**, 211–24.

GLASER, G. & CASHEL, M. (1979). *In vitro* transcripts from the *rrnB* ribosomal RNA cistron originate from two tandem promoter. *Cell*, **16**, 11–22.

GOFF, C. G. (1974). Chemical structure of a modification of the *Escherichia coli* ribonucleic acid polymerase α polypeptides induced by bacteriophage T4 infection. *Journal of Biological Chemistry*, **19**, 6181–90.

GOLDTHWAIT, D. A., ANTHONY, D. D. & WU, C-W. (1970). Studies with RNA polymerase, In *RNA Polymerase and Transcription,* ed. L. Silvestri, pp. 10–27. Amsterdam: North-Holland.

HANSEN, M. T., PAO, M. L., MOLIN, S., FIIL, N. P. & VON MEYENBURG, K. (1975). Simple downshift and resulting lack of correlation between ppGpp pool size and ribonucleic acid accumulation. *Journal of Bacteriology*, **122**, 585–91.

HAWLEY, D. K. & MCCLURE, W. R. (1980). *In vitro* comparison of the initiation properties of bacteriophage lambda wild type P_R and *x3* mutant promoters. *Proceedings of the National Academy of Sciences, USA*, (in press).

HSIEH, T. & WANG, J. C. (1978). Physicochemical studies on interactions between DNA and RNA polymerase. Ultraviolet absorption methods. *Nucleic Acids Research*, **5**, 3337–45.

ISHIHAMA, A. & SAITOH, T. (1979). Subunits of RNA polymerase in function and structure. IX. Regulation of RNA polymerase activity by stringent starvation protein (SSP). *Journal of Molecular Biology*, **129**, 517–30.

MAALØE, O. (1978). Regulation of the protein synthesising machinery – ribosomes, tRNA, factors etc. In *Biological Regulation and Development*, ed. R. F. Goldberger. New York: Plenum.

MCCLURE, W. R. (1980). Rate limiting steps in RNA chain initiation. *Proceedings of the National Academy of Sciences, USA*, (in press).

MAJORS, J. (1978). Ph.D. Thesis, Harvard University.

NIERLICH, D. P. (1978). Regulation of bacterial growth, RNA and protein synthesis. *Annual Review of Microbiology*, **32**, 393–432.

PONGS, O. & ULBRICH, N. (1976). Specific binding of formylated initiator tRNA to *Escherichia coli* RNA polymerase. *Proceedings of the National Academy of Sciences, USA*, **73**, 3064–7.

PRIBNOW, D. (1975). Bacteriophage T7 early promoters: nucleotide sequences of two RNA polymerase binding sites. *Journal of Molecular Biology*, **99**, 419–43.

PRIMAKOFF, P. & BERG, P. (1970). Stringent control of transcription of phage ø80 psu$_3$. *Cold Spring Harbor Symposium on Quantitative Biology*, **35**, 391–6.

REEH, S., PEDERSEN, S. & FRIESEN, J. D. (1976). Biosynthetic regulation of individual proteins in *relA*$^+$ and *relA* strains of *Escherichia coli* during amino acid starvation. *Molecular and General Genetics*, **149**, 279–89.

RICHARDSON, J. P. (1966). Some physical properties of RNA polymerase. *Proceedings of the National Academy of Sciences, USA*, **55**, 1616–20.

ROSENBERG, M. & COURT, D. (1979). Regulatory sequences involved in the promotion and termination of RNA transcription. *Annual Review of Genetics*, **13**, 319–53.

RYAN, M. J., BELAGAJE, R., BROWN, E. L., FRITZ, H-J. & KHORANA, H. G. (1979). A synthetic tyrosine suppressor tRNA gene with an altered promoter sequence. Its cloning and relative expression *in vivo*. *Journal of Biological Chemistry*, **254**, 10803–10.

SCHALLER, H., GRAY, C. & HERRMAN, K. (1975). Nucleotide sequence of an RNA polymerase binding site from the DNA of bacteriophage fd. *Proceedings of the National Academy of Sciences, USA*, **72**, 737–41.

SEEBURG, P., NÜSSLEIN, C. & SCHALLER, H. (1977). Interaction of RNA polymerase with promoters from bacteriophage fd. *European Journal of Biochemistry*, **74**, 107–13.

SIEBENLIST, U. (1979). RNA polymerase unwinds an 11-base pair segment of phage T7 promoter. *Nature, London*, **279**, 651–2.

SIEBENLIST, U., SIMPSON, R. & GILBERT, W. (1980). *E. coli* RNA polymerase interacts homologously with two different promoters. *Cell*, (in press).

SILVERSTONE, A. E., GOMAN, M. & SCAIFE, J. G. (1972). ALT: a new factor involved in the synthesis of RNA by *Escherichia coli. Molecular and General Genetics*, **118**, 223–34.

STOMATO, T. D. & PETTIJOHN, D. E. (1971). Regulation of ribosomal RNA synthesis in stringent bacteria. *Nature, London*, **234**, 99–102.

TAKANAMI, M., SUGIMOTO, K., SUGISAKI, H. & OKAMOTO, T. (1976). Sequence of promoter for coat protein of bacteriophage fd. *Nature, London*, **260**, 297–307.

TRAVERS, A. (1976a). Modulation of RNA polymerase specificity by ppGpp. *Molecular and General Genetics*, **147**, 225–32.

TRAVERS, A. (1976b). RNA polymerase specificity and the control of growth. *Nature, London*, **263**, 641–6.

TRAVERS, A. A. (1980a). Promoter sequence for stringent control of bacterial ribonucleic acid synthesis. *Journal of Bacteriology*, **141**, 973–6.

TRAVERS, A. A. (1980b). A tRNA^{Tyr} promoter with an altered response to ppGpp. *Journal of Molecular Biology*, (in press).

TRAVERS, A. & BUCKLAND, R. (1973). Heterogeneity of *E. Coli*. RNA polymerase. *Nature New Biology*, **243**, 257–60.

TRAVERS, A. A., BUCKLAND, R. & DEBENHAM, P. G. (1979). Regulation of promoter selection by RNA polymerases. In *From Gene to Protein: Information Transfer in Normal and Abnormal Cells*, ed. T. R. Russell, K. Brew, H. Faber & J. Schulz, pp. 271–96. New York: Academic Press.

TRAVERS, A., BUCKLAND, R. & DEBENHAM, P. (1980). Functional heterogeneity of *Escherichia coli* ribonucleic acid polymerase holoenzyme. *Biochemistry*, **19**, 1656–62.

TRAVERS, A. A., BUCKLAND, R., GOMAN, M., LE GRICE, S. S. G. & SCAIFE, J. G. (1978). A mutation affecting the σ subunit of RNA polymerase changes transcriptional specificity. *Nature, London*, **273**, 354–8.

TRAVERS, A. A., DEBENHAM, P. G. & PONGS, O. (1980). Translation initiation factor 2 alters transcriptional specificity of *Escherichia coli* ribonucleic acid polymerase. *Biochemistry*, **19**, 1651–6.

TRAVERS, A., KAMEN, R. & CASHEL, M. (1970). The *in vitro* synthesis of ribosomal RNA. *Cold Spring Harbor Symposium on Quantitative Biology*, **35**, 415–18.

VAN OOYEN, A. J. J., GRUBER, M. & JORGENSEN, P. (1976). The mechanism of action of ppGpp on rRNA synthesis *in vitro*. *Cell*, **8**, 123–8.

WALTER, G., SEIFERT, W. & ZILLIG, W. (1968). Modified DNA-dependent RNA polymerase from *E. coli* infected with bacteriophage T4. *Biochemical and Biophysical Research Communications*, **30**, 240–7.

YANG, H-L., ZUBAY, G., URM, E., REINESS, G. & CASHEL, M. (1974). Effects of guanosine tetraphosphate, guanosine pentaphosphate and β-γ methylenyl-guanosine pentaphosphate on gene expression of *Escherichia coli in vitro*. *Proceedings of the National Academy of Sciences, USA*, **71**, 63–7.

YOUNG, F. S. & NEIDHARDT, F. C. (1978). Effect of inhibitors of elongation factor Tu on the metabolic regulation of protein synthesis in *Escherichia coli*. *Journal of Bacteriology*, **135**, 675–86.

YOUNG, R. A. & STEITZ, J. A. (1979). Tandem promoters direct *E. coli* ribosomal RNA synthesis. *Cell*, **17**, 225–38.

ASPECTS OF THE CONTROL OF GENE EXPRESSION IN FUNGI

HERBERT N. ARST, JR.

Department of Genetics, Ridley Building, The University, Newcastle upon Tyne NE1 7RU, UK

INTRODUCTION

Amongst the lower eukaryotes, the fungi are perhaps the most popular group of organisms for the study of control of gene expression. Such studies have been largely confined to three ascomycetes – the yeast *Saccharomyces cerevisiae* and the filamentous fungi *Neurospora crassa* and *Aspergillus nidulans*. I shall concentrate on *Aspergillus nidulans* because it tends to be neglected in reviews of fungal gene regulation and because it is the organism I know best. In fact, I shall not even attempt to cover adequately all aspects of gene regulation in *A. nidulans* but shall concentrate almost exclusively on those areas where I feel that the existing literature needs to be clarified or augmented. Readers requiring a more general treatment of the field might well begin by consulting the recent edition of *Fungal Genetics* (Fincham, Day & Radford, 1979) and the symposium volume entitled *Genetics and Physiology of Aspergillus* (Smith & Pateman, 1977). Also note that Hopwood (this volume) discusses the regulation of antibiotic synthesis by *A. nidulans* and other fungi.

The virtues of *A. nidulans* for genetical and biochemical analysis have been extolled elsewhere (e.g. Pontecorvo *et al.*, 1953; Clutterbuck, 1974; Smith & Pateman, 1977), but several of these are so crucial to studies of gene regulation that they bear emphasising here. Firstly, *A. nidulans* is catabolically extremely versatile. It utilises a remarkable number of chemical compounds, including some rather improbable ones. Whilst microbial taxonomists have long exploited extensive nutritional screening, microbial geneticists have been much slower to appreciate that it can give enormous resolution to studies of catabolism. As nutritional testing is one of the least laborious methods of analysing gene roles, *A. nidulans* has a considerable advantage over more fastidious microorganisms for the study of gene regulation. Secondly, nutritional testing of *A. nidulans* frequently yields more information than nutritional testing of other microorganisms. On solid media *A. nidulans* has a colonial habitat and undergoes a distinctive degree of differentiation. Thus growth responses can be characterised not simply by growth or lack of growth, but by the colony diameter, height of aerial hyphae, extent of conidiation, formation of conidial pigment and other morphological criteria. This gives a precision to growth testing which in many ways exceeds that obtained using unicellular organisms such as *S. cerevisiae* or non-colonial organisms such as *N. crassa*. Thirdly, the existence of both sexual and parasexual cycles greatly facilitates genetic manipulation and analysis. Fourthly, the possibility of comparing growth responses of heterokaryons with those of corresponding diploids can sometimes yield information about regulatory systems (e.g. see Scazzocchio & Arst, 1978; for a general discussion see Cove, 1977).

Many of the basic elements involved in gene regulation in fungi would seem, at least on the basis of formal genetic studies, to be those already familiar from studies of gene regulation in prokaryotes. Some possible exceptions involve the regulation of mating type in *S. cerevisiae* (for a brief review see Leupold, 1980; some very recent references are Hicks, Strathern & Klar, 1979; Klar & Fogel, 1979; Strathern, Blair & Herskowitz, 1979; Strathern, Newlon, Herskowitz & Hicks, 1979; Kushner, Blair & Herskowitz, 1979; Nasmyth & Tatchell, 1980) mating type effects on *cis*-acting regulatory mutations in *S. cerevisiae* (Rothstein & Sherman, 1980; Lemoine, Dubois & Wiame, 1978; Dubois, Hiernaux, Grenson & Wiame, 1978; Deschamps & Wiame, 1979), and regulatory roles associated with introns in split genes (Church, Slonimski & Gilbert, 1979; Dujon, 1979; Jacq, Lazowska & Slonimski, 1980; Slonimski, 1980; Halbreich *et al.*, 1980; Church & Gilbert, 1980). As *A. nidulans* is homothallic and as no regulatory phenomena can as yet be ascribed to introns in *A. nidulans* (although Lazarus *et al.*, 1980) have identified an intervening sequence in the mitochondrial cistron specifying the large mitochondrial ribosomal RNA), I shall not discuss these topics here. In *A. nidulans* a number of *trans*-acting regulatory genes specifying diffusible positive or negative acting regulatory molecules have been identified. These regulatory genes can be divided into three categories, pathway-specific, wide domain, and integrator. Their products can apparently bind in a specific fashion to *cis*-acting receptor sites adjacent to structural genes under their respective control. As in other eukaryotes, structural genes whose products participate in the same pathway or process are seldom clustered. Nevertheless, there are a few examples of clustering of functionally related genes and the possibility of operon-type organisation remains open. In several but by no means all cases, these gene clusters have turned out to be 'cluster-genes' encoding polyfunctional polypeptides. One further similarity to prokaryotic regulatory systems is that enzyme induction and repression seem to involve *de novo* synthesis and to be exerted at the level of transcription. The evidence for this has been summarised elsewhere (Arst & Bailey, 1977; Fincham *et al.*, 1979) and, as all of the more recent findings have supported it, I shall simply assume it to be the case for all of the regulatory phenomena discussed here.

REGULATORY GENES

Pathway-specific regulatory genes

A number of pathway-specific regulatory genes have been identified in *A. nidulans*. Most of these are positive acting. A stimulating discussion of why positive control systems might predominate in eukaryotic cells is given by Metzenberg (1979). The only probably negative acting pathway-specific regulatory gene in *A. nidulans* to have been identified is *su*A*meth* which controls repression of a number of activities involved in sulphate assimilation and cysteine and methionine biosynthesis (Paszewski & Grabski, 1975; Lukaszkiewicz & Paszewski, 1976). The most extensively studied of the positive acting pathway-specific regulatory genes are *nir*A whose product mediates induction of the syntheses of the enzymes of nitrate assimilation and certain other activities (reviewed by Cove, 1979) and *ua*Y whose product mediates induction of the syntheses of a number of enzymes and permeases involved in purine catabolism (reviewed by Scazzocchio & Gorton, 1977 but a more recent and comprehensive account is given by Scazzocchio, Sdrin & Ong, 1981). The probable product of the *ua*Y gene has been recently isolated (Philippides & Scazzocchio, 1981).

Wide domain regulatory genes

There are at least two regulatory phenomena in *A. nidulans* which involve a large number of pathways and regulatory genes with correspondingly wide domains. These are nitrogen metabolite repression of enzymes and permeases involved in the utilisation of various nitrogen sources and carbon catabolite repression of enzymes and permeases involved in the utilisation of various carbon sources. There is as yet little information on how the products of wide domain regulatory genes act in concert with each other or with pathway-specific regulatory gene products (but see Rand & Arst, 1978).

Nitrogen metabolite repression is mediated by a positive acting regulatory gene designated *are*A and described in detail by Arst & Cove (1973) with some further characterisation by Hynes (1975b), Arst & Scazzocchio (1975) and Rand & Arst (1977). In confirmation of the positive nature of the role played by the *are*A product, Rand & Arst (1977) described a complete loss of function allele, associated with a translocation and unable to revert to *are*A$^+$. The implication was that it resulted from a translocation break-point occurring within the *are*A gene. More recent work (D. Tollervey & H. N. Arst, Jr., unpublished) indicates that partial, apparently intracistronic revertants but no fully wild-type revertants can be obtained, but even the partial revertants are very rare. The translocation involved is a non-reciprocal one in which a portion of linkage group IV including the *pal*C, *paba*B and *pyro*A loci is translocated to linkage group III. One translocation break-point is quite near to *pal*C. A likely, but as yet unproven, explanation is that this region of linkage group IV is inserted into the *are*A gene in linkage group III fairly near to the C-terminal end such that most of the more critical regions of the *are*A gene product are coded 'upstream' from the translocation break-point. Thus a single mutation might result in a reasonably acceptable C-terminal sequence in translocation strains and a partially active *are*A product. The phenotypes of these partial revertants seem more compatible with an explanation in which they have a structurally altered *are*A product rather than a reduced amount of normal *are*A product.

Mutations designated *cre*A^d, leading to relief of carbon catabolite derepression, were first selected and characterised by Arst & Cove (1973) and Bailey & Arst (1975). The *cre*A gene is likely to be a negative acting regulatory gene directly involved in the regulation of the syntheses of many carbon catabolite repressible enzymes and permeases (Arst & Bailey, 1977; see also Bailey, Penfold & Arst, 1979). Hynes & Kelly (1977) and Kelly & Hynes (1977) have shown that mutations in two other genes, designated *cre*B and *cre*C, can also lead to derepressed expression of certain of the genes under *cre*A control whilst pleiotropically lowering levels of other activities under *cre*A control such as the L-proline permease as well as activities such as L-glutamate uptake whose control *vis-à-vis cre*A is unclear.

Selection and characterisation of *cre*B and C alleles has also been done by my group, principally by Dr C. R. Bailey and, to a lesser extent, by myself (but we have benefited considerably from discussions of this work with Dr K. N. Rand). Dr Hynes kindly provided standard mutant alleles of both genes isolated in his laboratory. A full publication of this work will be made in due course but it is worth summarising this work and our conclusions about the role of *cre*B and C genes here because, if correct, they enable an adequate model of carbon catabolite repression in *A. nidulans* to be confined to a direct involvement of only a single regulatory gene, *viz. cre*A.

The first point to note about *cre*B and C is that the two genes are virtually certain to specify products involved in the same process (such as subunits of a heteropolymer or two enzymes of the same pathway) because non-leaky mutations in both genes have the same phenotype and are not even slightly additive in double mutants (C. R. Bailey, unpublished data; see also Hynes & Kelly, 1977). The second point to note is that the published phenotype of *cre*B and C mutations almost certainly represents the loss of function phenotype. This conclusion is based on the ease with which such mutations are selected using a variety of methods (Hynes & Kelly, 1977; Rand, 1978; Penfold, 1979; C. R. Bailey & H. N. Arst, Jr., unpublished data) as well as the fact that such mutations are recessive (Hynes & Kelly, 1977; C. R. Bailey, unpublished data).

Although it is clear that *cre*B$^-$ and C$^-$ mutations do lead to derepression of a number of carbon catabolite repressible activities, it is far from clear that the primary effect of these mutations is on a component of the regulatory system mediating carbon catabolite repression. Postulating a primary role in carbon catabolite repression for the *cre*B and C gene products seems particularly unattractive when the total phenotypes of *cre*B$^-$ and C$^-$ mutations are contrasted with those of *cre*A^d mutations. We believe that the evidence is much more compatible with a model in which the *cre*B and C gene products play an important role in determining the structure of the cell membrane, by coding either for a major membrane protein or for enzymes involved in the synthesis of a major membrane component. The loss of this component in *cre*B$^-$

or C^- mutants would affect a wide range of membrane functions, including permeability to a number of compounds. Reduced uptake of carbon catabolite repressing sugars in *cre*B^- and C^- mutants could therefore be responsible for derepression of activities such as acetamidase, alcohol dehydrogenase, extracellular protease and α-glucosidase, reported by Hynes & Kelly (1977). Similarly their observation that these mutants have reduced uptake capacity for L-proline and L-glutamate could result from a comparable inability of mutant membrane to incorporate or allow the proper functioning of particular permeases. Their finding of lower induced levels of D-quinate dehydrogenase and β-galactosidase in these mutants could reflect a similar effect on the uptake of the respective co-inducers.

The evidence for a primary role of the *cre*B and C gene products in membrane structure rather than carbon catabolic repression can be summarised as follows:

(1) Unlike *cre*A^d mutations, *cre*B^- and C^- mutations confer resistance (albeit at a low level) to a number of toxic sugars and sugar derivatives including 2-deoxy-D-glucose, L-sorbose, 2-deoxy-D-galactose, D-fucose, glucuronamide and D-glucosamine. In fact, the lack of effect of *cre*A^d mutations on L-sorbose and 2-deoxy-D-glucose toxicities was an important step in the screening process which allowed them to be distinguished from mutations affecting sugar transport (Arst & Cove, 1973; Bailey & Arst, 1975).

(2) *cre*B^- and C^-, but not *cre*A^d, mutations result in reduced ability to utilise a number of carbon sources including some which are carbon catabolite repressing (e.g., D-glucose, D-mannose, D-fructose). The utilisation of certain other carbon catabolite repressing carbon sources (e.g. D-xylose, acetate, sucrose) is largely unaffected.

(3) Whereas double mutants carrying a *cre*A^d mutation along with a mutation resulting in toxicity of a sugar are actually more sensitive to toxicity by the sugar than corresponding single (*cre*A^+) mutants (Bailey & Arst, 1975), *cre*B^- and C^- mutations in certain combinations of double mutations actually protect against sugar toxicity (in the case of D-glucose, D-fructose, D-mannose, L-arabinose, D-galactose and maltose but not in the cases of D-xylose, D-sorbitol and sucrose). One very useful and rather sensitive *in vivo* measurement of the degree to which carbon catabolite derepressing mutations result in carbon catabolite derepression of acetamidase in the presence of various carbon sources is

the extent to which they allow suppression of the inability of loss of function mutations in the *are*A gene to utilise acetamide in double mutants (Arst & Cove, 1973; Bailey & Arst, 1975; Arst & Bailey, 1977). (Loss of function mutations in the *are*A gene result in inability to utilise nitrogen sources other than ammonium under carbon catabolite repressing conditions but allow utilisation of a few nitrogen sources, including acetamide, when the carbon source is non-repressing (Arst & Cove, 1973; Arst & Bailey, 1977).) There is a reasonably good correlation between the degree to which *cre*B$^-$ and C$^-$ mutations can protect against the toxicity of carbon catabolite repressing sugars in strains where one or more of these is toxic and the extent to which, in strains lacking a functional *are*A allele, they allow utilisation of acetamide as a nitrogen source in the presence of the same carbon catabolite repressing sugars. For example, *cre*B$^-$ and C$^-$ mutations allow strong utilisation of acetamide in the presence of D-glucose, D-fructose and D-mannose but only weak utilisation on sucrose and none at all on D-xylose. *cre*A^d mutations allow strong acetamide utilisation in strains with defective *are*A alleles in the presence of all of these carbon sources, including sucrose and D-xylose.

(4) When present in triple mutants carrying a *sor*A$^-$ mutation, resulting in reduced D-glucose uptake (Elorza & Arst, 1971; C. R. Bailey, unpublished data), and a *mal*A$^-$ mutation, resulting in loss of the ability to take up maltose (Bailey, 1976), *cre*B$^-$ and C$^-$ mutations are additive, resulting in extremely weak growth on D-glucose as carbon source and very strong resistance to 2-deoxy-D-glucose and L-sorbose toxicities. *cre*A^d mutations have no comparable effect in *cre*A^d *sor*A$^-$ *mal*A$^-$ triple mutants. It would appear that D-glucose can be taken up by the *sor*A permease, the *mal*A permease and by at least one other mechanism absent in *cre*B$^-$ and C$^-$ strains.

(5) *cre*B$^-$ and C$^-$ mutations affect the toxicities of a wide range of compounds for which no involvement of carbon catabolite repression is to be expected and, where tested, *cre*A^d mutations have no such effect. Indeed, the first *cre*B$^-$ mutations to be selected were studied because they confer resistance to toxic concentrations of molybdate (Arst, MacDonald & Cove, 1970; Arst & Cove, 1970). The *mol*B gene of those studies has turned out to be the same gene as *cre*B (C. R. Bailey, unpublished data). It is also possible that amongst the lesser characterised mutants selected by Arst *et al.* (1970) were some carrying mutations allelic to *cre*C$^-$ mutations. In

any case, both *cre*B⁻ and C⁻ mutations confer resistance to both molybdate and tungstate toxicities (Arst *et al.*, 1970; C. R. Bailey & H. N. Arst, Jr., unpublished data). Their behaviour *in vivo* (but not *in vitro*) suggests that this is accompanied by reduced levels of the molybdoenzymes nitrate reductase and purine hydroxylase I (formerly xanthine dehydrogenase I – see Lewis, Hurt, Sealy-Lewis & Scazzocchio, 1978) because they reduce utilisation of nitrate (but not nitrite) and hypoxanthine (but not uric acid) as nitrogen sources and confer slight resistance to chlorate (see Cove, 1976) (Arst *et al.*, 1970). They lead to hypersensitivity to acriflavine, malachite green and ethidium bromide (Arst & Cove, 1970; C. R. Bailey, unpublished results). They result in hypersensitivity to the succinate dehydrogenase inhibitor carboxin (see Gunatilleke, Arst & Scazzocchio, 1975) on glucose as carbon source but not on acetate (C. R. Bailey, unpublished data). On the other hand they confer resistance to cycloheximide (Rand, 1978) and to L-glutamic γ-monohydroxamate (see Kinghorn & Pateman, 1977) as well as slight resistance to iodoacetate, D-serine and seleno-DL-methionine (C. R. Bailey, unpublished data).

(6) It is likely that D-galactose can be added to the list of sugars for which a direct effect of *cre*B⁻ and C⁻, but not *cre*A^{d}, mutations on utilisation can be observed. Arst & Cove (1970) reported that *mol*B⁻ mutations enhance radial growth rate (although the mycelial density is less) on solid media containing galactose as carbon source. This is also true of *cre*C⁻ mutations and sugar uptake mutations in the *sor*A gene (C. R. Bailey & H. N. Arst, Jr., unpublished data). D-galactose is toxic to a number of organisms which can nevertheless metabolise it such as *Lemna gibba* (DeKock, Cheshire, Mundie & Inkson, 1979), sugarcane cells (Maretzki & Thom, 1978) and *Dictyostelium discoideum* (DeMeglio & Friedman, 1978). It is therefore likely that for D-galactose as for uric acid (Arst & Scazzocchio, 1975), γ-amino-*n*-butyrate (Bailey *et al.*, 1979) and L-proline (Arst, MacDonald & Jones, 1980), a rate of uptake which exceeds the rate of catabolism is somewhat toxic, reducing radial growth rate, in *A. nidulans*. This is probably why mutations which reduce galactose uptake actually appear to enhance its utilisation. In support of this conclusion, *cre*B⁻ and C⁻ mutations protect strains lacking galactose-l-phosphate uridyl transferase (i.e. *gal*D⁻, Roberts, 1970) against galactose toxicity.

(7) Kelly & Hynes (1977) reported that the effects of *cre*B⁻ and C⁻, but not those of *cre*A^{d}, mutations on acetamide utilisation are

mediated mainly by their effects on induction by acetyl coenzyme A (see Hynes, 1977). In other words, the higher acetamidase activities of *cre*B$^-$ and C$^-$ mutants when grown in the presence of acetamide and a repressing carbon source (Hynes & Kelly, 1977) might be a secondary consequence of their derepressed levels of acetyl coenzyme A synthetase (Kelly & Hynes, 1977) whereas *cre*A^d mutations, although also derepressing acetyl coenzyme A synthetase, have a direct effect on the regulation of acetamidase.

In contrast to the wealth of *in vivo* data suggesting that *cre*B$^-$ and C$^-$ mutations reduce sugar uptake, direct uptake measurements have failed to demonstrate any convincing effect on the uptake of either D-glucose (Hynes & Kelly, 1977; C. R. Bailey, unpublished data) or 2-deoxy-D-glucose (C. R. Bailey & H. N. Arst, Jr., unpublished data). It is, however, possible that the measurements were not carried out at a development stage which would enable differences to be seen. Development regulation of transport activity has been described in a number of fungal systems, including sugar uptake in *A. nidulans* (Kurtz, 1979; Kurtz & Champe, 1979). It is possibly relevant and even analogous to consider the phenotype of *cys*-13 mutants of *N. crassa* (Marzluf, 1970a, b; Roberts & Marzluf, 1971). *cys*-13 mutants are resistant to the toxic sulphate analogue chromate and in double mutants carrying a *cys*-14 mutation, leading to loss of a mycelial sulphate permease (Marzluf, 1970a, b; Roberts & Marzluf, 1971), the ability to utilise sulphate is lost (whereas both *cys*-13 and *cys*-14 single mutants are able to utilise sulphate). However, when sulphate or chromate transport is measured in mycelia, *cys*-13 strains are indistinguishable from wild-type. When measurements are carried out using conidiospores, *cys*-13 strains show a very substantial reduction in uptake capacity for both sulphate and chromate. In view of the obvious parallels between *cys*-13 mutations of *N. crassa* and *cre*B$^-$ and C$^-$ mutations of *A. nidulans,* work is in hand to measure sugar uptake by conidiospores.

In spite of these difficulties in demonstrating differences in sugar uptake levels, there seems little question but that a primary role for *cre*B and C gene products in membrane structure remains the most attractive hypothesis. For example, hypersensitivity to malachite green and/or acriflavine and/or related compounds is a consequence of certain mutations affecting the cell surface in *Escherichia coli* (e.g. Ennis, 1971; Normark & Westling, 1971; Nakamura & Suganuma, 1972; Otsuji, Higashi & Kawamata, 1972; Coleman & Leive, 1979). In both *E. coli* (Anderson, Wilson & Oxender, 1979) and

Salmonella typhimurium (Postma, Cordaro & Roseman, 1977) highly pleiotropic membrane mutations affecting toxicities of a number of compounds and numerous transport functions have been described. Pleiotropic cross resistance and collateral hypersensitivity to a wide range of inhibitors results from a mutation modifying the cell membrane in *S. cerevisiae* (Rank, Robertson & Phillips, 1975a, b; Rank, Gerlach & Robertson, 1976). Mutational loss of specific outer membrane proteins of gram-negative bacteria affects a number of membrane functions including uptake systems (for references and review see Mäkelä & Stocker (this volume) and DiRienzo, Nakamura & Inouye, 1978; McIntosh, Pickett, Chenault & Earhart, 1978; Lo, 1979; Chai & Foulds, 1979; Nikaido, 1979; Alphen, Selm & Lugtenberg, 1978; Krieger-Brauer & Braun, 1980). Of course, syntheses of certain permeases and other membrane components are subject to carbon catabolite repression and it is therefore hardly surprising that *cre*A^{d} mutations affect colony morphology (Bailey & Arst, 1975; Arst & Bailey, 1977). Similarly, mutations affecting carbon catabolite repression in *E. coli* and *S. typhimurium* have effects on membrane composition, permease synthesis and cell morphology (Kumar, 1976; Alper & Ames, 1978; Kumar, Prakash & Agarwal, 1979; Alderman, Dills, Melton & Dobrogosz, 1979; Mallick & Herrlich, 1979). However, unlike the *cre*A^{d} phenotypes, the phenotype of *cre*B^{-} and C^{-} mutations seems pleiotropic well beyond the realm of carbon catabolite repression.

Even if the direct involvement of the *cre*B and C gene products is with membrane structure, it is nevertheless possible that the cell membrane plays some role in carbon catabolite repression in *A. nidulans* beyond uptake of co-repressors and co-inducers (or their precursors) and that this role is altered in *cre*B^{-} and C^{-} mutants. There is evidence for a membrane involvement in the regulation of adenylate cyclase and thus carbon catabolite repression in *E. coli* and *S. typhimurium* (Seto, Nagata & Maruo, 1975a, b; Postma & Roseman, 1976; Saier, 1977; Bolshakova, Gabrielyan, Bourd & Gershanovitch, 1978; Peterkofsky & Gazdar, 1979; Feucht & Saier, 1980). As in many other organisms, adenylate cyclase is bound to the plasma or intracellular membranes in a number of fungi including *N. crassa* (Flawiá & Torres, 1972a, b; Scott, 1976), *Blastocladiella emersonii* (Gomes, Mennucci & Maia, 1978), *Saccharomyces fragilis* (Sy & Richter, 1972); *S. cerevisiae* (Londesborough & Nurminen, 1972; Wheeler *et al.*, 1974) and possibly *Aspergillus niger* (Wold & Suzuki, 1974). In *N. crassa, S. cerevisiae* and *Mucor* *racemosus,* depolarisation of the membrane can increase cyclic AMP levels, presumably by activating membrane-bound adenylate cyclase (Trevillyan & Pall, 1979). There is, however, no evidence for an involvement of cyclic AMP in carbon catabolite repression in *A. nidulans* (Arst & Bailey, 1977; see also discussion by Trevillyan & Pall, 1979).

One final point concerns the possibility of functionally related genes located close to the *cre*B and C genes. The *cre*-34 mutation has in many respects a phenotype opposite to that of *cre*B$^-$ and C$^-$ mutations and shows linkage to the *cre*C gene (3–6 cM), which is probably not sufficiently tight for it to be a *cre*C allele (Kelly & Hynes, 1977). Moreover, the effects of *cre*-34 are roughly the same in double mutants with *cre*B-15 as with *cre*C-27. This apparent lack of interaction between the *cre*-34 mutation and the *cre*C-27 mutation makes it extremely unlikely that these two mutations lie in the same cistron. A curiously parallel situation exists involving *cre*B (C. R. Bailey, unpublished data). *cre*B is quite closely linked (2–4 cM) to the *acr*B gene on the extreme centromere distal region of the right arm of linkage group II. *acr*B mutations confer resistance to acriflavine and malachite green (Roper & Käfer, 1957). In this and certain other respects they have a phenotype which is opposite to that of *cre*B$^-$ and C$^-$ mutations. Allelism is again unlikely because the map distance is too great, an *acr*B$^-$/*cre*B$^-$ diploid is phenotypically wild-type, and *acr*B$^-$ *cre*B$^-$ double mutants resemble *acr*B$^-$ *cre*C$^-$ double mutants.

Integrator genes

An integrator gene is a positive acting regulatory gene which mediates the induction and concomitant expression of two or more non-contiguous structural genes of which at least one is also capable of independent expression mediated by another positive acting regulatory gene. It is an element of the model for gene regulation proposed by Britten & Davidson (1969, 1971; Davidson & Britten, 1973). One such gene, designated *int*A has been described in some detail by Arst (1976), Arst & Bailey (1977) and Arst, Penfold & Bailey (1978). It mediates the induction, by ω-amino acids, of acetamidase, γ-amino-n-butyrate permease, γ-amino-n-butyrate transaminase and an activity necessary for lactam utilisation, presumably a lactamase. The acetamidase is also induced by other structurally unrelated co-inducers (Hynes, 1978a) mediated by at

Table 1. *Relative activities of γ-amino-*n*-butyrate (GABA) permease, GABA transaminase and acetamidase in strains of* Aspergillus nidulans *carrying various* int*A alleles*

Relevant genotype	GABA uptake		GABA transaminase		Acetamidase	
	uninduced	induced	uninduced	induced	uninduced	induced
wild-type (*int*A$^+$)	41	100	14	100	28	100
*int*A$^-$-101	35	21	12	13	16	27
*int*A$^-$-500	52	45	18	67	27	25
*int*A^c-2	83	247	142	134	121	193
*int*A^c-304	85	70	209	280	64	75
*int*A^c-305	40	108	11	182	23	187

Activities are expressed as a percentage of the induced wild-type activity. 5 mM β-alanine was used to induce. Data are collated from Arst *et al.* (1978), Bailey *et al.* (1979), Penfold (1979), Bailey, Arst & Penfold (1980) and C. R. Bailey (unpublished data). See published references for details of growth conditions and assays.

least one other positive acting regulatory gene, designated *amd*A and acting in parallel with *int*A (Arst, 1976). As I have discussed integrator genes in general and *int*A in particular extensively elsewhere (*vide supra*), there are only two subjects I wish to discuss here.

The first point concerns the evidence for a direct involvement of the *int*A gene in the regulation of structural genes under its control. Arst & Bailey (1977) have described how allelic heterogeneity can be used to distinguish regulatory genes indirectly involved, where allelic heterogeneity is of a hierarchical nature, from those directly involved, where non-hierarchical allelic heterogeneity can be observed. Evidence *in vivo* for non-hierarchical heterogeneity of *int*A phenotype has been presented previously (Arst, 1976, 1978; Arst *et al.*, 1978), but data in Table 1 extend this evidence to quantitative measurements. In particular, it should be noted that *int*A mutations affect both the degree of inducibility and the maximal level of activities under *int*A control and that these effects are largely independent. Evidence of heterogeneity can be seen both when comparing levels of the same activity in two *int*A mutants and when comparing relative levels of different activities in the same *int*A mutant. For example, amongst *int*A^c mutations, *int*A^c-2 raises the maximal level of GABA uptake activity 2½-fold without appreciably altering the degree of inducibility whereas *int*A^c-304 results in fully constitutive expression but actually lowers the maximal level slightly. *int*A^c-2, -304, and -305 all raise the maximal level of GABA transaminase but to varying extents, and, whereas *int*A^c-2 results in full constitutivity, *int*A^c-304 lowers and *int*A^c-305 raises the degree of inducibility. *int*A^c-2 and *int*A^c-305 raise the maximal level of acetamidase to a comparable extent, but *int*A^c-2 reduces whilst *int*A^c-305 increases the degree of inducibility. *int*A^c-304 raises the maximal level of GABA transaminase whilst lowering maximal levels of GABA transport activity and acetamidase. *int*A^c-2 results in full constitutivity for GABA transaminase, a lowered degree of inducibility for acetamidase, and an unaffected degree of inducibility for GABA uptake. The possibilities for observing allelic heterogeneity are less when examining *int*A$^-$ mutations than when examining *int*A^c mutations. However, note that *int*A$^-$-500 retains a degree of inducibility for GABA transaminase but not for the other two activities. Although these data are too limited in genotypes and growth conditions examined to permit a full-scale comparison with *in vivo* phenotypes, some comparison can be made in the case of GABA transaminase. GABA transaminase can catalyse the synthesis of N^α-acetyl-L-ornithine, thereby substituting for the enzyme specified by the *orn*B gene (Arst, 1976) and the synthesis of β-alanine, thereby substituting for the enzyme specified by the *panto*C gene and possibly other enzymes of the same pathway (Arst, 1978). *int*A^c-2 and -304 are strong suppressors of *orn*B$^-$ (Arst, 1976) and *panto*C$^-$ (Arst, 1978) mutations whereas *int*A^c-305 has little effect. This correlates well with results in Table 1 showing that *int*A^c-2 and -304 but not *int*A^c-305 lead to high uninduced levels of GABA transaminase. *int*A$^-$-101 eliminates supplementation of *orn*B$^-$ (Arst, 1976) and *pantoC*$^-$ (Arst, 1978) auxotrophies by ω-amino acids acting as co-inducers of GABA transaminase whereas *int*A$^-$-500 reduces but does not eliminate such supplementation. This correlates well with results in Table 1 showing that *int*A$^-$-101 eliminates inducibility of GABA transaminase but that *int*A$^-$-500 only reduces its inducibility.

The other point which I believe requires some discussion concerns the relevance of studies of short repetitive DNA sequences (200–400 nucleotide base pairs) to the applicability of at least the integrator gene circuitry of the Britten & Davidson (1969) model to fungal gene regulation. The model postulates the presence of tandemly arranged receptor sites for integrator gene specified activators near the 5′ end of structural genes on the coding strand of the DNA. Partly for simplicity and partly to assign a role to the interspersed short repetitive sequences which have been found in the genomes of a number of higher eukaryotes, the early development of the model largely assumed that these activators specified by integrator genes would be RNA (and they might, in a sense be so in some organisms, see Davidson & Britten, 1979). Nevertheless, it was noted that a formally equivalent model could be constructed involving activator proteins as the products of integrator genes (Britten & Davidson, 1969, 1971). As it became increasingly clear that protein–DNA interactions account for the bulk of phenomena involving regulation of gene expression in prokaryotes, Davidson & Britten (1973) developed further the activator protein branch of their model. Whilst a receptor site of the order of 300 nucleotide base pairs might be a reasonable size for the binding of an activator RNA, it is clearly more than an order of magnitude longer than would be required to bind an activator protein. So if integrator genes in fungi code for activator proteins, then studies such as those of Timberlake (1978) in *A. nidulans* and Krumlauf & Marzluf (1979, 1980) in *N. crassa* demonstrating a lack of interspersed short repetitive DNA sequences are probably irrelevant to the question of applicability of the Britten–Davidson model or at least of integrator genes. Krumlauf & Marzluf (1980) note that their studies would not have detected very short (≤25 nucleotide base pairs) repetitive sequences which are almost certainly present in the *N. crassa* genome as receptor sites for positive acting regulatory proteins. Not only are protein receptor sites probably too short to be detected in any case, but it is very likely that minor sequence variations correlating with sensitivity to regulation occur within them (see Arst, 1976), making their detection as repetitive sequences even more difficult. A pertinent illustration is provided by the catabolite activator protein specified by the *crp* gene of *Escherichia coli* (for a review see Pastan & Adhya, 1976). *E. coli* does not contain repetitive DNA by the criteria applied to eukaryotes and yet there might be as many as 100 catabolite activator protein receptor sites in the *E. coli* genome.

RECEPTOR SITES

Cis-acting regulatory mutations identifying receptor sites for diffusible regulatory products have been identified in *A. nidulans* adjacent to the *amd*S gene specifying acetamidase (Hynes, 1975a, 1978b, 1979), the *s*B gene specifying sulphate permease (Lukaszkiewicz &

Paszewski, 1976), the *uap*A gene specifying a xanthine-uric acid permease (Arst & Scazzocchio, 1975; Scazzocchio & Arst, 1978), the *prn*B gene specifying L-proline permease (Arst & Cove, 1973; Arst & MacDonald, 1975; Arst *et al.*, 1980) and the *gab*A gene specifying γ-amino-n-butyrate permease (Bailey *et al.*, 1979). In addition a translocation has been obtained which apparently fuses the *nii*A gene specifying nitrite reductase to a new promoter, identifying the probable position of a *cis*-acting regulatory region for *nii*A (Rand & Arst, 1977; Arst, Rand & Bailey, 1979).

'CLUSTER-GENES'

Although 'cluster-genes' undoubtedly occur in *A. nidulans,* they have been much more extensively studied in other fungi. An interested reader might begin by consulting the reviews by Giles (1978) and Metzenberg (1979) and papers by Lumsden & Coggins (1977), Makoff, Buxton & Radford (1978) and Keesey, Bigelis & Fink (1979).

GENE CLUSTERS – DO FUNGAL OPERONS EXIST?

When the products of two or more structural genes are required simultaneously and in a constant stoichiometry, their co-ordinate expression would have a selective advantage. Co-ordinate expression can be achieved by an operon type of organisation in which the functionally related structural genes are expressed on the same polycistronic transcript. Alternatively, co-ordinate expression would be achieved if the structural genes in question had separate but structurally identical *cis*-acting regulatory regions, provided that their positions in the genome are such that the expression of some is not favoured relative to the expression of others. Even in prokaryotes such equivalence of genome position for scattered genes cannot be assumed, but it probably poses more difficulty in eukaryotes where the complexity of chromosome structure and replication is greater.

Completely co-ordinate expression of functionally related genes might not, however, always be advantageous. If a structural gene needs to be expressed in more than one physiological context and is present in the genome as only a single copy, then its inclusion in an operon-type structure might necessitate needless expression of other structural genes. This is potentially a problem in the *gal* operon of *E. coli* where one of the genes codes for an enzyme involved in synthesis of cell surface components as well as in galactose utilisation. How this particular conflict of interest is resolved (if it be resolved) is at present unclear (but see citation to personal communication by Adhya in Rosenberg & Paterson, 1979, as well as Musso, DiLauro, Adhya & de Crombrugghe, 1977; Adhya & Miller, 1979; DiLauro, Taniguchi, Musso & de Crombrugghe, 1979; Taniguchi, O'Neill & de Crombrugghe, 1979). One way to avoid such conflicts of interest is a system of overlapping transcripts differing in length by one or more cistrons, as occurs in the *deo* operon of *E. coli* (Albrechtsen, Hammer-Jespersen, Munch-Petersen & Fiil, 1976; Buxton, Albrechtsen & Hammer-Jespersen, 1977; Valentin-Hansen, Svenningsen, Munch-Petersen & Hammer-Jespersen, 1978; Valentin-Hansen, Hammer-Jespersen & Buxton, 1979). This allows greater regulatory versatility by enabling both independent and co-ordinate expression of individual cistrons, the choice being dependent on the regulatory stimulus. In evolutionary terms, an operon with overlapping transcripts might be an intermediate stage between clustered and non-clustered organisation (Arst & MacDonald, 1978).

Although co-ordinate expression of functionally related genes in enteric bacteria is achieved principally through operon organisation, in fungi and other eukaryotes there is considerable doubt whether polycistronic messenger RNA occurs and whether an eukaryotic protein-synthesising apparatus is capable of initiating translation of a second or subsequent polypeptide from a polycistronic messenger (see, for example, Watson, 1976). Petersen & McLaughlin (1973) showed that most, possibly all, messenger RNA in *S. cerevisiae* is monocistronic, or, more precisely, they were unable to find any evidence for more than one translational initiation per messenger in a polysome preparation. For anyone familiar with the genetic map of *S. cerevisiae* this result should come as no surprise, for once we have distinguished between 'cluster-genes' and gene clusters we find very few gene clusters. Moreover, the rich growth medium used by Petersen & McLaughlin (1973) probably prevented expression of the few gene clusters where a polycistronic messenger might be synthesised. Specifically, the *arg*BC cluster (Minet *et al.*, 1979) would have been repressed whilst the *dur* gene cluster (Whitney & Cooper, 1972; Lemoine, Dubois & Wiame, 1978; but this might be a 'cluster-gene', see citation to Cooper *et al.*, in preparation in Cooper, Gorski & Turoscy, 1979), the *dal* gene cluster (Lawther, Riemer, Chojnacki & Cooper, 1974; Cooper *et al.*, 1979), and the *gal* gene cluster (Douglas & Hawthorne, 1964; Bassel & Mortimer, 1971) would not have been induced (and might also have been repressed). Subsequently, Hopper & Rowe (1978) demonstrated the existence of separate messengers for the two genes at either end of the *gal* cluster. Their data establish that co-ordinate regulation of the expression of the genes of the *gal* cluster is at the level of transcription, and they note that these data do not rule out post-transcriptional cleavage of a single transcript leading to separate messengers.

The fact that polycistronic mRNA is, at the very least, rare in eukaryotes does not, however, preclude its existence. If it cannot be fully translated, its existence (or, in any case, its evolutionary perseverance) would be precluded. Sherman & Stewart (1975) reverted a strain of *S. cerevisiae* carrying an ochre mutation in the codon for the second amino acid of iso-1-cytochrome *c* but were unable to recover a single revertant resulting from a 'downstream' initiation site amongst over 130 intragenic revertants. Yet, when they reverted another strain in which the AUG initiation codon was mutated, they obtained revertants in which the AAG codon for the lysine in position 4 had mutated to an AUG initiation codon. They concluded that reinitiation could not occur after a chain termination codon. Rosenberg & Paterson (1979), using a wheat-germ cell-free protein-synthesising system, were able to demonstrate efficient translation of the first cistron (i.e. the 5′-proximal coding sequence) of polycistronic prokaryotic mRNAs provided the 5′ end had been 'capped' with a 7-methylguanosine residue so as to resemble an eukaryotic mRNA. In contrast, they were unable to detect translation of the second or subsequent cistrons from these mRNAs and concluded that proximity of the initiation codon to the 'cap' structure might be crucial for its recognition by an eukaryotic translational apparatus. Of course, anyone fond of analogies might argue that just as the 'cap' modification is necessary for translation of the first cistron so some other modification is necessary for translation of a second or subsequent cistron. After all, the limitations might be imposed by the heterology of the system rather than by the wheat-germ translational system alone.

Exinger & Lacroute (1979) constructed a strain of *S. cerevisiae* carrying one of each of the three kinds of chain termination mutations in the N-terminal portion (encoding carbamoyl phosphate synthase) of the 'cluster-gene' *ura*2. They then reverted this strain to prototrophy, which, given the presence of a separate carbamoyl phosphate synthase participating in arginine biosynthesis, requires only the expression of the 'downstream' portion of *ura*2 encoding aspartate carbamoyltransferase. The reversion frequency was very low but 17 revertants were obtained of which three apparently resulted from deletion of the region containing the three chain termination mutations without a shift of 'reading-frame'. The remaining 14 revertants would appear to result from creation of a reinitiation site with the *ura*2 gene but 'downstream' from the three chain termination mutations. If sequencing studies can confirm this interpretation, this elegant experiment will have gone a long way towards demonstrating the feasibility of translation of polycistronic mRNA in yeast.

If eukaryotic operons do exist, it might be predicted that polypeptide chain termination mutations would not exert a polar effect on the expression of distal cistrons as they do in prokaryotes, probably due to premature transcription termination (for a review see Adhya & Gottesman, 1978). In eukaryotes the existence of a nuclear membrane probably ensures the separation in space and time of the processes of transcription (in the nucleus) and translation (in the cytoplasm). Nevertheless, polar mutations have been sought in several gene clusters and, interestingly, have been found in the *arg*BC cluster of *S. cerevisiae* (Minet *et al.*, 1979). The authors suggested several possible explanations, but the correct one might have been provided by the finding of Losson & Lacroute (1979) that chain termination mutations interfere with mRNA stability in *S. cerevisiae.* Whether this finding can be extrapolated to other organisms depends upon how closely the control characteristics, localisations and specificities of their ribonucleases resemble those of yeast. Nevertheless, no polar mutations have been obtained in the *qa* cluster of *N. crassa* (Chaleff, 1974; Giles, 1978), the *prn* cluster of *A. nidulans* (Arst & MacDonald, 1978; S. A. Jones & H. N. Arst, Jr., unpublished data), or the *crn*A–*nii*A–*nia*D gene cluster of *A. nidulans* (D. J. Cove, unpublished data). Nevertheless, chain termination mutations have not been definitively identified in any of these three clusters although in each case the number of mutations selected is sufficiently large that it would be very surprising if a number of chain termination mutations had not been selected in each gene. Chain termination suppressors have not yet been defini-

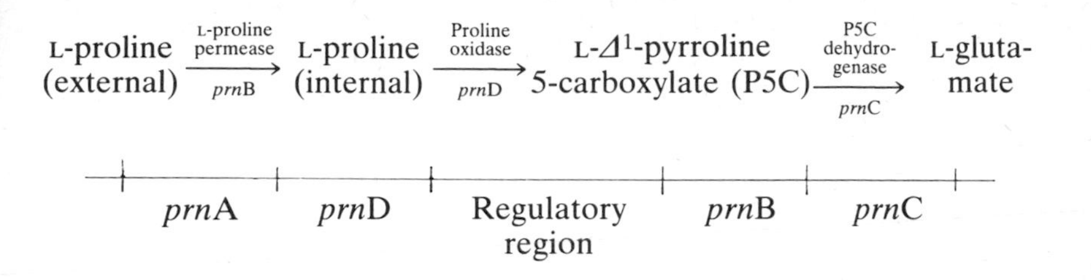

Fig. 1. The proline catabolic pathway in *Aspergillus nidulans:* gene roles and gene order within the *prn* gene cluster. Details of the pathway are given by Arst & MacDonald (1975, 1978) and Arst, MacDonald & Jones (1977, 1980). *prn*B, *prn*D and *prn*C are the structural genes for proline permease, proline oxidase and P5C dehydrogenase, respectively, whilst *prn*A is a positive acting regulatory gene mediating proline induction and necessary for the expression of *prn*D, *prn*C and, to a lesser extent, *prn*B (S. A. Jones, D. W. MacDonald & H. N. Arst, Jr., unpublished data and references cited above).

tively identified in *A. nidulans,* but two groups have identified allele specific, locus non-specific suppressor mutations (Bal, Maciejko, Kajtaniak & Gajewski, 1978; Bal, Kowalska, Maciejko & Weglenski, 1979; Roberts, Martinelli & Scazzocchio, 1979). Amongst the suppressible mutations selected by Roberts *et al.* (1979), there is a *nia*D^{-} allele. As there is evidence that transcription of *nii*A proceeds from the *nia*D-proximal side (Arst *et al.*, 1979), the lack of a polar effect of this mutation might already indicate a lack of polarity in this cluster. In fact, other evidence indicates that *nii*A cannot be expressed exclusively from a transcript which includes the *nia*D gene (*vide infra*).

The two most extensively studied gene clusters in *A. nidulans* are the *prn* gene cluster involved in L-proline catabolism and *crn*A–*nii*A–*nia*D gene cluster involved in nitrate assimilation. Gene orders and gene roles for these two clusters are given in Figs. 1 and 2. Deletion mutations extending from within *prn*D to within *prn*B reduce, but do not abolish, expression of *prn*C in *cis* (Arst & MacDonald, 1978). As all eight such deletions which have been isolated (Arst & MacDonald, 1978; S. A. Jones, H. N. Arst, Jr. & D. W. MacDonald, unpublished data) have this effect, it is probably not associated with the positions of the deletion endpoints but is more probably a consequence of loss of the regulatory region located between *prn*D and *prn*B. Arst & MacDonald (1978) have proposed that a *prn*B *prn*C dicistronic messenger is initiated in this region where *cis*-acting regulatory mutations affecting *prn*B expression also map (Arst & MacDonald, 1975; Arst *et al.*, 1980) but that an overlapping transcript(s) initiated elsewhere accounts for a

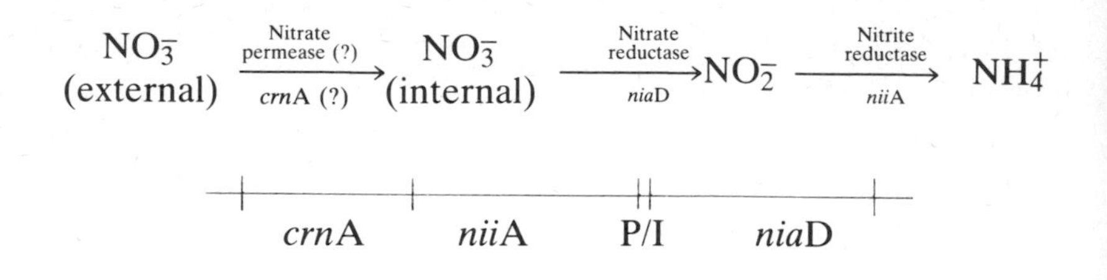

Fig. 2. The nitrate assimilation pathway in *Aspergillus nidulans:* gene roles and gene order within the *crn*A–*nii*A–*nia*D gene cluster. See Cove (1979) for a review of the pathway and Tomsett & Cove (1979) for deletion mapping. *nia*D and *nii*A are the structural genes for nitrate and nitrite reductases, respectively. Recent evidence (A. G. Brownlee & H. N. Arst, Jr., unpublished data) supports the suggestion of D. J. Cove (personal communication) that the *crn*A gene is involved in nitrate uptake, but it cannot specify the sole mechanism of nitrate uptake. P/I denotes the position of a promoter/initiator region for the *nii*A gene as proposed by Arst *et al.* (1979).

proportion of *prn*C expression. This overlapping transcript(s) might be mono-, tri- or tetracistronic. Although the behaviour of these deletions clearly implies that the direction of transcription of *prn*B and *prn*C is outward from the central regulatory region, there are as yet no data concerning the direction of transcription of *prn*A or *prn*D. It is intriguing that a situation possibly analogous to the effect of *prn*D to *prn*B deletions on *prn*C expression occurs in humans. Deletion of the γ- and δ-globin genes, responsible for γ-β-thalassaemia, reduces expression of the adjacent but apparently intact β-globin gene (Van der Ploeg *et al.*, 1980).

A rather unique insight into the regulation of the *crn*A–*nii*A–*nia*D gene cluster is afforded by an insertional translocation which fuses the *nii*A gene to a new promoter normally located in linkage group II and inserts a considerable portion of the right arm of linkage group II between the *nii*A and *nia*D genes in linkage group VIII (Rand & Arst, 1977; Arst *et al.*, 1979). The translocation does not affect expression of *nia*D but does reduce maximal expression of *nii*A. As *nii*A expression in translocation strains has become only partially independent of its normal regulatory responses (to nitrate induction and nitrogen metabolite repression), Arst *et al.* (1979) propose that the normal *nii*A promoter/initiator is still present, tandemly located to the translocated promoter. This interpretation implies that *nii*A is transcribed from the *nia*D-proximal side. The fact that the inserted segment does not abolish *nii*A expression via its normal promoter/initiator demonstrates that *nii*A expression cannot occur solely by means of a di- or polycistronic transcript initiated on the *nii*A-distal side of *nia*D. One species of *nii*A transcript must be initiated between *nii*A and *nia*D. The reduction in maximal *nii*A expression resulting from the translocation can be interpreted in several ways (see Arst *et al.*, 1979). Amongst these is the possibility that it prevents *nii*A expression via an overlapping di- or polycistronic transcript which includes *nia*D.

GENETICS AND ENZYMOLOGY

Although it is not directly related to a consideration of regulation of gene expression, I should like in closing to call attention to the elegant work of Scazzocchio & Sealy-Lewis (1978) in which study of the kinetics and substrate specifities of mutant forms of purine hydroxylase I of *A. nidulans* enabled them to propose a model for the geometry of the active site.

I am grateful to Chris Bailey and Claudio Scazzocchio for critically reading the manuscript and to Chris Bailey for permission to quote his unpublished data extensively. Unpublished work from this laboratory was supported by the Science Research Council through grants to Professor J. M. Thoday and myself and by the Royal Society to whom I am grateful to have been the Smithson Research Fellow.

REFERENCES

Adhya, S. & Gottesman, M. (1978). Control of transcription termination. *Annual Review of Biochemistry*, **47**, 967–96.

Adhya, S. & Miller, W. (1979). Modulation of the two promoters of the galactose operon of *Escherichia coli*. *Nature, London*, **279**, 492–4.

Albrechtsen, H., Hammer-Jespersen, K., Munch-Petersen, A. & Fiil, N. (1976). Multiple regulation of nucleoside catabolizing enzymes: effects of a polar *dra* mutation on the *deo* enzymes. *Molecular and General Genetics*, **146**, 139–45.

Alderman, E. M., Dills, S. S., Melton, T. & Dobrogosz, W. (1979). Cyclic adenosine 3′,5′-monophosphate regulation of bacteriophage T6/colicin K receptor in *Escherichia coli*. *Journal of Bacteriology*, **140**, 369–76.

Alper, M. D. & Ames, B. A. (1978). Transport of antibiotics and metabolic analogs by systems under cyclic AMP control: positive selection of *Salmonella typhimurium cya* and *crp* mutants. *Journal of Bacteriology*, **133**, 149–57.

Alphen, W. van, Selm, N. van & Lugtenberg, B. (1978). Pores in the outer membrane of *Escherichia coli* K12. Involvement of proteins b and e in the functioning of pores for nucleotides. *Molecular and General Genetics*, **159**, 75–83.

Anderson, J. J., Wilson, J. M. & Oxender, D. L. (1979). Defective transport and other phenotypes of a periplasmic 'leaky' mutant of *Escherichia coli* K-12. *Journal of Bacteriology*, **140**, 351–8.

Arst, H. N., Jr. (1976). Integrator gene in *Aspergillus nidulans*. *Nature, London*, **262**, 231–4.

Arst, H. N., Jr. (1978). GABA transaminase provides an alternative route of β-alanine synthesis in *Aspergillus nidulans*. *Molecular and General Genetics*, **163**, 23–7.

Arst, H. N., Jr. & Bailey, C. R. (1977). The regulation of carbon metabolism in *Aspergillus nidulans*. In *Genetics and Physiology of Aspergillus*, ed. J. E. Smith & J. A. Pateman, pp. 131–46. London: Academic Press.

Arst, H. N., Jr. & Cove, D. J. (1970). Molybdate metabolism in *Aspergillus nidulans*. II. Mutations affecting phosphatase activity or galactose utilization. *Molecular and General Genetics*, **108**, 146–53.

Arst, H. N., Jr. & Cove, D. J. (1973). Nitrogen metabolite repression in *Aspergillus nidulans*. *Molecular and General Genetics*, **126**, 111–41.

Arst, H. N., Jr. & MacDonald, D. W. (1975). A gene cluster in *Aspergillus nidulans* with an internally located *cis*-acting regulatory region. *Nature, London*, **254**, 26–31.

Arst, H. N., Jr. & MacDonald, D. W. (1978). Reduced expression of a distal gene of the *prn* gene cluster in deletion mutants of *Aspergillus nidulans*: genetic evidence for a dicistronic messenger in an eukaryote. *Molecular and General Genetics*, **163**, 17–22.

Arst, H. N., Jr., MacDonald, D. W. & Cove, D. J. (1970). Molybdate metabolism in *Aspergillus nidulans*. I. Mutations affecting nitrate reductase and/or xanthine dehydrogenase. *Molecular and General Genetics*, **108**, 129–45.

Arst, H. N., Jr., MacDonald, D. W. & Jones, S. A. (1977). The *prn* gene cluster in *Aspergillus nidulans*. *Heredity*, **39**, 434.

Arst, H. N., Jr., MacDonald, D. W. & Jones, S. A. (1980). Regulation of proline transport in *Aspergillus nidulans*. *Journal of General Microbiology*, **116**, 285–94.

Arst, H. N., Jr., Penfold, H. A. & Bailey, C. R. (1978). Lactam utilisation in *Aspergillus nidulans:* evidence for a fourth gene under the control of the integrator gene *int*A. *Molecular and General Genetics*, **166**, 321–7.

Arst, H. N., Jr., Rand, K. N. & Bailey, C. R. (1979). Do the tightly linked structural genes for nitrate and nitrite reductases form an operon? Evidence from an insertional translocation which separates them. *Molecular and General Genetics*, **174**, 89–100.

Arst, H. N., Jr. & Scazzocchio, C. (1975). Initiator constitutive mutation with an 'up-promoter' effect in *Aspergillus nidulans*. *Nature, London*, **254**, 31–4.

Bailey, C. R. (1976). Carbon catabolite repression in *Aspergillus nidulans*. Ph.D. thesis. University of Cambridge.

Bailey, C. R. & Arst, H. N., Jr. (1975). Carbon catabolite repression in *Aspergillus nidulans*. *European Journal of Biochemistry*, **51**, 573–7.

Bailey, C. R., Arst, H. N., Jr. & Penfold, H. A. (1980). A third gene affecting GABA transaminase levels in *Aspergillus nidulans*. *Genetical Research,* **36,** 167–80.

Bailey, C. R., Penfold, H. A. & Arst, H. N., Jr. (1979). *Cis*-dominant regulatory mutations affecting the expression of GABA permease in *Aspergillus nidulans*. *Molecular and General Genetics*, **169**, 79–83.

Bal, J., Kowalska, I. E., Maciejko, D. M. & Weglenski, P. (1979). Allele specific and locus non-specific suppressors in *Aspergillus nidulans*. *Journal of General Microbiology*, **115**, 457–70.

Bal, J., Maciejko, D. M., Kajtaniak, E. M. & Gajewski, W. (1978). Supersuppressors in *Aspergillus nidulans*. *Molecular and General Genetics*, **159**, 227–8.

Bassel, J. & Mortimer, R. (1971). Genetic order of the galactose structural genes in *Saccharomyces cerevisiae*. *Journal of Bacteriology*, **108**, 179–83.

BOLSHAKOVA, T. N., GABRIELYAN, T. R., BOURD, G. I. & GERSHANOVITCH, V. N. (1978). Involvement of the *Escherichia coli* phospho*enol*pyruvate phosphotransferase system in regulation of transcription of catabolic genes. *European Journal of Biochemistry*, **89**, 483–90.

BRITTEN, R. J. & DAVIDSON, E. H. (1969). Gene regulation for higher cells: a theory. *Science*, **165**, 349–57.

BRITTEN, R. J. & DAVIDSON, E. H. (1971). Repetitive and non-repetitive DNA sequences and a speculation on the origins of evolutionary novelty. *The Quarterly Review of Biology*, **46**, 111–38.

BUXTON, R. S., ALBRECHTSEN, H. & HAMMER-JESPERSEN, K. (1977). Overlapping transcriptional units in the *deo* operon of *Escherichia coli* K-12. Evidence from phage Mu-1 insertion mutants. *Journal of Molecular Biology*, **114**, 287–300.

CHAI, T.-J. & FOULDS, J. (1979). Isolation and partial characterization of protein E, a major protein found in certain *Escherichia coli* K-12 mutant strains: relationship to other outer membrane proteins. *Journal of Bacteriology*, **139**, 418–23.

CHALEFF, R. S. (1974). The inducible quinate-shikimate catabolic pathway of *Neurospora crassa:* genetic organization. *Journal of General Microbiology*, **81**, 337–55.

CHURCH, G. M. & GILBERT, W. (1980). Yeast mitochondrial intron products required in trans for RNA splicing. In *Mobilization and Reassembly of Genetic Information*, ed. D. R. Joseph, J. Schultz, W. A. Scott & R. Werner, in press. New York: Academic Press.

CHURCH, G. M., SLONIMSKI, P. P. & GILBERT, W. (1979). Pleiotropic mutations within two yeast mitochondrial cytochrome genes block mRNA processing. *Cell*, **18**, 1209–15.

CLUTTERBUCK, A. J. (1974). *Aspergillus nidulans.* In *Handbook of Genetics*, vol. 1, ed. R. C. King, pp. 447–510. New York: Plenum Press.

COLEMAN, W. G., JR. & LEIVE, L. (1979). Two mutations which affect the barrier function of the *Escherichia coli* K-12 outer membrane. *Journal of Bacteriology*, **139**, 899–910.

COOPER, T. G., GORSKI, M. & TUROSCY, V. (1979). A cluster of three genes responsible for allantoin degradation in *Saccharomyces cerevisiae. Genetics*, **92**, 383–96.

COVE, D. J. (1976). Chlorate toxicity in *Aspergillus nidulans.* Studies of mutants altered in nitrate assimilation. *Molecular and General Genetics*, **146**, 147–59.

COVE, D. J. (1977). The genetics of *Aspergillus nidulans.* In *Genetics and Physiology of Aspergillus*, ed. J. E. Smith & J. A. Pateman, pp. 81–95. London: Academic Press.

COVE, D. J. (1979). Genetic studies of nitrate assimilation in *Aspergillus nidulans. Biological Reviews*, **54**, 291–327.

DAVIDSON, E. H. & BRITTEN, R. J. (1973). Organization, transcription, and regulation in the animal genome. *The Quarterly Review of Biology*, **48**, 565–613.

DAVIDSON, E. H. & BRITTEN, R. J. (1979). Regulation of gene expression: possible role of repetitive sequences. *Science*, **204,** 1052–9.

DEKOCK, P. C., CHESHIRE, M. W., MUNDIE, C. M. & INKSON, R. H. E. (1979). The effect of galactose on the growth of *Lemna. The New Phytologist*, **82**, 679–85.

DEMEGLIO, D. C. & FRIEDMAN, T. B. (1978). Galactose metabolism in *Dictyostelium discoideum.* Regulation of galactose-1-phosphate uridyl transferase during growth and development. *Journal of Biochemistry, Tokyo*, **83**, 693–8.

DESCHAMPS, J. & WIAME, J.-M. (1979). Mating-type effect on *cis* mutations leading to constitutivity of ornithine transaminase in diploid cells of *Saccharomyces cerevisiae. Genetics*, **92**, 749–58.

DILAURO, R., TANIGUCHI, T., MUSSO, R. & DE CROMBRUGGHE, B. (1979). Unusual location and function of the operator in *Escherichia coli* galactose operon. *Nature, London*, **279**, 494–500.

DiRienzo, J. M., Nakamura, K. & Inouye, M. (1978). The outer membrane proteins of Gram-negative bacteria: biosynthesis, assembly, and functions. *Annual Review of Biochemistry*, **47**, 481–532.

Douglas, H. C. & Hawthorne, D. C. (1964). Enzymatic expression and genetic linkage of genes controlling galactose utilization in Saccharomyces. *Genetics*, **49**, 837–44.

Dubois, E., Hiernaux, D., Grenson, M. & Wiame, J.-M. (1978). Specific induction of catabolism and its relation to repression of biosynthesis in arginine metabolism of *Saccharomyces cerevisiae. Journal of Molecular Biology*, **122**, 383–406.

Dujon, B. (1979). Mutants in a mosaic gene reveal functions for introns. *Nature, London*, **282**, 777–8.

Elorza, M. V. & Arst, H. N., Jr. (1971). Sorbose resistant mutants of *Aspergillus nidulans. Molecular and General Genetics*, **111**, 185–93.

Ennis, H. L. (1971). Mutants of *Escherichia coli* sensitive to antibiotics. *Journal of Bacteriology*, **107**, 486–90.

Exinger, F. & Lacroute, F. (1979). Genetic evidence for the creation of a reinitiation site by mutation inside the yeast *ura*2 gene. *Molecular and General Genetics*, **173**, 109–13.

Feucht, B. U. & Saier, M. H., Jr. (1980). Fine control of adenylate cyclase by the phosphoenolpyruvate: sugar phosphotransferase system in *Escherichia coli* and *Salmonella typhimurium. Journal of Bacteriology*, **141**, 603–10.

Fincham, J. R. S., Day, P. R. & Radford, A. (1979). *Fungal Genetics*, 4th edn. Oxford: Blackwell Scientific Publications.

Flawiá, M. M. & Torres, H. N. (1972a). Adenylate cyclase activity in *Neurospora crassa.* I. General properties. *Journal of Biological Chemistry*, **247,** 6873–9.

Flawiá, M. M. & Torres, H. N. (1972b). Adenylate cyclase activity in lubrol-treated membranes from *Neurospora crassa. Biochimica et Biophysica Acta*, **289**, 428–32.

Giles, N. H. (1978). The organization, function, and evolution of gene clusters in eucaryotes. *The American Naturalist*, **112**, 641–57.

Gomes, S. L., Mennucci, L. & Maia, J. C. C. (1978). Adenylate cyclase activity and cyclic AMP metabolism during cytodifferentiation of *Blastocladiella emersonii. Biochimica et Biophysica Acta*, **541**, 190–8.

Gunatilleke, I. A. U. N., Arst, H. N., Jr. & Scazzocchio, C. (1975). Three genes determine the carboxin sensitivity of mitochondrial succinate oxidation in *Aspergillus nidulans. Genetical Research*, **26**, 297–305.

Halbreich, A., Pajot, P., Foucher, M., Grandchamp, C. & Slonimski, P. (1980). A pathway of cytochrome b mRNA processing in yeast mitochondria: specific slicing steps and an intron-derived circular RNA. *Cell*, **19**, 321–9.

Hicks, J., Strathern, J. N. & Klar, A. J. S. (1979). Transposable mating type genes in *Saccharomyces cerevisiae. Nature, London*, **282**, 478–83.

Hopper, J. E. & Rowe, L. B. (1978). Molecular expression and regulation of the galactose pathway genes in *Saccharomyces cerevisiae.* Distinct messenger RNAs specified by the GAL1 and GAL7 genes in the GAL7–GAL10–GAL1 cluster. *Journal of Biological Chemistry*, **253**, 7566–9.

Hynes, M. J. (1975a) A *cis*-dominant regulatory mutation affecting enzyme induction in the eukaryote *Aspergillus nidulans. Nature, London*, **253,** 210–12.

Hynes, M. J. (1975b). Studies on the role of the *are*A gene in the regulation of nitrogen catabolism in *Aspergillus nidulans. Australian Journal of Biological Sciences*, **28**, 301–13.

HYNES, M. J. (1977). Induction of the acetamidase of *Aspergillus nidulans* by acetate metabolism. *Journal of Bacteriology*, **131**, 770–5.

HYNES, M. J. (1978a). Multiple independent control mechanisms affecting the acetamidase of *Aspergillus nidulans. Molecular and General Genetics*, **161**, 59–65.

HYNES, M. J. (1978b). An 'up-promoter' mutation affecting the acetamidase of *Aspergillus nidulans. Molecular and General Genetics*, **166**, 31–6.

HYNES, M. J. (1979). Fine-structure mapping of the acetamidase structural gene and its controlling region in *Aspergillus nidulans. Genetics*, **91**, 381–92.

HYNES, M. J. & KELLY, J. M. (1977). Pleiotropic mutants of *Aspergillus nidulans* altered in carbon metabolism. *Molecular and General Genetics*, **150,** 193–204.

JACQ, C., LAZOWSKA, J. & SLONIMSKI, P. P. (1980). Sur un nouveau mécanisme de la régulation de l'expression génétique. *Comptes Rendus Hebdomadaires des Séances de l'Académie des Sciences de Paris Série D Sciences Naturelles*, **290**, 89–92.

KEESEY, J. K., BIGELIS, R. & FINK, G. R. (1979). The product of the *his*4 gene cluster in *Saccharomyces cerevisiae.* A trifunctional polypeptide. *Journal of Biological Chemistry*, **254**, 7427–33.

KELLY, J. M. & HYNES, M. J. (1977). Increased and decreased sensitivity to carbon catabolite repression of enzymes of acetate metabolism in mutants of *Aspergillus nidulans. Molecular and General Genetics*, **156**, 87–92.

KINGHORN, J. R. & PATEMAN, J. A. (1977). Nitrogen metabolism. In *Genetics and Physiology of Aspergillus*, ed. J. E. Smith & J. A. Pateman, pp. 147–202. London: Academic Press.

KLAR, A. J. S. & FOGEL, S. (1979). Activation of mating type genes by transposition in *Saccharomyces cerevisiae. Proceedings of the National Academy of Sciences, USA*, **76**, 4539–43.

KRIEGER-BRAUER, H. J. & BRAUN, V. (1980). Functions related to the receptor protein specified by the *tsx* gene of *Escherichia coli. Archives of Microbiology*, **124**, 233–42.

KRUMLAUF, R. & MARZLUF, G. A. (1979). Characterization of the sequence complexity and organization of the *Neurospora crassa* genome. *Biochemistry*, **18**, 3705–13.

KRUMLAUF, R. & MARZLUF, G. A. (1980). Genome organization and characterization of the repetitive and inverted repeat DNA sequences in *Neurospora crassa. Journal of Biological Chemistry*, **255**, 1138–45.

KUMAR, S. (1976). Properties of adenyl cyclase and cyclic adenosine 3′,5′-monophosphate receptor protein-deficient mutants of *Escherichia coli. Journal of Bacteriology*, **125**, 545–55.

KUMAR, S., PRAKASH, N. & AGARWAL, K. N. (1979). Cyclic AMP control of the envelope growth in *Escherichia coli*: envelope morphology of the mutants in *cya* and *crp* genes. *Indian Journal of Experimental Biology*, **17**, 325–7.

KURTZ, M. B. (1979). Changes in fructose transport during growth of *Aspergillus nidulans. Federation Proceedings*, **38**, 240.

KURTZ, M. B. & CHAMPE, S. P. (1979). Genetic control of transport loss during development of *Aspergillus nidulans. Development Biology*, **70**, 82–8.

KUSHNER, P. J., BLAIR, L. C. & HERSKOWITZ, I. (1979). Control of yeast cell types by mobile genes: a test. *Proceedings of the National Academy of Sciences, USA*, **76**, 5264–8.

LAWTHER, R. P., RIEMER, E., CHOJNACKI, B. & COOPER, T. G. (1974). Clustering of the genes for allantoin degradation in *Saccharomyces cerevisiae*. *Journal of Bacteriology*, **119**, 461–8.

LAZARUS, C. M., LÜNSDORF, H., HAHN, U., STĘPIEŃ, P. P. & KÜNTZEL, H. (1980). Physical map of *Aspergillus nidulans* mitochondrial genes coding for ribosomal RNA: an intervening sequence in the large rRNA cistron. *Molecular and General Genetics*, **177**, 389–97.

LEMOINE, Y., DUBOIS, E. & WIAME, J.-M. (1978). The regulation of urea amidolyase of *Saccharomyces cerevisiae*. Mating type influence on a constitutivity mutation acting in cis. *Molecular and General Genetics*, **166**, 251–8.

LEUPOLD, U. (1980). Transposable mating-type genes in yeasts. *Nature, London*, **283**, 811–12.

LEWIS, N. J., HURT, P., SEALY-LEWIS, H. M. & SCAZZOCCHIO, C. (1978). The genetic control of the molybdoflavoproteins in *Aspergillus nidulans*. IV. A comparison between purine hydroxylase I and II. *European Journal of Biochemistry*, **91**, 311–16.

LO, T. C. Y. (1979). The molecular mechanisms of substrate transport in gram-negative bacteria. *Canadian Journal of Biochemistry*, **57**, 289–301.

LONDESBOROUGH, J. C. & NURMINEN, T. (1972). A manganese-dependent adenyl cyclase in baker's yeast, *Saccharomyces cerevisiae*. *Acta Chemica Scandinavica*, **26**, 3396–8.

LOSSON, R. & LACROUTE, F. (1979). Interference of nonsense mutations with eukaryotic messenger RNA stability. *Proceedings of the National Academy of Sciences, USA*, **76**, 5134–7.

LUKASZKIEWICZ, Z. & PASZEWSKI, A. (1976). Hyper-repressible operator-type mutant in sulphate permease gene of *Aspergillus nidulans*. *Nature, London*, **259**, 337–8.

LUMSDEN, J. & COGGINS, J. R. (1977). The subunit structure of the *arom* multienzyme complex of *Neurospora crassa*. A possible pentafunctional polypeptide chain. *Biochemical Journal*, **161**, 599–607.

MCINTOSH, M. A., PICKETT, C. L., CHENAULT, S. S. & EARHART, C. F. (1978). Suppression of iron uptake deficiency in *Escherichia coli* K-12 by loss of two major outer membrane proteins. *Biochemical and Biophysical Research Communications*, **81**, 1106–12.

MAKOFF, A. J., BUXTON, F. P. & RADFORD, A. (1978). A possible model for the structure of the *Neurospora* carbamoyl phosphate synthase-aspartate carbamoyl transferase complex enzyme. *Molecular and General Genetics*, **161**, 297–304.

MALLICK, U. & HERRLICH, P. (1979). Regulation of synthesis of a major outer membrane protein: cyclic AMP represses *Escherichia coli* protein III synthesis. *Proceedings of the National Academy of Sciences, USA*, **76**, 5520–3.

MARETZKI, A. & THOM, M. (1978). Characteristics of a galactose-adapted sugarcane cell line grown in suspension culture. *Plant Physiology*, **61**, 544–8.

MARZLUF, G. A. (1970a). Genetic and biochemical studies of distinct sulfate permease species in differing developmental stages of *Neurospora crassa*. *Archives of Biochemistry and Biophysics*, **138**, 254–63.

MARZLUF, G. A. (1970b). Genetic and metabolic controls for sulfate metabolism in *Neurospora crassa*: isolation and study of chromate-resistant and sulfate transport-negative mutants. *Journal of Bacteriology*, **102**, 716–21.

METZENBERG, R. L. (1979) Implications of some genetic control mechanisms in *Neurospora*. *Microbiological Reviews*, **43**, 361–83.

MINET, M., JAUNIAUX, J.-C., THURIAUX, P., GRENSON, M. & WIAME, J.-M. (1979). Organization and expression of a two-gene cluster in the arginine biosynthesis of *Saccharomyces cerevisiae*. *Molecular and General Genetics*, **168**, 299–308.

MUSSO, R. E., DILAURO, R., ADHYA, S. & DE CROMBRUGGHE, B. (1977). Dual control for transcription of the galactose operon by cyclic AMP and its receptor protein at two interspersed promoters. *Cell*, **12**, 847–54.
NAKAMURA, H. & SUGANUMA, A. (1972). Membrane mutation associated with sensitivity to acriflavine in *Escherichia coli. Journal of Bacteriology,* **110,** 329–35.
NASMYTH, K. A. & TATCHELL, T. (1980). The structure of transposable yeast mating type loci. *Cell*, **19**, 753–64.
NIKAIDO, H. (1979). Permeability of the outer membrane of bacteria. *Angewandte Chemie International Edition in English*, **18**, 337–50.
NORMARK, S. & WESTLING, B. (1971). Nature of the penetration barrier in *Escherichia coli* K-12: Effect of macromolecular inhibition on penetrability in strains containing the *env*A gene. *Journal of Bacteriology*, **108**, 45–50.
OTSUJI, N., HIGASHI, T. & KAWAMATA, J. (1972). Genetic and physiological analysis of mitomycin C-sensitive mutants of *Escherichia coli* K12. *Biken Journal*, **15**, 49–59.
PASTAN, I. & ADHYA, S. (1976). Cyclic adenosine 5′-monophosphate in *Escherichia coli. Bacteriological Reviews*, **40**, 527–51.
PASZEWSKI, A. & GRABSKI, J. (1975). Enzymatic lesions in methionine mutants of *Aspergillus nidulans:* role and regulation of an alternative pathway for cysteine and methionine synthesis. *Journal of Bacteriology*, **124**, 893–904.
PENFOLD, H. A. (1979). Omega-amino acid catabolism in *Aspergillus nidulans.* Ph.D. thesis. University of Cambridge.
PETERKOFSKY, A. & GAZDAR, C. (1979). *Escherichia coli* adenylate cyclase complex: regulation by the proton electrochemical gradient. *Proceedings of the National Academy of Sciences, USA*, **76**, 1099–103.
PETERSEN, N. S. & MCLAUGHLIN, C. S. (1973). Monocistronic messenger RNA in yeast. *Journal of Molecular Biology*, **81**, 33–45.
PHILIPPIDES, D. & SCAZZOCCHIO, C. (1981). Positive regulation in a eukaryote, a study of the *ua*Y gene of *Aspergillus nidulans:* II. Identification of the effector binding protein. *Molecular and General Genetics,* in press.
PONTECORVO, G., ROPER, J. A., HEMMONS, L. A., MACDONALD, K. D. & BUFTON, A. W. J. (1953). The genetics of *Aspergillus nidulans. Advances in Genetics*, **5**, 141–238.
POSTMA, P. W., CORDARO, J. C. & ROSEMAN, S. (1977). Sugar transport. A pleiotropic membrane mutant of *Salmonella typhimurium. Journal of Biological Chemistry*, **252**, 7862–76.
POSTMA, P. W. & ROSEMAN, S. (1976). The bacterial phospho*enol*pyruvate:sugar phosphotransferase system. *Biochimica et Biophysica Acta*, **457**, 213–57.
RAND, K. N. (1978). Aspects of the control of nitrogen metabolism in *Aspergillus nidulans*. Ph.D. thesis. University of Cambridge.
RAND, K. N. & ARST, H. N. (1977). A mutation in *Aspergillus nidulans* which affects the regulation of nitrite reductase and is tightly linked to its structural gene. *Molecular and General Genetics*, **155**, 67–75.
RAND, K. N. & ARST, H. N., JR. (1978). Mutations in *nir*A gene of *Aspergillus nidulans* and nitrogen metabolism. *Nature, London*, **272**, 732–4.
RANK, G. H., GERLACH, J. H. & ROBERTSON, A. J. (1976). Some physiological alterations associated with pleiotropic cross resistance and collateral sensitivity in *Saccharomyces cerevisiae. Molecular and General Genetics*, **144**, 281–8.
RANK, G. H., ROBERTSON, A. J. & PHILLIPS, K. (1975a). Reduced plasma membrane permeability in a multiple cross-resistant strain of *Saccharomyces cerevisiae. Journal of Bacteriology*, **122**, 359–66.

RANK, G. H., ROBERTSON, A. J. & PHILLIPS, K. L. (1975b). Modification and inheritance of pleiotropic cross resistance and collateral sensitivity in *Saccharomyces cerevisiae. Genetics*, **80**, 483–93.
ROBERTS, C. F. (1970). Enzyme lesions in galactose non-utilising mutants of *Aspergillus nidulans. Biochimica et Biophysica Acta*, **201**, 267–83.
ROBERTS, K. R. & MARZLUF, G. A. (1971). The specific interaction of chromate with the dual sulfate permease systems of *Neurospora crassa. Archives of Biochemistry and Biophysics*, **142**, 651–9.
ROBERTS, T., MARTINELLI, S. & SCAZZOCCHIO, C. (1979). Allele specific, gene unspecific suppressors in *Aspergillus nidulans. Molecular and General Genetics*, **177**, 57–64.
ROPER, J. A. & KÄFER, E. (1957). Acriflavine-resistant mutants of *Aspergillus nidulans. Journal of General Microbiology*, **16**, 660–7.
ROSENBERG, M. & PATERSON, B. M. (1979). Efficient cap-dependent translation of polycistronic prokaryotic mRNAs is restricted to the first gene in the operon. *Nature, London*, **279**, 696–701.
ROTHSTEIN, R. J. & SHERMAN, F. (1980). Dependence on mating type for the overproduction of iso-2-cytochrome *c* in the yeast mutant *CYC7-H2. Genetics,* **94,** 891–8.
SAIER, M. H., JR. (1977). Bacterial phospho*enol*pyruvate:sugar phosphotransferase systems: structural, functional, and evolutionary interrelationships. *Bacteriological Reviews*, **41**, 856–71.
SCAZZOCCHIO, C. & ARST, H. N., JR. (1978). The nature of an initiator constitutive mutation in *Aspergillus nidulans. Nature, London*, **274**, 177–9.
SCAZZOCCHIO, C. & GORTON, D. (1977). The regulation of purine breakdown. In *Genetics and Physiology of Aspergillus*, ed. J. E. Smith & J. A. Pateman, pp. 255–65. London: Academic Press.
SCAZZOCCHIO, C., SDRIN, N. & ONG, G. (1981). Positive regulation in a eukaryote, a study of the *ua*Y gene of *Aspergillus nidulans*: I. Characterisation of alleles, dominance and complementation studies, and a preliminary fine structure map of the *ua*Y-*oxp*A cluster, submitted for publication.
SCAZZOCCHIO, C. & SEALY-LEWIS, H. M. (1978). A mutation in the xanthine dehydrogenase (purine hydroxylase I) of *Aspergillus nidulans* resulting in altered specificity. Implications for the geometry of the active site. *European Journal of Biochemistry*, **91**, 99–109.
SCOTT, W. A. (1976). Adenosine 3′:5′-cyclic monophosphate deficiency in *Neurospora crassa. Proceedings of the National Academy of Sciences, USA*, **73**, 2995–9.
SETO, H., NAGATA, Y. & MARUO, B. (1975a). Two types of glucose effects on beta-galactosidase synthesis in a membrane fraction of *Escherichia coli:* correlation with repression observed in intact cells. *Journal of Bacteriology*, **122**, 660–8.
SETO, H., NAGATA, Y. & MARUO, B. (1975b). Effect of glucose and its analogues on the accumulation and release of cyclic adenosine 3′5′-monophosphate in a membrane fraction of *Escherichia coli*: relation to beta-galactosidase synthesis. *Journal of Bacteriology*, **122**, 699–75.
SHERMAN, F. & STEWART, J. W. (1975). The use of iso-1-cytochrome *c* mutants of yeast for elucidating the nucleotide sequences that govern initiation of translocation. In *Organization and Expression of the Eukaryotic Genome; Biochemical Mechanisms of Differentiation in Prokaryotes and Eukaryotes,* ed. G. Bernardi & F. Gros, pp. 175–91. *Proceedings of the Tenth FEBS Meeting,* vol. 38. Amsterdam: North Holland, Press.
SLONIMSKI, P. (1980). Éléments hypothétiques de l'expression des génes morcelés: protéines messagères de la membrane nucléaire. *Comptes Rendus Hebdomadaires des Séances de l'Académie des Sciences de Paris Série D Sciences Naturelles*, **290**, 331–4.

SMITH, J. E. & PATEMAN, J. A. (EDS.) (1977). *Genetics and Physiology of Aspergillus*. London: Academic Press.
STRATHERN, J. N., BLAIR, L. C. & HERSKOWITZ, I. (1979). Healing of *mat* mutations and control of mating type interconversion by the mating type locus in *Saccharomyces cerevisiae. Proceedings of the National Academy of Sciences, USA*, **76**, 3425–9.
STRATHERN, J. N., NEWLON, C. S., HERSKOWITZ, I. & HICKS, J. B. (1979). Isolation of a circular derivative of yeast chromosome III: implication for the mechanism of mating type interconversion. *Cell*, **18,** 309–19.
SY, J. & RICHTER, D. (1972). Content of cyclic 3′,5′-adenosine monophosphate and adenylyl cyclase in yeast at various growth conditions. *Biochemistry*, **11**, 2788–91.
TANIGUCHI, T., O'NEILL, M. & DE CROMBRUGGHE, B. (1979). Interaction site of *Escherichia coli* cyclic AMP receptor protein on DNA of galactose operon promoters. *Proceedings of the National Academy of Sciences, USA*, **76**, 5090–4.
TIMBERLAKE, W. E. (1978). Low repetitive DNA content in *Aspergillus nidulans. Science*, **202**, 973–5.
TOMSETT, A. B. & COVE, D. J. (1979). Deletion mapping of the *nii*A *nia*D gene region of *Aspergillus nidulans. Genetical Research*, **34**, 19–32.
TREVILLYAN, J. M. & PALL, M. L. (1979). Control of cyclic adenosine 3′,5′-monophosphate levels by depolarizing agents in fungi. *Journal of Bacteriology*, **138**, 397–403.
VALENTIN-HANSEN, P., HAMMER-JESPERSEN, K. & BUXTON, R. S. (1979). Evidence for the existence of three promoters for the *deo* operon of *Escherichia coli* K12 *in vitro. Journal of Molecular Biology*, **133**, 1–17.
VALENTIN-HANSEN, P., SVENNINGSEN, B. A., MUNCH-PETERSEN, A. & HAMMER-JESPERSEN, K. (1978). Regulation of the *deo* operon in *Escherichia coli*. The double negative control of the *deo* operon by the *cyt*R and *deo*R repressors in a DNA directed *in vitro* system. *Molecular and General Genetics*, **159**, 191–202.
VAN DER PLOEG, L. H. T., KONINGS, A., OORT, M., ROOS, D., BERNINI, L. & FLAVELL, R. A. (1980). γ-β-thalassaemia studies showing that deletion of the γ- and δ-genes influences β-globin gene expression in man. *Nature, London*, **283**, 637–42.
WATSON, J. D. (1976). *Molecular Biology of the Gene*, 3rd edn. Menlo Park, California: W. A. Benjamin.
WHEELER, G. E., SCHIBECI, A., EPAND, R. M., RATTRAY, J. B. M. & KIDBY, D. K. (1974). Subcellular localization and some properties of the adenylate cyclase activity of the yeast, *Saccharomyces cervisiae. Biochimica et Biophysica Acta,* **372**, 15–22.
WHITNEY, P. A. & COOPER, T. G. (1972). Urea carboxylase and allophanate hydrolase. Two components of adenosine triphosphate:urea amido-lyase in *Saccharomyces cerevisiae. Journal of Biological Chemistry*, **247**, 1349–53.
WOLD, W. S. M. & SUZUKI, I. (1974). Demonstration in *Aspergillus niger* of adenyl cyclase, a cyclic adenosine 3′,5′-monophosphate-binding protein, and studies on intracellular and extracellular phosphodiesterases. *Canadian Journal of Microbiology*, **20**, 1567–76.

Note added in proof. Additional evidence that an *E. coli* outer membrane protein (the λ receptor) can facilitate the diffusion of some compounds whilst excluding others has recently appeared (T. Nakae & J. Ishii, *Journal of Bacteriology,* **142,** 735–40 (1980)). A *cis*-dominant regulatory mutation in the *arg*B *arg*C gene cluster of *S. cerevisiae* has been described (P. Jacobs, J.-C. Jauniaux & M. Grenson, *Journal of Molecular Biology,* **139,** 691–704 (1980)). Hynes (*Journal of Bacteriology,* **142,** 400–406 (1980)) has isolated a mutation apparently in the *int*A receptor site for the acetamidase structural gene which abolishes induction of acetamidase by ω-amino acids. Evidence that the *dur* loci of *S. cerevisiae* are part of a single 'cluster-gene' has now been published (T. G. Cooper, C. Lam & V. Turoscy, *Genetics,* **94,** 555–80 (1980)). An elegant study using mutations to alter the position of the translation initiation codon for iso-l-cytochrome *c* of *S. cerevisiae* (F. Sherman, J. W. Stewart & A. M. Schweingruber, *Cell,* **20,** 215–22 (1980)) will also interest readers.

GENETIC CONTROL OF NITROGEN FIXATION

RAY DIXON, CHRISTINA KENNEDY AND MIKE MERRICK

ARC Unit of Nitrogen Fixation, University of Sussex, Brighton BN1 9RQ, UK

INTRODUCTION

The ability to fix atmospheric nitrogen is a property which is widely distributed among a number of phylogenetically unrelated prokaryotes (Postgate, 1974). The enzyme complex termed nitrogenase, which catalyses nitrogen reduction, shows a remarkable structural uniformity among these organisms, even though they are found in diverse habitats and fix nitrogen under different physiological conditions. In addition, recent DNA–DNA hybridisation experiments have shown that structural gene sequences for nitrogenase proteins are conserved in both bacteria and blue-green algae (Nuti *et al.*, 1979; Mazur, Rice & Haselkorn, 1980; Ruvkun & Ausubel, 1980). However, other genes required for the process of nitrogen fixation, for example those determining proteins that are specifically involved in electron transport to nitrogenase, are most likely to have adapted to cope with the general metabolic requirements of a given organism.

Several characteristics of the nitrogen fixation process must be borne in mind when considering genetic regulation of this system. Firstly, nitrogen fixation requires a large input of energy; estimates of the energy requirement *in vivo* vary from 4–29 moles of ATP per mole of nitrogen reduced, according to the organism and the growth conditions, and studies *in vitro* with nitrogenase suggest a minimal ATP/N_2 ratio of 12–15 (for review see Andersen, Shanmugam & Valentine, 1977). Secondly, both the nitrogenase components are redox proteins which are irreversibly damaged by exposure to oxygen (see Mortenson & Thorneley, 1979). As a consequence of these facts, it is not surprising that nitrogenase synthesis is tightly controlled. Synthesis of a nitrogen-fixing system is only essential to an organism under conditions of nitrogen limitation and therefore, as expected, synthesis of nitrogenase is repressed by ammonia in all free-living nitrogen-fixing organisms so far examined. Likewise, in order to avoid synthesis of the enzyme in conditions which would lead to its inactivation, nitrogenase synthesis is also subject to regulation by oxygen. This chapter will be primarily concerned with genetic regulation of nitrogen fixation (*nif*) in *Klebsiella pneumoniae* and it is hoped that studies on the *nif* system in this organism will serve as a useful basis for work on other nitrogen-fixing organisms which are less amenable to genetic analysis.

NIF GENES IN *KLEBSIELLA PNEUMONIAE*

All known genes specific for nitrogenase synthesis and expression of its activity are clustered close to the histidine operon in *K. pneumoniae*. The order of genes in this region is *gnd rfb his nif shiA* (MacNeil, Supiano & Brill, 1979). The close-linkage of the *nif* and *his* genes has been utilised extensively for strain construction and for the derivation of conjugative plasmids which carry the *his-nif* region of the *K. pneumoniae* genome (Cannon, Dixon & Postgate, 1976; Dixon, Cannon & Kondorosi, 1976). The plasmid pRD1, which has the antibiotic resistance genes and the transfer regions derived from the P-incompatibility group plasmid RP4 and carries chromosomal genes *gnd rfb his nif* and *shiA*, has been particularly useful for complementation analysis of *nif* mutations. A similar plasmid, pMF100, which lacks drug resistance genes (Filser, 1980), has also been used for genetic analysis and transposon-induced mutagenesis of the *nif* genes.

Mutations in the *nif* gene cluster have been generated with a variety of chemical mutagens or by insertion of various translocatable genetic elements into the *nif* region. These mutations have been allocated to 14 cistrons on the basis of complementation tests; *nifQ nifB nifA nifF nifM nifV nifS nifU nifN nifE nifK nifD nifH* and *nifJ* (MacNeil, *et al.*, 1978; Merrick *et al.*, 1978, 1980). Although most mutations in the *nifA* region form a single complementation group, the phenotypic properties of mutations within this group suggest that this region is comprised of two genes, *nifA* and *nifL* (Kennedy, 1977; MacNeil *et al.*, 1978). At least 15 *nif* genes can therefore be recognised by genetic analysis.

Genetic and physical mapping

The order of mutations within the *nif* region was initially determined by three-factor transductional crosses with phage P1 (St John

nif

his Q B A L F M V S U X N E Y K D H J *shi*A

Fig. 1. Map of the *nif* gene cluster in *Klebsiella pneumoniae* as determined by genetic and physical mapping techniques (Merrick *et al.*, 1980; Reidel, Ausubel & Cannon, 1979). The *nifX* and *nifY* genes have been recently identified on the basis of 'protein mapping' of the *nif* region by W. Klipp and A. Pühler (unpublished). The lines beneath the map indicate transcriptional units and the arrows show the direction of transcription (where known).

et al., 1975; Kennedy, 1977). More recently, it has been possible to exploit the properties of translocatable genetic elements to isolate a large number of overlapping deletions with end-points in all the known *nif* cistrons both in the plasmid pRD1 and on the *K. pneumoniae* chromosome (Bachhuber, Brill & Howe, 1976; Elmerich *et al.*, 1978; MacNeil, Brill & Howe, 1978; Merrick *et al.*, 1978). The availability of these deletions has facilitated fine-structure mapping of *nif* mutations (MacNeil *et al.*, 1978; Merrick *et al.*, 1980) and the order of *nif* genes has been unambiguously determined as *his . . . nifQBALFMVSUNEKDHJ* (Fig. 1).

The cloning of *nif* DNA fragments derived by restriction digestion of plasmid pRD1 on small amplifiable plasmids (Cannon, Reidel & Ausubel, 1977, 1979; Pühler, Burkhardt & Klipp, 1979) has allowed the assignment of *nif* genes to individual restriction fragments and the establishment of a physical map of the *nif* gene cluster. *Nif* insertion mutations induced by bacteriophage Mu or by various transposons (Merrick *et al.*, 1980) have also been assigned physical locations on particular restriction fragments by determining alterations in the size of the fragment due to insertion of the element (Reidel, Ausubel & Cannon, 1979), providing a complete correlation between the genetic and physical maps.

Identification of proteins synthesised by cloned *nif* fragments has demonstrated the existence of additional *nif* genes which have not been detected by genetic analysis. *Nif* sequences have been cloned into vectors which allow expression of the inserted DNA from a strong constitutive promoter, thus facilitating detection of *nif*-specific polypeptides in *E. coli* minicells. 'Saturation mutagenesis' of the cloned fragment by random insertion of transposon Tn5 *in vivo* has allowed an unambiguous allocation of nitrogen fixation proteins to particular *nif* genes and a 'protein map' of the *nif* region has been established (W. Klipp & A. Pühler, unpublished). Two additional *nif* genes have been identified using these techniques (Fig. 1); *nifX*, located between *nifU* and *nifN*, and *nifY*, located between *nifE* and *nifK*. *NifY* encodes a 19 000 dalton polypeptide which was initially detected in strains carrying a cloned *Eco*R1 restriction fragment extending from *nifE* through *nifH* on plasmid pSA30 by Cannon, Reidel & Ausubel (1979). The *nif* region is 23 kb in length and the physical size of the 17 *nif* genes and the molecular weights of their products suggest that the map is now completely saturated.

Transcriptional organisation

The operon structure of the 17 *nif* genes has been determined by analysis of polar mutations induced by translocatable genetic elements. The transcriptional polarity of a large number of such mutations has been examined and the data at present indicate that the genes are organised into seven or eight transcriptional units of which five or six are polycistronic and the remainder are monocistronic (Elmerich *et al.*, 1978; MacNeil *et al.*, 1978; Merrick *et al.*, 1980; W. Klipp & A. Pühler, unpublished results). All polycistronic operons are transcribed in the same direction as the histidine operon, as shown in Fig. 1. Although *nifM* was initially thought to be in the same operon as *nifUSV* (MacNeil *et al.*, 1978) present evidence suggests that *nifM* is probably on a separate transcript (Merrick *et al.*, 1980; P. McLean, unpublished results). It is not clear whether *nifF* and *nifM* are in the same transcriptional unit. The majority of insertion mutations in *nifM* are not polar on *nifF* in genetic complementation tests (Merrick *et al.*, 1980) and the *nifF* product can be detected in extracts of *nifM*::Mu insertion strains on two-dimensional polyacrylamide gels (Roberts, MacNeil, MacNeil & Brill, 1978; Houmard, Bogusz, Bigault & Elmerich, 1980). However some Tn5 insertions in *nifM* show a polar effect on *nifF* in complementation tests (Filser, 1980) and the *nifF* gene product is absent in strains carrying a *nifM*::Tn5 insertion on a cloned DNA fragment containing *nifF* and *nifM*, where transcription of the inserted DNA is determined by a promoter on the plasmid vector (W. Klipp & A. Pühler, unpublished results). These ambiguities could possibly result from the presence of a strong secondary promoter located between *nifF* and *nifM*.

The nif*LA operon*

Strains carrying mutations in *nifA* and *nifL* can be distinguished from other Nif$^-$ strains since they give a white colour on plates containing 6-cyanopurine, in contrast to Nif^+ colonies and strains with mutations in other cistrons which give a purple colour on this medium in nitrogen-fixing conditions (MacNeil & Brill, 1978). However, the significance of this colour reaction is unknown at present. All Nif^- mutations mapping in the *nifLA* region form a single complementation group, which suggested initially that this region comprised a single gene, *nifA*. However, this region was sub-divided into two genes, *nifA* and *nifL*, by MacNeil *et al.* (1978) on the basis that almost all mutations which mapped in the *his*-distal portion of this region were either amber-suppressible point mutations or Mu insertions and hence likely to be polar on *nifA*. Moreover, Mu insertions in *nifL*, unlike those in any other *nif* gene, were found to revert to a Nif^+ phenotype at frequencies between 10^{-6} and 10^{-8}. Since Mu insertions do not normally revert, except when suppressed by mutation in another gene, MacNeil *et al.* (1978) suggested that such revertants arose either by a polarity-relief mutation or by creation of a reinitiation site for RNA polymerase between *nifA* and the *nifL* mutation. These data suggest that *nifL* may not be essential for growth on dinitrogen. Identification of the proteins encoded by this region confirms that *nifA* and *nifL* determine separate polypeptides and that Tn5 insertions in *nifL* are polar with respect to the synthesis of *nifA* product (W. Klipp & A. Pühler, unpublished results).

Biochemical phenotypes of nif *mutations*

Nitrogenase is an enzyme complex consisting of two component proteins; Kp1, a molybdenum and iron-containing protein comprising two types of sub-unit (α and β) and Kp2, a dimeric iron-sulphur protein with identical sub-units (for review see Eady & Smith, 1979; Mortenson & Thorneley, 1979). Some of the iron and sulphur atoms and all the molybdenum atoms in Kp1 are arranged in a co-factor called FeMoco which is essential for Kp1 activity and may contain the active site of the enzyme (Shah & Brill, 1977).

Examination of cell-free extracts from mutant strains for reactivation with purified Kp1, Kp2 and FeMoco, as well as the identification of *nif*-specific polypeptides on SDS polyacrylamide gels, has given an indication of the functions of most of the *nif* genes (St John *et al.*, 1975; Dixon *et al.*, 1977; Roberts *et al.*, 1978; Houmard *et al.*, 1980). The structural genes for nitrogenase are *nifH*, determining Kp2, and *nifD* and *nifK*, coding for the α and β sub-units of Kp1 respectively. Expression of these genes alone cannot determine active nitrogenase, and there is considerable evidence to suggest that their products are nascent polypeptides which require processing or modification by other *nif*-specific gene products. The *nifM* gene may be involved in the maturation of Kp2 since strains carrying mutations in this gene lack Kp2 activity (Roberts *et al.*, 1978). Strains carrying a *nifV* mutation synthesise Kp1 protein with altered properties with respect to substrate reduction (P. McLean, unpublished), implying that this cistron is involved in processing of Kp1. It is also possible that the *nifS* and *nifU* products are also involved in maturation of this protein (Houmard *et al.*, 1980). At least three genes, *nifB*, *nifN* and *nifE*, are required for the processing or synthesis of FeMoco. Strains carrying mutations in these cistrons lack Kp1 activity which can be restored by the addition of FeMoco to mutant crude extracts (Roberts *et al.*, 1978).

Nitrogenase requires a source of reducing potential for activity both *in vivo* and *in vitro* and two genes, *nifF* and *nifJ*, have been implicated in electron transport to nitrogenase in *K. pneumoniae* (St John *et al.*, 1975; Roberts *et al.*, 1978). *niF* and *nifJ* mutants have low nitrogenase activity *in vivo* but crude extracts have significant levels of activity *in vitro* with sodium dithionite as the reductant. The latter is an artificial electron donor which presumably by-passes the normal electron transport pathway. However both mutant classes lack nitrogenase activity *in vitro* with pyruvate as electron donor and pyruvate-supported nitrogenase activity can be restored to extracts of *nifF* but not *nifJ* mutants by addition of partially purified *Azotobacter chroococcum* flavodoxin, indicating that *nifF* may code for an electron transport protein (Hill & Kavanagh, 1980). The *nifJ* product may not only have a role in electron input to nitrogenase, it may also have a secondary role in maintaining the stability or activity of Kp1 protein since extracts of *nifJ* mutants have only 20% of the wild-type level of active Kpl (Hill & Kavanagh, 1980).

The *nifA* gene product seems to be involved in regulation of the *nif* gene cluster since strains carrying mutations in *nifA* lack all identifiable *nif*-specific polypeptides (Roberts *et al.*, 1978; Houmard *et al.*, 1980). Complementation tests suggest that the *nifA* product can activate *nif* derepression in *trans*, suggesting that *nifA* determines a positive activator for *nif* transcription (Dixon *et al.*, 1977). Little is known about the function of the *nifL* product, although there is evidence to suggest that it is involved in repression of the *nif* gene cluster in response to oxygen (see later). The roles of the remaining genes, *nifQ*, *nifX* and *nifY*, are unknown at present.

DEREPRESSION OF THE *NIF* GENE CLUSTER

Nitrogenase synthesis in *K. pneumoniae* occurs in anaerobic, nitrogen-limiting conditions. The enzyme is synthesised in the absence of nitrogen in an inert gas phase or in a vacuum, suggesting that a specific inducer is not required and that nitrogenase synthesis is a derepression rather than an induction phenomenon (Parejko & Wilson, 1970). Amino acids at low concentration stimulate derepression (Tubb & Postgate, 1973) but at high concentrations are repressive (Shanmugam & Morandi, 1976). It is important to note that nitrogen starvation in bacteria results in a stringent response (for review, see Gallant, 1979) and it is therefore likely that ppGpp may be involved directly or indirectly in regulation of nitrogen fixation, although such an effect has not so far been demonstrated.

Eady *et al.* (1978) observed that nitrogenase polypeptides were synthesised for about 1 hour before enzyme activity could be detected following a shift-down from growth on ammonia to nitrogen-fixing conditions, suggesting either that derepression of the *nif* operons is not co-ordinate or that there is a rate-limiting process in establishing a functional nitrogen-fixing system. Recently Casabadan & Cohen (1979) have developed an elegant technique for fusing any promoter to the *E. coli lac* operon. Gene fusions in which the *E. coli lac* operon is fused to a given *nif* promoter have been used to monitor derepression of individual transcriptional units in the *nif* gene cluster. Derepression of all seven *nif* transcriptional units is co-ordinate and in all cases is initiated at about 1 hour before nitrogenase activity is detectable, although the initial rate of synthesis of β-galactosidase directed by the *nifLA* and *nifBQ* promoters is slower than for the other transcriptional units (Dixon *et al.*, 1980). The slow derepression of the *nifBQ* operon may explain the late appearance of nitrogenase activity and is consistent with the finding that addition of purified FeMoco to crude extracts prepared from cultures in the initial stages of derepression results in a stimulation of nitrogenase activity (R. R. Eady, unpublished).

Requirement for nifA *product*

All *nifA* mutations examined so far have a pleiotropic effect on the synthesis of detectable *nif*-specific polypeptides (Dixon *et al.*, 1977;

Roberts *et al.*, 1978; Houmard *et al.*, 1980), which is consistent with the proposed role of the *nifA* product as a specific positive activator for *nif* transcription. The involvement of *nifA* in derepression has been further investigated using *nif-lac* fusions by constructing double mutants with an insertion mutation in *nifA* and a *lac* fusion in another *nif* transcriptional unit. In all cases tested the *nifA* mutation resulted in a considerable decrease in *lac* expression, indicating that the *nifA* product is required for transcription of other *nif* operons. Furthermore, introduction of plasmids carrying the *nifA*, *nif::lac* double mutations into a $nifA^+$ background restored β-galactosidase activity, indicating that the *nifA* product can act in *trans* (Dixon *et al.*, 1980). However, the *nifA* product is not required for its own synthesis (i.e. for transcription at the *nifLA* promoter) since both *nifA::lac* and *nifL::lac* fusions (which are polar on *nifA*) show similar levels of β-galactosidase activity in both *nif* deletion and $nifA^+$ backgrounds (Dixon *et al.*, 1980). These results suggest that the presence of the $nifA^+$ allele has an insignificant effect on the expression of the *nifA::lac* fusion, indicating that the *nifLA* operon is not autoregulated. In comparison with other systems which are subject to regulation by a positive regulator gene, *nifA* is therefore similar to *malT*, the positive activator of the maltose regulon which is also not autoregulated (Debarbouille & Swartz, 1979), rather than to the regulatory gene of the arabinose operon *araC* which *is* subject to autoregulation (Casabadan, 1976). It is not yet known whether the *nifA* product can act both as a repressor and an activator as has been shown for *araC* (Englesberg, Irr, Power & Lee, 1965; Englesberg, Squires & Meronk, 1969).

Regulation of the nifHDKY *operon*

Studies with *nif-lac* fusions have indicated that the *nif* structural genes, in contrast to the other *nif* transcriptional units, are autogenously regulated since levels of β-galactosidase synthesised in a diploid $nif^+/nifH::lac$ strain are considerably higher than those in a haploid *nifH::lac* fusion strain (Dixon *et al.*, 1980). There is a marked difference in the derepression kinetics of β-galactosidase in these two strains: the $nif^+/nifH::lac$ fusion shows a linear increase in β-galactosidase synthesis from the onset of derepression for at least 7 hours after transfer from ammonium to nitrogen-free medium whereas the *nifH::lac* haploid strain shows a cut-off in β-galactosidase synthesis after 3.5 to 4 hours in nitrogen-free medium, indicating that in the latter case there are insufficient levels of a regulatory effector to maintain transcription from the *nifH* promoter. The *nifH::lac* fusion exerts a strong polar effect on the expression of distal genes so it is possible that any of the gene products of the operon could be involved in autoregulation. Studies with diploid strains carrying plasmids with various mutations in *trans* to a *nifH::lac* fusion on the *K. pneumoniae* chromosome suggest that the *nifH* gene product is not required for maximal expression of the operon and that the *nifK* and *nifD*, or possibly the *nifY* gene products, are involved (Dixon *et al.*, 1980).

A second line of evidence suggests that a molybdo-protein may be involved in regulation of the *nifHDKY* operon. Molybdenum is essential for nitrogenase activity since it is a component of the iron-molybdenum co-factor of Kp1 but there is also some evidence to suggest that molybdenum may have a regulatory role with respect to nitrogenase synthesis. Brill, Steiner & Shah (1974) detected only low levels of antigenic cross-reacting material to Kp1 and Kp2 proteins when *K. pneumoniae* cells were derepressed in a molybdate-deficient medium and suggested that molybdenum may be required for *nif* derepression. On the other hand, Kennedy & Postgate (1977) found normal levels of nitrogenase polypeptides in a molybdenum-deficient *E. coli narD* mutant carrying the *nif*$^+$ plasmid pRD1, suggesting that molybdenum may not be essential for *nif* derepression. Recently this question has been re-examined by following the derepression of *nif-lac* fusion strains in molybdate-depleted medium. The absence of molybdate had no significant effect on the ability of haploid *nif::lac* fusion strains to synthesise β-galactosidase, indicating that molybdate has no regulatory function *per se*. However, in only *one* case, a *nif*$^+$/*nifH::lac* diploid, was derepression of β-galactosidase synthesis affected in the absence of molybdate and this culture showed similar derepression kinetics to the haploid *nifH::lac* fusion strain, exhibiting a cut-off in β-galactosidase synthesis after 3.5 to 4 hours in N-free medium (Dixon *et al.*, 1980). This result suggests that both the presence of molybdate and products of the *nifHDKY* operon are required for maximal expression from the *nifH* promoter and it is therefore possible that a molybdenum-containing protein, presumably Kp1 protein, is actively involved in autogenous regulation of this operon.

THE ROLE OF GLUTAMINE SYNTHETASE IN REGULATION OF THE *NIF* GENE CLUSTER

Klebsiella pneumoniae, like all enteric bacteria so far studied, has two pathways for ammonia assimilation. Under conditions where the internal ammonia pool is high, ammonia is assimilated via glutamate dehydrogenase (GDH) which catalyses the reductive amination of 2-oxoglutarate by ammonia in a reversible reaction utilising NADPH. However, under conditions of ammonia limitation, such as those necessary for derepression of nitrogenase synthesis, GDH is ineffective in nitrogen assimilation and an alternative pathway using the coupled reactions of glutamine synthetase (GS) and glutamate synthase (GOGAT) is operative.

Positive regulation

Glutamine synthetase (purified from *E. coli* W) has been studied in depth biophysically and enzymatically by Stadtman and co-workers (Ginsburg & Stadtman, 1973). The enzyme has a molecular weight of 600 000 and contains 12 identical sub-units. When ammonia is plentiful the GS levels in the cell are low and the enzyme is highly adenylylated and relatively inactive, but when the available ammonia level drops the level of GS increases and the enzyme becomes deadenylylated, with a consequent increase in activity.

The complex cascade mechanism regulating adenylylation/deadenylylation of the enzyme requires the products of three genes, *glnB*, *glnD* and *glnE*, which have been identified in *Klebsiella aerogenes* by Magasanik and co-workers (for review see Tyler, 1978).

In addition to its catalytic activity in the biosynthesis of glutamine, the deadenylylated form of GS has been implicated as a positive activator protein for the synthesis of a number of enzymes concerned with assimilation of ammonia from poor nitrogen sources. This regulatory function of GS was originally studied in relation to control of GS itself and of the histidine utilisation (*hut*) genes in *K. aerogenes* but has since been extended to other systems including enzymes for proline utilisation, arginine utilisation, urease, asparaginase, tryptophan transaminase and nitrogenase (Magasanik, 1977). The diagnostic test for positive regulation of these enzymes by GS required that their activity should parallel that of GS in the following way: (1) it should be high in wild-type cells under nitrogen limitation, where the GS level is high; (2) it should be low, even under nitrogen limitation, in mutants having a defect in the structural gene (*glnA*) for GS which lack the GS protein; (3) it should be high in mutants that produce high levels of GS constitutively even when grown with excess ammonia and a good carbon source.

This correlation could also apply if a product of the reaction catalysed by GS was the regulator, rather than GS itself. However, purified deadenylylated GS was shown to stimulate transcription of the *hut* operon in an in-vitro system (Tyler, Deleo & Magasanik, 1974) where the catalytic activity of the enzyme was irrelevant. Further support for the involvement of the GS protein comes from GlnAC$^-$ mutants such as *glnA51* in *K. aerogenes* which synthesise high levels of enzymatically inactive but antigenically active GS (Streicher, Deleo & Magasanik, 1976). When grown under conditions which repress wild-type GS, i.e. in glucose-ammonia medium, *glnA51* continues to synthesise cross-reacting material (CRM) to GS but this CRM can be repressed in histidine-glutamine medium. Synthesis of CRM in *glnA51* correlates with the level of urease, an enzyme known to require GS for its formation and reinforces the suggestion that *glnA51* synthesises a GS-like protein which although catalytically inactive can still act as an activator protein (Bender & Magasanik, 1977).

A second GlnAC$^-$ mutant in *K. aerogenes* (*glnA10*) also retains regulatory activity in ammonia but in this case no CRM for GS can be detected (Streicher *et al.*, 1976; Bender & Streicher, 1979). It is, however, possible that the product of *glnA10* is very unstable and is not detectable antigenically.

Similar tests to those used in *K. aerogenes* have been applied in *K. pneumoniae* to investigate the involvement of GS in the regulation of nitrogenase. Streicher *et al.* (1974) transduced three independently-isolated mutations causing constitutive synthesis of GS (GlnC) from *K. aerogenes* to *K. pneumoniae*. The resulting strains were also GlnC and two of them continued to synthesise nitrogenase to some extent in the presence of concentrations of ammonia which completely repress nitrogenase synthesis in the wild-type. Glutamine auxotrophs (GlnA$^-$) were also shown to be unable to synthesise nitrogenase and complementation of the *glnA* mutation by an *E. coli* F-prime restored both GS and nitrogenase synthesis. In a similar experiment Tubb (1974) introduced the nitrogen-fixation genes of *K. pneumoniae* into a GlnC strain of *K. aerogenes* and obtained synthesis of nitrogenase in the presence of ammonia.

GlnAC$^-$ mutants of *K. pneumoniae* were derived by Shanmugam, Chan & Morandi (1975) and, as in *K. aerogenes*, these were of two types, some which produced CRM for GS and some which did not. This latter class led the authors to speculate that there may be an additional factor, previously unrecognised, which might itself be the primary regulator of nitrogenase synthesis or which might substitute for GS in its absence. However, as with the *glnA10* mutation in *K. aerogenes*, these mutants may still produce a regulatory protein which has lost the relevant antigenic determinants. Another class of GS regulatory mutants (GlnA$^+$R$^-$) was obtained as Gln$^+$ revertants of *glnA* mutants which were still Nif$^-$Hut$^-$ (Ausubel *et al.*, 1979). In both cases the mutation responsible for the altered regulatory phenotype was tightly linked to *glnA* but it was not possible to say whether the mutations were within the *glnA* gene.

The original proposal that certain operons are regulated by the deadenylylated form of GS has since been complicated by the description of other types of regulatory mutants involved with nitrogen control in *K. aerogenes*, *E. coli* and *S. typhimurium*. Mutations in *glnF*, first discovered in *S. typhimurium* (Garcia, Bancroft, Rhee & Kustu, 1977) and then in *K. pneumoniae* (Streicher *et al.*, 1977), *K. aerogenes* (Gaillardin & Magasanik, 1978) and *E. coli* (Pahel, Zelenetz & Tyler, 1978), result in a glutamine requirement reflecting the inability of the mutant to produce GS at a higher level than the fully repressed level.

The phenotype of *glnF* mutants can be suppressed by mutations closely linked to *glnA* (Garcia *et al.*, 1977; Pahel *et al.*, 1978; Gaillardin & Magasanik, 1978). These suppressor mutations, in a gene now designated *glnG* (*glnR* in *S. typhimurium*), result in production of GS at a level intermediate between fully repressed and derepressed, irrespective of the composition of the medium (Pahel & Tyler, 1979; Kustu *et al.*, 1979). The product of *glnG* appears to act as both a repressor and activator of GS synthesis and the product of *glnF* is necessary for the activation exerted by the *glnG* product. Hence, in the absence of *glnF* product, *glnG* product represses GS.

Mutations in both *glnF* and *glnG* are pleiotropic with respect to utilisation of poor nitrogen sources and in *K. pneumoniae* such mutations result in an inability to synthesise nitrogenase (Streicher *et al.*, 1977; G. Espin & M. Merrick, unpublished). The description of *glnF* and *glnG* considerably complicates the question of regulation by GS, and the possible involvement of the *glnG* product in activation of *nif* and other enzyme systems (Kustu *et al.*, 1979) must now be considered.

Whilst it was originally proposed that deadenylylated GS was the regulatory form of the enzyme, a number of studies have questioned the significance of the adenylylation state for regulation. Kustu and co-workers found that altering the degree of adenylylation of GS by using mutations in *glnD* or *glnE* in *S. typhimurium* did not significantly affect regulation of GS synthesis (Bancroft, Rhee, Neumann & Kustu, 1978). Similarly, in *K. pneumoniae* a study of regulatory GS mutations affecting nitrogenase synthesis showed no correlation between adenylylation state of GS and the extent of nitrogenase derepression (Ausubel *et al.*, 1979). Goldberg & Hanau (1979) measured GS levels and adenylylation state in cultures of *K. pneumoniae* grown aerobically or anaerobically and in conditions either of nitrogen limitations or excess. They also found no clear correlation of adenylylation state with expression of histidase, urease, glutamate dehydrogenase or nitrogenase and hence they proposed that whilst alterations in GS adenylylation state may be a means of rapidly regulating biosynthetic activity of the enzyme they are not a way of altering its regulatory properties.

Ausubel *et al.* (1976, 1977) isolated mutations (designated *nifT*) within the *nif* cluster which allow *nif* expression in a $GlnA^+R^-$ background. One *nifT* mutation was *cis*-dominant, suggesting that it carried a mutation in a regulatory site. Since such mutations occur presumably at a single locus, Ausubel *et al.* (1977) argued that GS acts at one site within *nif* (e.g. the *nifLA* promoter) and that other *nif* operons are positively controlled by a specific *nif* activator protein (e.g. *nifA* product). The demonstration, using *nif-lac* fusions, that the *nifLA* operon (and all other transcriptional units) are not transcribed in a *glnA* mutant strain (Dixon *et al.*, 1980) is in agreement with this suggestion.

The interaction of GS and RNA polymerase with the specific DNA of the *nif* gene cluster has been studied by Janssen, Reidel, Ausubel & Cannon (1980). The binding of GS to DNA was first reported by Streicher & Tyler (1976), and Janssen *et al.* (1980) examined the ability of GS to protect *nif* DNA from cleavage by specific restriction endonucleases. They used DNA from the amplifiable plasmids pCRA10 and pCM1 which carry *hisDG* and *hisDG nifQ-nifK* respectively (Cannon *et al.*, 1979). Glutamine synthetase protected most of the same sites on the DNA as did RNA polymerase, which might be expected if GS affected transcription by binding at or near the binding sites for RNA polymerase. However, the binding of RNA polymerase is sensitive to salt concentration such that RNA polymerase protects more sites as the salt concentration is decreased and hence correlation of binding sites with promoter sites is questionable. Nevertheless binding of GS was apparently specific for *nif* genes since GS was not seen to protect sites on pCRA10 which carries only *his* genes.

Negative regulation

Nitrogenase synthesis in *K. pneumoniae* is efficiently repressed by ammonia (Tubb & Postgate, 1973) and when the *Klebsiella nif* genes are transferred to *E. coli*, *S. typhimurium* or *Azotobacter vinelandii* they remain subject to repression by ammonia (Dixon *et al.*, 1976; Cannon *et al.*, 1976; Cannon & Postgate, 1976).

Nif-lac fusions to each of the eight *nif* promoters have been used to show that all the *nif* transcriptional units are repressed by ammonia (Dixon *et al.*, 1980). This is in agreement with previous observations that no *nif*-specific mRNAs from pCM1 can be detected in ammonia-grown cultures (Janssen *et al.*, 1980) and that all identifiable *nif*-specific polypeptides are absent in such cultures (Eady *et al.*, 1978; Roberts *et al.*, 1978).

The catalytic activities of glutamine synthetase and glutamate synthase from *Klebsiella* are inhibited by the glutamate analogues methionine sulfoximine and methionine sulfone (Brenchley, 1973; Gordon & Brill, 1974). These analogues can stimulate GS synthesis (possibly as a consequence of inhibition of the autoregulatory properties of GS) and repress synthesis of glutamate dehydrogenase (Brenchley, 1973). The presence of either analogue allows *K. pneumoniae* to continue to synthesise nitrogenase in the presence of excess ammonia (Gordon & Brill, 1974), perhaps as a consequence of an increased GS level induced by the analogue or as suggested by Gordon & Brill (1974) because glutamine itself or other metabolic products of glutamine, in addition to ammonia (or perhaps rather than ammonia), function as the effector for nitrogenase repression. In agreement with this, Shanmugam & Morandi (1976) showed that certain amino acids, especially glutamine in combination with aspartate or glutamate, were able to repress nitrogenase synthesis in mutants of *K. pneumoniae* which are not repressed by ammonia. They concluded that repression of nitrogenase by ammonia occurs via some amino acid product and that the negative regulatory function of GS is simply to provide that product.

NifT mutants which are apparently independent of GS regulation (Ausubel *et al.*, 1976, 1977) are still repressed by ammonia, again implying that repression by ammonia may not be mediated solely (if at all) by a regulatory form of GS. Studies of the amino acid pools of *K. pneumoniae* during repression and derepression (Kleiner, 1976) showed a correlation between the intracellular concentration of glutamine and nitrogenase activity. By contrast the glutamate pool remained fairly stable during repression and derepression and this led Kleiner (1979) to suggest that glutamine may be a co-repressor for the *nif* system.

None of the genes in the *nif* gene cluster has been identified as being a specific *nif* repressor so there is currently no evidence for a repressor/co-repressor system for ammonia control of the *nif* cluster. It is possible that there is *no* specific negative control but that repression is simply lack of positive activation by GS although the repression of *nifT* mutants by ammonia would contradict this. An alternative model for repression mediated by amino acids such as glutamine would be the presence of an attenuator-type control such as that responsible for regulation of the *his*, *phe*, *leu* and *trp* operons in *E. coli* (Keller & Calvo, 1979). Such a mechanism does not require a specific repressor protein but modulates transcription of a particular operon according to the level of one or more particular aminoacyl tRNAs. Such a mechanism may operate in *nif* but sequencing of the *nifLA* and other promoter regions will probably be necessary to determine this.

REGULATION BY OXYGEN

Physiology

Nitrogen fixation in *K. pneumoniae* occurs only under anaerobic conditions or at very low dissolved O_2 tension (DOT). The effects of O_2 are two fold: it inactivates both nitrogenase components and also prevents their synthesis.

Studies of the regulatory action of O_2 on *nif* expression were initially encumbered by the need of an assay for inactive nitrogenase components, by the lack of a means to select O_2 insensitive mutants, and by the difficulty in maintaining precise DOTs in cultures under investigation, a refinement whose importance is just beginning to emerge (see later).

The first suggestion of a regulatory role for O_2 came from experiments by St John, Shah & Brill (1974) who failed to detect cross-reacting material (CRM) to Kp1 or Kp2 in NH_4-grown organisms transferred to N-free medium and shaken in air. However, it was found that Kp1 CRM was degraded so rapidly in cultures exposed to air that synthesis of this component would not have been detected. On the other hand, preformed Kp2 CRM was more stable in air and it was concluded that Kp2 was not synthesised in cells exposed to O_2.

Oxygen repression was firmly established by Eady *et al.* (1978) who showed that in derepressing cultures, synthesis of all three nitrogenase polypeptides was curtailed after bubbling with either air or 5% O_2 in N_2. In these experiments, nitrogenase was detected by pulse-labelling cultures with a mixture of ^{14}C amino acids followed by SDS polyacryalmide gel electrophoresis (PAGE) and gel autoradiography. During derepression, nitrogenase polypeptides comprised 10–15% of total proteins being synthesised, and after exposure to air or 5% O_2 the rates of synthesis decreased by half in 12 or 33 minutes, respectively, and then continued to decline. The decrease occurred far more rapidly than did degradation of preformed nitrogenase in cells exposed to O_2 and thus it was concluded that O_2 repressed synthesis of the three nitrogenase polypeptides, now known to be encoded by the *nifHDKY* operon.

More recently, using the *nif::lac* fusion strains already described, Dixon *et al.* (1980) found synthesis of β-galactosidase to be inhibited in N-free cultures bubbled vigorously with air. Expression from all *nif* promoters was affected, but variably, with a *nifH::lac* strain reduced to 3% of its activity under N_2 while a *nifL::lac* culture showed 43% of the activity obtained with the same organism sparged with N_2. Significantly, in cultures maintained at constant DOT in an oxystat, synthesis of β-galactosidase from the *nifH* promoter was inhibited at DOT 0.5% and higher, while expression from the *nifL* promoter was diminished by 3 to 5% DOT but not by 0.5% DOT. Thus O_2 repression of *nifH* transcription at a DOT of 0.5% is not mediated by repression at the *nifLA* promoter. The *nifH::lac* fusion was repressed by O_2 whether in a *nif*$^+$ or a *nif* deletion background and therefore a product of the *nifHDKY* operon is not apparently required for O_2 repression.

Genetics

Mutants in which *nif* expression is not affected by O_2 cannot be selected for directly on N-free medium. Some success was obtained by screening Nif^+ revertants of various *nif* mutant strains for O_2 repression of nitrogenase by pulse-labelling oxystat grown cultures followed by SDS-PAGE (S. Hill & E. Kavanagh, in preparation). One revertant, obtained from a *nifL* point mutant *nifL2265*, had the desired phenotype. In this strain a second mutation, designated *nif2400*, responsible for the Nif^+ phenotype was found to lie between *his* and *nifF*, since His^+Nif^+ transductants obtained from a cross between a *his-nifF* deletion strain and the revertant were found to escape O_2 repression. The Nif^+ revertant strain responded to NH_4^+ repression, consistent with the idea that O_2 and NH_4^+ repression occur via different mechanisms, as initially proposed by Eady *et al.* (1978) who found that in an NH_4-constitutive mutant, *nif* was repressed by O_2.

As with other aspects of *nif* regulation, studies on *nif2400* and other mutations have been expedited by the use of *nif::lac* fusion strains. In the double chromosomal mutant *nifL2265*, *nifH::lac*, β-galactosidase was not made in N-free medium sparged with either N_2 or O_2, as predicted from the $NifA^-$ phenotype of *nifL2265*. Also as predicted, in a strain carrying *nifL2265* and *nif2400*, *lac* expression from the *nifH* promoter proceeded in cultures maintained at 0.5% DOT or sparged with air. Introduction of the nif^+ plasmid pRD1 to this strain restored O_2 inhibition of β-galactosidase synthesis, suggesting that *nif2400* is a recessive mutation in a gene whose product can be supplied in *trans* to mediate O_2 repression (C. Kennedy & S. Hill, in preparation).

Another feature of *nif::lac* fusion strains is that they have a phenotype that is easily scored on N-free plates containing the β-galactosidase substrate 5-bromo-4-chloro-3-indolyl-β-D-galactoside. In the *nifL2265*, *nif2400*, *nifH::lac* strain, blue colour developed in colony patches incubated in air while the $NifA/L^+$ *nif::lac* strain was colourless. By testing various *nif* deletion mutants to see which could restore the 'colourless' phenotype to the double mutant, it has been possible to tentatively map *nif2400* within the *his*-proximal end of *nifL*.

The *nifT* alleles described previously, as well as a Nif^+ revertant of a *nifL::Mu* insertion mutant, also allow *nifH::lac* expression to proceed in air. However, the map position and dominance relationship with pRD1 have not yet been determined so their analogy with *nif2400* may be limited. Nevertheless it seems likely that the *nifL* gene product is a negative effector that prevents expression of the other *nif* operons in organisms exposed to O_2, possibly by inactivating the *nifA* product or by binding directly to *nif* DNA to prevent transcription (although an effect of O_2 on translation rather than transcription has not been ruled out).

While this discussion has focused on *nif*-specific aspects of O_2 regulation, it cannot be assumed that O_2 interacts directly with, for example, the *nifL* gene product. Several enzymes are known to be synthesised only under anaerobic growth conditions, e.g. fumarate reductase (Spencer & Guest, 1973), nitrate reductase (Fimmel & Haddock, 1979) and formate hydrogen lyase (Pichinoty, 1962). Others, such as certain cytochromes (Harrison, 1976) and histidase (Goldberg & Hanau, 1980), are made only in aerobically-grown organisms. Thus anaerobic *nif* expression could be controlled by a general regulatory system for enzymes needed during anaerobic growth, by interaction of the *nifL* product with oxidised or reduced cell metabolites or perhaps by a combination of both which is reflected in the different effects of high and low DOT on *nifL::lac* expression. An important step to follow could be the isolation of mutants mapping outside *nif* that affect O_2 regulation.

Nitrate regulation

Tubb & Postgate (1973) reported that nitrate-grown cultures of *K. pneumoniae* had no acetylene-reducing activity while Kennedy & Postgate (1977) showed that in *E. coli* carrying the nif^+ plasmid pRD1 nitrogenase proteins were absent from organisms grown in either nitrate or nitrite. Repression by NO_3^- was not observed in *E. coli chl* mutants which lack nitrate reductase activity. However, NO_2^- did prevent nitrogenase synthesis in these mutants, indicating that NO_3^- reduction is necessary for repression to occur.

Recently, in a more thorough study of nitrate regulation, rates of nitrogenase formation measured after SDS-PAGE of pulse-labelled organisms were shown to decline following addition of NO_3^- or NO_2^- to derepressing cultures of *K. pneumoniae*. Nitrogenase synthesis was normal in several Chl^r mutants in the presence of NO_3^- and the repressive effect of NO_3^- was coupled to the appearance of NO_2^-, again demonstrating that NO_3^- metabolism is required for repression of *nif* (Hom, Hennecke & Shanmugam, 1980). While NH_4^+ is produced by NO_2^- reduction and is an obvious candidate for involvement in NO_3^- repression, Hom *et al.* (1980) demonstrated that in several *glnC*-type NH_4^+-constitutive mutants nitrogenase is not synthesised in the presence of nitrate. They suggested that NO_3^- repression occurred via its use as an anaerobic terminal electron acceptor and so may show metabolic and regulatory features in common with O_2 repression. In fact, preliminary experiments bear this out; in the *nifL2265*, *nif2400*, *nifH::lac* fusion strain β-galactosidase activity was less affected by NO_3^- than its *nifL*$^+$ *nifH::lac* isogenic counterpart (H. Hennecke, S. Hill & C. Kennedy, unpublished). Since NO_3^- and NO_2^- concentrations are easier to manipulate than DOT, a practical advantage of a common regulatory pathway is obvious. However, both types of repression need further detailed study in order to finally answer this important question.

CONCLUSIONS

The preceding examples have been concerned almost entirely with the genetic control of nitrogen fixation in *Klebsiella pneumoniae*. However, many physiological aspects of the regulation of nitrogenase apply in other organisms and it seems quite likely that the genetic control mechanisms in these organisms are similar. In a number of cases, there is preliminary evidence for ammonia regulation mediated by GS since glutamine auxotrophs are Nif$^-$ in *Azospirillum lipoferum* (Gauthier & Elmerich, 1977), *Rhizobium* sp. strain 32H1 (Ludwig & Signer, 1977) and *Rhizobium meliloti* (Kondorosi, Svab, Kiss & Dixon, 1977). The possible involvement of GS in *nif* regulation in *Rhizobium* is, however, complicated by the presence of two forms of GS in *Rhizobium* spp. (Darrow & Knotts, 1977) and a rather different pattern of ammonium assimilation to that found in enteric bacteria (Ludwig, 1978).

Oxygen has been shown to repress nitrogenase synthesis in *Azotobacter chroococcum* (Robson, 1979) and also in slow-growing *Rhizobium* strains. In *Rhizobium* sp. strain 32H1, chemostat cultures grown under a high DOT do not contain nitrogenase CRM (Bergersen, Turner, Gibson & Dudman, 1976) while in batch cultures of *R. japonicum* ongoing synthesis of nitrogenase is repressed after the oxygen supply is increased (Scott, Hennecke & Lin, 1979).

It will be of considerable interest to determine whether the widespread genetic homology recently demonstrated for the nitrogenase structural genes (Ruvkun & Ausubel, 1980) can be extended to the regulatory genes, *nifA* and *nifL*, as the demonstration of such homology would be extremely good evidence for common regulatory mechanisms.

We would especially like to thank our colleagues in the ARC Unit of Nitrogen Fixation for their contributions to the recent experimental work described in this review. We are also very grateful to Hauke Hennecke, Werner Klipp and Alf Pühler for permitting us to mention their unpublished results. Finally we wish to thank John Postgate for constructive criticism of the manuscript.

REFERENCES

Andersen, K., Shanmugam, K. T. & Valentine, R. C. (1977). Nitrogen fixation (*nif*) regulatory mutants of *Klebsiella*: Determination of the energy cost of nitrogen fixation *in vivo*. In *Genetic Engineering for Nitrogen Fixation*, ed. A. Hollaender, pp. 95–110. New York: Plenum Press.

Ausubel, F. M., Bird, S. C., Durbin, K. J., Janssen, K. A., Margolskee, R. F. & Peskin, A. P. (1979). Glutamine synthetase mutations which affect the expression of nitrogen-fixing genes in *Klebsiella pneumoniae. Journal of Bacteriology*, **140**, 597–606.

Ausubel, F. M., Margolskee, R. F. & Maizels, N. (1976). Mutants of *Klebsiella pneumoniae* in which expression of nitrogenase is independent of glutamine synthetase control. In *Recent Developments in Nitrogen Fixation*, ed. W. Newton, J. R. Postgate & C. Rodriguez-Barrueco, pp. 347–56. London: Academic Press.

Ausubel, F. M., Reidel, G., Cannon, F., Peskin, A. & Margolskee, R. (1977). Cloning nitrogen fixation genes from *Klebsiella pneumoniae in vitro* and the isolation of *nif* promoter mutants affecting glutamine synthetase regulation. In *Genetic Engineering for Nitrogen Fixation*, ed. A. Hollaender, pp. 111–28. New York: Plenum Press.

Bachhuber, M., Brill, W. J. & Howe, M. (1976). Use of bacteriophage Mu to isolate deletions in the *his-nif* region of *Klebsiella pneumoniae. Journal of Bacteriology*, **128**, 749–53.

Bancroft, S., Rhee, S. G., Neumann, C. & Kustu, S. (1978). Mutations that alter the covalent modification of glutamine synthetase in *Salmonella typhimurium. Journal of Bacteriology*, **134**, 1046–55.

Bender, R. A. & Magasanik, B. (1977). Regulatory mutations in the *Klebsiella aerogenes* structural gene for glutamine synthetase, *Journal of Bacteriology*, **132**, 100–5.

Bender, R. & Streicher, S. (1979). Glutamine synthetase regulation, adenylylation state and strain specificity analysed by polyacrylamide gel electrophoresis. *Journal of Bacteriology*, **137**, 1000–7.

Bergersen, F. J., Turner, G. L., Gibson, A. H. & Dudman, W. F. (1976). Nitrogenase activity and respiration of cultures of *Rhizobium* sp. with special reference to concentration of dissolved oxygen. *Biochimica et Biophysica Acta*, **444**, 164–74.

BRENCHLEY, J. E. (1973). Effect of methionine sulfoximine and methionine sulfone on glutamate synthesis in *Klebsiella aerogenes*. *Journal of Bacteriology*, **114**, 666–73.

BRILL, W. J., STEINER, A. L. & SHAH, V. K. (1974). Effect of molybdenum starvation and tungsten on the synthesis of nitrogenase components in *Klebsiella pneumoniae*. *Journal of Bacteriology*, **118**, 986–9.

CANNON, F. C., DIXON, R. A. & POSTGATE, J. R. (1976). Derivation and properties of F-prime factors carrying nitrogen-fixation genes from *Klebsiella pneumoniae*. *Journal of General Microbiology*, **93**, 111–25.

CANNON, F. C. & POSTGATE, J. R. (1976). Expression of *Klebsiella* nitrogen fixation genes (*nif*) in *Azotobacter*. *Nature*, **260**, 271–2.

CANNON, F. C., REIDEL, G. E. & AUSUBEL, F. M. (1977). A recombinant plasmid which carries part of the nitrogen fixation (*nif*) gene cluster of *Klebsiella pneumoniae*. *Proceedings of the National Academy of Sciences, USA*, **74**, 2963–7.

CANNON, F. C., REIDEL, G. E. & AUSUBEL, F. M. (1979). Overlapping sequences of *Klebsiella pneumoniae nif* DNA cloned and characterised. *Molecular and General Genetics*, **174**, 59–66.

CASADABAN, M. J. (1976). Regulation of the regulatory gene for the arabinose pathway *araC*. *Journal of Molecular Biology*, **104**, 557–66.

CASADABAN, M. J. & COHEN, S. N. (1979). Lactose genes fused to exogenous promoters in one step using a Mu-*lac* bacteriophage: *in vivo* probe for transcriptional control sequences. *Proceedings of the National Academy of Sciences, USA*, **76**, 4530–33.

DARROW, R. A. & KNOTTS, R. R. (1977). Two forms of glutamine synthetase in free-living root-nodule bacteria. *Biochemical and Biophysical Research Communications*, **78**, 554–9.

DEBARBOUILLE, M. & SWARTZ, M. (1979). The use of gene fusions to study the expression of *malT* the positive regulator gene of the maltose regulon. *Journal of Molecular Biology*, **132**, 521–34.

DIXON, R. A., CANNON, F. C. & KONDOROSI, A. (1976). Construction of a P plasmid carrying nitrogen fixation genes from *Klebsiella pneumoniae*. *Nature*, **260**, 268–71.

DIXON, R., EADY, R., ESPIN, G., HILL, S., IACCARINO, M., KAHN, D. & MERRICK, M. (1980). Analysis of regulation of the *Klebsiella pneumoniae* nitrogen fixation (*nif*) gene cluster with gene fusions. *Nature*, **286**, 128–32.

DIXON, R. A., KENNEDY, C., KONDOROSI, A., KRISHNAPILLAI, V. & MERRICK, M. (1977). Complementation analysis of *Klebsiella pneumoniae* mutants defective in nitrogen fixation. *Molecular and General Genetics*, **157**, 189–98.

EADY, R. R., ISSACK, R., KENNEDY, C., POSTGATE, J. R. & RATCLIFFE, H. (1978). Nitrogenase synthesis in *Klebsiella pneumoniae*: comparison of ammonium and oxygen regulation. *Journal of General Microbiology*, **104**, 277–85.

EADY, R. R. & SMITH, B. E. (1979). Physico-chemical properties of nitrogenase and its components. In *A Treatise on Dinitrogen Fixation, Sections I & II. Inorganic and Physical Chemistry and Biochemistry*, ed. R. W. F. Hardy, F. Bottemly & R. C. Burns, pp. 399–490. New York: John Wiley & Sons.

ELMERICH, C., HOUMARD, J., SIBOLD, L., MANHEIMER, I. & CHARPIN, N. (1978). Genetic and biochemical analysis of mutants induced by bacteriophage Mu DNA integration into *Klebsiella pneumoniae* nitrogen fixation genes. *Molecular and General Genetics*, **165**, 181–9.

ENGLESBERG, E., IRR, J., POWER, J. & LEE, N. (1965). Positive control of enzyme synthesis by gene C in the L-arabinose system. *Journal of Bacteriology*, **90**, 946–57.

Englesberg, E., Squires, C. & Meronk, F. (1969). The L-arabinose operon in *Escherichia coli* B/r: a genetic demonstration of two functional states of the product of a regulator gene. *Proceedings of the National Academy of Sciences, USA*, **62**, 1100–7.

Filser, M. M. K. (1980). Genetic analysis of the *Klebsiella pneumoniae* nitrogen fixation gene cluster: plasmid constructed and transposon mutagenesis. *D. Phil. Thesis*, University of Sussex.

Fimmel, A. L. & Haddock, B. A. (1979). Use of *chl* C-*lac* fusions to determine regulation of gene *chlC* in *Escherichia coli* K12. *Journal of Bacteriology*, **138**, 726–30.

Gaillardin, C. M. & Magasanik, B. (1978). Involvement of the product of the *glnF* gene in autogenous regulation of glutamine synthetase formation in *Klebsiella aerogenes. Journal of Bacteriology*, **133**, 1329–38.

Gallant, J. A. (1979). Stringent control in *E. coli. Annual Review of Genetics*, **13**, 393–415.

Garcia, E., Bancroft, S., Rhee, S. G. & Kustu, S. (1977). The product of a newly identified gene, *glnF* is required for synthesis of glutamine synthetase in *Salmonella. Proceedings of the National Academy of Sciences, USA*, **74**, 1662–6.

Gauthier, D. & Elmerich, C. E. (1977). Relationship between glutamine synthetase and nitrogenase in *Spirillum lipoferum. FEMS Microbiology Letters*, **2**, 101–4.

Ginsburg, A. & Stadtman, E. R. (1973). Regulation of glutamine synthetase in *Escherichia coli.* In *The Enzymes of Glutamine Metabolism*, ed. S. Prusiner & E. R. Stadtman, pp. 9–43. New York: Academic Press.

Goldberg, R. B. & Hannau, R. (1979). Relation between the adenylylation state of glutamine synthetase and expression of other genes involved in nitrogen metabolism. *Journal of Bacteriology*, **137**, 1282–9.

Goldberg, R. B. & Hanau, R. (1980). Regulation of the *hut* operons of *Klebsiella pneumoniae* by oxygen. *Journal of Bacteriology*, **141**, 745–50.

Gordon, J. K & Brill, W. J. (1974). Derepression of nitrogenase synthesis in the presence of excess NH_4^+. *Biochemical and Biophysical Research Communications*, **59**, 967–71.

Harrison, D. E. F. (1976). The regulation of respiration rate in growing bacteria. In *Advances in Microbiol Physiology*, vol. 14, ed. A. H. Rose & D. W. Tempest, pp. 243–314. London: Academic Press.

Hill, S. & Kavanagh, E. (1980). Roles of *nifF* and *nifJ* gene products in electron transport to nitrogenase in *Klebsiella pneumoniae. Journal of Bacteriology*, **141**, 470–5.

Hom, S. S. M., Hennecke, H. & Shanmugam, K. T. (1980). Regulation of nitrogenase biosynthesis in *Klebsiella pneumoniae*: effect of nitrate. *Journal of General Microbiology*, **117**, 169–79.

Houmard, J., Bogusz, D., Bigault, R. & Elmerich, C. (1980). Characterisation and kinetics of the biosynthesis of some nitrogen fixation (*nif*) gene products in *Klebsiella pneumoniae. Biochimie* (in press).

Janssen, K. A., Reidel, G. E., Ausubel, F. M. & Cannon, F. C. (1980). Transcriptional studies with cloned nitrogen fixation genes. In *Nitrogen Fixation*, vol. 1, ed. W. E. Newton & W. H. Orme-Johnson, pp. 85–93. Baltimore: University Park Press.

Keller, F. B. & Calvo, J. M. (1979). Alternative secondary structures of leader RNAs and the regulation of the *trp*, *phe*, *his*, *thr* and *leu* operons. *Proceedings of the National Academy of Sciences, USA,* **12,** 6186–90.

Kennedy, C. (1977). Linkage map of the nitrogen fixation (*nif*) genes in *Klebsiella pneumoniae. Molecular and General Genetics*, **157**, 199–204.

KENNEDY, C. & POSTGATE, J. R. (1977). Expression of *Klebsiella pneumoniae* nitrogen fixation genes in nitrate reductase mutants of *Escherichia coli. Journal of General Microbiology*, **98**, 551–7.

KLEINER, D. (1976). Ammonium uptake and metabolism by nitrogen fixing bacteria II. *Archives of Microbiology*, **111**, 85–91.

KLEINER, D. (1979). Regulation of ammonium uptake and metabolism by nitrogen fixing bacteria III. *Clostridium pasteurianum. Archives of Microbiology*, **120**, 263–70.

KONDOROSI, A., SVAB, Z., KISS, G. B. & DIXON, R. A. (1977). Ammonia assimilation and nitrogen fixation in *Rhizobium meliloti. Molecular and General Genetics*, **151**, 221–6.

KUSTU, S., BURTON, D., GARCIA, E., MCCARTER, L. & MCFARLAND, N. (1979). Nitrogen control in *Salmonella*: Regulation by the *glnR* and *glnF* gene products. *Proceedings of the National Academy of Sciences, USA*, **76**, 4576–80.

LUDWIG, R. (1978). Control of ammonium assimilation in *Rhizobium* 32Hl. *Journal of Bacteriology*, **135**, 114–23.

LUDWIG, R. A. & SIGNER, E. R. (1977). Glutamine synthetase and control of nitrogen fixation in *Rhizobium. Nature*, **267**, 245–8.

MACNEIL, D. & BRILL, W. J. (1978). 6-Cyanopurine, a color indicator useful for isolating deletion, regulatory and other mutations in the *nif* (nitrogen fixation) genes. *Journal of Bacteriology*, **136,** 247–52.

MACNEIL, D., BRILL, W. J. & HOWE, M. M. (1978). Bacteriophage Mu-induced deletions in a plasmid containing the *nif* (N_2-fixation) genes of *Klebsiella pneumoniae. Journal of Bacteriology*, **134**, 821–9.

MACNEIL, T., MACNEIL, D., ROBERTS, G. P., SUPIANO, M. A. & BRILL, W. J. (1978). Fine-structure mapping and complementation analysis of *nif* (nitrogen fixation) genes in *Klebsiella pneumoniae. Journal of Bacteriology*, **136,** 253–66.

MACNEIL, D., SUPIANO, M. & BRILL, W. J. (1979). Order of genes near *nif* in *Klebsiella pneumoniae. Journal of Bacteriology*, **138**, 1041–5.

MAGASANIK, B. (1977). Regulation of bacterial nitrogen assimilation by glutamine synthetase. *Trends in Biochemical Sciences*, **2**, 9–12.

MAZUR, B., RICE, D. & HASELKORN, R. (1980). Identification of blue-green algal nitrogen fixation genes by using heterologous DNA hybridisation probes. *Proceedings of the National Academy of Sciences, USA*, **77**, 186–90.

MERRICK, M., FILSER, M., DIXON, R., ELMERICH, S., SIBOLD, L. & HOUMARD, J. (1980). Use of translocatable genetic elements to construct a fine-structure map of the *Klebsiella pneumoniae* nitrogen fixation (*nif*) gene cluster. *Journal of General Microbiology*, **117**, 509–20.

MERRICK, M., FILSER, M., KENNEDY, C. & DIXON, R. (1978). Polarity mutations induced by insertion of transposons Tn5, Tn7 and Tn10 into the *nif* gene cluster of *Klebsiella pneumoniae. Molecular and General Genetics*, **165**, 103–11.

MORTENSON, L. E. & THORNELEY, R. N. F. (1979). Structure and function of nitrogenase. *Annual Review of Biochemistry*, **48**, 387–418.

NUTI, M. P., LEPIDI, A. A., PRAKASH, R. K., SCHILPEROORT, R. A. & CANNON, F. C. (1979). Evidence for nitrogen fixation (*nig*) genes on indigenous *Rhizobium* plasmids. *Nature*, **282**, 533–5.

PAHEL, G. & TYLER, B. (1979). A new *glnA*-linked regulatory gene for glutamine synthetase in *Escherichia coli. Proceedings of the National Academy of Sciences, USA*, **76**, 4544–8.

PAHEL, G., ZELENETZ, A. & TYLER, B. (1978). *gltB* gene and regulation of nitrogen metabolism by glutamine synthetase in *Escherichia coli. Journal of Bacteriology*, **133**, 139–48.

PAREJKO, R. A. & WILSON, P. W. (1970). Regulation of nitrogenase synthesis by *Klebsiella pneumoniae. Canadian Journal of Microbiology*, **16**, 681–5.

PICHINOTY, F. (1962). Inhibition par l'oxygène de la biosynthese et de l'activité de l'hydrogénase et de l'hydrogène-lyase chez les bacteries anaerobies facultatives. *Biochimica et Biophysica Acta*, **64**, 111–19.

POSTGATE, J. R. (1974). Evolution within nitrogen fixing systems. In *Evolution in the Microbial World*, ed. M. J. Carlile & A. H. Rose, *Symposia of the Society for General Microbiology*, *24*, pp. 263–92. Cambridge University Press.

PÜHLER, A., BURKHARDT, H. J. & KLIPP, W. (1979). Cloning of the entire region for nitrogen fixation from *Klebsiella pneumoniae* on a multicopy plasmid. *Molecular and General Genetics*, **176**, 17–24.

REIDEL, G. E., AUSUBEL, F. M. & CANNON, F. C. (1979). Physical map of chromosomal nitrogen fixation (*nif*) genes of *Klebsiella pneumoniae. Proceedings of the National Academy of Sciences, USA*, **76**, 2866–70.

ROBERTS, G. P., MACNEIL, T., MACNEIL, D., & BRILL, W. J. (1978). Regulation and characterisation of protein products coded by the *nif* (nitrogen fixation) genes of *Klebsiella pneumoniae. Journal of Bacteriology*, **136**, 267–79.

ROBSON, R. L. (1979). O_2 repression of nitrogenase synthesis in *Azotobacter chroococcum. FEMS Microbiology Letters*, **5**, 259–62.

RUVKUN, G. B. & AUSUBEL, F. M. (1980). Interspecies homology of nitrogenase genes. *Proceedings of the National Academy of Sciences, USA*, **77**, 191–5.

SCOTT, D. B., HENNECKE, H. & LIN, S. T. (1979). The biosynthesis of Mo-Fe protein polypeptides in free-living cultures of *Rhizobium japonicum. Biochimica et Biophysica Acta*, **565**, 365–78.

SHAH, V. & BRILL, W. J. (1979). Isolation of an iron-molybdenum co-factor from nitrogenase. *Proceedings of the National Academy of Sciences, USA*, **74**, 3249–53.

SHANMUGAM, K. T., CHAN, I. & MORANDI, C. (1975). Regulation of nitrogen fixation. Nitrogenase-derepressed mutants of *Klebsiella pneumoniae. Biochimica et Biophysica Acta*, **408**, 101–11.

SHANMUGAM, K. T. & MORANDI, C. (1976). Amino acids as repressors of nitrogenase biosynthesis in *Klebsiella pneumoniae. Biochimica et Biophysica Acta*, **437,** 322–32.

SPENCER, M. E. & GUEST, J. R. (1973). Isolation and properties of fumarate reductase mutants of *Escherichia coli. Journal of Bacteriology*, **114**, 563–70.

ST JOHN, R. T., JOHNSTON, H. M., SEIDMAN, C., GARFINKEL, D., GORDON, J. K., SHAH, V. K. & BRILL, W. J. (1975). Biochemistry and genetics of *Klebsiella pneumoniae* mutant strains unable to fix nitrogen. *Journal of Bacteriology*, **121**, 759–65.

ST JOHN, R. T., SHAH, V. K. & BRILL, W. J. (1974). Regulation of nitrogenase synthesis by oxygen in *Klebsiella pneumoniae. Journal of Bacteriology*, **119**, 266–9.

STREICHER, S., BLOOM, F. R., FOOR, F., LEVIN, M. & TYLER, B. (1977). *Klebsiella pneumoniae* and *Escherichia coli* mutants altered in nitrogen assimilation. *Abstracts of the Annual Meeting of the American Society for Microbiology*, Abstract K88, 200.

STREICHER, S., DELEO, A. & MAGASANIK, B. (1976). Regulation of enzyme formation in *Klebsiella aerogenes* by episomal glutamine synthetase of *Escherichia coli. Journal of Bacteriology*, **127**, 184–192.

STREICHER, S. L., SHANMUGAM, K. T., AUSUBEL, F., MORANDI, C. & GOLDBERG, R. B. (1974). Regulation of nitrogen fixation in *Klebsiella pneumoniae*: evidence for a role of glutamine synthetase as a regulator of nitrogenase synthesis. *Journal of Bacteriology*, **120**, 815–21.

Streicher, S. L. & Tyler, B. (1976). New methods for the purification of glutamine synthetase (GS) from *Klebsiella aerogenes* indicate the presence of a GS-DNA complex in the bacterial cell. *Federation Proceedings*, **35**, 1471.

Tubb, R. S. (1974). Glutamine synthetase and ammonium regulation of nitrogenase synthesis in *Klebsiella*. *Nature*, **251**, 481–5.

Tubb, R. S. & Postgate, J. R. (1973). Control of nitrogenase synthesis in *Klebsiella pneumoniae*. *Journal of General Microbiology*, **79**, 103–17.

Tyler, B. (1978). Regulation of the assimilation of nitrogen compounds. *Annual Review of Biochemistry*, **47**, 1127–62.

Tyler, B., Deleo, A. B. & Magasanik, B. (1974). Activation of transcription of *hut* DNA by glutamine synthetase. *Proceedings of the National Academy of Sciences, USA*, **71**, 225–9.

GENETIC STUDIES OF ANTIBIOTICS AND OTHER SECONDARY METABOLITES

D. A. HOPWOOD

John Innes Institute, Colney Lane, Norwich NR4 7UH, UK

INTRODUCTION

Microorganisms produce a wide variety of pigments, odours, antibiotics, enzyme inhibitors and pharmacologically active agents, the value of which to the producing organism is often obscure and which are often produced at particular stages in the life of a colony or culture, typically at a stage when an early 'trophophase' gives way to a later 'idiophase'. The idiophase often coincides with a morphological differentiation; thus the production of these metabolites is an aspect of differential gene expression and so of differentiation. For these reasons they have been designated *secondary* metabolites (Bu'lock, 1961; Turner, 1971; Drew & Demain, 1977), in contrast to the so-called primary metabolites – amino acids, sugars, vitamins, organic acids, nucleic acid bases, etc. – which are the clearly essential building blocks of basic cellular components, or co-factors in their assembly, or parts of energy transfer systems.

Amongst the secondary metabolites, those with anti-microbial properties, the antibiotics, have received the greatest attention because of their obvious economic and social importance. The examples in this chapter are all antibiotics, but we should remember that even this grouping includes a range of compounds which are extremely heterogeneous both chemically and biologically; chemically because of the enormous variety of structures found amongst antibiotics, which range from simple small molecules like chloramphenicol (MW322), through the polycyclic structures of β-lactams (penicillins and cephalosporin C), tetracyclines, macrolides (erythromycin, tylosin, etc.) and polyethers (such as monensin), to peptide antibiotics like the actinomycins or gramicidins, which contain ten or more amino acid residues as parts of their complex structures; and biologically because of the diverse targets which antibiotics attack – DNA replication (mitomycin C), transcription (actinomycin D, rifamycin), translation at its various stages (aminoglycosides, erythromycin, chloramphenicol, cycloheximide, puromycin), cell wall or membrane synthesis or function (β-lactams, nystatin, monensin), fatty acid synthesis (cerulenin), etc. – which in turn determine their range of application as anti-bacterial, anti-fungal, anti-coccidial, anti-helminthic, anti-viral, anti-tumour or growth-promoting agents.

It is the anti-microbial properties of these molecules which usually lead to their discovery and we therefore tend to assume that this property is the evolutionarily relevant one, selected because it confers a competitive advantage on an antibiotic-producer in its natural habitat, which may be a highly competitive ecological niche such as the surface of insoluble organic debris in the soil. While this conclusion may well be accurate, at least for some antibiotics, it is remarkably difficult to demonstrate antibiotic production and activity in soil (Williams & Khan, 1974; Gottlieb, 1976), although the idea that anti-microbial activity may be significant only over short distances within differentiating colonies, to prevent parasitism by competing organisms (Chater & Merrick, 1979), would satisfactorily account for the low overall concentrations of antibiotics in natural substrates. However, competition is unlikely, intuitively, to be the correct explanation for *all* antibiotics, of which more than 3000 were known in 1974 (Bérdy, 1974); the number is probably nearly twice as large today.

An intriguing possibility is that some secondary metabolites play a regulatory or hormone-like role in the complex cycles of morphological differentiation typically displayed by antibiotic producers, which are mostly spore-forming or differentiating eubacteria (bacilli or myxobacteria) or mycelial, sporulating actinomycetes and moulds, rather than simple unicellular eubacteria or yeasts. Some evidence exists for such a role in bacilli and actinomycetes and this will be discussed later in the chapter.

That antibiotics in particular, and secondary metabolites in general, do confer some selective advantage on their producers can hardly be doubted in view of the considerable amount of DNA dedicated to their genetic determination. A typical antibiotic requires, for its step-wise synthesis, some dozen or more enzymes which generally appear to have no other roles. Members of most antibiotic-producing microbial groups, especially actinomycetes, make several antibiotics, many of them chemically distinct. A single strain of *Streptomyces clavuligerus*, for example, is known to produce 15 antibiotics (L. J. Nisbet, personal communication). This may not be a completely atypical example, since antibiotic discovery is in part a matter of the intensity with which an organism is investigated: the first antibiotic produced by the genetically well-known *Streptomyces coelicolor* A3(2) was identified only in 1976 (Wright & Hopwood, 1976a), but within three years at least five (Rudd, 1978) and probably more (Troost, Danilenko & Lomovskaya, 1979) were known. Thus some, at least, of these organisms may carry more than 100 expressed genes dedicated to antibiotic-production – over 1% of the genome. Even if little energy is required to drive these pathways, because small amounts of antibiotics are made by wild microorganisms, the replication of all this DNA, generation after generation, must be a significant energy drain.

While antibiotic production has surely been adaptive on an evolutionary time scale, it does not follow that the synthesis of all antibiotics by all antibiotic-producing microorganisms is advantageous at all times (Zähner, 1978). Indeed, antibiotic synthesis may fall into the same category as characters such as resistance to heavy metal ions or to antibiotics in enteric bacteria or staphylococci, or the ability of pseudomonads to utilise uncommon carbon sources; such characters are clearly adaptive to the population, but their advantage is discontinuous, since the inhibitory agent or potential nutrient occurs intermittently in the environment. This may explain why genes determining such characteristics are often borne on plasmids present only in some individuals of a species but capable of spreading to other members of the population, by conjugation, transduction or transformation. That conjugal spread of a plasmid may be directly related to the adaptive nature of its genes is clearly shown in strains of *Agrobacterium tumefaciens* carrying Ti plasmids determining opine utilisation. The opine itself derepresses self-transfer of the plasmid, leading to its infectious spread to organisms lacking it (Petit *et al.*, 1978). In view of these considerations, it is highly significant that plasmids control the synthesis of some antibiotics, particularly in actinomycetes (Hopwood, 1978; Chater, 1979; Okanishi, 1979) and a growing body of genetical studies, reviewed below, should clarify such control, which may involve an interaction between chromosomal and plasmid-borne genes.

Whatever the natural roles of antibiotics, it is clear that the isolation and study of mutants or plasmid variants modified in their ability to produce and to respond to these metabolites, or to over-produce them, must be a powerful tool in understanding their significance to the producing organisms, and of course will also have immediate industrial relevance. Moreover, because the roles of these compounds are different from those of primary metabolites, attempts to unravel their genetic determination may well reveal new genetic mechanisms, rather than merely providing further examples of phenomena understood in detail as a result of genetic studies over the last 30 years of the biochemical genetics of primary metabolism in such 'model' subjects as *Escherichia coli*, *Salmonella typhimurium*, *Neurospora crassa*, *Aspergillus nidulans* or *Saccharomyces cerevisiae*. I use the future tense because, in spite of having both an academic and an obvious applied interest, the genetics of secondary metabolism is still comparatively neglected. The time is ripe for its expansion because appropriate analytical tools are now available, or could be developed in predictable ways. A discussion of these tools will underlie much of the rest of this chapter.

GENETIC TOOLS

The ways in which genetics can be used as a tool in understanding any biological process depend strongly on the 'sexual biology' of the organisms concerned. Therefore detailed consideration of these tools will be given in sections dealing with the three major groups of antibiotic-producing microbes, but a few general points may usefully be made.

Natural systems of gene exchange

Although some antibiotics are produced by non-spore forming eubacteria – for example the microcins of *E. coli* (Aguilar *et al.*, 1980), prodigiosin in *Serratia marcescens* (Williams & Hearn, 1967) and several antibiotics in *Pseudomonas* spp. (Bérdy, 1974) – the most important antibiotic-producing groups are the filamentous fungi, the bacilli and the actinomycetes. Natural systems of gene exchange exist in some members of all these groups (Hopwood & Merrick, 1977) and they could probably be discovered or developed in any strain producing an interesting antibiotic. However, considerable effort may be required to develop such a system to a point where even relatively simple but precise genetic analysis can be carried out. At the very least, an adequately marked chromosomal linkage map is needed. In a fungus, possessing a set of, say, 8–20 linkage groups, each more than 50 map units long, this represents a heavy commitment even if the genetic system is particularly tractable; only one antibiotic-producing fungus, *Aspergillus nidulans*, falls into this category (see below). The situation in a eubacterium is simpler, since a single circular linkage map is likely to represent the entire chromosomal genome, but its definition may depend on the chance discovery of sex plasmids able to mobilise appropriate chromosomal segments: the linkage map of a strain of *Pseudomonas aeruginosa* was closed into a circle very recently (Holloway, Krishnapillai & Morgan, 1979), more than 20 years after recombination was first detected in this species. In the streptomycetes, with their genetically short circular genome (about 260 map units in *Streptomyces coelicolor* A3(2): Hopwood, 1966) and the apparently widespread occurrence of sex plasmids promoting the transfer of essentially random genome fragments, the situation is more favourable, but even so, relatively intractable strains are encountered (Hopwood & Merrick, 1977). Any reasonably penetrating genetic analysis of antibiotic production, using the tools provided by natural gene exchange, will therefore be limited to very few examples.

Artificial procedures for genetic manipulation

Significant advances have recently been made in the development of protoplast systems as potential genetic tools for the analysis and manipulation of antibiotic production. Two conceptually different procedures are involved: protoplast fusion and protoplast transformation/transfection. Both involve the treatment with polyethylene glycol (PEG) of fungal or bacterial protoplasts, prepared by removal of the cell walls with polysaccharide-degrading enzymes or lysozyme respectively. However, in *S. coelicolor* and its relatives the optimal conditions are clearly different for PEG-induced fusion and for PEG-assisted DNA uptake (Bibb, Ward & Hopwood, 1978; Hopwood & Wright, 1979; Suarez & Chater, 1980a), suggesting the operation of different mechanisms.

The potential applications are certainly different. Protoplast fusion can greatly enhance the frequency with which genetically different fungal nuclei or bacterial genomes enter a common cytoplasm, overcoming barriers to hyphal fusion between different fungal strains (Ferenczy, this volume), or the requirement for plasmid sex factors to promote gene transfer in bacteria (Fodor & Alföldi, 1976; Schaeffer, Cami & Hotchkiss, 1976; Hopwood, Wright, Bibb & Cohen, 1977). This can result in relatively high recombination frequencies provided that the parent strains are sufficiently closely related to carry homologous DNA segments which can recombine. The latter criterion limits the scope of recombination between strains. On the other hand, the ability to introduce the products of *in vitro* DNA recombination – plasmid or phage chimaeras – into antibiotic producers overcomes limitations to inter-strain recombination, as well as providing the tools for the physical analysis and *in vitro* mutation of relevant DNA segments (Timmis, this volume). It also offers the possibility, in principle, of transferring, for analytical or even production purposes, a whole set of genes for synthesis of an interesting antibiotic from a genetically or physiologically 'difficult' strain into one more suitable for study, perhaps because it is nutritionally convenient, or because it possesses error-prone repair systems allowing for efficient mutagenesis and lacking in the original host (Bridges, 1976), or because useful genetic devices, such as characterised suppressor systems, are available.

GENETIC ANALYSIS OF ANTIBIOTIC BIOSYNTHESIS

The topic of regulation of antibiotic synthesis has recently been reviewed (Martín & Demain, 1980) and so here we can concentrate primarily on the structural genes.

Fungi

'Natural' genetic systems

Hopwood & Merrick (1977) listed six antibiotic-producing fungal species in which genetic recombination had been demonstrated but in three of these (two *Emericellopsis* spp. and *Penicillium patulum*) genetic studies are very preliminary. This leaves three species whose investigation continues, primarily because of the extreme industrial and medical importance of their β-lactam antibiotics, penicillin G and cephalosporin C. (In each case the side chain of the natural antibiotic is artificially removed by hydrolysis, leaving a nucleus (6-aminopenicillanic acid or 7-aminocephalosporanic acid) to which artificial side chains can be attached, generating a huge variety of 'semi-synthetic' antibiotics with improved pharmacological or anti-microbial properties.) The industrial penicillin-producing strains are all mutational descendants of a single strain of *Penicil-*

lium chrysogenum, NRRL 1951, while cephalosporin-producers apparently trace back to a single isolate of *Cephalosporium acremonium*, the Brotzu strain (Abraham, 1974).

Both these fungi lack a sexual cycle, but in principle genetic analysis is possible by means of a series of unco-ordinated genetic events described and characterised as the 'parasexual cycle' by Pontecorvo and his colleagues in their pioneer studies of a strain of *Aspergillus nidulans* (NRRL 194), the 'Glasgow' strain (Pontecorvo *et al.*, 1953). This cycle (Sermonti, 1969; Hopwood, 1972) begins with fusion of hyphae of genetically different strains to give, after nuclear mixing, a heterokaryon; this tends to break down irregularly to the original homokaryons as branching occurs, and regularly at asexual spore formation since such spores (conidia) are uninucleate; thus heterokaryons can only be maintained as vegetative growth and by the use of at least one selective marker in each parental strain.

Occasional nuclear fusion within the heterokaryon produces heterozygous diploid nuclei. These can divide rather stably so that pure heterozygous diploid strains can be picked as vegetative sectors or by plating individual conidia. Either heterokaryons or, better, diploids allow dominance or complementation testing of mutant genes. Occasional mis-division of diploid nuclei give rise to a range of chromosome numbers between haploid and diploid; usually only the fully haploid nuclei, with their 'balanced' set of chromosomes, survive. This process of 'haploidisation', which can be enhanced by treatment with *p*-fluorophenylanine or other agents, allows easy assignment of mutations to linkage groups (corresponding to chromosomes) because alleles of genes on different chromosomes recombine freely as the chromosomes assort into haploid sets, while those on the same chromosome re-assort only rarely. Such rare recombination of linked genes is due to another infrequent event, mitotic crossing-over. Such crossing-over, not associated as it would be at meiosis with haploidisation, leads to the segregation of homozygous diploid genotypes from the heterozygote. Markers distal to a crossover point, with respect to the centromere, normally become homozygous co-incidentally, so the frequencies of occurrence of different homozygous genotypes serve to map a series of linked genes with respect to each other and to the centromere.

Together with this genetically convenient system, *A. nidulans* has a sexual cycle with regular nuclear fusion and meiosis which lends itself to genetic mapping, on both a gross and a fine level of analysis. Thus this organism is one of the few microbes which satisfy the criteria for a really effective and versatile genetic analysis (Clutterbuck, 1974). The realisation that *A. nidulans* produces penicillin G opened the way to a genetic analysis of its synthesis by both Mendelian and biometrical techniques. This system is well worthy of an intensive multi-disciplinary attack, involving chemistry, biochemistry, genetics and molecular biology.

P. chrysogenum is genetically less versatile than *A. nidulans*, lacking a sexual cycle and having a parasexual cycle which took considerable effort to develop into a workable system (Pontecorvo & Sermonti, 1954; Sermonti, 1969; Macdonald & Holt, 1976), but genetic analysis of penicillin biosynthesis by the organism has recently made significant progress (Normansell, Normansell & Holt, 1979). On the other hand, the experimental development of a parasexual cycle in *C. acremonium* has met with severe problems, at least in part caused by the normally uninucleate nature of the hyphal compartments which hinders heterokaryon selection. Diploidisation, mitotic crossing-over and haploidisation have been detected (Nüesch, Treichler & Liersch, 1973) but no genetic mapping of antibiotic genes has apparently been done.

Genetic analysis of the biosynthetic pathways of β-lactam synthesis

Although recombinational analysis of *C. acremonium* has not progressed, biochemical studies of mutants blocked in cephalosporin C production (reviewed by Hopwood & Merrick, 1977; Queener, Sebek & Vezina, 1978; Demain, 1981) have greatly aided elucidation of the biosynthetic pathway.

There seems little doubt that an early step in the pathway involves the synthesis of a tripeptide (δ-L-α-aminoadipyl)-L-cysteinyl-D-valine:ACV) from a molecule each of L-α-aminoadipic acid, L-cysteine and L-valine, followed by ring closures to generate the β-lactam and sulphur-containing (thiazolidine) rings of isopenicillin N (Fawcett, Usher & Abraham, 1976; Demain, 1981). Penicillin N (which differs from iso-penicillin N in having the D configuration instead of the L in the α-aminoadipyl moiety of the molecule) has long been known to be produced by *C. acremonium* and it was originally proposed that it lay on a side branch of the pathway to cephalosporin C. It now seems clear (Demain, 1981) that this is not the case but that a racemisation of iso-penicillin N to penicillin N is a step on the main pathway. This step is followed by a poorly understood expansion of the sulphur-containing ring from the five-membered ring of penicillin N to the six-membered cephalosporin ring. Studies of blocked mutants (Fujisawa, Kitano & Kanzaki, 1975; Liersch, Nüesch & Treichler, 1976) have been instrumental in showing that the first recognisable cephalosporin in the pathway is deacetoxycephalosporin C, followed by deacetylcephalosporin C and finally cephalosporin C.

Corresponding biochemical studies of mutants of *P. chrysogenum* (or *A. nidulans*) are less complete so that the pathway of biosynthesis of penicillin G is less clearly defined (Macdonald & Holt, 1976). However it is most probable that the pathways of penicillin G and cephalosporin C synthesis are essentially the same as far as isopenicillin N (Demain, 1981). Then, instead of the racemisation to penicillin N, the penicillin G producers carry out a transacylation reaction which exchanges the α-aminoadipyl side chain of isopenicillin N for a phenylacetyl side chain.

Some of the genes involved in β-lactam synthesis in both *P. chrysogenum* and *A. nidulans* have been identified by the isolation and genetic analysis of mutants (now called *npe* for non-penicillin producers) which fail to produce penicillin (in practice mutants producing less than 10% of the parental titre are put in this category: Edwards, Holt & Macdonald, 1974; Normansell *et al.*, 1979). For *P. chrysogenum*, such studies began by the isolation of nine *npe* mutants which were classified by the study of doubly heterozygous diploids into two complementation groups, one containing eight of the nine mutations (Sermonti, 1956). This work has recently been extended by Normansell *et al.* (1979). Twelve *npe* mutations were grouped in five loci (*npe* V, W, X, Y and Z) by complementation in heterozygous diploids and by haploidisation analysis; W, Y and Z were located by haploidisation from diploids to the same linkage group, while V and X were each mapped to another group. One of the five classes (*npe* Y) contained seven of the 12 mutants, just as seven of the eight mutations of Sermonti (1956) fell into a single group (presumably the same as *npe* Y: Normansell *et al.*, 1979).

In *A. nidulans*, 28 *npe* mutations were classified by Edwards *et al.* (1974) and Holt, Edwards & Macdonald (1976) into four genes (*npe* A, B, C and D) by complementation studies in diploids. The genes were mapped, by a combination of parasexual haploidisation and meiotic recombination analysis, to positions on linkage groups VI, III, IV and II respectively (J. F. Makins, G. Holt and K. D. Macdonald, unpublished results). As in *P. chrysogenum*, a large majority of the *npe* mutations (20 out of 28) fell into a single complementation group, *npe*A.

Recent results of Holt and co-workers indicate that the majority class of mutants in each organism (*npe*A in *A. nidulans* and *npe*Y in *P. chrysogenum*) is defective in synthesis of the ACV tripeptide. In *P. chrysogenum*, Normansell *et al.* (1979) found that *npe*Y mutants (and also members of the small classes *npe*X and *npe*Z) failed to accumulate tripeptide, whereas *npe*V and *npe*W mutants apparently did so and were presumably blocked at a later step in the pathway. *npe*A mutants of *A. nidulans* were also reported to be defective in tripeptide synthesis (Makins, Holt & Macdonald, 1980). These findings are interesting in relation to the idea (Martín, Luengo, Revilla & Villanueva, 1979) that formation of the ACV tripeptide from the three L-amino acid moieties – L-α-aminoadipic acid, L-cysteine and L-valine – with racemisation of the L-valine moiety to the D configuration, is likely to involve a multi-enzyme complex, just as does the synthesis of the peptide antibiotics of bacilli (Lipmann, 1973) and of the polyketide skeleton of many streptomycete antibiotics (see later). A possible explanation for the disproportionately frequent isolation of mutants defective in the activity of multi-enzyme complexes is discussed later in the context of streptomycete antibiotics.

Genetic analysis of penicillin yield

A complete genetic analysis of antibiotic yield, since it is a quantitative character, requires biometrical rather than Mendelian techniques. However, some mutations give a large enough increment of yield to be studied individually in sexual or parasexual crosses, and some fairly precise map locations have been obtained for several mutant genes in *A. nidulans*, with its versatile genetic system. Even so, following the segregation of these quantitative characters required particular analytical approaches (Macdonald & Holt, 1976; Ditchburn, Holt & Macdonald, 1976). The subject, reviewed by Hopwood & Merrick (1977), need not be covered again here since few new results have been published. No attempt appears to have been made to identify the biochemical basis of these increased titre mutations.

The total analysis of penicillin yield is really only possible by means of the sexual cycle of *A. nidulans*, since classical biometrical techniques require unbiased segregation and recovery of all possible genotypes from diploids (Caten & Jinks, 1976). The first studies by

Merrick and Caten (reviewed by Merrick, 1976a; Hopwood & Merrick, 1977) were of natural variation in penicillin titre between different wild isolates of *A. nidulans*. Interestingly, isolates having very similar titres were shown, by suitable crosses, to carry different alleles of a large series of genes affecting titre. Most such alleles interacted additively, so that higher titres could be achieved predictably by combining different series of 'plus' alleles from different wild isolates by successive crosses. Some wild *A. nidulans* strains produce no detectable penicillin. In many cases, such strains are apparently naturally occurring *npe*A mutants which carry, unexpressed, genes for the later steps of the biosynthetic pathway, as shown by the generation of penicillin-producing recombinants in crosses with *npe*B, C or D mutants of the Glasgow strain but not with *npe*A mutants (Cole, Holt & Macdonald, 1976).

The finding not only that most wild isolates of *A. nidulans* produce penicillin, and many at a rather similar titre, but that the 'plus' alleles which contribute to that titre are different in different isolates (Merrick, 1976a), lends weight to the idea that antibiotic production is adaptive, although not of course revealing the nature of its selective advantage. If antibiotic production had been non-adaptive it seems unlikely that so precise a phenotype as a capacity to produce a particular yield of the antibiotic would be found in a wide variety of genetically rather isolated strains belonging to different heterokaryon incompatibility groups (Merrick & Caten, 1975).

More recently (Simpson & Caten, 1981), a biometrical analysis of induced variation for penicillin titre has been carried out.

Artificial genetic manipulation

Protoplast fusion has produced hybrids between various *Penicillium* spp. and between *Aspergillus* spp. (reviewed by Ferenczy, this volume) but there is so far little information on penicillin production by such hybrids. As expected, fusion between closely related species produced heterokaryons at high frequency, but interactions between genetically different nuclei were relatively infrequent. In *C. acremonium*, intraspecific protoplast fusion led, unexpectedly, not to heterokaryons or diploids, but to frequent haploid recombinants (Hamlyn & Ball, 1979). Only an empirical analysis of cephalosporin C yield was made; it demonstrated the potential of protoplast fusion to combine yield-enhancing genes from divergent selection lines, but no estimate was made of the number of genes involved or the nature of their interactions.

No serious attempt has apparently been made to develop DNA vectors and uptake systems for antibiotic-producing fungi, although their successful development in yeast (Beggs, 1978; Hinnen, Hicks & Fink, 1978; Botstein *et al.*, 1979) suggested that comparable procedures could work in filamentous fungi (Case, Schweizer, Kushner & Giles, 1979). Particularly promising is the finding that a piece of yeast chromosome including the *trp*-1 locus can apparently act, when cleaved from the genome, as a separate replicon into which foreign DNA can be inserted (Struhl, Stinchcomb, Scherer & Davis, 1979). This result suggests a general means of generating cloning vectors for any eukaryotic host.

A very interesting approach has recently been taken by Makins *et al.* (1980) in the biochemical study of *npe* mutants of *A. nidulans*. Although such mutations fell into four complementation groups revealed by penicillin synthesis by doubly heterozygous diploids (see above), no co-synthesis of penicillin occurred when mycelia of mutants belonging to different complementation groups were grown together. The same was true of members of the five *npe* complementation groups of *P. chrysogenum* (Normansell *et al.*, 1979). Such results, which are by no means uncommon in studies of antibiotic non-producing mutants in actinomycetes also, may reflect the instability of biosynthetic intermediates, or their conversion to 'shunt products' before they are able to enter the biosynthetic pathway of a mutant blocked at an earlier point in the biosynthesis which should, in principle, be capable of converting the intermediate to the end-product of the pathway.

A more common explanation for a failure of co-synthesis may be the impermeability of the cell membranes to the intermediates so that they are unable to move from the cells of one mutant strain to those of the other. Makins *et al.* (1980) found a way round this obstacle by using osmotically fragile mycelia, prepared by the digestion of the cell walls under osmotically stabilising conditions, and incubation of the resulting (partially) wall-less cells in the presence of inhibitors of cell-wall synthesis to prevent wall regeneration. By this approach, four out of six pairwise combinations of *npe*A, B, C and D mutants of *A nidulans* synthesised penicillin, the three combinations involving an *npe*A partner producing a good yield, perhaps reflecting donation of ACV tripeptide from *npe*B, C or D mutants to the tripeptide non-producing *npe*A strain.

That *npe*A mutants were indeed defective in tripeptide synthesis was shown by feeding tripeptide to an *npe*A mutant. Normally such experiments fail, presumably because of a barrier to uptake of the tripeptide, but this was ingeniously circumvented by enclosing the tripeptide in lipid vesicles, which were fused with protoplasts of the *npe*A strain; penicillin synthesis was then observed (J. F. Makins, G. Holt and K. D. Macdonald, unpublished results).

Eubacteria

Study of the genetics of antibiotic production by eubacteria has been severely hampered by the fact that the antibiotics which have been seen to be of scientific or commercial interest are not produced by the strains in which genetic analysis is far advanced. It is striking that the handful of organisms with a really sophisticated genetics, notably *Escherichia coli* K-12 (Bachmann & Low, 1980), *Salmonella typhimurium* LT2 (Sanderson & Hartman, 1978), *Pseudomonas aeruginosa* PAO and PAT (Holloway *et al.*, 1979) and *Bacillus subtilis* 168 (Henner & Hock, 1980), produce either no antibiotics or, in the case of *B. subtilis*, antibiotics which have been considered unworthy of detailed investigation. Thus the potential for genetic analysis by one or more of the natural systems of gene exchange – conjugation and generalised transduction for the Gram-negative *E. coli*, *S. typhimurium* and *P. aeruginosa*, or transformation and generalised transduction for *B. subtilis* – has not been exploited for an understanding of antibiotic synthesis. Genetic analysis by conjugation could probably be extended relatively easily to the Gram-negative *Serratia marcescens*, in which considerable progress has been made in the isolation and classification, biochemically and by co-synthesis tests, of mutants blocked in prodigiosin biosynthesis (Williams & Hearn, 1967). This follows from the availability of broad host-range sex plasmids of the P-1 incompatibility group which can be transferred from *E. coli* or *P. aeruginosa* strains, in which they have normally been discovered, to members of any of a wide range of Gram-negative genera, including *Serratia*. In the new hosts, the natural plasmids, or variants of them such as R68.45, can mobilise segments of the chromosome in a way convenient for mapping purposes (Holloway, 1979). There are in fact R plasmids to be found in strains of *Serratia marcescens* itself, capable of promoting chromosomal recombination (Hedges, 1978). Gene–enzyme relations and regulation in prodigiosin biosynthesis could be interesting since a convergent pathway is apparently involved, (Williams & Hearn, 1967) and, although the antibiotic contains three pyrrole rings, biosynthesis of each proceeds from different components (Gerber, 1975). Moreover, prodigiosin production by *S. marcescens* provides a classical case of genetic instability (Bunting, 1946), well worthy of investigation by modern molecular biological approaches.

There are some examples of plasmid involvement in antibiotic synthesis amongst various eubacterial strains (reviewed by Hopwood, 1978), but the roles of the plasmids are unclear. It might be thought self-evident that they would carry structural genes for antibiotic biosynthetic enzymes, but in view of the good evidence that plasmids in streptomycetes sometimes control antibiotic production in other ways (see below), this cannot be assumed.

The oligopeptide antibiotics of bacilli

From several points of view the peptide antibiotics produced by various species of the genus *Bacillus* (Katz & Demain, 1977) are extremely interesting. (1) Although these antibiotics consist of peptide chains (sometimes with other moieties), of eight or more amino acid residues, some linear (for example gramicidins A, B and C) and others cyclic (gramicidin S, tyrocidines, bacitracin), they are not synthesised by translation of an RNA message on ribosomes but by a unique stepwise assembly process on multi-enzyme thiotemplate complexes (Lipmann, 1973; Frøyshov, 1975). (2) There is some evidence, and much argument, that the antibiotics play a role in the development and/or outgrowth of the endospores typical of the genus *Bacillus*. (3) A few of the antibiotics – bacitracin, colistin, the gramicidins and tyrocidines – have some commercial importance.

Genetic studies can play a key part in illuminating the synthesis and roles of these antibiotics. Antibiotic non-producing mutants have already contributed to ideas on the possible role of these antibiotics as developmental signals and this topic is dealt with in a subsequent section. As for analysis of the biosynthetic processes, understanding of gramidicin S synthesis by *B. brevis* has greatly benefited from the isolation and biochemical study of blocked mutants. Such mutants were classified into three groups by *in vitro* complementation (Iwaki *et al.*, 1972; Kambe, Imae & Kurahashi, 1974). The gramicidin S synthetase consists of two components, the 'light' sub-unit which activates and racemises L-phenylalanine, and the 'heavy' sub-unit which activates L-proline, L-valine, L-ornithine and L-leucine. The groups of mutants contained those in which the light but not the heavy sub-unit was functional, those with a functional heavy but no functional light sub-unit, and those in which neither sub-unit was functional. Later, 20 of the mutants were classified further in studies of the activation of individual amino acids by the synthetase sub-units (Shimura *et al.*, 1974). In particular, various mutants defective in heavy sub-unit function were found to lack the ability to activate a single amino acid while retaining the capacity to activate the other three; presumably they represented missense mutations in structural genes for individual polypeptides of the multi-enzyme heavy sub-unit. This conclusion has been strongly supported by purification of the heavy sub-unit from the wild-type and from those mutants failing to activate prolinc, valine or leucine respectively; the mutant sub-units all had apparently the same size as that from the wild-type strain (Hori *et al.*, 1978). In the case of mutants defective in light sub-unit function, studies of the purified enzyme showed that, while it was the same size as that of the wild-type and could still activate L-phenylalanine, it could not racemise it to the D-isomer; it was suggested that the mutations affected the thiol site of the normal enzyme (Kanda *et al.*, 1978).

These elegant studies demonstrate the use of mutants as powerful tools in unravelling an intricate biosynthetic process involving a series of gene products which interact to form a multi-functional complex. It would be very interesting to know how the corresponding genes are arranged and regulated. However, in the continuing absence of convenient systems of genetic analysis in *B. brevis* (and in *B. licheniformis* and *B. polymyxa* which also make well-known peptide antibiotics), we are still totally ignorant of the locations and organisation of any of the genes involved.

What are the prospects for developing useful genetic systems for the analysis of antibiotic production in bacilli by artificial means? They should be reasonably good, although no results of studies with this declared objective appear to have been published. In *B. subtilis* (Schaeffer, Cami & Hotchkiss, 1976), recombination analysis was carried out by protoplast fusion and the results could be compared with those of the more traditional transformation and transduction techniques, while in *B. megaterium* (Fodor & Alföldi, 1976) protoplast fusion provided a new possibility for carrying out recombinational analysis in an otherwise asexual organism. The *B. subtilis* system has recently been studied in some detail; the frequencies of recombination, initially apparently rather low when assayed by prototroph selection, have been found to be quite high (up to 1%), when recombinants carrying recessive markers were sought under conditions allowing better protoplast regeneration (Gabor & Hotchkiss, 1979; Hotchkiss & Gabor, 1980), but anomalies in marker segregation emerged which, if they were to be found in a new organism, might make linkage analysis difficult. Nevertheless, an attempt to develop such a system, particularly for the gramicidin S-producing strain of *B. brevis*, would surely be worthwhile.

There are possibilities for the development of suitable systems for the application of recombinant DNA techniques to antibiotic-producing bacilli. Naturally competent *B. subtilis* 168 can be transformed with plasmid DNA, but apparently only dimeric molecules (or higher multimers) are effective so that direct gene cloning on plasmid vectors in such a system fails, unless the strain contains a corresponding resident plasmid into which incomplete vector molecules can be recombined (Gryczan, Contente & Dubnau, 1980). However, since the interesting antibiotic-producing strains may well not have a naturally competent state, the potentially easily generalisable technique of plasmid transformation of *Bacillus* protoplasts in the presence of PEG (Chang & Cohen, 1979; Vorobjeva, Khmei & Alföldi, 1980) is likely to be more useful. DNA vectors for bacilli are being developed (Erhlich, 1978; Young & Wilson, 1978).

Actinomycetes

'Natural' systems of gene exchange

The actinomycetes in general, and the streptomycetes in particular, are antibiotic-producing organisms in which natural systems of gene exchange are widespread, if not ubiquitous. To the numerous examples of conjugation in mesophilic species and the transformation system in the thermophilic *Thermoactinomyces vulgaris* (reviewed by Hopwood & Merrick, 1977) has recently been added a well-documented case of generalised transduction (Stuttard, 1979). In only a few strains have these genetic systems been exploited to a stage where a reasonably well-developed chromosomal linkage map is available. There is most information for *Streptomyces coelicolor* A3(2) (Hopwood, Chater, Dowding & Vivian, 1973), while two strains of *S. rimosus* also have fairly well-marked linkage maps (Friend & Hopwood, 1971; Alačević, 1976). Amongst other examples of linkage maps in antibiotic-producing strains are those of *S. glaucescens* (Baumann, Hütter & Hopwood, 1974), *S. venezuelae* (Akagawa, Okanishi & Umezawa, 1975), *S. bikiniensis* var. *zor-*

bonensis (Coats & Roeser, 1971) and *Nocardia mediterranei* (Schupp, Hütter & Hopwood, 1975).

The methodology of genetic analysis by conjugation, developed in *S. coelicolor* A3(2) and applied, with minor modifications, in the other species, has been reviewed elsewhere (Hopwood, 1967; Sermonti, 1969; Hopwood *et al.*, 1973; Hopwood & Chater, 1974; Hopwood & Merrick, 1977). In *S. coelicolor* A3(2), two sex plasmids, called SCP1 and SCP2, and capable of interacting with the chromosome in various ways (Hopwood & Wright, 1976; Bibb, Freeman & Hopwood, 1977) allow generalised or more specific marker transfer between parental strains. A comparable plasmid, SRP1, occurs in *S. rimosus* (Friend, Warren & Hopwood, 1978) and it seems likely that, at least in those other species in which genetic recombination occurs at a frequency of 10^{-6} or above, sex plasmids are at work.

In streptomycetes, gross genetic mapping can often be carried out simply by selecting recombinants which inherit one marker from each of two parent strains and analysing such recombinants for the segregation of a series of other, non-selected, markers. Since the total linkage map is short, linkage of a new mutation to existing markers is almost invariably found (Hopwood, 1972).

Although these genetic systems are suitable for mapping genes controlling antibiotic synthesis to chromosomal regions, or for demonstrating extrachromosomal inheritance, they could usefully be extended. What is particularly lacking is a really good system for fine genetic mapping. Although conjugation has been used to order some groups of fairly closely linked mutations (Harold & Hopwood, 1970; Chater, 1972; Carere, Russi, Bignami & Sermonti, 1973) the resolution attainable is not very high and the 'classical' problems of negative interference and gene conversion may tend to make sequencing ambiguous. In principle, transduction should lend itself to fine mapping but the only system so far known is apparently confined to strains of *S. venezuelae* (Stuttard, 1979).

Involvement of plasmid-borne and chromosomal genes in antibiotic biosynthesis

Amongst the actinomycetes there are examples of antibiotics whose synthesis is apparently controlled entirely by chromosomal genes, others in which the activity of both plasmid-borne and chromosomal genes is required for significant antibiotic production, and at least one in which all those genes which appear to be required solely for antibiotic synthesis are carried on a plasmid.

A single strain, *S. coelicolor* A3(2), provides examples of at least two of these patterns of genetic control for its five or more antibiotics. These antibiotics include three chemically characterised compounds: methylenomycin A (Wright & Hopwood, 1976a); actinorhodin (Wright & Hopwood, 1976b); and a red compound (Rudd & Hopwood, 1980), recently identified as a derivative of prodigiosin, a 'prodiginine' (S. W. Lee, C.-J. Chang, B. A. M. Rudd & H. G. Floss, personal communication). Genetic studies (Kirby, Wright & Hopwood, 1975; Kirby & Hopwood, 1977) established that the structural genes for the enzymes of methylenomycin biosynthesis are borne on the SCP1 plasmid, since point mutations abolishing antibiotic synthesis invariably mapped to SCP1, while transfer of SCP1 to another species of *Streptomyces*, *S. parvulus*, caused it to produce methylenomycin A. Only the last step in the pathway of methylenomycin A synthesis is known (Hornemann & Hopwood, 1978), but probably at least five or six SCP1-coded steps are involved (Hornemann & Hopwood, 1980). No chromosomal mutations specifically abolishing methylenomycin synthesis have been isolated. In contrast, all mutations leading to lack of actinorhodin synthesis or that of the prodiginine were mapped to the chromosome, with no evidence of plasmid involvement in the biosynthesis. These mutations were classified, on the basis primarily of co-synthesis studies, into a minimum of seven phenotypic classes for actinorhodin (Rudd & Hopwood, 1979) and five classes for the prodiginine (Rudd & Hopwood, 1980). For each antibiotic, members of all phenotypic classes of mutations were closely linked, suggesting clustering of the structural genes.

There are some examples of clustered chromosomal genes for antibiotic synthesis in other actinomycetes. In *S. rimosus*, mutations in at least six genes controlling steps in the pathway of oxytetracycline synthesis, and three others for steps in the biosynthesis of a co-factor required for a step in the main pathway, mapped to two clusters on opposite sides of the circular linkage group (Rhodes, Winskill, Friend & Warren, 1981). In *Nocardia mediterranei*, three chromosomal genes controlling the last three steps in the biosynthesis of rifamycin B are closely linked (T. Schupp, personal communication).

There appear to be no other clear examples resembling the situation with methylenomycin A in which structural biosynthetic genes for antibiotic synthesis are plasmid-borne, but there are several other well-established cases in which plasmids play a role in antibiotic synthesis, as well as many others in which such a conclusion should be regarded as tentative (reviewed by Hopwood, 1979). Amongst the well-established examples is that of chloramphenicol in *S. venezuelae* in which a plasmid appears to modulate the expression of chromosomal structural genes for antibiotic biosynthesis (Akagawa, Okanishi & Umezawa, 1975, 1979). The mechanism of this postulated control system is completely unknown, but there is no reason to believe that the plasmid is directly and specifically concerned with antibiotic synthesis, especially since loss of antibiotic production is commonly observed as part of a spectrum of phenotypic changes in other systems, whether they arise from pleiotropic chromosomal mutations (Piggot & Coote, 1976; Merrick, 1976b) or from some less well-characterised genetic change (Pogell, 1979).

Involvement of multi-enzyme synthetases in polyketide formation

As was noted in connection with the genetics of penicillin synthesis in *P. chrysogenum* and *A. nidulans*, mutations apparently abolishing the production of the ACV tripeptide, an intermediate whose synthesis may require the activity of a multi-enzyme complex, seem to constitute a high proportion of total *npe* mutations. An analogous situation may be presented by actinomycete antibiotics whose biosynthesis involves the successive condensation of activated acetate or propionate units: the so-called polyketide mode of synthesis. There is considerable evidence that such synthesis occurs by iteration of a series of six or seven reactions, as in fatty-acid synthesis (Turner, 1971; Packter, 1980), catalysed by multi-enzyme complexes. The macrolide tylosin represents a good example of such an antibiotic and it is significant that, amongst a collection of mutants blocked in tylosin biosynthesis, a very high proportion (59 out of 72) appeared to be defective in assembly of the lactone ring (the polyketide-derived aglycone), with many fewer mutants blocked at later stages in the pathway, involving specific oxidations of the aglycone and the attachment to it and O-methylation of the two sugar residues (Baltz & Seno, 1981; Seno & Baltz, 1981). It has been suggested (R. H. Baltz, personal communication) that, as in the case of typical single enzymes of primary metabolism (Clarke & Johnston, 1976), many missense mutations in genes coding for single enzymes of tylosin synthesis go undetected but that, in the case of genes coding for polypeptides which need to form a spatially accurate multi-enzyme complex, not only nonsense, but also most missense mutations abolish enzyme function and are therefore detected as antibiotic-negative mutations.

A similar situation, although less striking numerically, may apply for actinorhodin (Rudd & Hopwood, 1979) and for the prodiginine pigment (Rudd & Hopwood, 1980) in which large classes of mutants occurred which acted as converters in co-synthesis with other mutants, but not as secretors; this is the phenotype expected of mutants blocked early in the biosynthetic pathway, although the nature of the blocks has not been determined.

Artificial genetic manipulation

Recent advances in the handling of protoplasts make the actinomycetes promising organisms for the development of artificial systems which will both aid the genetic analysis of antibiotic synthesis and allow its experimental manipulation.

Protoplast fusion under the influence of PEG can lead to very efficient homologous recombination of chromosomal genes in *Streptomyces* (reviewed by Baltz, 1980; Hopwood, 1980a). This allows the generation of multiply marked strains which can be used for subsequent linkage mapping, either through protoplast fusion itself (Baltz, 1980), or by the more conventional conjugational analysis. Again, the loss of plasmids from *S. coelicolor* by the simple expedient of forming and regenerating protoplasts suggests that this treatment may provide presumptive evidence for plasmid involvement in antibiotic synthesis in a new situation, on the view that it is less likely than the use of chemicals like acridines or ethidium bromide to lead to structural changes in DNA, and so should be less ambiguous than such 'curing' treatments (Hopwood, 1980a). However, the most powerful application of protoplasts in the analysis of antibiotic production must surely be their use as hosts in recombinant DNA experiments.

Two approaches to the development of *Streptomyces* host–vector systems for *in vitro* generated recombinants have so far succeeded. These involve, respectively, the use of phages or of plasmids, but they have in common the induced uptake of DNA into *Streptomyces* protoplasts under the influence of PEG, followed by protoplast regeneration. For phages, primary transfectants are detected as plaques in soft agar overlays containing the transfection mixture together with an excess of untreated protoplasts or spores (Suarez &

Chater, 1980a). In the case of plasmids, transformants were originally detected by allowing populations of PEG-treated protoplast-DNA mixtures to regenerate, and then re-plating the spores from such confluent cultures in the presence of an excess of untransformed spores in order to detect transformants as 'pocks' (circular zones of inhibition of the plasmid-negative population by plasmid-positive individuals: Bibb *et al.*, 1978). More recently, primary transformants have been routinely detected as 'pocks' on the regeneration plates themselves (Thompson, Ward & Hopwood, 1980).

To develop a phage vector, a viable deletion mutant of the temperate *Streptomyces* phage φC31 was selected by its resistance to a chelating agent and into it was cloned the whole of the *E. coli* plasmid pBR322 to produce a bifunctional replicon capable of reproducing as a phage in *Streptomyces* and as a plasmid in *E. coli* and containing unique sites for several restriction endonucleases in dispensible regions (Suarez & Chater, 1980b); the vector has been further developed by the selection of additional deletions which will allow the insertion of DNA segments of up to 7 000 base pairs (K. F. Chater, personal communication). To produce plasmid vectors, antibiotic resistances from various *Streptomyces* spp. have been cloned into the SLP1.2 plasmid of *S. lividans* (Hopwood, Bibb, Ward & Westpheling, 1979; Bibb, Ward & Hopwood, 1980) to provide it with directly selectable markers, or into a reduced replicon derived from the SCP2* plasmid of *S. coelicolor* (Bibb, Schottel & Cohen, 1980; Thompson *et al.*, 1980). These vectors are, of course, still comparatively primitive but can now be developed in predictable ways to allow the efficient cloning of genes directly concerned with antibiotic synthesis, both structural and regulatory. Apart from cloning chromosomal genes, and the structural biosynthetic genes for methylenomycin synthesis carried on SCP1, it will be particularly interesting to analyse the functions of segments of putative regulatory plasmids, both in their natural hosts and in other strains, producing other antibiotics, whose synthesis might conceivably be regulated by plasmid-borne genes from other species (Hopwood, 1980b).

POSSIBLE ROLES OF ANTIBIOTICS AS MORPHOGENETIC REGULATORS

For both sporulating bacilli and streptomycetes, there are studies suggesting a role for some antibiotics, in some strains, in the regulation of normal spore development or in the outgrowth of the spores but it is difficult to deduce valid generalisations, especially since different groups working with the same strains have not always obtained the same results. Certainly not all antibiotics play such a role; some antibiotics are produced by non-sporulating organisms (Demain, 1974), while in others, apparently normal sporulation occurs in mutants, or plasmid-negative variants, lacking the ability to make particular antibiotics (Haavik & Thomassen, 1973; Chater & Merrick, 1979). Probably further studies of mutants unable to produce antibiotics and, if they really act as morphogenetic regulators, to respond to them, will be the single most powerful tool in understanding the mechanisms involved.

Bacilli

Most studies have involved *B. brevis*, but with a variety of strains, making different antibiotics. Strain ATCC 8185 makes tyrothricin (a mixture of the cyclic tyrocidine and the linear gramicidins A, B and C), while strain Nagano makes the cyclic gramicidin S. Some groups have reported an effect of antibiotics on spore maturation, while others find effects only on spore outgrowth.

In strain ATCC 8185, Mukherjee & Paulus (1977) isolated antibiotic non-producing mutants and found, as well as presumed pleiotropic mutations similar to those reported in other systems (they produced neither tyrocidine nor linear gramicidins and failed to sporulate), two interesting mutants producing tyrocidine but not gramicidin. They sporulated, but the spores were low in dipicolinic acid and were less heat resistant than wild-type spores. Gramicidin added to the mutant culture, at a critical period prior to sporulation, restored heat resistance of the spores nearly to the wild-type level. It was suggested, in view of earlier results indicating that gramicidin could alter the activity and specificity of RNA polymerase, that the antibiotic might be influencing sporulation by affecting the transition between the transcription of vegetative and sporulation genes. This view was supported by the finding (Sarkar, Langley & Paulus, 1977) that gramicidin apparently inhibits transcription at the level of promoter recognition and is selective for certain classes of genes. In the same *B. brevis* wild-type strain, Ristow and co-workers found interactions between the effects of tyrocidine and linear gramicidins in transcription, suggesting that normal sporulation may be regulated by a balanced antagonism between the two types of antibiotic

(Ristow, Schazschneider, Bauer & Kleinkauf, 1975; Ristow, Schazschneider, Vater & Kleinkauf, 1975). Early vegetative cell populations were induced to sporulate by the addition of tyrothricin, but only in a medium lacking a nitrogen source (Ristow, Pschorn, Hansen & Winkel, 1979), emphasising the complex interaction of nutritional and other factors involved in the sporulation response.

In strain Nagano, gramicidin S-negative mutants apparently produced spores with the same heat resistance and dipicolinic acid content as wild-type spores. Gramicidin S seemed to regulate spore outgrowth, rather than sporulation itself. Wild-type spores, which contained appreciable amounts of gramicidin S, outgrew only at spore concentrations below a certain level (10^8 spores ml^{-1}), unless the antibiotic was removed from the spores, while mutant spores outgrew even at very high concentrations, unless gramicidin S was added to the mutant culture (Nandi & Seddon, 1978; Seddon & Nandi, 1978). Gramicidin S restored wild-type outgrowth properties to mutant spores only if it was added to late vegetative cultures (not to mature spores), when it was incorporated into the spores (Lazaridis *et al.*, 1980). It appears that in this organism gramicidin S acts as a regulator of spore outgrowth, controlling the density-dependent outgrowth of a population of spores. There are analogies with fungal and higher plant systems in which specific metabolites (which may not have been studied as antibiotics) act to control spore or seed dormancy, probably to prevent simultaneous germination of the whole of a population in response to the first sign of favourable conditions (Sussman & Halvorson, 1966).

The idea that the differences in antibiotic effects in the two *B. brevis* strains – on spore maturation and outgrowth respectively – might simply reflect real differences between the strains and the antibiotics involved is shaken by some conflicting results. J. Piret and A. L. Demain (personal communication) obtained similar results on spore outgrowth in *B. brevis* Nagano to those discussed above, but could not repeat the observations of Mukherjee & Paulus (1977) on spore maturation; the spores of the linear gramicidin-negative mutants appeared to be indistinguishable from wild-type spores. Evidently much interesting work still remains to be done in these *Bacillus* systems.

Streptomycetes

As in the case of bacilli, roles for antibiotics in both the maturation and the germination of streptomycete spores have been suggested, but the genetic and biochemical analysis of these systems is still at a very early stage.

The first report of an endogenously produced compound affecting streptomycete differentiation, but not apparently an antibiotic, was the discovery by A. S. Khokhlov and his colleagues of a chemically characterised metabolite, the A-factor (see Queener *et al.*, 1978). It was produced in trace amounts by wild-type *Streptomyces griseus* but not by non-sporulating mutants, and could restore sporulation ability to the mutants.

Pogell (1979) summarised the results of studies in *Streptomyces alboniger* (Redshaw, McCann, Pentella & Pogell, 1979) in which a chemically partially characterised antibiotic, pamamycin, was found to be produced by the wild-type strain but not by a class of variants (Amy$^-$) which had lost the ability to produce aerial mycelium (apparently sporulation still occurred but the evidence for this was not given). Pamamycin stimulated aerial mycelium production by the wild-type but not by the Amy$^-$ variants. Two inhibitors of aerial mycelium production were also detected, both in the wild-type and in the Amy$^-$ variants; they showed no antimicrobial activity on test organisms but reversed that of pamamycin. They were not characterised, but they appeared to be chemically related to pamamycin. Thus the possibility exists that a step in differentiation is regulated by a balance between the effects of more than one secondary metabolite (we should recall the similar suggestion for tyrocidine and linear gramicidin in *B. brevis*). Unfortunately for the analysis of this system (though interestingly in other ways) the Amy$^-$ variants did not apparently represent simple point mutations but had properties – lack of reversion to Amy$^+$ and pleiotropic loss of several properties – consistent with loss of a segment of genetic material, possibly a plasmid. Perhaps a deliberate search for simple revertible mutations affecting the synthesis of pamamycin and the two putative differentiation inhibitors would provide strains capable of clarifying the roles of these compounds. There was no evidence that the best known antibiotic of this strain of *S. alboniger*, puromycin, was involved in differentiation; it was produced normally by both Amy$^+$ and Amy$^-$ strains.

In *Streptomyces viridochromogenes*, an antibiotic activity resembling one known to be produced by this species, nonactin, was found to be released from germinating spores and to inhibit the germination of other spores of the strain (Ensign, 1978; Hirsch & Ensign, 1978). It was suggested that the role of this antibiotic could be to control dormancy of the spores (Ensign, 1978); this is reminiscent of the situation for gramicidin S in *B. brevis* Nagano spores. No mutants were used in the study of the *S. viridochromogenes* antibiotic; they should provide a useful clue as to its true function.

CONCLUSION: POSSIBILITIES FOR GENETIC APPROACHES TO THE DEVELOPMENT OF INDUSTRIAL STRAINS

As well as the scientific interest of genetic studies of antibiotic-producing microorganisms, which I hope to have demonstrated in this chapter, it is clear that there are considerable possibilities for the use of genetics in the construction of industrial strains which make more of a known antibiotic, or make it more cheaply, or produce new compounds with useful new properties. The artificial techniques, now being perfected, for the genetic manipulation of antibiotic-producing strains are particularly promising in this regard. Some of the possibilities have recently been discussed elsewhere (Hopwood, 1980b; Queener & Baltz, 1979; Hopwood & Chater, 1980).

REFERENCES

Abraham, E. P. (1974). Some aspects of the development of the penicillins and cephalosporins. *Developments in Industrial Microbiology,* **15**, 3–15.

Aguilar, A., Duro, A. F., Candela, A., Rodriguez, M., Fernández-Jorge, D., Asensio, C. & Baquero, F. (1980). On the mechanisms of action of microcins. In *Current Chemotherapy and Infectious Disease*, ed. R. Day & G. Pollack. Washington DC: American Society for Microbiology, pp. 556–7.

Akagawa, H., Okanishi, M. & Umezawa, H. (1975). A plasmid involved in chloramphenicol production in *Streptomyces venezuelae*: evidence from genetic mapping. *Journal of General Microbiology*, **90**, 336–46.

Akagawa, H., Okanishi, M. & Umezawa, H. (1979). Genetics and biochemical studies of chloramphenicol-nonproducing mutants of *Streptomyces venezuelae* carrying plasmid. *Journal of Antibiotics*, **32**, 610–20.

Alačević, M. (1976). Recent advances in *Streptomyces rimosus* genetics. In *Second International Symposium on the Genetics of Industrial Microorganisms*, ed. K. D. Macdonald, pp. 513–19. London: Academic Press.

Bachmann, B. J. & Low, K. B. (1980). Linkage map of *Escherichia coli* K-12, edition 6. *Microbiological Reviews*, **44**, 1–56.

Baltz, R. H. (1980). Genetic recombination by protoplast fusion in *Streptomyces. Developments in Industrial Microbiology,* **21,** 43–54.

BALTZ, R. H. & SENO, E. T. (1981). Biosynthesis of tylosin: properties of *Streptomyces fradiae* mutants blocked in specific biosynthetic steps. *Journal of General Microbiology* (submitted).

BAUMANN, R., HÜTTER, R. & HOPWOOD, D. A. (1974). Genetic analysis in a melanin-producing streptomycete, *Streptomyces glaucescens*. *Journal of General Microbiology*, **81**, 463–74.

BEGGS, J. D. (1978). Transformation of yeast by a replicating hybrid plasmid. *Nature, London*, **275**, 104–9.

BÉRDY, J. (1974). Recent developments of antibiotic research and classification of antibiotics according to chemical structure. *Advances in Applied Microbiology*, **18**, 309–406.

BIBB, M. J., FREEMAN, R. F. & HOPWOOD, D. A. (1977). Physical and genetical characterisation of a second sex factor, SCP2, for *Streptomyces coelicolor* A3(2). *Molecular and General Genetics*, **154**, 155–66.

BIBB, M. J., SCHOTTEL, J. L. & COHEN, S. N. (1980). A DNA cloning system for interspecies gene transfer in antibiotic-producing *Streptomyces*. *Nature, London*, **284**, 526–31.

BIBB, M. J., WARD, J. M. & HOPWOOD, D. A. (1978). Transformation of plasmid DNA into *Streptomyces* at high frequency. *Nature, London*, **274**, 398–400.

BIBB, M. J., WARD, J. M. & HOPWOOD, D. A. (1980). The development of a cloning system for *Streptomyces*. *Developments in Industrial Microbiology*, **21**, 55–64.

BOTSTEIN, D., FALCO, S. C., STEWART, S. E., BRENNAN, M., SCHERER, S., STINCHROMB, D. T., STRUHL, K. & DAVIS, R. W. (1979). Sterile host yeasts (Shy): a eukaryotic system of biological containment for recombinant DNA experiments. *Gene*, **8**, 17–24.

BRIDGES, B. A. (1976). Mutation induction. In *Second International Symposium on the Genetics of Industrial Microorganisms*, ed. K. D. Macdonald, pp. 7–14. London: Academic Press.

BU'LOCK, J. D. (1961). Intermediate metabolism and antibiotic synthesis. *Advances in Applied Microbiology*, **3**, 293–342.

BUNTING, M. I. (1946). The inheritance of color in bacteria, with special reference to *Serratia marcescens*. *Cold Spring Harbor Symposium on Quantitative Biology*, **11**, 25–31.

CARERE, A., RUSSI, S., BIGNAMI, M. & SERMONTI, G. (1973). An operon for histidine biosynthesis in *Streptomyces coelicolor*. I. Genetic evidence. *Molecular and General Genetics*, **123**, 219–24.

CASE, M. E., SCHWEIZER, M., KUSHNER, S. R. & GILES, N. H. (1979). Efficient transformation of *Neurospora crassa* by utilizing hybrid plasmid DNA. *Proceedings of the National Academy of Sciences, USA*, **76**, 5259–63.

CATEN, C. E. & JINKS, J. L. (1976). Quantitative genetics. In *Second International Symposium on the Genetics of Industrial Microorganisms*, ed. K. D. Macdonald, pp. 93–111. London: Academic Press.

CHANG, S. & COHEN, S. N. (1979). High frequency transformation of *Bacillus subtilis* protoplasts by plasmid DNA. *Molecular and General Genetics*, **168**, 111–15.

CHATER, K. F. (1972). A morphological and genetic mapping study of white colony mutants of *Streptomyces coelicolor*. *Journal of General Microbiology*, **72**, 9–28.

CHATER, K. F. (1979). Some recent developments in *Streptomyces* genetics. In *Genetics of Industrial Microorganisms*, ed. O. K. Sebek & A. I. Laskin, pp. 123–33. Washington DC: American Society for Microbiology.

CHATER, K. F. & MERRICK, M. J. (1979). Streptomycetes. In *Developmental Biology of Prokaryotes*, ed. J. H. Parish, pp. 93–114. Oxford: Blackwell.

CLARKE, C. H. & JOHNSTON, A. W. B. (1976). Intragenic mutational spectra and hot spots. *Mutation Research*, **36**, 147–64.

CLUTTERBUCK, A. J. (1974). *Aspergillus nidulans.* In *Handbook of Genetics*, vol. 1, ed. R. C. King, pp. 447–510. New York: Plenum.

COATS, J. H. & ROESER, J. (1971). Genetic recombination in *Streptomyces bikiniensis* var. *zorbonensis*. *Journal of Bacteriology*, **105**, 880–5.

COLE, D. C., HOLT, G. & MACDONALD, K. D. (1976). Relationship of the genetic determination of impaired penicillin production in naturally occurring strains of *Aspergillus nidulans*. *Journal of General Microbiology*, **96**, 423–6.

DEMAIN, A. L. (1974). How do antibiotic producing microorganisms avoid suicide? *Annals of the New York Academy of Sciences,* **235,** 601–2.

DEMAIN, A. L. (1981). β-lactam Antibiotics. In *Handbook of Experimental Pharmacology,* ed. A. L. Demain. Berlin; Springer (In press).

DITCHBURN, P., HOLT, G. & MACDONALD, K. D. (1976). The genetic location of mutations increasing penicillin yield in *Aspergillus nidulans.* In *Second International Symposium on the Genetics of Industrial Microorganisms*, ed. K. D. Macdonald, pp. 213–27. London: Academic Press.

DREW, S. W. & DEMAIN, A. L. (1977). Effect of primary metabolites on secondary metabolism. *Annual Review of Microbiology*, **31**, 343–56.

EDWARDS, G. F. ST. L., HOLT, G. & MACDONALD, K. D. (1974). Mutants of *Aspergillus nidulans* impaired in penicillin biosynthesis. *Journal of General Microbiology*, **84**, 420–2.

ENSIGN, J. C. (1978). Formation, properties and germination of actinomycete spores. *Annual Review of Microbiology*, **32**, 185–219.

ERHLICH, S. D. (1978). Cloning in *Bacillus subtilis. Proceedings of the National Academy of Sciences, USA*, **75**, 1433–6.

FAWCETT, P. A., USHER, J. J. & ABRAHAM, E. P. (1976). Aspects of cephalosporin and penicillin biosynthesis. In *Second International Symposium on the Genetics of Industrial Microorganisms*, ed. K. D. Macdonald, pp. 129–38. London: Academic Press.

FODOR, K. & ALFÖLDI, L. (1976). Fusion of protoplasts of *Bacillus megaterium. Proceedings of the National Academy of Sciences, USA*, **73**, 2147–50.

FRIEND, E. J. & HOPWOOD, D. A. (1971). The linkage map of *Streptomyces rimosus. Journal of General Microbiology*, **68**, 187–97.

FRIEND, E. J., WARREN, M. & HOPWOOD, D. A. (1978). Genetic evidence for a plasmid controlling fertility in an industrial strain of *Streptomyces rimosus. Journal of General Microbiology*, **106**, 201–6.

FRØYSHOV, Ø. (1975). Enzyme-bound intermediates in the biosynthesis of bacitracin. *European Journal of Biochemistry*, **59**, 201–6.

FUJISAWA, Y., KITANO, K. & KANZAKI, T. (1975). Accumulation of deacetoxycephalosporin C by a deacetylcephalosporin C negative mutant of *Cephalosporium acremonium. Agricultural and Biological Chemistry*, **39**, 2049–55.

GABOR, M. H. & HOTCHKISS, R. D. (1979). Parameters governing bacterial regeneration and genetic recombination after fusion of *Bacillus subtilis* protoplasts. *Journal of Bacteriology*, **137**, 1346–53.

GERBER, N. (1975). Prodigiosin-like pigments. *Critical Reviews in Microbiology*, **3**, 469–85.

GOTTLIEB, D. (1976). The production and role of antibiotics in soil. *Journal of Antibiotics*, **29**, 987–1000.

GRYCZAN, T., CONTENTE, S. & DUBNAU, D. (1980). Molecular cloning of heterologous chromosomal DNA by recombination between a plasmid vector and a homologous resident plasmid in *Bacillus subtilis. Molecular and General Genetics*, **177**, 459–67.

Haavik, H. & Thomassen, S. (1973). A bacitracin-negative mutant of *Bacillus licheniformis* which is able to sporulate. *Journal of General Microbiology*, **76**, 451–4.

Hamlyn, P. F. & Ball, C. (1979). Recombination studies with *Cephalosporium acremonium*. In *Genetics of Industrial Microorganisms*, ed. O. K. Sebek & A. I. Laskin. Washington DC: American Society for Microbiology.

Harold, R. J. & Hopwood, D. A. (1970). Ultraviolet-sensitive mutants of *Streptomyces coelicolor*. II. Genetics. *Mutation Research*, **10**, 439–48.

Hedges, R. W. (1978). The R factors of Serratia. In *The Genus Serratia*, ed. A. von Graevenitz & S. J. Rubin. Cleveland: CRC Uniscience Press.

Henner, D. J. & Hoch, J. A. (1980). The *Bacillus subtilis* chromosome. *Microbiological Reviews*, **44**, 57–82.

Hinnen, A., Hicks, J. B. & Fink, G. R. (1978). Transformation of yeast. *Proceedings of the National Academy of Sciences, USA*, **75**, 1929–33.

Hirsch, C. F. & Ensign, J. C. (1978). Some properties of *Streptomyces viridochromogenes* spores. *Journal of Bacteriology*, **134**, 1056–63.

Holloway, B. W. (1979). Plasmids that mobilize bacterial chromosome. *Plasmid*, **2**, 1–19.

Holloway, B. W., Krishnapillai, V. & Morgan, A. F. (1979). Chromosomal genetics of *Pseudomonas*. *Microbiological Reviews*, **43**, 73–102.

Holt, G., Edwards, G. F. St L. & Macdonald, K. D. (1976). The genetics of mutants impaired in the biosynthesis of penicillin. In *Second International Symposium on the Genetics of Industrial Microorganisms*, ed. K. D. Macdonald, pp. 199–211. London: Academic Press.

Hopwood, D. A. (1966). Lack of constant genome ends in *Streptomyces coelicolor*. *Genetics*, **54**, 1177–84.

Hopwood, D. A. (1967). Genetic analysis and genome structure in *Streptomyces coelicolor*. *Bacteriological Reviews*, **31**, 373–403.

Hopwood, D. A. (1972). Genetic analysis in microorganisms. In *Methods in Microbiology*, vol. 7B, ed. J. R. Norris & D. W. Ribbons, pp. 29–159. London: Academic Press.

Hopwood, D. A. (1978). Extrachromosomally determined antibiotic production. *Annual Review of Microbiology*, **32**, 373–92.

Hopwood, D. A. (1979). Genetics of antibiotic production by actinomycetes. *Journal of Natural Products*, **42**, 596–602.

Hopwood, D. A. (1980a). Uses of protoplasts in the genetic manipulation of streptomycetes. In *International Symposium on Actinomycete Biology*, ed. K. P. Schaal. Stuttgart: Gustav Fischer-Verlag (in press).

Hopwood, D. A. (1980b). Possible applications of genetic recombination in the discovery of new antibiotics in actinomycetes. In *Antibiotics of the Future*, ed. L. Ninet. London: Academic Press (in press).

Hopwood, D. A., Bibb, M. J., Ward, J. M. & Westpheling, J. (1979). Plasmids in *Streptomyces coelicolor* and related species. In *Plasmids of Medical, Environmental and Commercial Importance*, ed. K. N. Timmis & A. Pühler, pp. 245–58. Amsterdam: Elsevier-North Holland.

Hopwood, D. A. & Chater, K. F. (1974). *Streptomyces coelicolor*. In *Handbook of Genetics*, vol. 1, ed. R. C. King, pp. 237–55. New York: Plenum Press.

Hopwood, D. A. & Chater, K. F. (1980). Fresh approaches to antibiotic production. *Philosophical Transactions of the Royal Society*, **B, 290,** 313–28.

Hopwood, D. A., Chater, K. F., Dowding, J. E. & Vivian, A. (1973). Advances in *Streptomyces coelicolor* genetics. *Bacteriological Reviews*, **37**, 371–405.

Hopwood, D. A. & Merrick, M. J. (1977). Genetics of antibiotic production. *Bacteriological Reviews*, **41**, 595–635.

HOPWOOD, D. A. & WRIGHT, H. M. (1976). Interactions of the plasmid SCP1 with the chromosome of *Streptomyces coelicolor* A3(2). In *Second International Symposium on the Genetics of Industrial Microorganisms*, ed. K. D. Macdonald, pp. 607–19. London: Academic Press.

HOPWOOD, D. A. & WRIGHT, H. M. (1979). Factors affecting recombinant frequency in protoplast fusions of *Streptomyces coelicolor. Journal of General Microbiology*, **111**, 137–43.

HOPWOOD, D. A., WRIGHT, H. M., BIBB, M. J. & COHEN, S. N. (1977). Genetic recombination through protoplast fusion in *Streptomyces. Nature, London*, **268**, 171–4.

HORI, K., KUROTSU, T., KANDA, M., MIURA, S., NOZOE, A. & SAITO, Y. (1978). Studies on gramicidin S synthetase. Purification of the heavy enzyme obtained from some mutants of *Bacillus brevis. Journal of Biochemistry (Tokyo)*, **84**, 425–34.

HORNEMANN, U. & HOPWOOD, D. A. (1978). Isolation and characterisation of desepoxy-4,5-didehydro-methylenomycin A, a precursor of methylenomycin A in SCP1$^+$ strains of *Streptomyces coelicolor* A3(2). *Tetrahedron Letters*, 2977–8.

HORNEMANN, U. & HOPWOOD, D. A. (1980). Biosynthesis of methylenomycin A: a plasmid-determined antibiotic. In *Antibiotics, Biosynthesis*, vol. IV, ed. J. W. Corcoran. Berlin, Heidelberg, New York: Springer. (In press.)

HOTCHKISS, R. D. & GABOR, M. H. (1980). Biparental products of bacterial protoplast fusion showing unequal parental chromosome expression. *Proceedings of the National Academy of Sciences, USA*, **77**, 3553–7.

IWAKI, M., SHIMURA, K., KANDA, M., KAJI, E. & SAITO, Y. (1972). Some mutants of *Bacillus brevis* deficient in gramicidin S formation. *Biochemical and Biophysical Research Communications*, **48**, 113–18.

KAMBE, M., IMAE, Y. & KURAHASHI, K. (1974). Biochemical studies on gramicidin S non-producing mutants of *Bacillus brevis* ATCC 9999. *Journal of Biochemistry (Tokyo)*, **75**, 481–93.

KANDA, M., HORI, K., KUROTSU, T., MIURA, S., NOZOE, A. & SAITO, Y. (1978). Studies of gramicidin S synthetase. Purification and properties of the light enzyme obtained from some mutants of *Bacillus brevis. Journal of Biochemistry (Tokyo)*, **84**, 435–41.

KATZ, E. & DEMAIN, A. L. (1977). The peptide antibiotics of *Bacillus*: chemistry, biogenesis and possible functions. *Bacteriological Reviews*, **41**, 449–74.

KIRBY, R. & HOPWOOD, D. A. (1977). Genetic determination of methylenomycin synthesis by the SCP1 plasmid of *Streptomyces coelicolor* A3(2). *Journal of General Microbiology*, **98**, 239–52.

KIRBY, R., WRIGHT, L. F. & HOPWOOD, D. A. (1975). Plasmid-determined antibiotic synthesis and resistance in *Streptomyces coelicolor. Nature, London*, **254**, 265–7.

LAZARIDIS, I., FRANGON-LAZARIDIS, M., MACCUISH, F. C., NANDI, S. & SEDDON, B. (1980). Gramicidin S content and germination and outgrowth of *Bacillus brevis* Nagano spores. *FEMS Microbiology Letters*, **7**, 229–32.

LIERSCH, M., NÜESCH, J. & TREICHLER, H. J. (1976). Final steps in the biosynthesis of cephalosporin C. In *Second International Symposium on the Genetics of Industrial Microorganisms*, ed. K. D. Macdonald, pp. 179–95. London: Academic Press.

LIPMANN, F. (1973). Nonribosomal polypeptide synthesis on polyenzymal templates. *Accounts of Chemical Research*, **6**, 361–7.

MACDONALD, K. D. & HOLT, G. (1976). Genetics of biosynthesis and overproduction of penicillin. *Science Progress (London)*, **63**, 547–73.

MAKINS, J. F., HOLT, G. & MACDONALD, K. D. (1980). Cosynthesis of penicillin following treatment of mutants of *Aspergillus nidulans* impaired in antibiotic production with lytic enzymes. *Journal of General Microbiology*, **119**, 397–404.

MARTÍN. J. F. & DEMAIN, A. L. (1980). Control of antibiotic biosynthesis. *Microbiological Reviews*, **44**, 230–51.

MARTÍN, J. F., LUENGO, J. M., REVILLA, G. & VILLANUEVA, J. R. (1979). Biochemical genetics of the β-lactam antibiotic biosynthesis. In *Genetics of Industrial Microorganisms*, ed. O. K. Sebek & A. I. Laskin, pp. 83–9. Washington DC: American Society for Microbiology.

MERRICK, M. J. (1976a). Hybridization and selection for penicillin production – a biometrical approach to strain improvement. In *Second International Symposium on the Genetics of Industrial Microorganisms*, ed. K. D. Macdonald, pp. 229–42. London: Academic Press.

MERRICK, M. J. (1976b). A morphological and genetic mapping study of bald colony mutants of *Streptomyces coelicolor*. *Journal of General Microbiology*, **96**, 299–315.

MERRICK, M. J. & CATEN, C. E. (1975). The inheritance of penicillin titre in wild-type isolates of *Aspergillus nidulans*. *Journal of General Microbiology*, **86**, 283–93.

MUKHERJEE, P. K. & PAULUS, H. (1977). Biological function of gramicidin: Studies on gramicidin-negative mutants. *Proceedings of the National Academy of Sciences, USA*, **74**, 780–4.

NANDI, S. & SEDDON, B. (1978). Evidence for gramicidin S functioning as a bacterial hormone specifically regulating spore outgrowth in *Bacillus brevis* Nagano. *Biochemical Society Transactions*, **6**, 409–11.

NORMANSELL, P. J. M., NORMANSELL, I. D. & HOLT, G. (1979). Genetic and biochemical studies of mutants of *Penicillium chrysogenum* impaired in penicillin production. *Journal of General Microbiology*, **112**, 113–26.

NÜESCH, J., TREICHLER, H. J. & LIERSCH, M. (1973). The biosynthesis of cephalosporin C. In *Genetics of Industrial Microorganisms: Actinomycetes and Fungi*, ed. Z. Vaněk, Z. Hošťálek & J. Cudlin, pp. 309–34. Prague: Academia.

OKANISHI, M. (1979). Plasmids and antibiotic synthesis in streptomycetes. In *Genetics of Industrial Microorganisms*, ed. O. K. Sebek & A. I. Laskin, pp. 134–40. Washington, DC: American Society for Microbiology.

PACKTER, N. M. (1980). Biosynthesis of acetate-derived phenols (polyketides). In *The Biochemistry of Plants*, vol. 4, ed. P. K. Stumpf, pp. 535–70. New York: Academic Press.

PETIT, A., TEMPÉ, J., KERR, A., HOSTERS, M. VAN MONTAGUE, M. & SCHELL, J. (1978). Substrate induction of conjugative activity of *Agrobacterium tumefaciens* Ti plasmids. *Nature, London*, **271**, 570–2.

PIGGOTT, P. J. & COOTE, J. G. (1976). Genetic aspects of bacterial endospore formation. *Bacteriological Reviews*, **40**, 908–62.

POGELL, B. M. (1979). Regulation of aerial mycelium formation in streptomycetes. In *Genetics of Industrial Microorganisms*, ed. O. K. Sebek & A. I. Laskin, pp. 218–34. Washington DC: American Society for Microbiology.

PONTECORVO, G., ROPER, J. A., HEMMONS, L. M., MACDONALD, K. D. & BUFTON, A. (1953). The genetics of *Aspergillus nidulans*. *Advances in Genetics*, **5**, 141–253.

PONTECORVO, G. & SERMONTI, G. (1954). Parasexual recombination in *Penicillium chrysogenum*. *Journal of General Microbiology*, **11**, 94–104.

QUEENER, S. W. & BALTZ, R. H. (1979). Genetics of industrial microorganisms *Annual Reports on Fermentation Processes*, **3**, 5–45.

Queener, S. W., Sebek, O. K. & Vezina, C. (1978). Mutants blocked in antibiotic synthesis. *Annual Review of Microbiology*, **32**, 593–636.

Redshaw, P. A., McCann, P. A., Pentella, M. A. & Pogell, B. M. (1979). Simultaneous loss of multiple differentiated functions in aerial mycelium-negative isolates of streptomycetes. *Journal of Bacteriology*, **137**, 891–9.

Rhodes, P. M., Winskill, N., Friend, E. J. & Warren, M. (1981). Biochemical and genetic characterisation of *Streptomyces rimosus* mutants impaired in oxytetracycline biosynthesis. *Journal of General Microbiology* (submitted).

Ristow, H., Pschorn, W., Hansen, J. & Winkel, U. (1979). Induction of sporulation in *Bacillus brevis* by peptide antibiotics. *Nature, London*, **280**, 165–6.

Ristow, H., Schazschneider, B., Bauer, K. & Kleinkauf, H. (1975). Tyrocidine and the linear gramicidins. Do these peptide antibiotics play an antagonistic regulative role in sporulation? *Biochimica et Biophysica Acta*, **390**, 246–52.

Ristow, H., Schazschneider, B., Vater, J. & Kleinkauf, H. (1975). Some characteristics of the DNA tyrocidine complex and a possible mechanism of the gramicidin action. *Biochimica et Biophysica Acta*, **414**, 1–8.

Rudd, B. A. M. (1978). Genetics of Pigmented Secondary Metabolites in *Streptomyces coelicolor*. Ph.D. Thesis, University of East Anglia, Norwich.

Rudd, B. A. M. & Hopwood, D. A. (1979). Genetics of actinorhodin biosynthesis in *Streptomyces coelicolor* A3(2). *Journal of General Microbiology*, **114**, 35–43.

Rudd, B. A. M. & Hopwood, D. A. (1980). A pigmented mycelial antibiotic in *Streptomyces coelicolor*: control by a chromosomal gene cluster. *Journal of General Microbiology,* **119,** 333–40.

Sanderson, K. E. & Hartmann, P. E. (1978). Linkage map of *Salmonella typhimurium*, edition V. *Microbiological Reviews*, **42**, 471–519.

Sarkar, N., Langley, D. & Paulus, H. (1977). Biological function of gramicidin: Selective inhibition of RNA polymerase. *Proceedings of the National Academy of Sciences, USA*, **74**, 1478–82.

Schaeffer, P., Cami, B. & Hotchkiss, R. D. (1976). Fusion of bacterial protoplasts. *Proceedings of the National Academy of Sciences, USA*, **73**, 2151–5.

Schupp, T., Hütter, R. & Hopwood, D. A. (1975). Genetic recombination in *Nocardia mediterranei. Journal of Bacteriology*, **121**, 128–36.

Seddon, B. & Nandi, S. (1978). Biochemical aspects of germination and outgrowth of *Bacillus brevis* Nagano and control by gramicidin S. *Biochemical Society Transactions*, **6**, 412–13.

Seno, E. T. & Baltz, R. H. (1981). Biosynthesis of tylosin: Properties of S-adenosyl-L-methionine: macrocin O-methyltransferase in extracts of *Streptomyces fradiae* strains which produce normal or elevated levels of tylosin. *Journal of General Microbiology* (submitted).

Sermonti, G. (1956). Complementary genes which affect penicillin yields. *Journal of General Microbiology*, **15**, 599–608.

Sermonti, G. (1969). *Genetics of Antibiotic-Producing Microorganisms*. London: Wiley-Interscience.

Shimura, K., Iwaki, M., Kanda, M., Hori, K., Kaji, E. & Saito, Y. (1974). On the enzyme system obtained from some mutants of *Bacillus brevis* deficient in gramicidin S formation. *Biochimica et Biophysica Acta*, **338**, 577–87.

Simpson, I. N. & Caten, C. F. (1981). Genetics of penicillin titre in lines of *Aspergillus nidulans* selected through recurrent mutagenesis. *Journal of General Microbiology* (in press).

Struhl, K., Stinchcomb, D. T., Scherer, S. & Davis, R. W. (1979). High frequency transformation of yeast: autonomous replication of hybrid DNA molecules. *Proceedings of the National Academy of Sciences, USA,* **76,** 1035–9.

STUTTARD, C. (1979). Transduction of auxotrophic markers in a chloramphenicol-producing strain of *Streptomyces*. *Journal of General Microbiology*, **110**, 479–82.
SUAREZ, J. E. & CHATER, K. F. (1980a). Polyethylene glycol-assisted transfection of *Streptomyces* protoplasts. *Journal of Bacteriology*, **142**, 8–14.
SUAREZ, J. E. & CHATER, K. F. (1980b). DNA cloning in *Streptomyces*: a bifunctional replicon comprising pBR322 cloned into a *Streptomyces* phage. *Nature, London,* **286,** 527–9.
SUSSMAN, A. S. & HALVORSON, H. O. (1966). *Spores, their Dormancy and Germination*. New York: Harper & Row.
THOMPSON, C. J., WARD, J. M. & HOPWOOD, D. A. (1980). DNA cloning in *Streptomyces*: resistance genes from antibiotic-producing species. *Nature, London,* **286,** 525–7.
TROOST, T. R., DANILENKO, V. N. & LOMOVSKAYA, N. D. (1979). Fertility properties and regulation of antimicrobial substance production by SCP2 plasmid of *Streptomyces coelicolor* A3(2). *Journal of Bacteriology*, **140**, 359–68.
TURNER, W. B. (1971). *Fungal Metabolites*. London and New York: Academic Press.
VOROBJEVA, I. P., KHMEI, I. A. & ALFÖLDI, L. (1980). Transformation of *Bacillus megaterium* protoplasts by plasmid DNA. *FEMS Microbiology Letters* (in press).
WILLIAMS, R. P. & HEARN, W. R. (1967). Prodigiosin. In *Antibiotics: Biosynthesis*, vol. 2, ed. D. Gottlieb & P. D. Shaw, pp. 410–32. Berlin, Heidelberg and New York: Springer-Verlag.
WILLIAMS, S. T. & KHAN, M. R. (1974). Antibiotics – A soil microbiologist's view. *Postepy Higieny i Medycyny Doswiadczalnej*, **28**, 395–408.
WRIGHT, L. F. & HOPWOOD, D. A. (1976a). Identification of the antibiotic determined by the SCP1 plasmid of *Streptomyces coelicolor* A3(2). *Journal of General Microbiology*, **95**, 96–106.
WRIGHT, L. F. & HOPWOOD, D. A. (1976b). Actinorhodin is a chromosomally determined antibiotic in *Streptomyces coelicolor* A3(2). *Journal of General Microbiology*, **96**, 289–97.
YOUNG, F. E. & WILSON, G. A. (1978). Development of *Bacillus subtilis* as a cloning system. In *Genetic Engineering*, ed. A. M. Chakrabarty, pp. 145–7. CRC press: West Palm Beach.
ZÄHNER, H. (1978). The search for new secondary metabolites. In *Antibiotics and other Secondary Metabolites*, ed. R. Hütter, T. Leisinger, J. Nüesch, W. Wehrli, pp. 1–17. London: Academic Press.

GENETICS OF THE BACTERIAL CELL SURFACE

P. HELENA MÄKELÄ* AND B. A. D. STOCKER†

Central Public Health Laboratory, SF–00280 Helsinki 28, Finland

†Department of Medical Microbiology, Stanford University School of Medicine, Stanford, California 94305, USA

INTRODUCTION

The surface of a bacterium exposed to the outer world determines the outcome of many interactions of the bacterium and its environment–in particular interaction of pathogenic bacteria with vertebrate hosts–and for this reason has been studied since the earliest days of bacteriology. Observations on bacterial surface structures thus accumulated have made possible many important discoveries in genetics. In fact Griffith's work in 1928 which led to the identification of DNA as the material basis of the gene was on transformation of the capsular serotypes of pneumococci.

Since then, genetic methods have been extensively applied to the study of bacterial surface structures. This work has led to a considerable body of information about the determination of these structures and enabled us to study structure–function relationships in the interactions of bacteria and environment. We review here this genetic information, showing how the genetic approach has been used to study structures, biosynthetic pathways and functions. Because of limitations of space we will deal only cursorily with some areas discussed in recent reviews (Braun, 1978; DiRienzo, Nakamura & Inouye, 1978; Halegoua & Inouye, 1979; Hazelbauer & Parkinson, 1977; Hazelbauer, 1979; Iino, 1977; Mäkelä & Mayer, 1976; Mäkelä & Stocker, 1969; Manning & Achtman, 1979; Markovitz, 1977; Mayer & Schmidt, 1979; Ørskov, Ørskov, Jann & Jann, 1977; Osborn, 1979; Reeves, 1979; Silverman & Simon, 1977; Stocker & Mäkelä, 1971; Stocker & Mäkelä, 1978). Because much more is known of the genetics of Gram-negative organisms, especially of the *Enterobacteriaceae* family, the review will be mostly concerned with them, even if the genetic work started in Gram-positive bacteria.

GENETIC PRINCIPLES AND METHODS

The surface characteristics – like any other phenotypic trait – of an individual bacterium are determined by the interaction of its genotype and the environment in which it developed. The genetic information of a bacterium is embodied in the base sequence of its DNA; all or most of this DNA in the bacteria so far studied forms a single closed circular molecule. This is called the chromosome, even though it differs from the eucaryotic chromosome both in composition and organization. However, additional genetic information may be present in smaller DNA molecules called plasmids, which are of great importance in the determination of some bacterial surface features. Both the chromosomal and plasmid DNA sequences include (i) structural genes, which can be sequences both transcribed and translated into polypeptides, or sequences only transcribed into non-messenger RNA such as ribosomal or transfer RNA and (ii) non-transcribed sequences, many of which serve as punctuation marks for genetic messages. These include sequences that interact with proteins (such as repressors and positive regulatory proteins, or RNA polymerase), with RNA (as in initiation or termination of transcription or translation), or with other DNA sequences to form special structures such as the loops perhaps involved in excision, insertion and inversion of DNA segments. The genetic analysis of bacterial surface structures starts from the study of the effects of changes in all these different kinds of genetic elements (= 'genes'), either arising by mutation or pre-existing in different isolates of related bacteria. Then the genes identified by these changes are mapped by any of the various systems available – conjugation, resulting in chromosomal recombination or plasmid transfer, DNA-mediated transformation, phage-mediated transduction, lysogenization and conversion, protoplast fusion or deletion mapping. In the last few years the traditional methods have been supplemented by the use of cloning in plasmid or phage vectors followed by direct sequencing of DNA and the protein products. The final goal is to define the type of gene (in terms described under i–ii above), its immediate product and its mode of action; this may require the integrated use of a variety of methods.

Mapping of genes provides a useful guide for their identification and techniques for their further manipulation. The 'chromosomes' of *E. coli* and *Salmonella typhimurium* have been extensively mapped (Bachmann & Low, 1980; Sanderson & Hartman, 1978)

Table 1. *Means of creating frequent genotypic variation*

Mechanism	Frequency of reverse change	Examples	Section of this review or other reference
Loss of plasmid	zero	*Shigella sonnei* LPS[a] *E. coli*, OMP[b] *Yersinia*, V and W antigens	LPS, Antigenic Conversion Major OMP, Regulation Other OMP, Effect of Plasmids
Invertible DNA segments ('site-specific' recombination by virtue of inverted repeat sequences)	low or high	*Salmonella* flagellar antigens	Flagella and Phase Variation
		Possibly many others: *Salmonella* O antigens	LPS, Regulation of O Side chains
		Some capsular polysaccharides	Capsular Polysaccharides
		Fimbriation and colony forms in enteric bacteria, *Neisseria* etc.	Fimbriae and Pili
Tandem duplication of genes (legitimate or *rec* dependent recombination)	low or high	*his* operon of *Salmonella* *ampA* gene of *E. coli*	Anderson & Roth (1977) Normark *et al.* (1977)
Mutational hot spots	usually low	*r*II gene in phage T4 *lacI* gene of *E. coli*	Benzer (1961) Coulondre *et al.* (1978)

[a] LPS, lipopolysaccharide.
[b] OMP, outer membrane proteins.

and form a firm basis for positioning new genes. For convenience, the chromosome is in both cases divided into 100 units, by which the position of each gene is defined. In *E. coli* they correspond closely to minutes of entry data derived from Hfr-mediated conjugation; in *Salmonella* the time of entry data must be modified to fit the 100 unit map.

The genotype of bacteria is normally stable, with mutations causing infrequent changes at the average estimated rate of 10^{-6} per gene per cell generation. More rapid alterations are, however, often seen in bacterial surface properties. Several mechanisms by which such more rapid changes can occur have now been identified (Table 1). Variation has perhaps been easier to detect in surface properties than in many other characteristics, and it seems that these mechanisms may be of general importance. In the following, examples and details of these mechanisms will be given at appropriate places.

THE SURFACE STRUCTURES

Nearly all eubacteria possess, external to the cytoplasmic membrane, a rigid cell wall made of the characteristic peptidoglycan or murein. In Gram-positive bacteria this cell wall is usually interlaced and covered by teichoic acids, polysaccharides of various sorts, and/or proteins. Although these often form a microscopically visible capsule around the cells they do not form organized structures and apparently do not efficiently prevent solutes from reaching the underlying structures

By contrast, the Gram-negative bacteria have, outside the peptidoglycan, an outer membrane that can be seen by electron microscopy as a separate structure and isolated from other cell components by physical or chemical separation methods. The outer membrane constitutes an important permeability barrier, preventing the entry of large water-soluble and in many Gram-negative groups also of hydrophobic molecules (Nikaido, 1979). The major components of the outer membrane are lipopolysaccharide (LPS), phospholipids and a limited number of protein species. In wild-type strains of *E. coli* or *Salmonella*, all or nearly all the LPS appears to be located in the outer leaflet of the outer membrane, leaving nearly all the phospholipids in the inner leaflet (Nikaido & Nakae, 1979). Thus, about half of the surface is made of LPS, and the rest, of the major outer membrane proteins. In mutant cells with defective LPS this arrangement may be disturbed, with phospholipid exposed and the permeability properties of the membrane altered. Polysaccharide and sometimes protein capsules may be present outside the outer membrane.

Both Gram-positive and Gram-negative bacteria may have special structures extending from the cell wall into the environment. These include pili and fimbriae, which seem to function in recognition of and adhesion to environmental structures, and motility-conferring flagella.

Many surface components are recognized by antisera and are therefore often described as surface antigens. Thus, the main antigens used for the identification and/or typing of Gram-negative bacteria in clinical bacteriology are capsular structures (K antigens), LPS (O antigen) or flagella (H antigens) (Kauffmann, 1954). Bacteriophages and bacteriocins likewise recognize structures exposed on the bacterial surface and use them as receptors (Lindberg, 1977; Konisky, 1979; Braun & Hantke 1977). All these properties provide generally applicable methods for isolating mutants resistant to the action of the specific agents.

GENETICS OF CELL SURFACE POLYSACCHARIDES

Lipopolysaccharide

Structural features

A wide range of Gram-negative prokaryotes, including blue-green algae, manufacture a macromolecular material, lipopolysaccharide or LPS, of the same general structure (Fig. 1, p. 228; Galanos, Lüderitz, Rietschel & Westphal, 1977). In the best investigated bacteria, *Salmonella* species and *E. coli*, each LPS molecule has an innermost part, called lipid A and made up of a β1–6 linked D-glucosamine disaccharide. This disaccharide is substituted at both its amino and hydroxy groups with fatty acids, which characteristically include β-hydroxymyristic (C14) acid, and, at both ends, with phosphate, phosphoryl-ethanolamine, and 4-aminoarabinose. These substitutions are often not stoichiometric and may be subject to variation e.g. according to the growth conditions of the bacteria. The main effects of LPS on mammalian hosts – such as lethality, pyrogenicity and mitogenic activity – reside in the amphiphilic lipid A. These endotoxic effects are shared by the LPS of widely different species, with the structurally unusual LPS of some photosynthetic bacteria as an exception.

To lipid A is attached an oligosaccharide usually made up of three units of 3-deoxy-D-*manno*-octulosonic acid (KDO) which in turn bears a short chain of several units of heptose, glucose, galactose and/or *N*-acetylglucosamine. Monosaccharide units of this core may be substituted e.g. by phosphate. All salmonellas have the same core structure (called Ra), whereas in *E. coli* several related cores (R_1 to R_4 recognized so far) are found. In wild-type – termed smooth because of their colony morphology – *Salmonella* and *E. coli* many of the LPS core molecules bear sidechains, each a linear polymer of an oligosaccharide repeat unit; the composition of the repeat unit differs widely between different serogroups and is the determinant of the O antigenic character of these bacteria.

Some Gram-negative bacteria – e.g. *Neisseria* species – have LPS that apparently consists of only lipid A and core (Adams, 1971; Perry *et al.*, 1978; Stead, Main, Ward & Watt, 1975). In bacteria that normally have O sidechains, these can be missing because of mutation in genes determining the biosynthesis of LPS. Such mutants are called rough or R because of their frequently rough colony form, and hence genes of LPS biosynthesis, defined by these mutations, are termed *rf*. It may be worth noting that many laboratory strains of *E. coli* such as K-12 and B are such R mutants; the presumed smooth ancestors of these strains are no longer available and thus their original O unit character (O serotype) is not known.

Biosynthesis

Knowledge of how the LPS is synthesized, like knowledge of its structure, was obtained to a very large extent by study of mutants with defects in its biosynthesis. The reactions involved – so far as known or guessed – can be grouped into five general classes (Nikaido, 1973; Osborn, 1979; Osborn & Rothfield, 1971; Robbins & Wright, 1971):

(i) Synthesis of water-soluble precursors, which where known are high-energy sugar nucleotides. Some of the enzymes involved have other functions as well, such as UDP-galactose epimerase used both for the conversion of UDP-glucose to UDP-galactose, the precursor of the galactose units of LPS, and for the utilization of galactose as energy source.

(ii) Transfer of monosaccharide, phosphate or other unit from the precursor to an acceptor, a lipid which may carry the more or less incomplete oligo- or polysaccharide under assembly. This lipid can be a precursor or complete form of lipid A for the assembly of the LPS core, or a special isoprenol carrier lipid for the assembly of O units. Two of the core transferases have been extensively purified (Endo & Rothfield, 1969; Müller, Hinckley & Rothfield, 1972).

(iii) Polymerization = transfer of a previously synthesized lipid-borne O unit or polymer of O units to the distal, non-reducing end of a newly synthesized O unit, attached to the lipid molecule on which it was built.

(iv) Transfer of a chain of one or (usually) several O units from the carrier lipid to the core-LPS. The enzymes probably involved have not been investigated.

(v) Modification of the polysaccharide. The addition of glucosyl branches to the O sidechain probably takes place when the O unit polymer is still attached to its carrier lipid (Takeshita & Mäkelä,

1971). The glucose is first transferred to its own carrier lipid and from this to the O unit polymer in a linkage specified by the transferase (Nikaido & Nikaido, 1971; Nikaido, Nikaido, Nakae & Mäkelä, 1971). The steps involved in other modifications are less well known. These modifications are often subject to variation by form variation and/or antigenic conversion by phages (see below).

Genetics of lipid A

Little has been learnt about the biosynthesis of lipid A, perhaps because of lack of relevant mutants (Osborn, 1979). *S. typhimurium* mutants with temperature-sensitive mutations affecting steps in the synthesis of KDO (Lehmann, Rupprecht & Osborn, 1977; Rick & Osborn, 1972) synthesize at restrictive temperature an incomplete lipid A lacking ester-linked fatty acids (Lehmann, 1977; Lehmann & Rupprecht, 1977; Rick, Fung, Ho & Osborn, 1977). One of the genes has been located at the appr. position 38 on the *S. typhimurium* linkage map (Fig. 2, *see* page 240; Rick & Osborn, 1977).

The KDO mutations are lethal, and the cells, when transferred to high temperature (42 °C) stop LPS synthesis almost immediately, and DNA and protein synthesis after one generation (Rick & Osborn, 1977). This, and the inability so far to isolate mutants defective in earlier steps of lipid A synthesis suggest that lipid A is essential for outer membrane assembly and for survival of the cells (Osborn, 1979). Recently, two classes of mutations phenotypically hypersensitive to detergents (*acrA*, at position 10 in the *E. coli* K-12 map, and *pmrA* for resistance to polymyxin, at position 95 in the *S. typhimurium* map) have been suggested to be altered in the KDO-lipid A part of LPS (Nakamura, 1968; Coleman & Leive, 1979; Mäkelä, Sarvas, Calcagno & Lounatmaa, 1978; Vaara, Vaara & Sarvas, 1979; M. Vaara, personal communication). These mutations do not prevent the synthesis of complete smooth-type LPS.

Genetics of LPS core

An *E. coli* B mutant has been described that lacks both the distal KDO unit and the phosphoryl ethanolamine substituent of the branch KDO (Prehm *et al.*, 1976a). The enzymic defect was not further specified nor the gene mapped. More is known about the more distal part of the LPS core. In *S. typhimurium* many mutants with different defects in core structure have been identified and the affected *rfa* genes mapped, most of them as a cluster between *cysE* (at position 79) and *pyrE*. The genetic analysis in *E. coli* is less complete, but the same rules seem to apply.

We first consider genes in the main *rfa* cluster. Mutations of *rfaD* or *rfaC* affect the formation of the proximal heptose unit. The presence of an abnormal heptose in the LPS of the *rfaD* mutants (Lehmann *et al.*, 1973) suggests that this gene is involved in the epimerization step in the synthesis of the heptose precursor. The nature of the *rfaC* function is not known. (In addition to these two genes in the main *rfa* cluster, a third gene, *rfaE*, located outside it, in the general region of 63–68, prevents by an unknown mechanism the addition of any heptose to the core.)

In *S. minnesota,* mutants called *rfaP* lack ability to transfer phosphate from ATP to LPS core and, presumably in consequence of this, fail to add LPS core units distal to glucose I (Mühlradt, Risse, Lüderitz & Westphal, 1968). Similar mutants have not been found in *S. typhimurium.*

Mutations at *rfaF* and *rfaG* prevent the addition of, respectively, the heptose II and glucose I units to the core. Extracts of *rfaG* mutants lack normal ability to transfer glucose from UDP-glucose to glucose-deficient LPS (of chemotype Rd1) – such as is made by *rfaG* mutants or by *galU* mutants, unable to synthesize UDP glucose (Osborn, 1968). Thus *rfaG* is presumed to be the structural gene for this glucosyl transferase. A Carbon plasmid carrying the corresponding *E. coli* chromosomal region restores normal LPS phenotype to *S. typhimurium* with a mutation at *rfaG* (Creeger & Rothfield, 1979) or at several other closely linked *rfa* loci (E. S. Creeger, personal communication).

Mutation at *rfaI* (formerly called *rfa*(R-res-2)) leads to the synthesis of LPS of type Rb3, with only the branch galactose present (Lindberg & Hellerqvist, 1980). On this basis *rfaI* would have to be the structural gene of galactose I transferase. However, results of tests of *in vitro* transferase activity in extracts of these mutants contradicted this hypothesis (M. J. Osborn & L. Rothfield, personal communication); the discrepancy remains unexplained.

Mutants unable to add the branch galactose unit, mentioned as an expected class in our previous review (Stocker & Mäkelä, 1978), have now been identified by structural analysis of the LPS made by one subset of *rfa*(FOR) mutants (Hudson, Lindberg & Stocker, 1978). FOR mutants (MacPhee, Krishnapillai, Roantree & Stocker, 1975) are smooth by the usually accepted criteria for the normal presence of O sidechains but, because of their resistance to phage

FO and concomitant increase of antibiotic sensitivity, are inferred to have defects in LPS core structure. The absence of the galactose branch apparently blocks core synthesis only partially, allowing a considerable proportion of the LPS molecules to be completed. Several mutations leading to the FOR phenotype and absence of the galactose branch unit were mapped in the main *rfa* cluster; on the assumption that they all are affected in the same gene we propose the symbol *rfaB* for it.

Genes *rfaJ* and *rfaK* are concerned with the addition, respectively, of the glucose II and *N*-acetylglucosamine branch unit to the complete core. Although the presumed transferase activities have not been measured, it seems reasonable to suppose that *rfaJ* and *rfaK* specify these enzymes since both the precursors (UDP-glucose, UDP-*N*-acetylglucosamine) and the acceptor LPS are present in the mutant cells.

Mutation at *rfaL* prevents transfer of O chains to the apparently complete core. Because certain mutations (*rfbT*) in the main *rfb* cluster (see below) have a similar phenotype, it is assumed that the products of both these genes participate in the transfer of O chains from the carrier lipid to complete core.

The *rfa* genes listed above cover most of the transfer reactions required to build the basic *Salmonella* core structure shown in Fig. 1. However, additional units are known or suspected to be present in the *Salmonella* core LPS – these include a branch heptose, and an unknown substituent that protects the galactose branch from attack by galactose oxidase (Funahara & Nikaido, 1980). Genes concerned with the information of such structures remain to be discovered, perhaps amongst the *rfa*(FOR) mutants whose core defects are not yet identified (Hudson, Lindberg & Stocker, 1978).

A probable regulatory gene is located outside the main *rfa* cluster. Initially, the phage sensitivity pattern and LPS monosaccharide composition of mutants of a class called *rfaH* suggested that they were deficient in addition of the α1–3 linked galactose I unit of the core (Wilkinson & Stocker, 1968). The *rfaH* gene was inferred to specify the single polypeptide chain of the purified transferase (Endo & Rothfield, 1969) because such transferase activity was lacking in extracts of these mutants, and their LPS accepted galactose when incubated with extracts of wild-type cells (Osborn, 1968; Kuo & Stocker, 1972). Later results make this, however, unlikely (Lindberg & Hellerqvist, 1980). Chemical and immunochemical analyses of the LPS of *rfaH* mutants indicate that their core

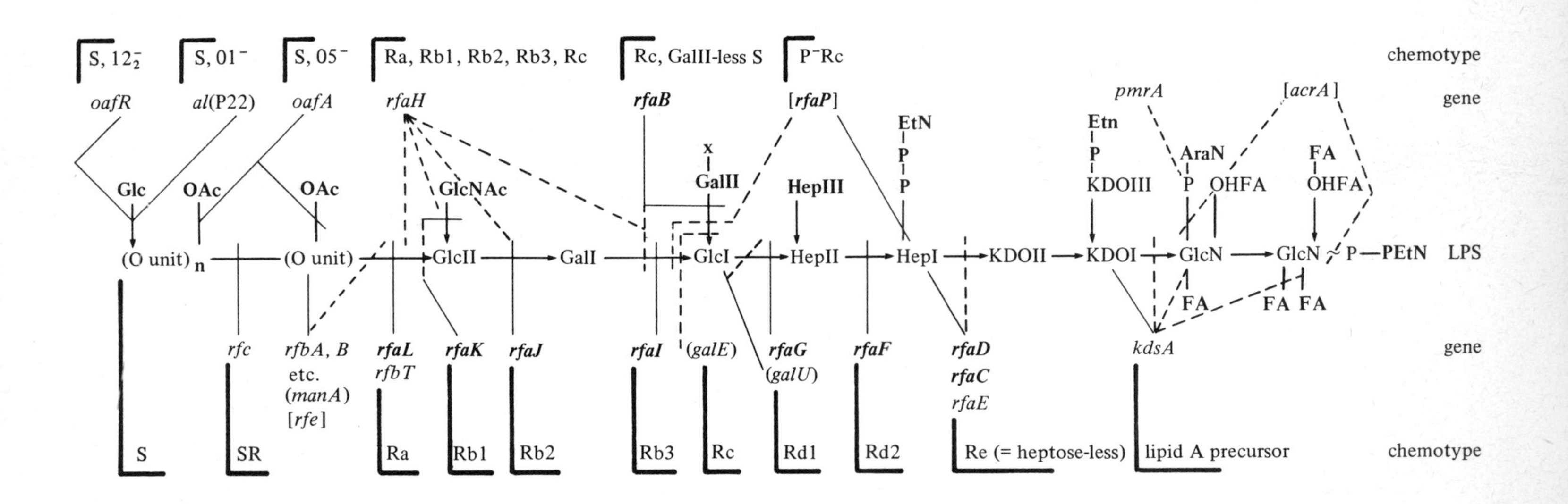
S, 12_2^-
S, 01^-
S, 05^-
Ra, Rb1, Rb2, Rb3, Rc
Rc, GalII-less S
P⁻Rc
chemotype
oafR
al(P22)
oafA
rfaH
rfaB
[rfaP]
pmrA
[acrA]
gene
Glc
OAc
OAc
GlcNAc
x
GalII
HepIII
EtN
P
P
Etn
P
KDOIII
AraN
P
OHFA
FA
OHFA
(O unit)
n
(O unit)
GlcII
GalI
GlcI
HepII
HepI
KDOII
KDOI
GlcN
GlcN
P
PEtN
LPS
FA
FA FA
rfc
rfbA, B
etc.
(manA)
[rfe]
rfaL
rfbT
rfaK
rfaJ
rfaI
(galE)
rfaG
(galU)
rfaF
rfaD
rfaC
rfaE
kdsA
gene
S
SR
Ra
Rb1
Rb2
Rb3
Rc
Rd1
Rd2
Re (= heptose-less)
lipid A precursor
chemotype

LPS molecules are heterogeneous, some being of type Rc, others of types Rb3, Rb2, Rb1 and even Ra, the complete core. Thus the block in the addition of the galactose I caused by the *rfaH* mutation is only partial, while several other reactions for addition of more distal core units, and probably also for addition of O chains, are partly blocked, too. Genetic analysis showed that all but one of five mutations of this sort were located between *metE* and *pepQ* (at position 84), presumably in a single *rfaH* gene (Stocker, Males & Takano, 1980). The fifth, phenotypically similar but less completely characterized, maps elsewhere.

Two possible explanations for the core defects of *rfaH* mutants can be proposed: (i) that the *rfaH* gene product is a polypeptide needed for efficient action of several transferases in core biosynthesis, or (ii) that it is a regulatory substance needed for complete turning-on of several *rfa* genes. It has recently been reported that in *E. coli* K-12 mutation at *sfrB,* also near *metE,* not only prevents the expression of several *tra* genes of the F factor but also causes defective LPS core structure, as indicated by changes in phage sensitivity (Beutin & Achtman, 1979). Expression of several functions of an F′*lac*$^+$*finP*$^-$ plasmid was found to be greatly reduced in any of several *rfaH* mutants of *S. typhimurium* although the expression of the same plasmid was normal in smooth cells of several other classes of *rfa* mutants, including *rfa-648* whose phenotype in other respects is *rfaH*-like (Sanderson & Stocker, manuscript in preparation). *E. coli* K-12 plasmids carrying the *metE* region restored the smooth LPS character to *rfaH* mutants of *Salmonella* (Stocker, Males & Takano, 1980; Creeger, Chen & Rothfield, 1979). These observations speak for alternative ii above. Thus the function of the *rfaH* gene of *Salmonella*, and its K-12 equivalent, *sfrB,* would be needed for full expression of several *rfa* genes as well as of several *tra* genes of the F factor. One can only speculate as to the possible advantage of such co-regulation. Furth-

Fig. 1. The structure of the lipopolysaccharide of *Salmonella typhimurium* and genes concerned in its synthesis. The genes are connected to the links or structures whose formation they prevent by continuous (more or less direct action) or dotted (indirect action) lines. Genes in the main *rfa* cluster at position 79 are printed in bold face. Genes that participate in LPS synthesis only secondarily are in parentheses. Genes not demonstrated in *Salmonella typhimurium* (but in other *Salmonellas* or in *E. coli*) are in square brackets.

In the LPS structure, groups for which heterogeneity is demonstrated or strongly suspected are printed in bold face. Abe, abequose; Ac, acetyl; EtN, ethanolamine; FA fatty acid; Gal, galactose; Glc, glucose; GlcN, glucosamine; Hep, heptose; KDO, 3-deoxy-D-manno-octulosonic acid; LPS, lipopolysaccharide; OHFA, hydroxy fatty acid; P, phosphate; X, unknown substituent.

ermore, because it has been too laborious to measure the activity of the *rfa*-determined transferases we do not know whether or not the genes of the main *rfa* cluster are grouped into operons under the possible regulation of *rfaH*.

Genetics of the O sidechain synthesis in Salmonella

Salmonella species are the best investigated in respect of O chain synthesis and genetics; among them two general patterns have been found. One pattern is seen in the serogroups A, B, D and E, in all of which the O repeat unit is a heterosaccharide with the backbone structure mannosyl-rhamnosyl-galactose. Each new O unit is synthesized on the carrier lipid by the sequential addition of its constituents, followed by polymerization of the ready-made units. Mutants defective in the polymerization step have only one O unit in their final LPS and are in many characters intermediate between smooth (with complete O chains) and rough forms (with no O chains), therefore termed SR, for 'semirough' (Naide *et al.*, 1965).

The second pattern of O chain assembly is seen in the *Salmonella* O groups C1 and L, in which the backbone of the O unit consists of mannose and *N*-acetylglucosamine (C1) or galactose and *N*-acetylgalactosamine (L), and probably also in *E. coli* 08 and 09 (both mannans). In this pattern, no pre-assembly and subsequent polymerization of repeat units has been shown, and no SR mutants with defective polymerization found. The synthesis of the O chain then appears to proceed through sequential addition of the monosaccharides to the non-reducing end of the growing chain attached to a carrier lipid (Flemming & Jann, 1978a, b). It is not clear whether the carrier lipid is the same isoprenol as in the first pattern. Unexpected features in this pattern of O chain synthesis include the start of the mannan chain in *E. coli* 08 and 09 by glucose (Flemming & Jann, 1978a, b) and the reported presence of glyceraldehyde in the corresponding position in *Salmonella* group C (Gmeiner, 1975). The function of a gene or genes of a *rfe* locus close to *ilv* at position 83 is required for the O chain assembly of this second type (Mäkelä, Jahkola & Lüderitz, 1970; Schmidt, Mayer & Mäkelä, 1976), as well as for the synthesis of the T1 antigen and the glycolipid ECA (see below), but its mode of action is still not known.

Some monosaccharide constituents of the O side chain are sugars used also for other purposes in the bacteria. In such cases, the enzymes for the synthesis of the sugar nucleotides are determined by genes *galE* (for making UDP-galactose), *galU* (for UDP-glucose) or *manA* (phosphomannoisomerase) located separately from the LPS genes. All other genes for the synthesis of the components of the O chains and for their assembly are specified by genes in the main *his*-linked *rfb* cluster at position 44–45 in both the *Salmonella* and *E. coli* map, or *rfc* (linked to *trp* at approximately position 32) in some cases. This applies to all the several O groups of *Salmonella* and *E. coli* studied. For example, at least nine enzymes are known to be required for the synthesis of TDP-rhamnose, GDP-mannose and CDP-abequose, the precursors of the O unit of *Salmonella* group B. All these enzymes, as well as at least four transferases, are determined by closely linked genes at *rfb*. The order of most of these genes has been revealed by study of enzyme deficiencies in *his-rfb* deletion strains (Nikaido, Levinthal, Nikaido & Nakane, 1967). Reduced activity of all enzymes determined by genes to one side of the site of various presumed point *rfb* mutations indicated a polar effect of these mutations and, by inference, that at least six of the mapped *rfb* genes are part of a single transcription unit or operon (Levinthal & Nikaido, 1969).

Polymerization of O units in *Salmonella* groups B and D requires the polymerase function determined at the *trp*-linked *rfc* site. In group E a similar polymerase function is specified by a gene in the main *rfb* cluster (Nyman, Plosila, Howden & Mäkelä, 1979).

The enzymes responsible for O sidechain modification are often determined by genes outside *rfb* and *rfc*. For example, O-acetylation of the abequose in the O unit of *Salmonella* group B, detected serologically as the O factor 5 requires the function of an *oafA* gene at position 46 (Mäkelä, 1965). Glucosylation of the O unit galactose has been shown to depend on the gene *oafR* in *Salmonella* group B or *oafC* in group E, both at approximately position 12–13 (Mäkelä, 1973; Plosila & Mäkelä, 1972).

Antigenic conversion

LPS modification genes can also be carried by non-chromosomal elements such as phage or plasmid. Introduction of these into the cell can result in a change of the LPS structure, known as antigenic conversion. Thus, when bacteria of the *Salmonella* group E1 (O antigens 3,10) are lysogenized by the converting phage ε15, the action of the bacterial O polymerase is inhibited by an inhibitory protein coded by a gene on the phage genome (Robbins, Keller,

Wright & Bernstein, 1965). In its place, a new, phage-specified polymerase effecting polymerization in β-linkage instead of the original α-linkage, appears, and the LPS subsequently synthesized has an altered structure recognized by new antigenic and phage receptor properties (Robbins & Uchida, 1962).

In salmonellas of groups A, B and D a very similar conversion is exerted by phage P27, changing the linkage between O units from galactose-(1→2)-mannose to galactose-(1→6)-mannose (Bagdian & Mäkelä, 1971; Lindberg, Hellerqvist, Bagdian-Motta & Mäkelä, 1978). It is presumed that the action of the bacterial polymerase was completely repressed because no old, 1→2 linkages remained in the lysogenized *S. bredeney*. That the new, 1→6 linkage polymerase is phage-coded, is suggested by the ability of the phage to restore smooth-type LPS with polymerized side chains to strains that had one O unit only because of lack of bacterial *rfc*-coded polymerase function (Mäkelä, Bagdian-Motta & Nurminen, manuscript in preparation).

Several *Salmonella* phages cause glucosylation of O side chains, recognizable as new serological O antigen factors. The phage genes involved have been studied for phage ε34 in group E and phage P22 in group B (Wright & Barzilai, 1971; Mäkelä, 1973 and unpublished). In both cases, the phage was shown to carry information both for the formation of the glucosyl-carrier-lipid and for the transfer of the glucose to the LPS. The former function could replace the missing glucosyl-carrier-lipid formation in several bacterial mutants unable to perform their own (*oafR*-determined) LPS glucosylation.

In *Shigella flexneri,* phage-determined acetylation results in serologically recognized conversion (Gemski, Koeltzow & Formal, 1975). A different kind of conversion appears to be caused by a plasmid in *Shigella sonnei.* It has long been known that *Shigella sonnei* strains give rise at high frequency to a rough-looking colony variant, termed phase-2; the sugar composition of the LPS of phase-2 strains suggests that they are indeed rough so that their LPS lacks O side chains. The recent experiments of D. Kopecko (personal communication) indicate that change to phase-2 in *Shig. sonnei* results from loss of a plasmid and that the plasmid, if transferred to *S. typhi* or various other hosts, makes them develop the phase-1 antigen of *Shig. sonnei,* not previously known to occur in any other species or genus. It will indeed be of interest if this converting plasmid supplies the total genetic information for assembly and polymerization of an O unit composed of two unusual monosaccharides (Kenne *et al.*, 1980). The same may be true for the drastic alteration of LPS composition of *S. typhimurium* caused by the acquisition of a derepressed plasmid ColIb *drd2* (Hoffman *et al.*, 1980). The original smooth group B O side chain was apparently completely replaced by a chain composed of new pentasaccharide repeat units. These differed from the original e.g. by lack of abequose and presence of *N*-acetyl-glucosamine. The genetic mechanisms involved remain to be determined.

Regulation of O side chains: form variation

The synthesis of the O chains is not subject to any recognized regulation by environmental manipulation or mutation of regulatory genes. However, in *S. anatum* of O group E1 growth at 20 °C results in the production of LPS with fewer but longer O chains than formed at 37 °C (McConnell & Wright, 1979) with resulting increased sensitivity to the phage FO which absorbs to the complete core LPS. The mechanism is not known but it might be change in relative efficiency of the reactions for transfer of O chains to LPS versus polymerization.

The modification and polymerization steps are in many cases subject to variation called form-variation. Thus a *Salmonella typhimurium* culture is likely to contain bacteria whose LPS is fully glucosylated and others with non-glucosylated LPS. This is the glucosyl-(1→4)-galactose modification determined by *oafR* genes at position 12–13, and recognized serologically as the antigen factor 12_2. Furthermore, bacteria in the 'on' state (glucosylated LPS, and therefore 12_2+) give rise with relatively high frequency to descendants in the 'off' state (12_2−, non-glucosylated LPS), and vice versa. This 12_2 variation was shown to be determined at the *oafR* site, which controls both the on or off state and whether variation occurs or not (Mäkelä & Mäkelä, 1966). The molecular mechanism has not been determined but may well be similar to that of flagellar phase variation (see below).

Phage-coded glucosylations, such as antigen factor 1 determined by P22, undergo phenotypically similar form variation in established lysogens; newly lysogenized bacteria are always antigen 1-positive. The same applies to the antigen factor 27 that corresponds to the phage P27-determined O unit polymerization: *S. typhimurium* strains lysogenized with P27 alternate between 27+ and 27− forms, and LPS extracted from a mass culture shows variable proportions of the chromosomally-determined 1→2 and the phage-coded 1→6 linkages (Lindberg *et al.*, 1978).

Unusual LPS types

Polysaccharides other than the O chains are in some circumstances attached to the LPS core. T1 chains, polymers of ribose and galactose in the unusual furanose form (Berst, Lüderitz & Westphal, 1971) have been found in some isolates of *S. paratyphi B* unable to make O chains. By genetic methods the T1 determinant genes (which presumably derepress the previously unexpressed biosynthetic genes) called *rft* (close to *purE* at position 12) can be transferred to other strains including smooth ones. In such recombinants, both O and T1 chains can be linked to the core LPS, both requiring the complete core for attachment, and the function of the *rfaL* gene (Sarvas, 1967). Strains with the T1 antigen become T1-negative at a high frequency by mutation at or near *rft*; the reverse change does not occur with detectable frequency.

Another T antigen, T2, has been described. Its chemical determinant appears to be an incompletely characterized *N*-acylglucosamine attached to the complete *Salmonella* core (Bruneteau, Volk, Singh & Lüderitz, 1974). Its expression is determined by *rfu* genes not close to *rfb* (Valtonen, Sarvas & Mäkelä, 1976).

Capsular polysaccharides

Exopolysaccharides are found on a large variety of bacteria, both Gram-positive and Gram-negative (Sutherland, 1977; Jann & Westphal, 1975). The polysaccharide may be found as slime in the medium surrounding the bacteria, or as capsules attached to the bacterial cell wall. The mode of attachment remains, however, unknown even if linkage to a membrane lipid (which is not lipid A) has been suggested. The distinction between slime and capsular polysaccharides is not sharp, and the same bacteria may switch from producing one type to producing the other (Sutherland, 1977). Even the distinction between LPS and capsular polysaccharide is not always sharp. Normally LPS contains neutral heteropolysaccharide chains linked to lipid A, while capsular polysaccharides are almost always acidic homo- or heteropolysaccharides containing phosphate residues and/or uronic acids, and not linked to lipid A. However, some enteric bacteria have acidic LPS, or a lipid A-linked homopolysaccharide (polymannan) as LPS and a complex heteropolysaccharide as capsular polysaccharide (Ørskov, Ørskov, Jann & Jann, 1977).

Features of the polysaccharide structure can be identified by antisera, and with their use tens of capsular serotypes have been identified among bacteria such as *Streptococcus pneumoniae, E. coli* and *Klebsiella pneumoniae.* Mutants can be easily isolated by phage or antiserum selection methods. Thorough genetic and biochemical analysis of such mutants has, however, been performed in only a few cases.

In bacteria pathogenic for animals or man, the capsule appears to shield the bacteria from the antibacterial defences of the host – bactericidal action of serum plus complement, and phagocytosis; however, if antibodies to the capsular polysaccharide are present, the shielding function is prevented (Robbins *et al.*, 1980). The capsular polysaccharide can also mediate the adhesion of the bacteria to surfaces as suggested both for the caries-associated oral streptococci (Gibbons & van Houte, 1975) and for adherent aquatic bacteria (Dudman, 1977). It has also been suggested that the acidic polysaccharides may be important for binding of water and of divalent cations, but these ideas have not been tested e.g. by using mutants devoid of these polysaccharides.

Pneumococcal capsules

Pneumococcal capsular polysaccharides were the first polysaccharides whose function and synthesis were studied by genetic methods. In fact, the first experiment of bacterial genetics, that identified the chemical nature of the gene, concerned pneumococcal capsules, and the selection in the transformation was based on the necessity of the capsule for infectivity (Griffith, 1928). Studies of biosynthesis established that the structure of the polysaccharide of each type is determined by the action of specific transferases, determined by type-specific genes. In many cases, specific enzymes are also required to synthesize the monosaccharide components of the capsular polysaccharide, and are then determined by type-specific genes (Taylor, 1949; Mills & Smith, 1965). The same principles were later demonstrated and expanded in the studies of LPS, as described above.

The early transformation experiments with pneumococcal capsules are still the best demonstration of the role of the capsules as virulence factors (Griffith, 1928). Encapsulated bacteria of serotype

II were pathogenic (killed mice after sub-cutaneous injection); a non-capsulated mutant of serotype II was non-pathogenic, but could kill mice when injected together with heat-killed smooth bacteria (=DNA) of another serotype (III); the bacteria recovered from these dead mice were encapsulated transformants with the donor serotype III. MacLeod & Krauss (1950) showed that also the quantity and the quality of the capsular polysaccharide were important determinants of virulence.

Capsules of Gram-negative bacteria

The continuous transitions between LPS and capsular polysaccharide also apply to their function. Thus, many virulent bacteria have polysaccharide capsules (apparently as a shielding device) together with either smooth LPS (many *E. coli* and *Klebsiella* strains, *Salmonella typhi*) or rough-type LPS (*Neisseria meningitidis*), while others have only smooth LPS (most *Salmonella*). Epidemiological evidence suggests that the quality of the capsular polysaccharide is also important in pathogenicity. Thus, almost all *E. coli* strains causing neonatal meningitis in man have the capsular serotype K1, which is α2–8-linked poly-*N*-acetylneuraminic acid (Robbins *et al.*, 1974). *N*-acetylneuraminic acid is also present in the capsules of many other pathogenic bacteria, which has led to speculations of its special ability to promote virulence (Robbins *et al.*, 1980). However, the associations may be coincidental, as suggested by the relatively few O-K types seen in human meningitis (Robbins *et al.*, 1974; Sarff *et al.*, 1975). It seems that the question could only be settled by preparing isogenic bacterial strains differing in their capsular polysaccharides and measuring their virulence in a suitable animal model.

Genetically, the capsular polysaccharides (K antigens) of *E. coli* fall into two classes. Several K antigens, all occurring together with the neutral heteropolysaccharidic LPS, have been shown to be determined by *kps* genes at position 61 close to *serA* (Ørskov & Nyman, 1974; Ørskov, Sharma & Ørskov, 1976). These include the disease-associated type K1. The enzymes determined at this locus have not been identified. The presence of a regulatory locus at some other site has been suggested (Ørskov & Nyman, 1974; Ørskov & Ørskov, 1962) but its nature is unknown. The other class is formed by several K antigens that all occur mainly in combination with LPS of the serotypes 08 and 09 that contain only mannose. The genes determining these K antigens are found closely linked to *his* and *rfb,* or linked to *trp* (Schmidt *et al.,* 1977). The genetic data suggest that in this case the serological classification has been misinterpreted, and that these K antigens are in reality LPS antigens, requiring the function of both *rfb* and *rfc* genes, like the O antigens of *Salmonella* groups B and D. Chemical data at least partly support this view; some K antigens of this type are structurally similar to some O antigens and have been found linked to lipid A (Ørskov, Ørskov, Jann & Jann, 1977).

A high-frequency variation resembling the form variation of *Salmonella* LPS has been described for some capsular polysaccharides. The K1 antigen in *E. coli* undergoes a variation recognized by differential reactivity with anti-K antisera. The K1− and + forms correspond to *O*-acetylated and non-*O*-acetylated forms of the capsular polysaccharide, poly-*N*-acetylneuraminic acid (Ørskov *et al.,* 1979).

Salmonella typhi is one of the few salmonellas that have a capsular polysaccharide, the Vi antigen; this is a polymer of *O*- and *N*-acetylated galactosaminuronic acid (Clark, McLaughlin & Webster, 1958). Two gene loci, at about 92 (*viaB*) and 46 (*viaA*) have been demonstrated by conjugational transfer to *S. typhimurium* that does not naturally have a capsule (Johnson, Krauskopf & Baron, 1965, 1966). The opaque and translucent colonies regularly seen in some Vi-containing *S. typhi* cultures correspond to strong or weak development of the Vi antigen. Both colony forms on sub-culture give rise to variants of the other form (Kauffmann, 1954). This variation has been analysed in *Citrobacter ballerup*, which has a similar Vi polysaccharide and also shows rapid alternation between Vi-synthesizing and non-synthesizing forms. Conjugational analysis showed that this variation was determined by a gene at or closely linked to *viaB* (Snellings, Johnson & Baron, 1979).

Colanic acid = M antigen

All *E. coli, Salmonella* and *Enterobacter cloacae* strains can synthesize an additional, mucoid, surface polysaccharide, which is antigenically very similar in all these strains. This M antigen, also called colanic acid, consists of hexasaccharide repeating units. Colanic acid isolated from different strains can vary in its *O*-acetylation and in the pyruvyl substituents peculiar to this molecule (Sutherland, 1977). In wild-type strains, the synthesis of large amounts of the M antigen can be turned on by environmental conditions such as low temperature, high concentration of salts and excess of fermentable sugars. Mutants permanently derepressed for the synthesis of the M antigen are easily obtained, for example by selection for resistance to many phages. Many genes involved in the synthesis of the monosaccharide precursors of this molecule have been identified. The regulation of the synthesis is complex and involves several *cap* genes. Because it is not yet completely understood we will not discuss it here but refer to a recent review (Markovitz, 1977).

Enterobacterial common antigen = ECA

A glycolipid whose saccharide part is composed of *N*-acetylmannosaminuronic acid and *N*-acetylglucosamine is also a component shared by many bacteria. It is the antigen common to members of *Enterobacteriaceae,* hence termed ECA. Mutants defective in its synthesis have been found accidentally or by indirect selection methods. The genes involved are close to *ilv* (*rfe* and *rff* at position 83) or in the *rfb* cluster at position 44. The *rff* genes determine enzymes required for the synthesis of UDP-*N*-acetylmannosaminuronic acid (Lew, Nikaido & Mäkelä, 1978); the function of the two other gene groups is not known. For reviews see Mäkelä & Mayer (1976) and Mayer & Schmidt (1979).

GENETICS OF CELL SURFACE PROTEINS

Major outer membrane proteins

A fairly small number of the proteins of the outer membrane of Gram-negative bacteria occur in large amounts (10^5 molecules/cell) and are therefore called major outer membrane proteins (OMP). Their identification was initially based on the apparent molecular weight as determined by their mobility in SDS-polyacrylamide gel electrophoresis. This caused much confusion in nomenclature due to slightly differing gel systems and strain differences. Now we can define porins (generally in the 34 000 to 40 000 molecular weight region) by their hydrophilic pore function and certain common physicochemical characteristics, and lipoprotein (*ca* 7 000 molecular weight) by its covalently linked lipid and linkage of some of the molecules to peptidoglycan. The third major OMP, often called II*, with a molecular weight around 30 000, is required for F-mediated conjugation in a so-far unknown way. An exact way to identify the major proteins is by referring to them by their structural genes with the help of known mutants. Thus protein II* is, in the following, called the *ompA* protein, whereas the major porins are products of the *ompC*, *ompD*, and *ompF* genes; the lipoprotein structural gene is *mlp* (for map positions see Fig. 2).

Most work on the major OMP has been done in *E. coli* and *Salmonella* in which groups these proteins are closely homologous. The outer membrane protein profiles of many other Gram-negative organisms such as *Pseudomonas* and *Neisseria* have also been shown to display similar features (Hancock & Nikaido, 1978; Frasch & Mocca, 1978; Johnston, 1978).

Both the porins and the *ompA* protein are exposed on the surface in *E. coli* and *Salmonella* cells as shown by their accessibility to phages, bacteriocins and antibodies. Mutants deficient in any one of these proteins can be easily isolated by selecting for phage or bacteriocin resistance (Reeves, 1979). *ompA* mutants are also found because they are defective in conjugation (Havekes & Hoekstra, 1976). Porin mutants can be selected on the basis of their pore function as resistant to Cu^{2+} or chloramphenicol (Pugsley & Schnaitman, 1978a). Several previously isolated pleiotropic transport or 'cryptic' mutants have turned out to be defective in porin function (Bavoil, Nikaido & von Meyenburg, 1977; Beacham, Haas & Yagil, 1977). The porin mutants have been essential in ascertaining the physiological pore function of these proteins, initially suggested by the ability of the proteins to form pores when incorporated in artificial membranes (Nikaido, 1979). The lipoprotein is probably not exposed in bacteria with an intact outer membrane, and no phages or bacteriocins are known to use it as receptor. Consequently mutants of lipoprotein synthesis have been found only by the laborious means of random screening (Hirota, Suzuki, Nishimura & Yasuda, 1977) or complicated suicide selection (Wu & Lin, 1976).

Membrane localization

Most of the above described *omp* and *mlp* mutants have very much reduced or zero amounts of the corresponding protein in their outer membranes suggesting that an altered amino acid composition may easily prevent the incorporation of the protein in the outer membrane. The transport of outer membrane components to their

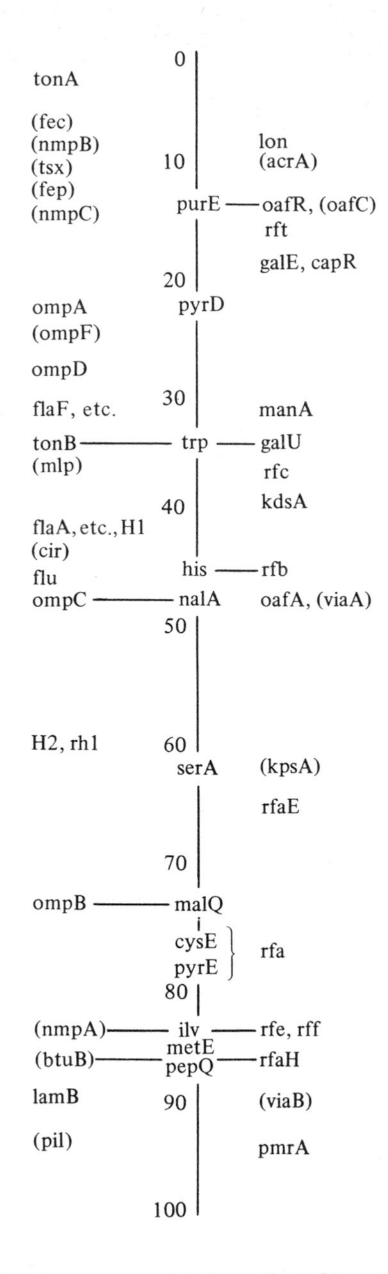

Fig. 2. Simplified map of the chromosome of *Salmonella typhimurium*. Points 0 and 100 are directly joined to form a circle. Genes not (yet) demonstrated in *S. typhimurium* (but in other *Salmonellas* or *E. coli*) are in parenthesis.

Cotransducible genes are joined by horizontal lines. The *E. coli* map is closely homologous, but differs by having a large inversion around *trp* (position 34 in *S. typhimurium*, 27 in *E. coli*) and by lacking the gene H2 for flagellar phase 2; also perhaps by lacking *ompD*.

final site outside the cytoplasmic membrane where they were synthesized has been an intriguing question that has eluded final clarification (Osborn, 1979). The proteins now appear to provide us with means to attack this problem; these studies rely heavily on the use of genetic methods, both mutants, gene fusions, and cloning (Halegoua & Inouye, 1979; Silhavy, Bassford & Beckwith, 1979).

The *E. coli* genes *ompA* and *mlp* have been cloned (Henning *et al.*, 1979; Nakamura *et al.*, 1980). The comparison of the DNA base sequence with the proteins synthesized *in vivo* and *in vitro* shows that the genes indeed are longer than the final membrane proteins. The extension in the immediate translational product is N-terminal and has the function of a signal or leader peptide previously demonstrated for certain eucaryotic membrane and secreted proteins (Blobel & Dobberstein, 1975) and essential for the transfer of the protein through the cytoplasmic membrane during synthesis. The newly synthesized lipoprotein requires both the removal of the leader peptide and further processing – covalent attachment of lipids and linkage to peptidoglycan. A mutant altered in an amino acid in the signal peptide was unable to complete any of these steps (Lin, Kanazawa & Wu, 1980a, b; Lin, Lai & Wu, 1980). Fusing the genes of an outer membrane protein (the *lamB* product) and β-galactosidase (for easy detection) has proved a very powerful tool for analysing what determines the localization of proteins in the outer membrane (Silhavy, Bassford & Beckwith, 1979; Emr *et al.*, 1980). Much still remains to be learned in this area; for example, mutants of all the proteins involved in the processing steps are still lacking.

Regulation

The major outer membrane proteins are subject to complex regulation which is so far only partially understood. Environmental control by osmolarity, availability of phosphate, or concentration of glucose has been shown to cause severe repression/derepression of the various porins (Kawaji, Mizuno & Mizushima, 1979; van Alphen & Lugtenberg, 1977). Such derepression can lead to the appearance of a new major porin protein that otherwise had been undetectable (Overbeeke & Lugtenberg, 1980). Genetic derepression has a similar result in mutants defective in the major porin species (Bavoil, Nikaido & von Meyenburg, 1977; Pugsley & Schnaitman, 1978b). The genes specifying the new proteins have been termed *nmp* for new membrane protein (Fig. 2). Many mutants selected as being defective in a major porin have the mutation in a regulatory gene *ompB*. Such mutations can cause the absence of either one or both of the *ompC* and *ompF* proteins (Sarma & Reeves, 1977; Verhoef *et al.*, 1979). Hall and Silhavy (1979) studied the *ompB* function using a strain in which the *lac* gene (for β-galactosidase) was fused to *ompC*; their findings strongly suggest that the *ompB* gene product is a diffusible positive regulator of *ompC*. The regulation by osmolarity was found independent of *ompB*.

Plasmids and bacteriophages are known that can change the major OMP composition of the strains they inhabit. The phenomenon appears in principle to be similar to antigenic conversion discussed above under lipopolysaccharide. Schnaitman, Smith & Forn de Salsas (1975) showed that lysogenization of *E. coli* K-12 by a phage PA-2 resulted in almost complete repression of the major porin and the production of a new protein of slightly larger size (Diedrich, Summers & Schaitman, 1977) which also functions as porin (Pugsley & Schnaitman, 1978a). Iyer (1979) has shown that plasmids of Inc group N similarly depress the synthesis of the major porin of *E. coli* B/r, and cause the appearance of new polypeptides of approximately the same size class. Whether these are modified forms of the original porin or new gene products could not be established. Different plasmids caused the production of different new polypeptides. Furthermore, the synthesis of the new polypeptides was variable (form variation? loss of plasmid genes?).

Other outer membrane proteins

Although the few species of major OMP are responsible for most of the protein content of the outer membrane, many other proteins are present. Some occur in few copies only; many can be increased greatly e.g. in response to environmental factors. Many of these proteins appear to serve specialized transport functions, although their exact mode of action is usually not known. The best known may be the protein that was termed *lamB* because it served as the receptor for phage lambda (Hofnung, Jezierska & Braun-Breton, 1976). It is now clear that this protein forms pores in the OM that are somewhat specific for maltose and higher oligosaccharides of the maltose series (Luckey & Nikaido, 1980). The *lamB* gene and especially the sequence determining the OM localization of its product have been studied in detail as discussed above under major OMP.

The OM proteins involved in the capture and transport of iron from low external concentrations have been extensively studied (Braun & Hantke, 1977; Braun, 1978). Several OM proteins (products of the genes *tonA, fepA, cir* and *fec*) are implicated. All these genes are derepressed by iron starvation, resulting in much increased amounts of the respective proteins. The mechanism of the control is, however, not known. The *tsx* gene product is needed for nucleoside transport, and the *btu* gene product for the transport of vitamin B_{12}. Mutants of both are easily obtained by phage (T6, BF23) or colicin (ColE) selection (Reeves, 1979).

Effect of plasmids

Conjugative plasmids determine several proteins located in either the cytoplasmic or the outer membrane (Manning & Achtman, 1979). Some of the latter are required to allow stable contacts to form between the conjugating cells and some of them also mediate exclusion of other related plasmids. Several conjugative plasmids reduce the sensitivity of their *E. coli* hosts to the bacteriolytic action of serum + complement (Taylor, Hughes & Robinson, 1979). The plasmid genes conferring serum resistance have been identified in two instances. Binns, Davies & Hardy (1979) identified a 5 300 base-pairs sequence in a cloned fragment of plasmid ColV as the determinant of enhanced chick-virulence and reduced serum-sensitivity of a smooth *E. coli* strain. Moll, Manning & Timmis (1980) and Timmis, Moll & Danbara (1979) report that the serum-resistance gene of the conjugative R plasmid R6–5 is *traT*, one of the two surface-exclusion genes of this plasmid; the protein specified by *traT* was present, about 21 000 copies per bacterium, in the outer membrane. Acquisition of a conjugative plasmid, especially a mutant derepressed for transmissibility, sometimes causes increased sensitivity to various antibacterial agents as well as slow growth. It is not clear whether this results from alteration in outer membrane protein composition or from some other mechanism, for example interference with LPS synthesis (see *rfaH* under Genetics of core synthesis).

Williams (1979) found that the virulence-promoting effect of colicin V (ColV) plasmids (Smith, 1974; Smith & Huggins, 1976) was associated with and probably dependent upon their ability to confer a new iron-acquisition ability on their *E. coli* host; the plasmid-determined iron-acquisition ability was independent of the enterochelin system, but, like other known systems, required host *tonB* function. The possible role of altered outer membrane protein composition in this phenomenon remains to be determined.

In *Yersinia enterocolitica* and *Y. pseudotuberculosis* virulence is associated with presence of the V and W (lipoprotein) antigens at the cell surface. Gemski, Lazere & Casey (1980) report that in each of these species presence of VW and invasiveness depends on plasmids, of about 42 megadaltons; the plasmids from the two species differed in endonuclease fragment pattern (P. Gemski, personal communication).

Many plasmids (R factors), both non-conjugative and conjugative, cause resistance to tetracycline, and such resistance, especially after exposure to an 'inducing' low concentration of tetracycline, is associated with the appearance of new proteins (Levy & McMurry, 1974) and of a new antigen in the envelope fraction of *E. coli* (Bolder & Sompolinsky, 1974). We are, however, not aware of any proof that a 'tetracycline-resistance' protein or antigen is detectable at the external surface of the cell.

Fimbriae and pili

Non-flagellar filamentous appendages occur in very many bacterial groups. In some they may determine twitching motility but in most groups there is no evidence for their involvement in bacterial movement. Bacterial conjugation is dependent on filaments which are specified by the conjugative plasmids (Achtmann & Skurray, 1977; Manning & Achtmann, 1979). We refer all discussion of their genetics to these reviews. In accordance with the proposal of Ottow (1975) we term them sex pili and use the term fimbriae (Duguid *et al.*, 1955) for all other non-flagellar filaments.

Type 1 fimbriae

In the *Enterobacteriaceae* fimbriae called type 1 are very commonly found (Jones, 1977). They are distinguished by the specific inhibition (or reversal) by mannose or mannosides of their adherence to a variety of surfaces. These include red cells and epithelial cells of various vertebrates, yeast cells and broth-air interfaces. Hence the commonly used term ('mannose-sensitive' agglutination) to describe the behaviour of type 1 fimbriae. The sticking of the fimbriate bacteria to the broth–air interface in unshaken broth cultures allows their continued growth as a pellicle after the end of the logarithmic phase of growth, and allows isolation of a fimbriate form from strains consisting initially only of non-fimbriate cells (Old *et al.*, 1968).

Several kinds of variation affecting type 1 fimbriation have been recognized. Their study is facilitated by the smaller size of colonies, characteristic of the fimbriate form in some strains (Brinton, Buzzel & Lauffer, 1954), and by a conspicuous difference in electrophoretic mobility between fimbriate and non-fimbriate cells. First, production of fimbriae may be prevented by certain environmental conditions. Thus Saier, Schmidt & Leibowitz (1978) showed that production of type 1 fimbriae in *S. typhimurium* is subject to catabolite repression by fermentable carbohydrates. In *cya* mutants it is dependent on provision of cyclic AMP. In some strains suspensions of heavily fimbriate bacteria (e.g. from pellicles from unshaken broth cultures) on incubation in fresh broth give rise to descendants with few or no fimbriae; this suggests that log-phase cells do not produce fimbriae.

Secondly, there is clear evidence for spontaneous alteration of some *E. coli, Shigella* and *Salmonella* strains between fimbriate and non-fimbriate 'phases'. In one of the earliest reports on type 1 fimbriae Brinton, Buzzell & Lauffer (1954) estimated the rates of change of phase in *E. coli* B as 10^{-3}/cell/generation, in each direction. In a more detailed investigation of another *E. coli* strain (Brinton, 1959) the rate of change from fimbriate to non-fimbriate phase was 4×10^{-4}/cell/generation and the generation time for non-fimbriate bacteria was 10% shorter than that for the fimbriate form. Both the rate of change of phase and difference in growth-rate were much altered by changes in temperature of incubation.

Thirdly, fimbriation may be affected by ordinary mutation. In an *E. coli* K-12 line mutations at *pil*, at position 98, caused irreversible loss of ability to make fimbriae; complementation tests indicated presence of at least two, perhaps three, *pil* genes at this locus (Swaney, Liu, Ippen-Ihler & Brinton, 1977b; Swaney *et al.*, 1977a). It is not known whether the fimbrial phases are also determined at *pil*. However, a gene called *flu* (for fluffy) close to *his* has recently been described in *E. coli* K-12: this gene appears to control frequent alternation in respect of cell aggregation, colony form and presence of type 1 fimbriae (Diderichsen, 1980). In *S. typhimurium* most strains can be grown to produce a fimbriate phase by static broth culture selection. Nearly all the exceptions observed among independent natural isolates belong to the sub-type or sub-species called

FIRN (Duguid, Anderson & Campbell, 1966); all FIRN strains appear to have an identical mutation (or mutations) preventing production of fimbriae, but the locus concerned has been only approximately mapped.

'Mannose-resistant' fimbriae

A protein antigen called K88 (K for capsular antigen) was found by Ørskov, Ørskov, Sojka & Leach (1961) to be associated with *E. coli* strains from pig diarrhoea. This turned out to be the first example of what now appears an important mechanism of pathogenicity; a fibrillar surface protein mediating adhesion of bacteria to surfaces with a distinct specificity of attachment (Jones, 1977; Swanson, 1980). It is still not clear whether this protein is equivalent to fimbriae as exemplified by the type 1 fimbriae. In neither case has the mode of attachment to the bacterial surface been characterized. It is therefore only for convenience that we here call the K88 and similar proteins fimbriae.

K88 fimbriae were soon shown to be plasmid-specified (Ørskov & Ørskov, 1966; Stirm, Ørskov & Ørskov, 1966). The number and organization of plasmid genes required has not been determined. The association of the K88 antigen with strains causing pig diarrhoea suggested that it had a role in the pathogenicity of these strains. By transferring the plasmid to other strains this was shown to be so: attachment to pig intestinal epithelium and ability to cause diarrhoea were dependent on the transfer of the K88-determinant (Smith & Linggood, 1971). The same plasmid also determined the synthesis of enterotoxin, another established factor of virulence. The need of attachment for disease production could be confirmed by showing that both were absent in pigs genetically resistant to diarrhoea caused by *E. coli* K88 (Rutter, Burrows, Sellwood & Gibbons, 1975). The receptor in the epithelial cells of pig intestine to which the K88 fimbriae bind has been studied by comparing the binding ability of various preparations from cells of K88-sensitive and -resistant pigs. Such binding was found in the glycolipid fraction, but further work is required before final identification of the receptor (Kearns & Gibbons, 1979). Mannose is not involved in the binding, which is thus 'mannose-resistant'.

Other apparently similar adhesion-mediating and plasmid-determined fimbriae are the K99 antigen of *E. coli* strains causing diarrhoea in calves and lambs (Smith & Linggood, 1972; Burrows,

Sellwood & Gibbons, 1976) and the several serologically separable Colonization Factor Antigens or CFA in enterotoxigenic *E. coli* strains causing travellers' diarrhoea (Evans & Evans, 1978; Evans *et al.*, 1975). The binding specificity for these fimbriae has not been established, but the binding is mannose-resistant. Preliminary evidence has implicated the disialylganglioside GM2 as the specificity determinant (Faris, Lindahl & Wadström, 1980). The synthesis of K99 and the CFA antigens has been shown to be regulated by environmental conditions but the mechanisms have not been elucidated (Graaf, Wientjes & Klaasen-Boor, 1980; Evans, Evans & Tjoa, 1977).

Adhesion to urinary tract epithelium has recently been demonstrated characteristic of *E. coli* strains causing urinary tract infections (Svanborg-Edén *et al.*, 1976, 1978). The adhesive property coincides with mannose-resistant hemagglutination and presence of fimbriae. These are different from type 1 fimbriae also present on most of these strains but also on others, not associated with urinary tract infections (T. K. Korhonen & C. Svanborg-Edén, personal communication; Korhonen, Edén & Svanborg-Edén, 1980; Källenius & Möllby, 1979; Ørskov, Ørskov & Birch-Andersen, 1980). The receptor for these fimbriae is present in the cells of most humans but lacking in individuals of the rare p̄ blood type to whose cells the bacteria do not adhere (Källenius *et al.*, 1980). From a comparison of the ability of glycolipid fractions of P^+ and p̄ cells to inhibit adhesion the receptor was concluded to be a glycolipid of the globoside series, thus different from the receptors of K99, CFA I or type 1 fimbriae (Källenius *et al.*, 1980; Leffler & Svanborg Edén, 1980). The genetic analysis of these fimbriae has only just been started with the isolation of a mutant lacking fimbriae and mannose-resistant agglutination properties (Hull, Hull, Minshaw & Falkow, 1980).

Fimbriae in Neisseria *species*

The virulence of *Neisseria gonorrhoeae* has been shown to be associated with the presence of fimbriae; among disease isolates many antigenically distinct types are found (Swanson, 1980; Buchanan *et al.*, 1978). The presence of fimbriae (as well as other factors, including outer membrane proteins) influence the colonial appearance (Swanson, 1980), on the basis of which colony types T1 to T4 with varying disease-causing properties have been identified (Kellogg *et al.*, 1963). The colony types are subject to marked variation both spontaneously and as a response to growth conditions (Swanson, 1980; James & Swanson, 1978; Penn *et al.*, 1978). The rate of variation would be compatible with a mechanism similar to that of form variation, but the mechanism has not been studied in the gonococcus although these bacteria would be amenable to genetic analysis (Roberts, Elwell & Falkow, 1978). Since the variation may influence the disease-causing properties of these bacteria and indeed be an essential part of the infectious process, such an analysis would seem of considerable importance. Fimbriation and its association with adhesion have likewise been shown in another pathogenic *Neisseria*, the meningococcus. Both variability and its association with disease have been suggested but not yet substantiated (Craven & Frasch, 1978).

Flagella and phase variation

The reader is referred to other reviews (Iino, 1977; Silverman & Simon, 1977; Hazelbauer & Parkinson, 1977; Hazelbauer, 1979) for accounts of the many genes concerned in formation and motor function of flagella and in chemotaxis (see also Parkinson, this volume). We will here consider only the flagellar components known to be exposed at the surface: the proximal hook portion, made of a protein specified by a *fla* gene at position 31 in one of the two main clusters of *fla* genes; and the main part of the flagellum, composed (in most investigated species) of a single protein, flagellin.

Most *Salmonella* species have two flagellin-specifying genes: *H1*, determining phase-1 H antigen, at position 41 (which corresponds to *hag* of *E. coli*); and *H2*, for phase-2 antigen, at position 59, without any *E. coli* homologue. Representatives of the many different *H1* and *H2* alleles determine flagellins differing in serological specificity and primary structure (McDonough, 1965). Several mutations altering the serological character of antigen *i* (of *S. typhimurium*) have been mapped in *H1* and shown to be amino acid substituents, (or short deletions) in the same tryptic peptide of the *i* flagellin (Joys & Martin, 1973). Other mutations at *H1* or *H2* cause straight or 'curly' flagella (Iino & Mitani, 1967; Iino, 1962). In many species the ε-amino residues of some lysines (presumably those exposed on the folded *H1*- or *H2*-specified flagellin molecule) are methylated. This *N*-methylation modifies the antigenic character, and is determined by gene *nml*, closely linked to *H1* (Stocker, McDonough & Ambler,

1961). All the gene regions mentioned above (*fla*, *H1* and *H2*) have been cloned in λ phage and small plasmid vectors (Silverman & Simon, 1977; Zieg, Silverman, Hilmen & Simon, 1977). This has allowed detailed identification of gene with product, and also led to the elucidation of the molecular basis of flagellar phase variation that had intrigued bacterial geneticists for so long.

Normally only the *H1* or the *H2* gene is expressed by an individual *Salmonella* bacterium, but bacteria in phase 1 (i.e. expressing *H1*) give rise to phase 2 descendants with probabilities in the range of 5×10^{-3} to 10^{-4}/cell/generation. The probability for the reverse change is somewhat lower (10^{-3} to 10^{-5}), so that populations in continuous growth tend towards a mutational equilibrium in which phase 2 predominates. Transduction of *H1* or *H2* showed that phase is determined at or very close to *H2* (Lederberg & Iino, 1956). Various observations led to the hypothesis of an *H2* operon, alternating stochastically between 'on' and 'off' states, and comprising the structural gene for phase 2 flagellin and a separate gene, *rh1*, specifying a repressor of the activity of *H1*. This explained the observed phenomena but said nothing as to the basis for the 'on' and 'off' states, presumed to be determined at or near the promoter of the postulated operon. One proposal (Lederberg & Stocker, 1970) was the existence of an inverted repeat, with an intervening sequence whose orientation would be reversed by a crossover between the synapsed homologous sequences of the repeats. Reduction in frequency of phase change caused by one *recA* mutant allele but lack of effect of two others neither confirmed nor excluded the hypothesis.

A comparison of the *H2* gene region in the 'on' and 'off' states, separately cloned in a λ vector, led to the discovery of the mechanism of the alternation between the 'on' and 'off' configurations (Zieg *et al*., 1977): each corresponds to one orientation of an invertible segment, of about 900 base pairs, which later was shown to be flanked by short inverted repeats. The co-ordinate control of the *H1* gene was shown to be exerted, as previously postulated, by a repressor protein specified by *rh1* in the same operon as *H2* (Silverman, Zieg & Simon, 1979).

As shown in Fig. 3, the right-hand side of the invertible segment (in the 'on' orientation) includes the promoter for the adjacent (to the right) *H2* operon (Silverman, Zieg, Hilmen & Simon, 1979, Silverman & Simon, 1980). Mutations affecting either copy of the inverted repeat entirely prevent change of phase. Mutations in the left half of the invertible segment (in 'on' configuration) reduce frequency of change of phase by about 10^{-3}; the low-level residual variation is abolished by a *recA* mutation. It thus seems likely that inversion occurs by 'site-specific' recombination of the two copies of the repeat, perhaps in some way mediated by a polypeptide specified by a part of the invertible segment. The experiments of Kutsukake & Iino (1980), involving the same cloned *H2* fragment, confirm this picture and show that $vh2^-$, an *H2*-linked gene which blocks phase variation, corresponds to a stable mutation in the part of the invertible segment needed for normal inversion frequency. A $vh2^+$ allele in *trans* restored normal frequency of inversion to a $vh2^-$ invertible segment, as also did presence of a Mu or P1 prophage (each of these two phage genomes includes an invertible segment whose inversion is dependent on a phage gene whose function can be replaced by the corresponding gene of the other). It is not yet known whether, or to what extent, the flanking inverted repeats of the three invertible segments resemble each other.

The greater frequency of change from 'on' to 'off' state than from 'off' to 'on' is not predicted by the mechanism so far revealed – but it is easy to propose possible explanations. It will be apparent that inversion of a segment bounded by inverted repeats mediated either by 'site-specific' recombination or by ordinary (*recA*-dependent, 'legitimate') recombination, provides a possible mechanism for stochastic alternation of any gene or operon between active and inactive forms, and so a possible explanation for several phenomena discussed in the preceding sections, such as form variation of O antigen factors, variation in expression or modification of the capsular polysaccharides, Vi and K1, and alternation between fimbriate and non-fimbriate phases.

FINAL REMARKS

The classical methods of bacterial genetics, supplemented with the new possibilities of cloning and *in vitro* recombination, make it possible to synthesize isogenic bacterial strains – especially of the *Enterobacteriaceae* family – differing only in precisely known ways in defined surface properties. The use of such pairs is most helpful in identifying the functional significance of corresponding structures. At the moment such studies are expected to increase our understanding of the structure–function relationships of biological membranes, and of synthesis and handling of molecules destined to be transported through membranes. Practical applications of this

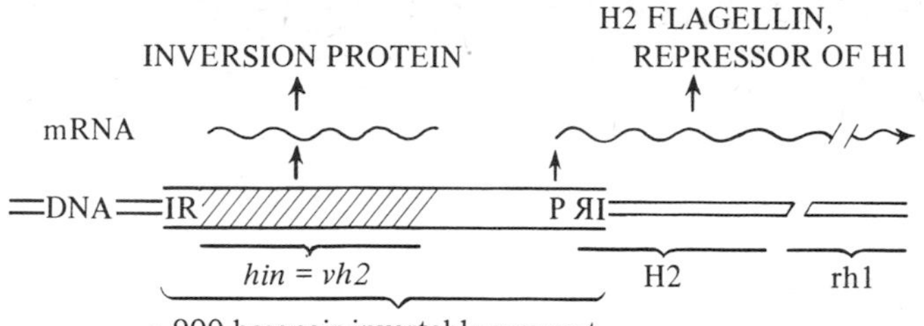

Fig. 3. Schematic drawing of the *Salmonella typhimurium* chromosome in the region determining flagellar phase variation (approximate position 60). The switch between phases is executed by inversion of the DNA segment flanked by short inverted repeats (IR, ЯI). The inversion requires the presence of the inverted repeats (in *cis*) and the inversion protein whose structural gene is *hin* (previously called *vh2*) within the invertible segment. The transcription of the operon containing the genes H2 (for flagellar protein in phase 2) and *rhl* (for repressor that turns off the H1 gene (at position 41 and determining the phase 1 flagellar protein) starts at promotor P within the invertible segment when this segment is in the 'on' orientation in which P comes close to the H2 operon.

knowledge are already visible, for example in biotechnology in the design of strains that export wanted proteins.

From a different angle, such predesigned bacterial strains have already been exploited to some extent to identify determinants of pathogenicity. As examples can be cited Griffiths' type transformation experiments as proof of the role of capsular polysaccharide production in the mouse-pathogenicity of pneumococci; the identification of the K88/K99 antigens as virulence factors in diarrhoea in domestic animals (section on Mannose-resistant fimbriae), and our use of *Salmonella* transductants differing only in their LPS antigens to test the effect of O repeat-unit character on mouse virulence (Mäkelä, Valtonen & Valtonen, 1973; Lyman, Stocker & Roantree, 1977) and to investigate the O-specificity or non-specificity of protection conferred by vaccination (Lyman, Stocker & Roantree, 1979; Eisenstein, 1975). The power of this approach can be further increased by taking advantage of the genetic variation of susceptibility in the host (examples in the Fimbriae section). Increased knowledge in this field will very probably lead to new ways of preventing and perhaps of treating infectious diseases.

We have above considered genetics as a tool for the study of the bacterial surface. It is perhaps self-evident that the bacterial surface also has been, and will continue to be, a most important tool in the study of genetic mechanisms in general. A recent example is the discovery of the material basis of the phase variation of flagellar antigens described above: reversible inversion of a specific DNA segment, mediated by site-specific or general recombination. It provides a possible mechanism not only for various other examples of alternate production and non-production of bacterial surface components (form variation of O antigen factors of *Salmonella*, or some capsular antigens in *S. typhi* and *E. coli*; variability of fimbriation in many bacteria but, with appropriate auxiliary hypotheses, (i) for alternation or variation between more than two antigenic states, both in prokaryotes such as *Borrelia recurrentis*, and in various parasitic eukaryotes, especially the trypanosomes (Turner, 1980); (ii) for the metastable turning on, or turning-off, of genes or groups of genes, during development in eukaryotes.

We thank Ms Marita Antila, Marianne Hovi and Kaija Helisjoki for their patient help in the preparation of this paper. The work of B.A.D.S. was supported in part by Grant No. AI 07/65 from the National Institute of Allergy and Infectious Disease.

REFERENCES

Achtman, M. & Skurray, R. (1977). A redefinition of the mating phenomenon in bacteria. In *Microbial Interactions*, series B, vol. 3, ed. J. L. Reissig, pp. 233–79. London: Chapman and Hall.

Adams, G. A. (1971). Structural investigations on a cell-wall lipopolysaccharide from *Neisseria sicca. Canadian Journal of Biochemistry*, **49**, 243–50.

Anderson, R. P. & Roth, J. R. (1977). Tandem genetic duplications in phage and bacteria. *Annual Reviews of Microbiology*, **31**, 473–505.

Bachmann, B. J. & Low K. B. (1980). Linkage map of *Escherichia coli* K-12, edition 6. *Microbiological Reviews*, **44**, 1–56.

Bagdian, G. & Mäkelä, P. H. (1971). Antigenic conversion by phage P27. I. Mapping of the prophage attachment site on the *Salmonella* chromosome. *Virology,* **43**, 403–11.

Bavoil, P., Nikaido, H. & von Meyenburg (1977). Pleiotropic transport mutants of *Escherichia coli* lack porin, a major outer membrane protein. *Molecular and General Genetics*, **158**, 23–33.

Beacham, I. R., Haas, D. & Yagil, E. (1977). Mutants of *Escherichia coli* 'cryptic' for certain periplasmic enzymes: evidence for an alteration of the outer membrane. *Journal of Bacteriology*, **129**, 1034–44.

Benzer, S. (1961). On the topography of the genetic fine structure. *Proceedings of the National Academy of Sciences*, *USA*, **47,** 403–15.

Berst, M., Lüderitz, O. & Westphal, O. (1971). Studies on the structure of T1 lipopolysaccharide. *European Journal of Biochemistry*, **18**, 361–8.

Beutin, L. & Achtman, M. (1979). Two *Escherichia coli* chromosomal cistrons, *sfrA* and *sfrB*, which are needed for expression of F factor *tra* functions. *Journal of Bacteriology*, **139**, 730–7.

Binns, M. M., Davies, D. L. & Hardy, K. G. (1979). Cloned fragments of the plasmid ColV, I-K94 specifying virulence and serum resistance. *Nature*, *London*, **279**, 778–81.

Blobel, G. & Dobberstein, B. (1975). Transfer or proteins across membranes. *Journal of Cell Biology*, **67**, 835–62.

Bolder, I. & Sompolinsky, D. (1974). Antigen specific for bacteria resistant to tetracycline. *Antimicrobial Agents and Chemotherapy*, **47**, 117–20.

BRAUN, V. (1978). Structure-function relationships of the Gram-negative bacterial cell envelope. In *Relations between Structure and Function in the Prokaryotic Cell*, ed. R. Y. Stanier, H. J. Rogers & J. B. Ward, pp. 111–38, *Symposium of the Society for General Microbiology*, 28. Cambridge University Press.

BRAUN, V. & HANTKE, K. (1977). Bacterial receptors for phages and colicins as constituents of specific transport system. In *Microbial Interactions*, series B, vol. 3, ed. J. L. Reissig, pp. 99–137. London: Chapman and Hall.

BRINTON, C. C. JR. (1959). Non-flagellar appendages of bacteria. *Nature, London*, **183**, 782–6.

BRINTON, C. C. JR., BUZZELL, A. & LAUFFER, M. A. (1954). Electrophoresis and phage susceptibility studies on a filament-producing variant of the *E. coli* bacterium. *Biochimica et Biophysica Acta*, **15**, 533–42.

BRUNETEAU, M., VOLK, W. A., SINGH, P. P. & LÜDERITZ, O. (1974). Structural investigation on the Salmonella T2 lipopolysaccharide. *European Journal of Biochemistry*, **43**, 501–8.

BUCHANAN, T. M., CHEN, K. C. S., JONES, R. B., HILLEBRAND, J. F., PEARCE, W. A., HERMODSON, M. A., NEWLAND, J. C. & LUCHTEL, D. C. (1978). Pili and principal outer membrane protein of *Neisseria gonorrhoeae*: Immunochemical, structural, and pathogenic aspects. In *Immunobiology of Neisseria gonorrhoeae*, ed. G. F. Brooks, E. C. Gotschlich, K. K. Holmes, D. W. Sawyer & F. E. Young, pp. 145–54. Washington DC: American Society for Microbiology.

BURROWS, M. R., SELLWOOD, E. & GIBBONS, R. A. (1976). Haemagglutinating and adhesive properties associated with the K99 antigen of bovine strains of *Escherichia coli. Journal of General Microbiology*, **96**, 269–75.

CLARK, W. R., MCLAUGHLIN, J. & WEBSTER, M. E. (1958). An aminohexuronic acid as the principal hydrolytic component of the Vi antigen. *Journal of Biological Chemistry*, **230**, 81–9.

COLEMAN, W. G. JR. & LEIVE, L. (1979). Two mutations which affect the barrier function of the *Escherichia coli* K-12 outer membrane. *Journal of Bacteriology*, **139**, 899–910.

COULONDRE, C., MILLER, J. H., FARABAUGH, P. J. & GILBERT, W. (1978). Molecular basis of base substitution hotspots in *Escherichia coli. Nature, London*, **247**2, 775–80.

CRAVEN, D. E. & FRASCH, C. E. (1978). Pili-mediated and -nonmediated adherence of *Neisseria meningitidis* and its relationship to invasive disease. In *Immunobiology of Neisseria gonorrhoeae*, ed. G. F. Brooks, E. C. Gotschlich, K. K. Holmes, D. W. Sawyer & F. E. Young, pp. 250–2. Washington DC: American Society for Microbiology.

CREEGER, E. S., CHEN, J. F. & ROTHFIELD, L. (1979). Cloning of genes for bacterial glycosyltransferases. II. Selection of a hybrid plasmid carrying the *rfaH* gene. *Journal of Biological Chemistry*, **254**, 811–15.

CREEGER, E. S. & ROTHFIELD, L. I. (1979). Cloning of genes for bacterial glycosyltransferases. I. Selection of hybrid plasmids carrying genes for two glucosyltransferases. *Journal of Biological Chemistry*, **254**, 804–10.

DIDERICHSEN, B. (1980). *flu*, a metastable gene controlling surface properties of *Escherichia coli. Journal of Bacteriology*, **141**, 858–67.

DIEDRICH, D. L., SUMMERS, A. O. & SCHNAITMAN, C. A. (1977). Outer membrane of *Escherichia coli.* V. Evidence that protein 1 and bacteriophage-directed protein 2 are different polypeptides. *Journal of Bacteriology*, **131**, 598–607.

DIRIENZO, J. M., NAKAMURA, K. & INOUYE, M. (1978). The outer membrane proteins of Gram-negative bacteria: biosynthesis, assembly, and functions. In *Annual Review of Biochemistry*, vol. 47, ed. E. E. Snell, P. D. Boyer, A. Meister & C. C. Richardson, pp. 481–532. Palo Alto, California: Annual Reviews Inc.

DUDMAN, W. F. (1977). The role of surface polysaccharides in natural environments. In *Surface Carbohydrates of the Prokaryotic Cell*, ed. I. Sutherland, pp. 357–414. London: Academic Press.

DUGUID, J. P., ANDERSON, E. S. & CAMPBELL, I. (1966). Fimbriae and adhesive properties in salmonellae. *Journal of Pathology and Bacteriology*, **92**, 107–38.

DUGUID, J. P., SMITH, I. W., DEMPSTER, G. & EDMUNDS, P. N. (1955). Nonflagellar filamentous appendages ('fimbriae') and hemagglutinating activity in *Bacterium coli. Journal of Pathology and Bacteriology*, **70**, 335–48.

EISENSTEIN, T. K. (1975). Evidence for O antigens as the antigenic determinants in 'ribosomal' vaccines prepared from *Salmonella. Infection and Immunity*, **12**, 364–77.

EMR, S. D., HEDGPETH, J., CLÉMENT, J.-M., SILHAVY, T. J. & HOFNUNG, M. (1980). Sequence analysis of mutations that prevent export of λ receptor, an *Escherichia coli* outer membrane protein. *Nature, London*, **285**, 82–5.

ENDO, A. & ROTHFIELD, L. (1969). Studies of a phospholipid-requiring bacterial enzyme. I. Purification and properties of uridine diphosphate galactose: lipopolysaccharide α-3-galactosyl transferase. *Biochemistry*, **9**, 3500–15.

EVANS, D. G. & EVANS, D. J. JR. (1978). New surface-associated heat-labile colonization factor antigen (CFA/II) produced by enterotoxigenic *Escherichia coli* of serogroups 06 and 08. *Infection and Immunity*, **21**, 638–47.

EVANS, D. G., EVANS, D. J. JR. & TJOA, W. (1977). Hemagglutination of human group A erythrocytes by enterotoxigenic *Escherichia coli* isolated from adults with diarrhoea: correlation with colonization factor. *Infection and Immunity*, **18**, 330–7.

EVANS, D. G., SILVER, R. P., EVANS, D. J. JR., CHASE, D. G. & GORBACH, S. L. (1975). Plasmid-controlled colonization factor associated with virulence in *Escherichia coli* enterotoxigenic for humans. *Infection and Immunity*, **12**, 656–67.

FARIS, A., LINDAHL, M. & WADSTRÖM, T. (1980). GM_2-like glycoconjugate as possible erythrocyte receptor for the CFA/I and K99 haemagglutinins of enterotoxigenic *Escherichia coli. FEMS Microbiology Letters*, **7**, 265–9.

FLEMMING, H.-C. & JANN, K. (1978a). Biosynthesis of the 09 antigen of *Escherichia coli* growth of the polysaccharide chain. *European Journal of Biochemistry*, **83**, 47–52.

FLEMMING, H.-C. & JANN, K. (1978b). Biosynthesis of the 08-antigen of *Escherichia coli* glucose at the reducing end of the polysaccharide and growth of the chain. *FEMS Microbiology Letters*, **4**, 203–5.

FRASCH, C. E. & MOCCA, L. F. (1978). Heat-modifiable outer membrane proteins of *Neisseria meningitidis* and their organization within the membrane. *Journal of Bacteriology*, **136**, 1127–34.

FUNAHARA, Y. & NIKAIDO, H. (1980). Asymmetric localization of lipopolysaccharides on the outer membrane of *Salmonella typhimurium. Journal of Bacteriology*, **141**, 1463–5.

GALANOS, C., LÜDERITZ, O., RIETSCHEL, E. T. & WESTPHAL, O. (1977). Newer aspects of the chemistry and biology of bacterial lipopolysaccharides, with special reference to their lipid A component. In *International Review of Biochemistry*, vol. 14, ed. T. W. Goodwin, pp. 239–335. Baltimore: University Park Press.

GEMSKI, P. JR., KOELTZOW, D. E. & FORMAL, S. B. (1975). Phage conversion of *Shigella flexneri* group antigens. *Infection and Immunity*, **11**, 685–91.

GEMSKI, P. LAZERE, J. R. & CASEY, T. (1980). Plasmid associated with pathogenicity and calcium dependency of *Yersinia enterocolitica. Infection and Immunity*, **27**, 682–5.

GIBBONS, R. J. & VAN HOUTE, J. (1975). Bacterial adherence in oral microbial ecology. In *Annual Review of Microbiology*, vol. 29, ed. M. P. Starr, J. L.

Ingraham & S. Raffel, pp. 19–44. Palo Alto, California: Annual Reviews Inc.
GMEINER, J. (1975). Glyceraldehyde phosphate at the reducing terminus of *Salmonella* O haptens. *European Journal of Biochemistry*, **51**, 449–57.
GRAAF, F. K., WIENTJES, F. B. & KLAASEN-BOOR, P. (1980). Production of K99 antigen by enterotoxigenic *Escherichia coli* strains of antigen groups 08, 09, 020, and 0101 grown at different conditions. *Infection and Immunity*, **27**, 216–21.
GRIFFITH, F. (1928). The significance of pneumococcal types. *Journal of Hygiene*, **27**, 8–159.
HALEGOUA, S. & INOUYE, M. (1979). Biosynthesis and assembly of the outer membrane proteins. In *Bacterial Outer Membranes*, ed. M. Inouye, pp. 67–113. New York: Wiley-Interscience.
HALL, M. N. & SILHAVY, T. J. (1979). Transcriptional regulation of *Escherichia coli* K-12 major outer membrane protein 1b. *Journal of Bacteriology*, **140**, 342–50.
HANCOCK, R. E. W. & NIKAIDO, H. (1978). Outer membranes of Gram-negative bacteria. XIX. Isolation from *Pseudomonas aeruginosa* PAO1 and use in reconstitution and definition of the permeability barrier. *Journal of Bacteriology*, **136**, 381–90.
HAVEKES, L. M. & HOEKSTRA, W. P. M. (1976). Characterization of an *Escherichia coli* K-12 F$^-$Con$^-$ mutant. *Journal of Bacteriology*, **126**, 593–600.
HAZELBAUER, G. L. (1979). The outer membrane and chemotoxis: indirect influences and secondary involvements. In *Bacterial Outer Membranes*, ed. M. Inouye, pp. 449–73. New York: Wiley-Interscience.
HAZELBAUER, G. L. & PARKINSON, J. S. (1977). Bacterial chemotaxis. In *Microbial Interactions*, series B, vol. 3, ed. J. L. Reissig, pp. 59–98. London: Chapman and Hall.
HENNING, U., ROYER, H.-D., TEATHER, R. M., HINDENNACH, I. & HOLLENBERG, C. P. (1979). Cloning of the structural gene (*ompA*) for an integral outer membrane protein of *Escherichia coli* K-12. *Proceedings of the National Academy of Sciences, USA*, **76**, 4360–4.
HIROTA, Y., SUZUKI, H., NISHIMURA, Y. & YASUDA, S. (1977). On the process of cellular division in *Escherichia coli*: A mutant of *E. coli* lacking a murein-lipoprotein. *Proceedings of the National Academy of Sciences, USA*, **74**, 1417–20.
HOFFMAN, J., LINDBERG, B., GLOWACKA, M. DERYLO, M. & LORKIEWICZ, Z. (1980). Structural studies of the lipopolysaccharide from *Salmonella typhimurium* 902 (ColIb drd2). *European Journal of Biochemistry*, **105**, 103–7.
HOFNUNG, M., JEZIERSKA, A. & BRAUN-BRETON, C. (1976). *lamB* mutations in *Escherichia coli* K12: growth of λ host range mutants and effect of nonsense suppressors. *Molecular and General Genetics*, **145**, 207–13.
HUDSON, H. P., LINDBERG, A. A. & STOCKER, B. A. D. (1978). Lipopolysaccharide core defects in *Salmonella typhimurium* mutants which are resistant to Felix O phage but retain smooth character. *Journal of General Microbiology*, **109**, 97–112.
HULL, S. I., HULL, R. A., MINSHAW, B. H. & FALKOW, S. (1980). Genetic and physical analysis of mannose resistant pili from a human UTI strain of *Escherichia coli*. In *Abstracts of the Annual Meeting of the American Society for Microbiology*, p. 25, B 51. Miami Beach, Florida, 11–16 May 1980.
IINO, H. (1962). Curly flagellar mutants in Salmonella. *Journal of General Microbiology*, **27**, 167–75.
IINO, T. (1977). Genetics of structure and function of bacterial flagella. *Annual Review of Genetics*, **11**, 161–82.
IINO, T. & MITANI, M. (1967). A mutant of *Salmonella* possessing straight flagella. *Journal of General Microbiology*, **49**, 81–8.

IYER, R. (1979). Variant forms of matrix protein in *Escherichia coli* B/r bearing N plasmids. *Biochemica et Biophysica Acta*, **556**, 86–95.

JAMES, J. F. & SWANSON, J. (1978). Color opacity colonial variants of *Neisseria gonorrhoeae* and their relationship to the menstrual cycle. In *Immunobiology of Neisseria gonorrhoeae*, ed. G. F. Brooks, W. D. Sawyer & F. E. Young, pp. 338–43. Washington DC: American Society for Microbiology.

JANN, K. & WESTPHAL, O. (1975). Microbial polysaccharides. In *The Antigens*, vol. III, pp. 1–125. New York: Academic Press.

JOHNSON, E. M., KRAUSKOPF, B. & BARON, L. S. (1965). Genetic mapping of Vi and somatic antigenic determinants in *Salmonella. Journal of Bacteriology*, **90**, 302–8.

JOHNSON, E. M., KRAUSKOPF, B. & BARON, L. S. (1966). Genetic analysis of the ViA-his chromosomal region in Salmonella. *Journal of Bacteriology*, **92**, 1457–63.

JOHNSTON, K. H. (1978). Antigenic profile of an outer membrane complex of *Neisseria gonorrhoeae* responsible for serotypic apecificity. In *Immunobiology of Neisseria gonorrhoeae*, ed. G. F. Brooks, E. C. Gotschlich, K. K. Holmes, W. D. Sawyer & F. E. Young, pp. 121–9. Washington DC: American Society of Microbiology.

JONES, G. W. (1977). The attachment of bacteria to the surfaces of animal cells. In *Microbial Interactions*, series B, vol. 3, ed. J. L. Reissig, pp. 139–76. London: Chapman and Hall.

JOYS, T. M. & MARTIN, J. F. (1973). Identification of amino acid changes in serological mutants of the i flagellar antigen of *Salmonella typhimurium. Microbios*, **7**, 71–3.

KÄLLENIUS, G. & MÖLLBY, R. (1979). Adhesion of *Escherichia coli* to human periurethal cells correlated to mannose-resistant agglutination of human erythrocytes. *FEMS Microbiology Letters*, **5**, 295–9.

KÄLLENIUS, G, MÖLLBY, R., SVENSON, S. B., WINBERG, J., LUNDBLAD, A., SVENSSON, S. & CEDERGREN, B. (1980). The p^k antigen as receptor for the haemagglutinin of pyelonephritic *Escherichia coli. FEMS Microbiology Letters*, (in press).

KAUFFMANN, F. (1954). *Enterobacteriaceae*, 2nd edn. Copenhagen: Ejnar Munksgaard Publisher.

KAWAJI, H., MIZUNO, T. & MIZUSHIMA, S. (1979). Influence of molecular size and osmolarity of sugars and dextrans on the synthesis of outer membrane proteins 0–8 and 0–9 of *Escherichia coli* K-12. *Journal of Bacteriology*, **140**, 843–7.

KEARNS, M. J. & GIBBONS, R. A. (1979). The possible nature of the pig intestinal receptor for the K88 antigen of *Escherichia coli. FEMS Microbiology Letters*, **6**, 165–8.

KELLOGG, D. JR., PEACOCK, W. L. JR., DEACON, W. E., BROWN, L. & PIRKLE, C. I. (1963). *Neisseria gonorrhoeae.* I. Virulence genetically linked to clonal variation. *Journal of bacteriology*, **85**, 1274–9.

KENNE, L., LINDBERG, B., PETERSSON, K., KATZENELLENBOGEN, E. & ROMANOWSKA, E. (1980). Structural studies of the O-specific side chains of the *Shigella sonnei* phase I lipopolysaccharide. *Carbohydrate Research*, **78**, 119–26.

KONISKY, J. (1979). Specific transport systems and receptors for colicins and phages. In *Bacterial Outer Membranes*, ed. M. Inouye, pp. 319–59. New York: Wiley-Interscience.

KORHONEN, T. K., EDÉN, S. & SVANBORG EDÉN, C. (1980). Binding of purified *Escherichia coli* poli to human urinary tract epithelial cells. *FEMS Microbiology Letters*, **7**, 237–40.

KUO, T.-T. & STOCKER, B. A. D. (1972). Mapping of *rfa* genes in *Salmonella typhimurium* by ES18 and P22 transduction and by conjugation. *Journal of Bacteriology*, **112**, 48–57.

KUTSUKAKE, K. & IINO, T. (1980). A *trans*-acting factor mediates inversion of a specific DNA segment in flagellar phase variation of *Salmonella. Nature, London*, **284**, 479–81.

LEDERBERG, E. & STOCKER, B. A. D. (1970). Phase variation in *rec*⁻ mutants of *Salmonella typhimurium. Bacteriological Proceedings*, 35.

LEDERBERG, J. & IINO, T. (1956). Phase variation in *Salmonella. Genetics*, **41**, 744–57.

LEFFLER, H. & SVANBORG EDÉN, C. (1980). Chemical identification of a glycosphingolipid receptor for *Escherichia coli* attaching to human urinary tract epithelial cells and agglutinating human erythrocytes. *FEMS Microbiology Letters*, (in press).

LEHMANN, V. (1977). Isolation, purification and properties of an intermediate in 3-deoxy-D-*manno*-octulosonic acid – lipid A biosynthesis. *European Journal of Biochemistry*, **75**, 257–66.

LEHMANN, V., HÄMMERLING, G., NURMINEN, M., MINNER, I., RUSCHMANN, E., LÜDERITZ, O., KUO, T.-T. & STOCKER, B. A. D. (1973). A new class of heptose-defective mutant of *Salmonella typhimurium. European Journal of Biochemistry*, **32**, 268–75.

LEHMANN, V. & RUPPRECHT, E. (1977). Microheterogeneity in lipid A demonstrated by a new intermediate in the biosynthesis of 3-deoxy-D-*manno*-octulosonic-acid – lipid A. *European Journal of Biochemistry*, **81**, 443–52.

LEHMANN, V., RUPPRECHT, E. & OSBORN, M. (1977). Isolation of mutants conditionally blocked in the biosynthesis of the 3-deoxy-D-*manno*-octulosonic-acid – lipid-A part of lipopolysaccharides derived from *Salmonella typhimurium. European Journal of Biochemistry*, **76**, 41–9.

LEVINTHAL, M. & NIKAIDO, H. (1969). Consequences of deletion mutations joining two operons of opposite polarity. *Journal of Molecular Biology*, **42**, 511–20.

LEVY, S. B. & MCMURRY, L. (1974). Detection of an inducible membrane protein associated with R-factor-mediated tetracycline resistance. *Biochemical and Biophysical Research Communications*, **56**, 1060–8.

LEW, H. C., NIKAIDO, H. & MÄKELÄ, P. H. (1978). Biosynthesis of uridine diphosphate N-acetylmannosaminuronic acid in *rff* mutants of *Salmonella typhimurium. Journal of Bacteriology*, **136**, 227–33.

LIN, J. J. C., KANAZAWA, H. & WU, H. C. (1980a). Assembly of outer membrane lipoprotein in an *Escherichia coli* mutant with a single amino acid replacement within the signal sequence of prolipoprotein. *Journal of Bacteriology*, **141**, 550–7.

LIN, J. J. C., KANAZAWA, H. & WU, H. C. (1980b). Purification and characterization of the outer membrane lipoprotein from an *Escherichia coli* mutant altered in the signal sequence of prolipoprotein. *Journal of Biological Chemistry*, **255**, 1160–3.

LIN, J. J. C., LAI, J.-S. WU, H. C. (1980). Characterization of murein-bound lipoprotein in an *Escherichia coli* mutant altered in the signal sequence of prolipoprotein. *FEBS Letters*, **109**, 50–4.

LINDBERG, A. A. (1977). Bacterial surface carbohydrates and bacteriophage adsorption. In *Surface Carbohydrates of the Prokaryotic Cell*, ed. I. Sutherland, pp. 289–356. London: Academic Press.

LINDBERG, A. A. & HELLERQVIST, C.-G. (1980). Rough mutants of *Salmonella typhimurium:* immunochemical and structural analysis of lipopolysaccharides from *rfaH* mutants. *Journal of General Microbiology*, **116**, 25–32.

LINDBERG, A. A., HELLERQVIST, C. G., BAGDIAN-MOTTA, G. & MÄKELÄ, P. H. (1978). Lipopolysaccharide modification accompanying antigenic conversion by phage P27. *Journal of General Microbiology*, **107**, 279–87.

LUCKEY, M. & NIKAIDO, H. (1980). Specificity of diffusion channels produced by λ phage receptor protein of *Escherichia coli. Proceedings of the National Academy of Sciences, USA*, **77,** 167–71.

LYMAN, M. B., STOCKER, B. A. D. & ROANTREE, R. J. (1977). Comparison of the virulence of O:9,12 and O:4,5,12 *Salmonella typhimurium his*$^+$ transductants for mice. *Infection and Immunity*, **15**, 491–9.

LYMAN, M. B., STOCKER, B. A. D. & ROANTREE, R. J. (1979). Evaluation of the immune response directed against the *Salmonella* antigenic factors O4,5 and O9. *Infection and Immunity*, **26**, 956–65.

MCCONNELL M. & WRIGHT, A. (1979). Variation in the structure and bacteriophage-inactivating capacity of *Salmonella anatum* lipopolysaccharide as a function of growth temperature. *Journal of Bacteriology*, **137,** 746–51.

MCDONOUGH, M. W. (1965). Amino acid composition of antigenically distinct *Salmonella* flagellar proteins. *Journal of Molecular Biology*, **12**, 342–55.

MACLEOD, C. M. & KRAUSS, M. R. (1950). Relation of virulence of pneumococcal strains for mice to the quantity of capsular polysaccharide formed *in vitro. Journal of Experimental Medicine*, **92,** 1–9.

MACPHEE, D. G., KRISHNAPILLAI, V., ROANTREE, R. J. & STOCKER, B. A. D. (1975). Mutations in *Salmonella typhimurium* conferring resistance to Felix O phage without loss of smooth character. *Journal of General Microbiology*, **87**, 1–10.

MÄKELÄ, P. H. (1965). Inheritance of the O antigens of *Salmonella* groups B and D. *Journal of General Microbiology*, **41**, 57–65.

MÄKELA, P. H. (1973). Glucosylation of lipopolysaccharide in *Salmonella*: mutants negative for O antigen factor $12_2{}^1$. *Journal of Bacteriology*, **116**, 847–56.

MÄKELÄ, P. H., JAHKOLA, M. & LUDERITZ, O. (1970). A new gene cluster *rfe* concerned with the biosynthesis of *Salmonella* lipopolysaccharide. *Journal of General Microbiology*, **60**, 91–106.

MÄKELÄ, P. H. & MÄKELÄ, O. (1966). Salmonella antigen 12_2: genetics of form variation. *Annales Medicinae Experimentalis Fenniae*, **44**, 310–17.

MÄKELÄ, P. H. & MAYER, H. (1976). Enterobacterial common antigen. *Bacteriological Reviews*, **40**, 591–632.

MÄKELÄ, P. H. SARVAS, M., CALCAGNO, S. & LOUNATMAA, K. (1978). Isolation and characterization of polymyxin resistant mutants of *Salmonella. FEMS Microbiology Letters*, **3**, 323–6.

MÄKELÄ, P. H. & STOCKER, B. A. D. (1969). Genetics of polysaccharide biosynthesis. *Annual Review of Genetics*, **3**, 291–322.

MÄKELÄ, P. H., VALTONEN, V. V. & VALTONEN, M. (1973). Role of O-antigen (lipopolysaccharide) factors in the virulence of *Salmonella. Journal of Infectious Diseases*, **128**, suppl., S81–S85.

MANNING, P. A. & ACHTMAN, M. (1979). Cell-to-cell interactions in conjugating *Escherichia coli*: the involvement of the cell envelope. In *Bacterial Outer Membranes*, ed. M. Inouye, pp. 409–47. New York: Wiley-Interscience.

MARKOVITZ, A. (1977). Genetics and regulation of bacterial capsular polysaccharide biosynthesis and radiation sensitivity. In *Surface Carbohydrates of the Prokaryotic Cell*, ed. I. Sutherland, pp. 415–62. London: Academic Press.

MAYER, H. & SCHMIDT, G. (1979). Chemistry and biology of the enterobacterial common antigen (ECA). In *Current Topics in Microbiology and Immunology*, vol. 85, ed. W. Arber, S. Falkow, W. Henle, P. H. Hofschneider, J. H. Humphrey, J. Klein, P. Koldovský, H. Koprowski, D. Maaløe, F. Melchers, R.

Rott, H. G. Schweiger, L. Styrucek & P. K. Vogt, pp. 99–153. Berlin: Springer-Verlag.

Mills, G. T. & Smith, E. E. B. (1965). Biosynthesis of capsular polysaccharides in the pneumococcus. *Bulletin de la Société de Chimie Biologique*, **47**, no. 10, 1751–65.

Moll, A., Manning, P. A. & Timmis, K. N. (1980). Plasmid-determined resistance to serum bactericidal activity. II. A major outer membrane protein. TraTp, is responsible for plasmid-specified serum resistance in *Escherichia coli. Infection and Immunity*, (in press).

Mühlradt, P., Risse, H. J., Lüderitz, O. & Westphal. O. (1968). Biochemical studies on lipopolysaccharides of *Salmonella* R mutants. 5. Evidence for a phosphorylating enzyme in lipopolysaccharide biosynthesis. *European Journal of Biochemistry*, **4**, 139–45.

Müller, E., Hinckley, A. & Rothfield, L. (1972). Studies of phospholipid-requiring bacterial enzymes. III. Purification and properties of uridine diphosphate glucose:lipopolysaccharide glucosyltransferase I. *Journal of Biological Chemistry*, **247**, 2614–22.

Naide, Y., Nikaido, H., Mäkelä, P. H., Wilkinson, R. G. & Stocker, B. A. D. (1965). Semirough strains of *Salmonella. Proceedings of the National Academy of Sciences, USA*, **52**, 147–53.

Nakamura, K. (1968). Genetic determination of resistance to acriflavive, phenethyl alcohol, and sodium dodecyl sulfate in *Escherichia coli. Journal of Bacteriology*, **96**, 987–96.

Nakamura, K., Pirtle, R. M., Pirtle, I. L., Takeishi, K. & Inouye, M. (1980). Messenger ribonucleic acid of the lipoprotein of the *Escherichia coli* outer membrane. *Journal of Biological Chemistry*, **255**, 210–16.

Nikaido, H. (1962). Studies on the biosynthesis of cell wall polysaccharide in mutant strains of Salmonella, II. *Proceedings of the National Academy of Sciences, USA,* **48**, 1542–8.

Nikaido, H. (1973). Biosynthesis and assembly of lipopolysaccharide and the outer membrane layer of Gram-negative cell wall. In *Bacterial Membranes and Walls*, vol. 1, ed. L. Leive, pp. 131–208. New York: Marcel Dekker.

Nikaido, H. (1979). Nonspecific transport through the outer membrane. In *Bacterial Outer Membranes*, ed. M. Iouye, pp. 361–407. New York: Wiley-Interscience.

Nikaido, H., Levinthal, M., Nikaido, K. & Nakane, K. (1967). Extended deletions in the histidine-rough-B region of the Salmonella chromosome. *Proceedings of the National Academy of Sciences, USA*, **57**, 1825–32.

Nikaido, H. & Nakae, T. (1979). The outer membrane of Gram-negative bacteria. *Advances in Microbial Physiology*, **20**, 163–250.

Nikaido, K. & Nikaido, H. (1971). Glucosylation of lipopolysaccharide in *Salmonella:* Biosynthesis of O antigen factor 12_2. II. Structure of the lipid intermediate. *Journal of Biological Chemistry*, **246**, 3912–19.

Nikaido, H., Nikaido, K., Nakae, T. & Mäkelä, P. H. (1971). Glucosylation of lipopolysaccharide in *Salmonella*: Biosynthesis of O antigen factor 12_2. I. Over-all reaction. *Journal of Biological Chemistry*, **246**, 3902–11.

Normark, S., Edlund, T., Grundström, T., Bergström, S. & Wolf-Watz, H. (1977). *Escherichia coli* K-12 mutants hyperproducing chromosomal beta-lactamase by gene repetitions. *Journal of Bacteriology*, **132**, 912–22.

Nyman, K., Plosila, M., Howden, L. & Mäkelä, P. H. (1979). Genetic determination of lipopolysaccharide: Locus of O-specific unit polymerase in group E of *Salmonella. Zentralblatt für Bakteriologie, Parasitenkunde, Infektioskrankheiten und Hygiene*, I. Abt. Orig. **A243**, 355–62.

OLD, D. C., CORNEIL, I., GIBSON, L., THOMPSON, A. & DUGUID, P. (1968). Fimbriation, pellicle formation and the amount of growth of *Salmonellas* in broth. *Journal of General Microbiology*, **51**, 1–16.

ØRSKOV, F. & ØRSKOV, I. (1962). Behaviour of *E. coli* antigens in sexual recombination. *Acta Pathologica et Microbiologica Scandinavica*, **55**, 99–109.

ØRSKOV, F., ØRSKOV, I., SUTTON, A., SCHNEERSON, R., LIN, W., EGAN, W., HOFF, G. E. & ROBBINS, J. B. (1979). Form variation in *Escherichia coli* K1: determined by O-acetylation of the capsular polysaccharide. *Journal of Experimental Medicine*, **149**, 669–85.

ØRSKOV, I. & NYMAN, K. (1974). Genetic mapping of the antigenic determinants of two polysaccharide K antigens, K10 and K54 in *Escherichia coli. Journal of Bacteriology*, **120**, 43–51.

ØRSKOV, I. & ØRSKOV, F. (1966). Episome-carried surface antigen K88 of *Escherichia coli.* I. Transmission of the determinant of the K88 antigen and influence on the transfer of chromosomal markers. *Journal of Bacteriology*, **91**, 69–75.

ØRSKOV, I. & ØRSKOV, F. (1977). Special O:K:H: serotypes among enterotoxigenic *E. coli* strains from diarrhoea in adults and children. Occurrence of the CF (colonization factor) antigen and of hemagglutinating abilities. *Medical Microbiology and Immunology*, **163**, 99–110.

ØRSKOV, I., ØRSKOV, F. & BIRCH-ANDERSEN, A. (1980). Comparison of *Escherichia coli* fimbrial antigen F7 with type 1 fimbriae. *Infection and Immunity*, **27**, 657–66.

ØRSKOV, I., ØRSKOV, F., JANN, B. & JANN, K. (1977). Serology, chemistry, and genetics of O and K antigens of *Escherichia coli. Bacteriological Reviews*, **41**, 667–710.

ØRSKOV, I., ØRSKOV, F., SOJKA, W. J. & LEACH, J. M. (1961). Simultaneous occurrence of *E. coli* B and L antigens in strains from diseased swine. *Acta Pathologica et Microbiologica Scandinavica*, **53**, 404–22.

ØRSKOV, I., SHARMA, V. & ØRSKOV, F. (1976). Genetic mapping of the K1 and K4 antigens (L) of *Escherichia coli.* Non-allelism of K (L) antigens with K antigens of 08:K27(A), 08:K8(L) and 09:K57(B). *Acta Pathologica et Microbiologica Scandinavica*, Sect. B **84**, 125–31.

OSBORN, M. J. (1968). Biochemical characterization of mutants of *Salmonella typhimurium* lacking glucosyl or galactosyl lipopolysaccharide transferases. *Nature, London*, **217**, 957–60.

OSBORN, M. J. (1979). Biosynthesis and assembly of lipopolysaccharide of the outer membrane. In *Bacterial Outer Membranes*, ed. M. Inouye, pp. 15–34. New York: Wiley-Interscience.

OSBORN, M. J. & ROTHFIELD, L. I. (1971). Biosynthesis of the core region of lipopolysaccharide. In *Microbial Toxins*, vol. IV, ed. G. Weinbaum, S. Kadis & S. J. Ajl, pp. 331–50. New York: Academic Press.

OTTOW, J. C. G. (1975). Ecology, physiology, and genetics of fimbriae and pili. *Annual Review of Microbiology*, **29**, 79–108.

OVERBEEKE, N. & LUGTENBERG, B. (1980). Expression of outer membrane protein e of *Escherichia coli* K12 by phosphate limitation. *FEBS Letters*, **112**, 229–32.

PENN, C. W., PARSONS, N. J., RITTENBERG, S. C., SANYAL, S. C., VEALE, D. R. & SMITH, H. (1978). Gonococci grown *in vivo* and *in vitro*: selection and phenotypic change in relation to pathogenesis and immunity. In *Immunobiology of Neisseria gonorrhoeae*, ed. G. F. Brooks, E. C. Gotschlich, K. K. Holmes, W. D. Sawyer & F. E. Young, pp. 356–9. Washington DC: American Society for Microbiology.

PERRY, M. B., DAOUST, V., JOHNSON, K. G., DIENNA, B. B. & ASHTON, F. E. (1978). Gonococcal R-type lipopolysaccharides. In *Immunobiology of Neisseria gonorrhoeae*, ed. G. F. Brooks, E. C. Gotschlich, K. K. Holmes, W. D. Sawyer & F. E. Young, pp. 101–7. Washington DC: American Society for Microbiology.

PLOSILA, M. & MÄKELÄ, P. H. (1972). Mapping of a gene *oafC* determining antigen 1 in *Salmonella* of group E. *Scandinavian Journal of Clinical and Laboratory Investigation*, **29**, suppl. 122, 55.

PREHM, P., JANN, B., JANN, K., SCHMIDT, G. & STIRM, S. (1976a). On a bacteriophage T3 and T4 receptor region within the cell wall lipopolysaccharide of *Escherichia coli* B. *Journal of Molecular Biology*, **101**, 277–81.

PREHM, P., STIRM, S., JANN, B., JANN, K. & BOMAN, H. G. (1976b). Cell-wall lipopolysaccharides of ampicillin-resistant mutants of *Escherichia coli* K-12. *European Journal of Biochemistry*, **66**, 369–77.

PUGSLEY, A. P. & SCHNAITMAN, C. A. (1978a). Outer membrane proteins of *Escherichia coli.* VII. Evidence that bacteriophage-directed protein 2 functions as a pore. *Journal of Bacteriology*, **133**, 1181–9.

PUGSLEY, A. P. & SCHNAITMAN, C. A. (1978b). Identification of three genes controlling production of new outer membrane pore proteins in *Escherichia coli* K-12. *Journal of Bacteriology*, **135**, 1118–29.

REEVES, P. (1979). The genetics of outer membrane proteins. In *Bacterial Outer Membranes*, ed. M. Inouye, pp. 255–91. New York: Wiley-Interscience.

RICK, P. D., FUNG, L. W.-M., HO, C. & OSBORN, M. J. (1977). Lipid A mutants of *Salmonella typhimurium.* Purification and characterization of a lipid A precursor produced by a mutant in 3-deoxy-D-mannooctulosonate-8-phosphate synthetase. *Journal of Biological Chemistry*, **252**, 4904–12.

RICK, P. D. & OSBORN, M. J. (1972). Isolation of a mutant of *Salmonella typhimurium* dependent on D-arabinose-5phosphate for growth and synthesis of 3-deoxy-D-mannooctulosonate (ketodeoxyoctonate). *Proceedings of the National Academy of Sciences, USA*, **69**, 3756–60.

RICK, P. D. & OSBORN, M. J. (1977). Lipid A mutants of *Salmonella typhimurium.* Characterization of a conditional lethal mutant in 3-deoxy-D-mannooctulosonate-8-phosphate synthetase. *Journal of Biological Chemistry*, **252**, 4895–903.

ROBBINS, J. B., MCCRACKEN, G. H., GOTSCHLICH, E. C., ØRSKOV, F., ØRSKOV, I. & HANSON, L. (1974). *Escherichia coli* K1 capsular polysaccharide associated with neonatal meningitis. *New England Journal of Medicine*, **290**, 1216–20.

ROBBINS, J. B., SCHNEERSON, R., EGAN, W. B., VANN, W. R. & LIU, D. T. (1980). Virulence properties of bacterial capsular polysaccharides – unanswered questions. In *The Molecular Basis of Microbial Pathogenicity*, ed. H. Smith, J. J. Skehel & M. J. Turner, pp. 115–32. Dahlem Konferenzen, 1980. Weinheim: Verlag Chemie.

ROBBINS, P. W., KELLER, J. M., WRIGHT, A. & BERNSTEIN, R. L. (1965). Enzymatic and kinetic studies on the mechanism of O-antigen conversion by bacteriophage ε^{15}. *Journal of Biological Chemistry*, **240**, 384–90.

ROBBINS, P. W. & UCHIDA, T. (1962). Studies on the chemical basis of the phage conversion of O-antigen in the E-group *Salmonellae. Biochemistry*, **1,** 323–34.

ROBBINS, P. W. & WRIGHT, A. (1971). Biosynthesis of O-antigens. In *Microbial Toxins*, vol. IV, ed. G. Weinbaum, S. Kadis & S. J. Ajl, pp. 351–68. New York: Academic Press.

ROBERTS, M., ELWELL, L. & FALKOW, S. (1978). Introduction to the mechanisms of genetic exchange in the gonococcus: plasmids and conjugation in *Neisseria gonorrhoeae.* In *Immunobiology of Neisseria gonorrhoeae*, ed. G. F. Brooks, E. C. Gotschlich, K. K. Holmes, W. D. Sawyer & F. E. Young, pp. 38–43. Washington DC: American Society for Microbiology.

RUTTER, J. M., BURROWS, M. R., SELLWOOD, R. & GIBBONS, R. A. (1975). A genetic basis for resistance to enteric disease caused by *E. coli. Nature, London*, **257**, 135–6.

SAIER, M. H. JR., SCHMIDT, M. R. & LEIBOWITZ, M. (1978). Cyclic AMP-dependent synthesis of fimbriae in *Salmonella typhimurium:* Effects of *cya* and *pts* mutations. *Journal of Bacteriology*, **134**, 356–8.

SANDERSON, K. E. & HARTMAN, P. E. (1978). Linkage map of *Salmonella typhimurium*, Edition V. *Microbiological Reviews*, **42**, 471–519.

SARFF, L. D., MCCRACKEN, G. H. JR., SCHIFFER, M. S., GLODE, M. P., ROBBINS, J. B., ØRSKOV, I. & ØRSKOV, F. (1975). Epidemiology of *Escherichia coli* K1 in healthy and diseased newborns. *The Lancet*, **I**, 1099–104.

SARMA, V. & REEVES, P. (1977). Genetic locus (*ompB*) affecting a major outer-membrane protein in *Escherichia coli* K-12. *Journal of Bacteriology*, **132**, 23–7.

SARVAS, M. (1967). Inheritance of *Salmonella* T1 antigen. *Annales Medicinae Experimentalis Fenniae*, **45**, 447–72.

SCHMIDT, G., JANN, B., JANN, K., ØRSKOV, I. & ØRSKOV, F. (1977). Genetic determinants of the synthesis of the polysaccharide capsular antigen K(A)27 of *Escherichia coli. Journal of General Microbiology*, **100**, 355–61.

SCHMIDT, G., MAYER, H. & MÄKELÄ, P. H. (1976). Presence of *rfe* genes in *Escherichia coli*: Their participation in biosynthesis of O antigen and enterobacterial common antigen. *Journal of Bacteriology*, **127**, 755–62.

SCHNAITMAN, C., SMITH, D. & FORN DE SALSAS, M. (1975). Temperate bacteriophage which causes the production of a new major outer membrane protein by *Escherichia coli. Journal of Virology*, **15**, 1121–30.

SILHAVY, T. J., BASSFORD, P. J. JR. & BECKWITH, J. R. (1979). A genetic approach to the study of protein localization in *Escherichia coli.* In *Bacterial Outer Membranes*, ed. M. Inouye, pp. 203–54. New York: Wiley-Interscience.

SILVERMAN, M. & SIMON, M. I. (1977). Bacterial flagella. *Annual Review of Microbiology*, **31**, 397–419.

SILVERMAN, M., ZIEG, J., HILMEN, M. & SIMON, M. (1979). Phase variation in *Salmonella*: Genetic analysis of a recombinational switch. *Proceedings of the National Academy of Sciences, USA*, **76**, 391–5.

SILVERMAN, M., ZIEG, J. & SIMON, M. (1979). Flagellar-phase variation: Isolation of the *rh1* gene. *Journal of Bacteriology*, **137**, 517–23.

SILVERMAN, M. & SIMON, M. (1980). Phase variation: genetic analysis of switching mutants. *Cell*, **19**, 845–54.

SMITH, H. W. (1974). A search for transmissible pathogenic characters in invasive of *Escherichia coli*: the discovery of a plasmid-controlled toxin and a plasmid-controlled lethal character closely associated, or identical with colicine V. *Journal of General Microbiology*, **83**, 95–111.

SMITH, H. W. & HUGGINS, M. B. (1976). Futher observations on the association of the colicine V plasmid of *Escherichia coli* with pathogenicity and with survival in the alimentary tract. *Journal of General Microbiology*, **92**, 335–50.

SMITH, H. W. & LINGGOOD, M. A. (1971). Observations on the pathogenic properties of the K88, hly and ent plasmids of *Escherichia coli* with particular reference to porcine diarrhoea. *Journal of Medical Microbiology*, **4**, 467–85.

SMITH, H. W. & LINGGOOD, M. A. (1972). Further observations on *Escherichia coli* enterotoxins with particular regard to those produced by atypical piglet strains and by calf and lamb strains: the transmissible nature of these enterotoxins and of a K antigen possessed by calf and lamb strains. *Journal of Medical Microbiology*, **5**, 243–50.

SNELLINGS, N. J., JOHNSON, E. M. & BARON, L. S. (1979). Genetic investigation of Vi antigen variation in *Citrobacter ballerup.* In *Abstracts of the Annual Meeting of the American Society for Microbiology*, Los Angeles, California, May 1979, p. 48.

STEAD, A., MAIN, J. S., WARD, M. E. & WATT, P. J. (1975). Studies on lipopolysaccharides isolated from strains of *Neisseria gonorrhoeae. Journal of General Microbiology*, **88**, 123–31.

STIRM, S., ØRSKOV, I. & ØRSKOV, F. (1966). K88, an episome-determined protein antigen of *Escherichia coli. Nature, London*, **209**, 507–8.

STOCKER, B. A. D. & MÄKELÄ, P. H. (1971). Genetics aspects of biosynthesis and structure of *Salmonella lipopolysaccharide. In Microbial Toxins*, vol. IV, ed. S. Ajl, pp. 396–438. New York and London: Academic Press.

STOCKER, B. A. D. & MÄKELÄ, P. H. (1978). Genetics of the (gram-negative) bacterial surface. *Proceedings of the Royal Society of London*, **B202**, 5–30.

STOCKER, B. A. D., MALES, B. M. & TAKANO, W. (1980). *Salmonella typhimurium* mutants of *RfaH*-phenotype: Genetics and antibiotic sensitivities. *Journal of General Microbiology*, **116**, 17–24.

STOCKER, B. A. D., MCDONOUGH, M. W. & AMBLER, R. P. (1961). A gene determining presence or absence of ε-N-methyl-lysine in *Salmonella* flagellar protein. *Nature, London*, **189**, 556–8.

SUTHERLAND, I. W. (1977). Bacterial exopolysaccharides – their nature and production. In *Surface Carbohydrates of the Prokaryotic Cell*, ed. I. Sutherland, pp. 27–96. London: Academic Press.

SVANBORG-EDÉN, C., ERIKSSON, B., HANSON, L. Å., JODAL, U., KAIJSER, B., JANSON, G. L., LINDBERG, U. & OLLING, S. (1978). Adhesion to normal human uroepithelial cells of *Escherichia coli* from children with various forms of urinary tract infection. *Journal of Pediatrics*, **93**, 398–403.

SVANBORG-EDÉN, C., JODAL, U., HANSON, L. Å., LINDBERG, U. & SOHL AKERLUND, A. (1976). Variable adherence to normal human urinary-tract epithelial cells of *Escherichia coli* strains associated with various forms of urinary-tract infection. *The Lancet*, **2**, 490–2.

SWANEY, L. M., LIU, Y.-P., TO, C.-M., TO, C.-C., IPPEN-IHLER, K. & BRINTON, C. C. JR. (1977a). Isolation and characterization of *Escherichia coli* phase variants and mutants deficient in type 1 pilus production. *Journal of Bacteriology*, **130**, 495–505.

SWANEY, L. M., LIU, Y.-P., IPPEN-IHLER, K. & BRINTON, C. C. JR. (1977b). Genetic complementation analysis of *Escherichia coli* type 1 somatic pilus mutants. *Journal of Bacteriology*, **130**, 506–11.

SWANSON, J. (1980). Adhesion and entry of bacteria into cells: a model of the pathogenesis of Gonorrhoea. In *The Molecular Basis of Microbial Pathogenicity*, ed. H. Smith, J. J. Skehel & M. J. Turner, pp. 17–39. Dahlem Konferenzen, Berlin 1979. Weinheim: Verlag Chemie.

TAKESHITA, M. & MÄKELÄ, P. H. (1971). Glucosylation of lipopolysaccharide in *Salmonella:* biosynthesis of O antigen factor 12_2. III. The presence of 12_2 determinants in haptenic polysaccharides. *Journal of Biological Chemistry*, **246**, 3920–7.

TAYLOR, H. E. (1949). Additive effects of certain transforming agents from some variants of pneumococcus. *Journal of Experimental Medicine*, **89**, 399–424.

TAYLOR, P. W., HUGHES, C. & ROBINSON, M. (1979). Plasmids and the serum resistance of enterobacteria. In *Plasmids of Medical Environmental and Commercial Importance*, vol. 1, ed. K. N. Timmis & A. Pühler, pp. 135–43. Amsterdam: Elsevier/North-Holland Biomedical Press.

TIMMIS, K. N., MOLL, A. & DANBARA, H. (1979). Plasmid gene that specifies resistance to the bactericidal activity of serum. In *Plasmids of Medical, Environmental and Commercial Importance*, vol. 1, ed. K. N. Timmis & A. Pühler, pp. 145–53. Amsterdam: Elsevier/North-Holland Biomedical Press.

TURNER, M. J. (1980). Antigenic variation. In *The Molecular Basis of Microbial Pathogenicity*, ed. H. Smith, J. J. Skehel & M. J. Turner, pp. 133–58. Dahlem Konferenzen, 1980. Weinheim: Verlag Chemie.

VAARA, M., VAARA, T. & SARVAS, M. (1979). Decreased binding of polymyxin by polymyxin-resistant mutants of *Salmonella typhimurium*. *Journal of Bacteriology*, **139**, 664–7.

VALTONEN, V. V., SARVAS, M. & MÄKELÄ, P. H. (1976). T2 lipopolysaccharide antigen of *Salmonella*: Genetic determination of T2 and properties of the T2, T2, S, and T2, SR forms. *Infection and Immunity*, **13**, 1647–53.

VAN ALPHEN, W. & LUGTENBERG, B. (1977). Influence of osmolarity of the growth medium on the outer membrane protein pattern of *Escherichia coli*. *Journal of Bacteriology*, **131**, 623–30.

VERHOEF, C., LUGTENBERG, B., VAN BOXTEL, R., DE GRAAF, P. & VERHEIJ, H. (1979). Genetics and biochemistry of the peptidoglycan-associates proteins b and c of *Escherichia coli* K12. *Molecular and General Genetics*, **169**, 137–46.

WILKINSON, R. G. & STOCKER, B. A. D. (1968). Genetics and cultural properties of mutants of *Salmonella typhimurium* lacking glucosyl or galactosyl lipopolysaccharide transferases. *Nature, London*, **217**, 955–7.

WILLIAMS, P. H. (1979). Novel iron uptake system specified by ColV plasmids: an important component in the virulence of invasive strains of *Escherichia coli*. *Infection and Immunity*, **26**, 925–32.

WRIGHT, A. & BARZILAI, N. (1971). Isolation and characterization of nonconverting mutants of bacteriophage ε^{34}. *Journal of Bacteriology*, **105**, 937–9.

WU, H. C. & LIN, J. J.-C. (1976). *Escherichia coli* mutants altered in murein lipoprotein. *Journal of Bacteriology*, **126**, 147–56.

ZIEG, J., SILVERMAN, M., HILMEN, M. & SIMON, M. (1977). Recombinational switch for gene expression. *Science*, **196**, 170–2.

GENETICS OF BACTERIAL CHEMOTAXIS

JOHN S. PARKINSON

Biology Department, University of Utah, Salt Lake City, Utah, USA 84112

INTRODUCTION

Many types of motile bacteria are capable of detecting and responding to changes in their environment. Phototactic, chemotactic and thermotactic movements in bacteria are similar to more complex behaviours seen in higher organisms, and constitute useful model systems for investigating the molecular events underlying sensory transduction phenomena. The best-studied of these systems is the chemotactic behaviour of *Escherichia coli* and *Salmonella typhimurium*. Extensive genetic and biochemical analyses of the chemotaxis machinery in these organisms has led to an intriguing picture of how bacteria detect and process sensory information. At the molecular level, the chemotactic apparatus of bacteria has proven to be surprisingly sophisticated, although many of the mechanistic details are still poorly understood. In this chapter I summarize current knowledge in this area, with particular emphasis on the various ways that genetic methods are being used to investigate the chemotactic behaviour of *E. coli* and *S. typhimurium*. This presentation is necessarily brief and rather speculative. More extensive discussion of this subject can be found in a number of recent reviews (Adler, 1975; Berg, 1975; Hazelbauer & Parkinson, 1977; Parkinson, 1977; Goy & Springer, 1978; Koshland, 1978; Macnab, 1978).

Locomotor behaviour

In the absence of chemical stimuli, wild-type *E. coli* and *S. typhimurium* swim about in random walk fashion consisting of smooth translational movements ('runs') and abrupt turning motions ('tumbles') (Berg & Brown, 1972). Tumbling episodes occur approximately once per second and enable the cell to change its swimming direction in an essentially random manner. In the presence of chemical gradients, chemotactic movements are accomplished by altering tumble probability in response to concentration changes encountered as the organism swims (Berg & Brown, 1972; Macnab & Koshland, 1972). For example, whenever an individual happens to swim toward an attractant source, tumble probability decreases, producing a somewhat longer average run length in the favourable direction. Since bacteria are sufficiently small to be knocked off course by Brownian motion, they spend a great deal of time moving in the 'wrong' direction during a chemotactic response: they must constantly assess their direction of movement relative to the gradient and can only achieve net migration in a biased random walk fashion.

Both runs and tumbles are produced by rotation of the cells' flagellar filaments, which are semi-rigid helical structures that function like propellers (Berg & Anderson, 1973; Silverman & Simon, 1974). A bacterium typically possesses about six flagella emanating from different sites on the cell surface which act together in a bundle at the rear of the cell to generate co-ordinated swimming movements. The hydrodynamic details of bacterial motility are complex, but in essence runs are produced by counter-clockwise rotation of the flagellar filaments, whose helical sense is normally left-handed. (The direction of rotation is defined by viewing along the bundle of filaments toward the cell body.) Clockwise rotation of one or more filaments causes disruption of the bundle and produces a tumbling episode (Macnab & Ornston, 1977). Flagellar rotation can be monitored by tethering a filament to a microscope slide with specific antibodies and observing rotation of the cell body (Silverman & Simon, 1974). Tethered wild-type cells exhibit frequent reversals in rotation which presumably correspond to episodes of runs and tumbles in free-swimming cells (Berg, 1974; Larsen *et al.*, 1974b).

Rotational motion is most likely generated at the base of the flagellum, which is anchored in the cell wall and membranes and has a complex structure in the electron microscope (DePamphilis & Adler, 1971). The flagellar base is composed of approximately ten different proteins (Silverman & Simon, 1977a). It seems likely that many additional proteins required for motility reside in the inner membrane, because the energy source for rotation is known to be derived from an electrochemical gradient of protons across the cytoplasmic membrane (Larsen, Adler, Gargus & Hogg, 1974a). Both electrical (i.e. membrane potential) and chemical (i.e. pH differential) gradients of protons can be harnessed to power flagellar rotation; ATP is not involved (Khan & Macnab, 1980a). Only a few motility-related membrane components have yet been identified biochemically, and it is not at all clear how proton motive force is coupled to flagellar rotation.

Stimulus detection

Chemotactic responses, i.e. changes in the rotational behaviour of the flagella, are triggered by *temporal* changes in attractant or repellent concentrations encountered by the swimming bacteria (Macnab & Koshland, 1972; Brown & Berg, 1974). Concentration measurements are made by means of specific chemoreceptors arrayed either on the external surface of the cytoplasmic membrane or in the periplasmic space between the inner and outer membranes. Receptors for the sugar attractants have been shown to be specific binding proteins. The extent of receptor occupancy therefore provides a measure of ligand concentration in the environment.

The galactose, maltose and ribose receptors are periplasmic binding proteins (Hazelbauer & Adler, 1971; Aksamit & Koshland, 1974; Hazelbauer, 1975); the glucose and other sugar receptors are binding proteins associated with the cytoplasmic membrane (Adler, & Epstein, 1974; Lengeler, 1975). These binding proteins function as specific recognition devices for both chemotaxis and active transport; however, sugar uptake is not required for chemotaxis, so the two processes are clearly different, even though they share a common component (Adler, 1969). The identity of the receptors for the amino acid attractants aspartate and serine is less clear, but they may also be specific binding proteins in the cytoplasmic membrane (Clarke & Koshland, 1979). Competition studies have provided indirect evidence for the existence of a number of repellent receptors as well, including compounds such as fatty acids, alcohols, hydrophobic amino acids, and several divalent cations (Tso & Adler, 1974). None of these repellent receptors has been identified biochemically.

Sensory transduction

In order to respond only to *changes* in receptor occupancy, and not to absolute chemical concentrations, the bacteria must have a means of adapting their response machinery to static levels of attractants and repellents. The existence of such an adaptation system has been convincingly demonstrated by subjecting cells to rapid, isotropic changes in attractant or repellent concentration. Attractant increases and repellent decreases suppress clockwise (CW) rotation, whereas stimuli of opposite sign tend to enhance CW rotation (Macnab & Koshland, 1972; Larsen *et al.*, 1974b; Berg and Tedesco, 1975; Spudich & Koshland, 1975). In the absence of further stimuli, these responses persist for a period of time that is directly related to the magnitude of the change in receptor occupancy. The return to a pre-stimulus behavioural pattern marks the completion of the adaptation process.

These sorts of temporal stimulation studies have served to define two phases to the chemotactic response. Upon encountering a change in attractant or repellent concentration, the cell undergoes an immediate change in flagellar behaviour. This constitutes the excitation phase of the response, during which the level of some sort of internal tumble-controlling signal is altered by chemotactic stimuli. During the adaptation phase of the response, this signal is returned to its pre-stimulus value. The nature of the tumble-controlling signal(s) is not known, but at some point sensory inputs from different receptors must be integrated by the response machinery, because the cell is able to sum multiple or conflicting stimuli algebraically (Tsang, Macnab & Koshland, 1973; Adler & Tso, 1974; Berg & Tedesco, 1975; Spudich & Koshland, 1975). Thus different receptor signals are functionally equivalent. Moreover, all receptors appear to use a common adaptation system because response times to combinations of stimuli are essentially additive.

GENETIC ANALYSIS OF CHEMOTAXIS

The chemotactic machinery of *E. coli* and *S. typhimurium* constitutes a network of signalling elements through which sensory information about the chemical environment is transmitted from chemoreceptors to flagella. These informational pathways are amenable to genetic dissection in much the same manner as more conventional biochemical pathways. The only substantial difference is that defects in biochemical pathways usually lead to accumulation of a discrete chemical intermediate which can be identified and used to reconstruct the sequence of reaction steps. The sequence of signalling steps in the chemotactic pathway is more difficult to establish because very few of the informational intermediates have been identified. Nevertheless, a probable pathway of information flow can be constructed by examining mutants with various sorts of chemotaxis defects. Since there are many chemoreceptors which ultimately transmit signals through a common transduction mechanism, the severity of a mutant phenotype should reflect the position of the defective component in the convergent signalling pathway.

In *E. coli*, approximately 50 loci are known to be required either

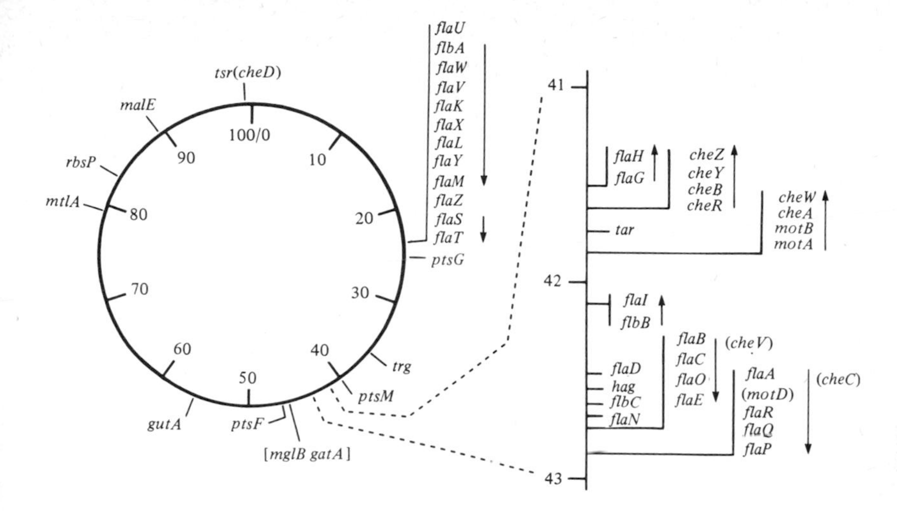

Fig. 1. Genetic map of chemotaxis loci in *E. coli*. Numbers within the circle indicate map position in minutes on the *E. coli* chromosome. In cases where a locus can give rise to more than one type of mutant defect, the primary or null type is shown first, followed by special types shown in parentheses. Arrows above the genes indicate the direction and extent of transcriptional units.

Locus	*Mutant phenotype*	*Probable function*
gatA, *gutA*, *malE*, *mglB*, *mtlA*, *ptsF*, *ptsG*, *ptsM*, *rbsP*	lack responses to a limited set of related compounds	chemoreception
tar, *tsr*, *trg*	multiple response defects	signalling
cheA, *B*, *C*, *D*, *R*, *V*, *W*, *Y*, *Z*	motile, but generally nonchemotactic	regulation of flagellar rotation
fla, *flb* (27 genes)	nonflagellate	structure and assembly of flagella
motA, *motB*	flagellar paralysis	energize rotation (?)
hag	filament defects	filament protein

for motility or for chemotaxis (Fig. 1). Mutational alterations in these genes can lead to four basic types of chemotaxis defects. Chemoreceptor mutants, which define the input end of the signalling pathway, have normal motility and lack responses to only a few structurally related compounds. Although there is physiological evidence for at least 20 different chemoreceptor species, many of them, notably the amino acid and repellent receptors, have not yet been identified by mutation. The output end of the chemotaxis machinery is defined by motility mutants, which are either unable to synthesize flagella (*fla* mutants) or which have aberrant filaments (*hag* mutants) or paralysed flagella that cannot rotate (*mot* mutants). Recent reviews on chemoreception (Hazelbauer &

Table 1. *Chemotaxis genes involved in signalling and information processing*

Locus	Gene product		Mutant phenotype			
	Size[a]	Location[b]	Flagellar rotation	Response to temporal stimuli[c]		
				Serine	Aspartate	Ribose
tar	~60 000	cyt. memb.	normal	+	−	+
tsr	~65 000	cyt. memb.	normal	−	+	+
trg	~55 000	cyt. memb.	normal	+	+	−
cheA	78 000 69 000	cytoplasm* cytoplasm	CCW	−	−	−
cheB	38 000	cytoplasm*	*CW*	±	(−)	±
cheC	?	?	*CCW or CW*	±	±	±
cheD	(65 000)	(cyt. memb.)	CCW	(−)	(−)	(−)
cheR	28 000	cytoplasm*				
cheV	?	?	CCW or CW	±	±	±
cheW	15 000	cytoplasm	CCW	−	−	−
cheY	11 000	cytoplasm	CCW	(−)	(−)	(−)
cheZ	28 000	cytoplasm	CW	±	+	+

[a] Approximate molecular weight of proteins based on sodium dodecyl sulphate-polyacrylamide gel electrophoresis; data are from Silverman & Simon, 1977b, 1977c and from Parkinson *et al.*, unpublished.
[b] Data of Ridgway, Silverman & Simon, 1977; asterisk indicates that a portion of the gene product was also found in the cytoplasmic membrane fraction.
[c] Ability of sudden changes in attractant concentration to elicit changes in flagellar rotation. Responses in brackets are weak or equivocal. Data of Parkinson, 1974, 1976 and unpublished; Springer, Goy & Adler, 1977b; Hazelbauer & Harayama, 1979.

Parkinson, 1977) and motility (Iino, 1977) should be consulted for further discussion of these topics.

Two groups of mutants, whose properties are summarized in Table 1, appear to define transduction elements that link the chemoreceptors and flagella. So-called signalling mutants (*tar, tsr, trg*) have normal motility, but lack a number of responses that are initiated by different chemoreceptors (Ordal & Adler, 1974; Reader *et al.*, 1979). Each signalling element therefore seems to handle inputs from a subset of chemoreceptors. Generally non-chemotactic (*che*) mutants (Armstrong, Adler & Dahl, 1967) are unable to carry out effective chemotactic responses to any compounds and must define transduction components common to all of the signalling pathways. Although capable of flagellar rotation, all *che* mutants exhibit aberrant swimming patterns characterized by too few or too frequent tumbling episodes. No *che* mutants with normal tumbling behaviour have been isolated, suggesting that all *che* functions are somehow involved in generating or regulating reversals in the direction of flagellar rotation. Many *che* strains are still capable of responding to chemotactic stimuli in temporal assays, but have high response thresholds or altered adaptation behaviour.

The central machinery of chemotaxis, as defined by signalling and *che* mutants, appears to be quite similar in *E. coli* and *S. typhimurium*. Nine *che* loci have been identified in both organisms (Armstrong & Adler, 1969a; Aswad & Koshland, 1975a; Collins & Stocker, 1976; Parkinson, 1976, 1978; Silverman & Simon, 1977b; Warrick, Taylor & Koshland, 1977), and the correspondence between *che* functions in the two species has been established by complementation analysis in which F-prime elements carrying *E. coli che* genes were used to correct *che* defects in *Salmonella* recipients (DeFranco, Parkinson & Koshland, 1979). Direct evidence was obtained for correspondence of seven *che* functions in the two organisms, and a uniform nomenclature of these seven loci has been adopted (see Table 1). Mutants of the *cheV* class have not yet been examined by interspecies complementation tests, but have similar phenotypes and map positions in both organisms and are tentatively assumed to be homologous (Warrick *et al.*, 1977; Parkinson, unpublished observations). Thus far *cheD* mutants are only known in *E. coli* and *cheS* mutants only in *S. typhimurium*. Both types of mutants are quite rare and probably arise through highly specific alterations of chemotaxis-related functions (Parkinson, 1976, 1980; Warrick *et al.*, 1977). Three types of signalling mutants have been isolated in *E. coli: tar* mutants are defective in aspartate and maltose taxis (Reader *et al.*, 1979); *tsr* mutants are defective in serine taxis and in a variety of repellent responses (Hazelbauer, Mesibov & Adler, 1969; Tso & Adler, 1974); *trg* mutants are defective in ribose and galactose responses (Ordal & Adler, 1974; Hazelbauer & Harayama, 1979). Only *trg* mutants have thus far been isolated in *Salmonella* (M. Fahnestock, cited in Strange & Koshland, 1976); however, biochemical and physiological evidence for signallers comparable to the *tar* and *tsr* products has been obtained in this organism. Much of the following discussion will draw primarily on the work in *E. coli*, although most of the conclusions will undoubtedly apply to *Salmonella* as well.

As shown in Table 1, the gene products associated with most of the signalling and *che* functions have been identified. The *tar*, *tsr*, *trg* and *cheD* products are cytoplasmic membrane proteins, whereas the *cheA*, *cheB*, *cheR*, *cheW*, *cheY* and *cheZ* products are primarily located in the cytoplasm (Ridgway *et al.*, 1977). A portion of the *cheB* and *cheR* proteins as well as the larger of the two *cheA* proteins, are also found associated with the cytoplasmic membrane. With the exception of *trg*, these product identifications were made by utilizing plasmids or specialized λ transducing phages to synthesize radio-labelled chemotaxis proteins in minicells or in heavily UV-irradiated whole cells (Silverman & Simon, 1977b; Matsumura, Silverman & Simon, 1977). The genetic content of different plasmids or phages, established by complementation analysis, could be compared with the proteins made by these strains to establish the identity of each gene product. Although λ transducing phage carrying the *cheC* and *cheV* loci have been constructed (Komeda, Shimada & Iino, 1977; Kondoh, 1977), these gene products have not yet been observed by these methods, perhaps because their promoters are less active than other *che* promoters under these experimental conditions.

The probable roles of the signalling and *che* components in chemotaxis are summarized in the two sections below on excitation and adaptation. Many of the transduction elements are important in both processes, but in order to simplify this discussion, the two response phases will be treated separately. Because a great deal of physiological, biochemical and genetic evidence is currently available, interpretive models are presented as a means of summarizing these diverse studies.

EXCITATION

Components involved in the excitation phase of the chemotactic response are shown in Fig. 2. The excitation pathway is clearly convergent, with many inputs at the receptor end but only a single output at the flagellum. The summation of conflicting signals can probably occur at several different points in the pathway depending on the applied stimuli. For example, aspartate and maltose stimuli, which are both handled by the *tar* signaller, must be integrated by the *tar* product. However, aspartate and ribose stimuli, which are channelled through different signallers, must be integrated by some common component later in the pathway. The eventual target of all incoming signals must be the flagellum, which appears to be controlled by some sort of switch.

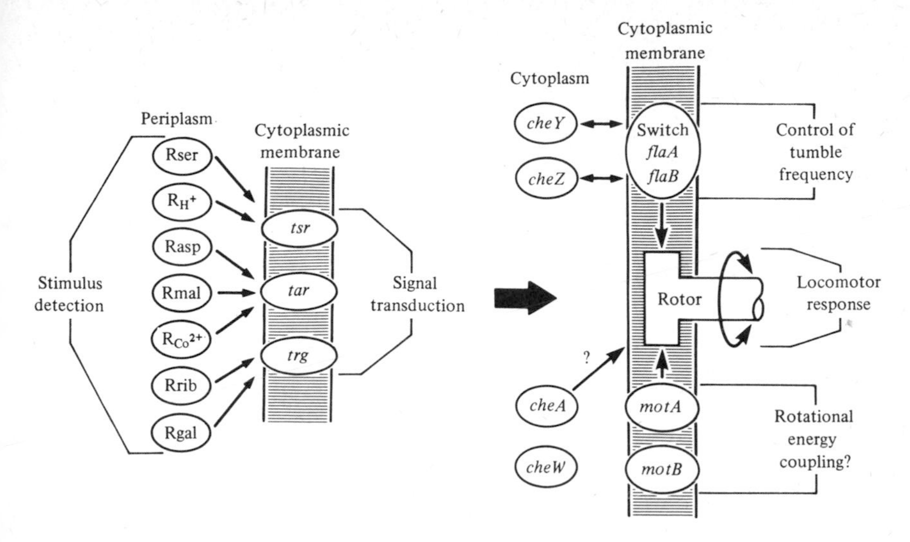

Fig. 2. Components of the excitation system. Occupied chemoreceptors are thought to interact with membrane proteins that in turn generate signals to control the direction of flagellar rotation. The nature of the excitatory signal(s) is unknown.

The flagellar switch

The *cheC* and *cheV* gene products may be the flagellar target site for incoming receptor signals. Mutants of the *cheC* and *cheV* classes are quite rare relative to other *che* strains. In fact, many of the available *cheC* and *cheV* mutations in *E. coli* were obtained as suppressors of other *che* defects during reversion analyses (discussed in a later section). In transductional crosses, *cheC* and *cheV* mutations map in the vicinity of several *fla* operons (Armstrong & Adler, 1969b; Silverman & Simon, 1973; Parkinson, 1976 and unpublished data), indicating that *cheC* and *cheV* mutants might have specific flagellar defects. The relationship of *cheC* and *cheV* to *fla* functions has been investigated with the aid of a specialized λ transducing phage that carries the *cheC–cheV* region of the *E. coli* chromosome (Fig. 3) (Parkinson, Parker & Houts, unpublished data). Deletions of the parental phage were obtained by selecting particles resistant to inactivation by chelating agents. These deletions were then used to construct a fine-structure map of the region by complementation and recombination analyses. As shown in Fig. 3, *cheC* mutations map at the *flaA* locus and *cheV* mutations at the *flaB* locus. These *fla* loci, which are essential for synthesis or assembly of flagella, can evidently be altered in subtle ways by *cheC* or *cheV* mutations to

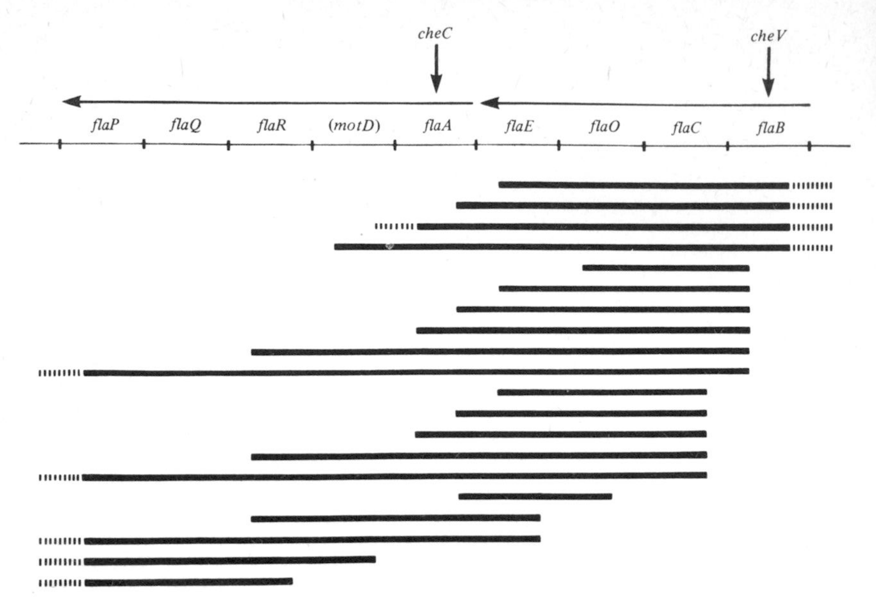

Fig. 3. Localization of *cheC* and *cheV* mutations by deletion mapping. Deletions were obtained by chelating agent inactivation (Parkinson & Huskey, 1971) of *λfla36*, a plaque-forming transducing phage that carries the *flaA* and *flaB* operons (Komeda *et al.*, 1977). Deleted phages were crossed to representative *fla* and *che* strains, as described by Parkinson (1978), to map *cheC* and *cheV* mutations (Parkinson *et al.*, unpublished data).

permit flagellar assembly while interfering with normal rotational behaviour, thereby causing a defect in chemotaxis.

Unlike other *che* loci, mutations at the *cheC* and *cheV* loci can result in either excessively high or low tumbling rates (Warrick *et al.*, 1977; Rubik & Koshland, 1978; Parkinson *et al.*, unpublished data). The fact that specific *flaA* (i.e., *cheC*) and *flaB* (i.e., *cheV*) defects can lead to predominately CW or counter-clockwise (CCW) rotation suggests that these flagellar proteins play a role in controlling the direction of flagellar rotation. The *flaA* and *flaB* products may constitute a 'switch' that is intimately associated with the flagellar basal complex and determines its direction of rotation. In the wild-type cell this switch is set to produce CW and CCW rotation with equal probabilities, but mutations of the right sort can introduce a bias in either direction.

The switch is thought to receive signals generated by the chemoreceptors during excitation. The switch also must respond to signals that set the spontaneous tumbling rate of the cell. The nature of these signals is poorly understood; however, genetic evidence discussed in later sections suggests that the two signal types may not be identical. In any case, the chemotaxis defects of *cheC* and *cheV* mutants are probably due to an inability to respond properly to these controlling signals. For example, a mutant with CW bias in the switch would require a correspondingly large CCW signal in order to initiate a change in flagellar rotation. Physiological studies of both CW and CCW switch mutants demonstrate that they can still respond to stimuli, but with higher than normal thresholds (Parkinson, 1976 and unpublished data; Rubik & Koshland, 1978). In fact, some switch mutants with only slight alterations in spontaneous tumbling rate exhibit considerable chemotactic ability under certain assay conditions (Parkinson *et al.*, unpublished data).

Regulation of tumbling rates

The *cheY* and *cheZ* products appear to play a role in regulating the rotational behaviour of the cell, probably by interacting directly with the flagellar switch. Mutants with *cheY* and *cheZ* defects resemble switch mutants in several respects: *cheY* strains exhibit extreme CCW bias in their flagellar behaviour, whereas *cheZ* strains have an extreme CW bias; both types of mutant can still respond to chemotactic stimuli, but with high thresholds (Parkinson, 1974, 1976, 1978; unpublished data). Unlike switch mutants, however, *cheY* and *cheZ* defects can arise through nonsense mutations or deletions which eliminate gene function entirely. Thus neither *cheY* nor *cheZ* function is essential for flagellar formation as the switch products are, but they do play a role in setting the cell's spontaneous tumbling rate.

The functional relationship between the *cheY* and *cheZ* products and the flagellar switch has been investigated by searching for direct interactions among the various gene products. Reversion analysis is a powerful method for detecting such interactions, particularly weak ones that would be difficult to observe by conventional biochemical methods (Parkinson & Parker, 1979). The reasoning behind this approach is that when two gene products interact, defects in one can often be alleviated by a compensating alteration of the other. This strategy has been applied to the *che* system by selecting revertants of *cheY* or *cheZ* mutants and then examining those revertants for the presence of suppressor mutations in other chemotaxis genes. When the original mutants carried missense mutations in *cheY* or *cheZ*, many of the chemotactic revertants acquired compensating mutations at the *cheC* and *cheV* loci (Parkinson *et al.*, unpublished data).

The nature of the interactions between the products of *cheY* and *cheZ* and the switch components has been deduced by studying the specificity of these suppression effects. In the first place, not all *cheZ* or *cheY* defects can be suppressed. For example, nonsense mutations, which do not make a complete gene product, cannot be corrected by any sort of switch mutation. This implies that the mutant *cheY* and *cheZ* gene products even though they are aberrant are still required for suppression to occur. Moreover, a mutant suppressed by one switch mutation may not be corrected by another. Finally, suppressors of *cheY* defects cannot correct *cheZ* defects and *vice versa.* These results indicate that the *cheY* and *cheZ* products probably interact directly with the switch proteins.

These interactions could play a role in controlling the cell's spontaneous tumbling rate. For example, the switch may have two alternative states or configurations, one that causes CCW rotation, and one that causes CW rotation. The spontaneous tumbling frequency is presumably determined by the rate at which transitions occur from one switch state to the other. The *cheY* and *cheZ* products could be involved in initiating these transitions or in stabilizing one form or the other of the switch. As a general rule, *cheZ* mutants, which are tumbly or excessively CW, are suppressed by CCW switch mutants; the converse is true for *cheY* mutants. It is possible that in the CCW state the switch has affinity only for *cheZ* protein, not *cheY* protein; in the CW state, the switch may bind *cheY* protein, but not *cheZ* protein.

If this model is true, the relative amount or activity of the *cheY* and *cheZ* proteins should influence the swimming behaviour of the cell. It seems unlikely that these proteins are directly involved in excitation, however they could influence the cell's responsiveness by altering the threshold characteristics of the switch. Moreover, the swimming patterns of other *che* strains, in particular *cheB* and *cheR* mutants (see the section on adaptation), may be due primarily to their effects on the *cheY* and *cheZ* proteins. One way to test these ideas is to alter the normal stoichiometry of the system by selectively changing the number of copies of *cheY* or *cheZ* genes in the cell. This can be readily accomplished with λ transducing phages, and such experiments are presently in progress.

Generation of CW rotation

Mutants defective in *cheA* or *cheW* function have a complete CCW bias in flagellar rotation and are unable to respond to even the very strongest sorts of CW stimuli (Warrick *et al.*, 1977; Smith & Parkinson, unpublished data). Reversion analyses indicate that *cheA* may interact with the switch and other flagellar components (R. Smith, personal communication). (*cheW* has not yet been examined.) It seems that *cheA* (and perhaps *cheW*) function is required to activate or potentiate the flagellar motor for CW rotation. It is known for example that CCW rotation is possible at lower values of the proton motive force than is CW rotation (Kahn & Macnab, 1980b). It may be that *cheA* and *cheW* are involved in coupling proton motive force to the switch or motor. In this regard it is interesting to note that the *cheA* and *cheW* genes form an operon with the two *mot* genes, which are probably involved in energizing flagellar rotation (Silverman & Simon, 1976).

The *cheA* locus directs the synthesis of two polypeptides, designated p[*cheA*]$_S$ and p[*cheA*]$_L$ (Silverman & Simon, 1977b). These two products are identical in sequence except that p[*cheA*]$_L$, the larger of the two, has an additional 90 or so amino acid residues at its amino terminal end. Examination of the polypeptides produced by various nonsense mutants of *cheA* indicates that both proteins are made from the same coding sequence and are translated in the same reading frame, but from different start sites (Smith & Parkinson, 1980). The *cheA* locus thus appears to be the first example of overlapping genes in bacteria because complementation tests indicate that both proteins are probably needed for chemotaxis, but the purpose in making two similar proteins in this manner is not yet clear. One attractive possibility is that p[*cheA*]$_L$ and p[*cheA*]$_S$ interact with different components of the chemotaxis machinery. For example, the amino terminal sequence of p[*cheA*]$_L$ may enable it to enter the cytoplasmic membrane, whereas p[*cheA*]$_S$ is restricted to the soluble portions of the cell.

Signalling

The *tar*, *trg* and *tsr* gene products are inner membrane proteins that appear to be required for the generation of chemoreceptor signals. Each of these signalling elements handles inputs from a set of chemoreceptors. Presumably, occupied receptor proteins are capable of interacting with the appropriate signaller species to elicit tumble-controlling signals. These signals are assumed to act on the flagellar switch; however, their nature and mode of transmission are not yet understood.

Mutants of the *cheD* class provide a handle of sorts on the signalling problem. These mutants are relatively rare and prove to be completely dominant to wild-type in complementation tests, implying that they produce an altered gene product that actively interferes with chemotaxis. Genetic studies strongly indicate that the *cheD* phenotype is generated by a special alteration of the *tsr* signaller (Parkinson, 1980). For example, *cheD* and *tsr* mutations are very tightly linked in a region of the genetic map that contains no other known chemotaxis loci. Moreover, the inhibitory product made by *cheD* strains can be inactivated by *tsr* lesions, suggesting that the inhibitor is an altered form of the *tsr* product. Both *tsr* and *cheD* defects could conceivably arise by changes in the same gene product. Null mutations such as deletions, which completely abolish function, would lead to a Tsr⁻ condition, whereas special missense mutations might convert the *tsr* protein into an inhibitor of chemotaxis. Since *cheD* mutants have low tumbling frequencies, they may synthesize a *tsr* signaller that is somehow locked in the CCW signalling mode. This model predicts that it should be possible to restore chemotaxis in *cheD* strains by selectively modifying the normal target of *tsr* signals so that it is no longer inhibited by aberrant signals due to the *cheD* defect.

External suppressors of *cheD* mutations have been obtained by isolating pseudorevertants from strains with two copies of the mutant *cheD* locus (Parkinson, unpublished data). Because *cheD* defects are dominant, two independent mutations, one in each *cheD* copy, are required to produce either true revertants or Tsr⁻ pseudorevertants of *cheD* diploids and these reversion types should prove to be relatively rare in this selection, which favours detection of external suppressors that can arise in a single step. These studies revealed that mutations at the *cheB*, *cheZ*, *flaA* and *flaB* loci were capable of suppressing *cheD* defects. All such suppressors exhibit considerable CW bias of flagellar rotation in the absence of the *cheD* lesion; in combination with *cheD*, flagellar rotation approximates wild-type behaviour. None of these interactions appears to be allele-specific, suggesting that restoration of a normal tumbling rate, by whatever means, is sufficient to restore chemotactic ability in *cheD* strains. This result implies that the excitation and adaptation machinery is still intact in *cheD* mutants, but cannot function properly unless the tumbling rate is increased.

If the *cheD* model is valid, the failure to obtain specific target mutants could mean that all receptor signals are qualitatively identical and therefore the component that receives those signals is unable to discriminate among them. Thus it may not be possible for the cell to ignore *cheD* signals and still detect other signal inputs; the only recourse for reversion is to reset the spontaneous tumbling frequency so that *cheD* signals no longer 'drown out' the other channels. A second conclusion from these studies is that the interaction between a signaller and its target may not be a direct protein–protein interaction. If the interaction were direct, it seems unlikely that all signallers would appear identical to the target since the *tar*, *tsr* and *trg* proteins are clearly not identical to one another. Two possible signal types are consistent with these findings. The signal could be a cytoplasmic molecule or ion whose concentration is regulated by the transduction machinery. Alternatively, the signallers might control changes in the cell's membrane potential to initiate flagellar responses.

ADAPTATION

A variety of studies, principally by Adler and his co-workers (reviewed by Springer, Goy & Adler, 1979), have shown that sensory adaptation is accompanied by a net change in the level of methylation of several inner membrane proteins known as 'methyl-accepting chemotaxis proteins' or MCPs (Kort, Goy, Larsen & Adler, 1975). The changes in MCP methylation elicited by various chemotactic stimuli are illustrated in Fig. 4. Tumble-suppressing stimuli such as attractant increases produce a net increase in MCP methylation; tumble-enhancing stimuli such as repellent increases cause a net decrease in methylation level. The kinetics of methylation or demethylation following application of a chemotactic stimulus are quite similar to the time-course of adaptation, suggesting that methylation–demethylation of MCP plays a causative role in the adaptation process (Goy, Springer & Adler, 1977).

The methyl groups on MCP are derived from methionine, via S-adenosylmethionine (SAM), and are attached as methyl esters to glutamic acid residues (Kleene, Toews & Adler, 1977; Van der Werf & Koshland, 1977). Removal of the methyl groups is readily accomplished by hydrolysis, yielding methanol (Toews & Adler, 1979). Each MCP appears to possess several methyl-accepting sites per molecule (Chelsky & Dahlquist, 1980; DeFranco & Koshland, 1980; Engström & Hazelbauer, 1980). Following stimulation, methyl groups may be added to or removed from these sites in a

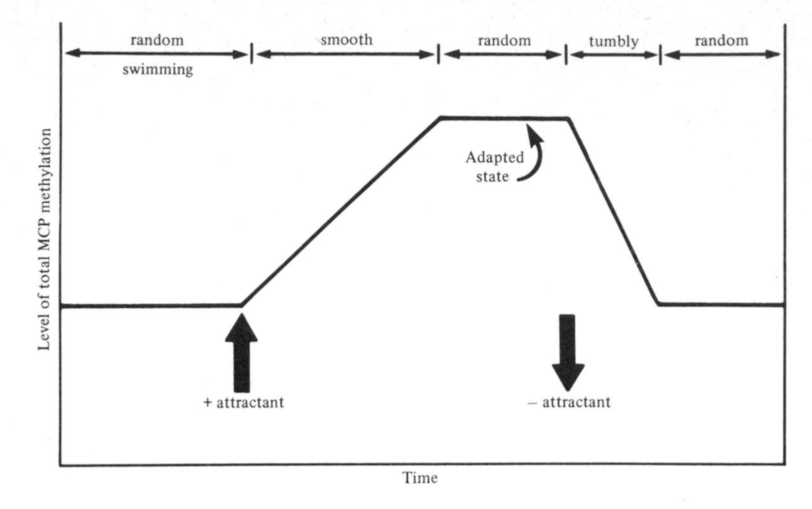

Fig. 4. Effects of chemotactic stimuli on swimming pattern and MCP methylation. Evidence for this scheme is summarized by Springer *et al.* (1979).

specific sequence. Cells deprived of methyl donors, either methionine or SAM, are still capable of responding to certain types of chemotactic stimuli, but are unable to adapt to those stimuli (Armstrong, 1972a, b; Aswad & Koshland, 1974, 1975b; Springer *et al.*, 1975; Springer, Goy & Adler, 1977a). Mutants lacking *cheR* function, discussed below, have a similar defect, indicating that MCP methylation is essential for adaptation, but apparently is not crucial for excitation.

Methylation substrates

Three different MCP structural genes have been identified by mutation, but there is reason to suspect that at least one additional MCP species may exist (Koiwai, Minoshima & Hayashi, 1980). MCPI activity is missing in *tsr* mutants; MCPII is missing in *tar* mutants; and MCPIII is missing in *trg* mutants (Springer *et al.*, 1977b; Silverman & Simon, 1977c; Kondoh, Ball & Adler, 1979). Lambda transducing phages carrying the *tsr* or *tar* loci synthesize methyl-accepting proteins upon infection that appear identical to MCPI and MCPII, respectively (Silverman & Simon, 1977c). These findings indicate that the *tsr* and *tar* gene products (and presumably the *trg* product as well) are MCPs involved in sensory adaptation. Since these gene products are also required for signal production during the excitation phase of the chemotactic response, they must

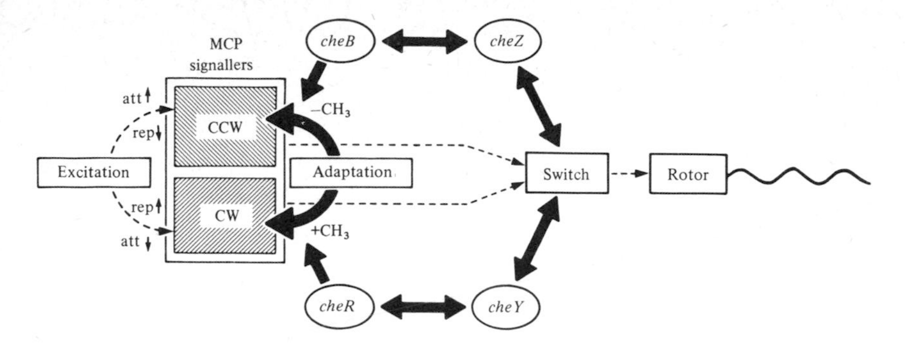

Fig. 5. Information flow during excitation and adaptation. The excitation pathway (dotted lines) is responsible for triggering changes in flagellar rotation in response to chemotactic stimuli. The adaptation system (large arrows) cancels those responses. Both systems appear to act on the MCP signallers in the cytoplasmic membrane. Gene product interactions that could play a role in feedback regulation of the adaptation process have been inferred from genetic studies (Parkinson & Parker, 1979; Parkinson *et al.*, unpublished data).

play a dual role in chemotaxis. They interact with occupied chemoreceptors to generate tumble-modulating signals and then are acted upon by the methylation–demethylation system to cancel or negate those signals to achieve adaptation.

A simple scheme that summarizes the dual function of MCPs in chemotaxis is shown in Fig. 5. This model postulates that the *tar*, *tsr*, and *trg* signallers can each exist in two conformational states, one that causes CCW rotation of the flagella and one that causes CW rotation. The pattern of flagellar rotation thus reflects the relative amount of signaller species in each form. Conversion of a signaller from one rotational state to the other is caused by stimuli, i.e. occupied chemoreceptors, and by adding or removing methyl groups.

Methylation enzymes

Mutants defective in *cheR* function have low spontaneous tumbling rates and can respond to tumble-enhancing stimuli, but do not show adaptation to those stimuli (Parkinson & Revello, 1978; Goy, Springer & Adler, 1978). This behaviour is essentially identical to that of methionine-deprived cells, which are unable to add methyl groups to MCP. *In vivo* studies of *cheR* strains have shown that they are defective in incorporating labelled methyl groups from methionine into the methyl-accepting proteins (Goy *et al.*, 1978). The MCPs present in *cheR* strains are nearly devoid of methyl groups (Hayashi, Koiwai & Kozuka, 1979) demonstrating that the inability of *cheR* mutants to turn over MCP methyl groups is due to an inability to add methyl groups to the MCP molecules.

Mutants defective in *cheB* function exhibit abnormally high tumbling rates, but are able to respond to certain tumble-suppressing stimuli (Parkinson, 1974, 1976). Usually, *cheB* strains exhibit transient responses to such stimuli; however, when subjected to combinations of stimuli that utilize several MCP species for signalling, *cheB* mutants exhibit prolonged responses indicative of some sort of adaptation defect (Rubik & Koshland, 1978). Methylation assays *in vivo* showed that *cheB* mutants do not turn over MCP methyl groups (Kort *et al.*, 1975; Stock & Koshland, 1978). The MCPs present in these strains appear to be fully methylated, (Hayashi *et al.*, 1979), implying that *cheB* function is required to remove MCP methyl groups.

Methylation assays in permeabilized cells or with crude membrane preparations indicate that wild-type cells contain soluble proteins that catalyse the addition or removal of methyl groups on MCP. Mutants defective in *cheR* function lack the methyltransferase (i.e. methyl-adding) activity (Springer & Koshland, 1977); mutants defective in *cheB* function lack the methylesterase (i.e. methyl-removing) activity (Stock & Koshland, 1978). The steady-state level of MCP methylation observed in wild-type strains is presumably due to a balance between these two enzymatic reactions. In *cheR* and *cheB* strains, where one of these activities is defective, a very low or very high steady-state level of MCP methylation results. The aberrant tumbling rates of *cheR* and *cheB* strains may be caused by the aberrant methylation levels: *cheR* mutants are undermethylated and show predominately CCW flagellar rotation; *cheB* mutants are overmethylated and show predominately CW rotation. These phenotypes are consistent with the data summarized in Fig. 4 which show for example that adaptation to CCW stimuli is accompanied by the addition of methyl groups to the MCPs. Thus methylated MCP is correlated with CW rotation and unmethylated MCP with CCW rotation.

Methylation control

In the absence of chemotactic stimuli there is a slow turnover of methyl groups on the MCPs. The resultant methylation level reflects the steady-state rates of the addition and removal reactions. Upon excitation, the activity of one or both of these reactions must be altered in order to achieve a net change in methylation level. When adaptation is complete, a new steady-state methylation level is maintained. This level is directly related to the abundance of occupied chemoreceptors; large proportional changes in receptor occupancy upon stimulation elicit correspondingly large changes in the steady-state level of MCP methyl groups.

How does the cell regulate the level of MCP methylation in response to chemotactic stimuli? *A priori*, control could be exerted either by altering the substrate proteins (MCPs) or by altering the methyltransferase or methylesterase enzymes. For example, changes in chemoreceptor occupancy probably cause conformational changes in the cognate signaller species during excitation. These conformational changes might in turn generate new methyl-accepting sites on the MCP molecule or expose previously protected methylated sites. If the methyltransferase and methylesterase activities were substrate-limited, these changes in substrate conformation would lead to new steady-state methylation levels. An alternative means of control would be through modulation of the catalytic activity of the enzymes themselves, rather than their substrates. The evidence discussed below suggests that both of these mechanisms may come into play during the adaptation process.

Following application of a stimulus, only one species of MCP undergoes a *net* change in methylation level. The MCP involved is always the one that is responsible for generating the excitatory signal for a particular stimulus. For example, serine stimuli which are processed by MCPI, the *tsr* product, cause a net change in methylation of only MCPI. This result implies that methylation levels are controlled by substrate conformation. However, such stimuli also cause *transient* methylation changes of non-signalling MCPs (M. S. Springer *et al.*, personal communication). For example, a serine increase causes a brief increase in the level of methyl groups on MCPII, but MCPII methylation returns to pre-stimulus levels upon completion of the adaptation process. This effect could be due to a transient change in the conformation of the MCPII substrate, even though MCPII is not required to process serine signals. Alternatively, this effect could be caused by transient stimulus-elicited changes in the catalytic activities of the methyltransferase and methylesterase enzymes. Several lines of indirect evidence suggest that this latter explanation may be correct.

Genetic studies have shown that both the methyltransferase

(*cheR* product) and the methylesterase (*cheB* product) interact with other chemotaxis components. For example, in interspecies complementation tests, *S. typhimurium* mutants defective in *cheR* or *cheY* activity can only be corrected by supplying *both* of the homologous functions from *E. coli* (DeFranco *et al.*, 1979). Similarly, *both cheB* and *cheZ* function from *E. coli* are needed to complement *Salmonella* mutants lacking either *cheB* or *cheZ* activity. These results suggest that there may be an interaction of some sort between the *cheR* and *cheY* products and between the *cheB* and *cheZ* products. Although these interacting complexes are evidently functionally interchangeable between species, the components of the complexes are not. Thus *cheR* product from *E. coli* can only interact productively with *cheY* product from *E. coli*; the *Salmonella cheY* product does not suffice. These interactions could play a role in modulating activity of the methylation–demethylation enzymes during adaptation. Neither *cheY* nor *cheZ* function is essential to either enzyme activity; however some *cheY* and *cheZ* strains exhibit defects in methylation of MCPs *in vivo*, implying that altered forms of these gene products may interfere with the enzymatic activities of the *cheR* and *cheB* proteins (Kort *et al.*, 1975). Methylesterase activity in extracts of some *cheZ* strains is quite low, whereas it is essentially normal in other *cheZ* strains (Stock & Koshland, 1978).

A summary of gene product interactions involved in excitation and adaptation is shown in Fig. 5. Since the flagellar switch interacts with the *cheY* and *cheZ* products, which in turn interact with the *cheR* and *cheB* products, respectively, it seems possible that the methyltransferase and methylesterase activities are modulated by feedback signals from the switch. The mechanism by which these interactions modulate enzymatic activity is not known, but one possibility is shown in Fig. 6. If the *cheR-cheY* and *cheB-cheZ* products form tight complexes, the switch could modulate enzyme availability by selectively binding one complex or the other. For example, a CCW signal from an MCP, upon interaction with the flagellum, might cause the switch to release *cheY* product and to bind *cheZ* product, thereby activating the methyltransferase and inhibiting the methylesterase. Methylation of the MCPs would proceed until the CCW signal had been cancelled. The switch would then return to its pre-stimulus behaviour and the activity of the two enzymes would return to steady-state values. MCP signals causing CW rotation would activate the methylesterase and inhibit the methyltransferase in an analogous manner.

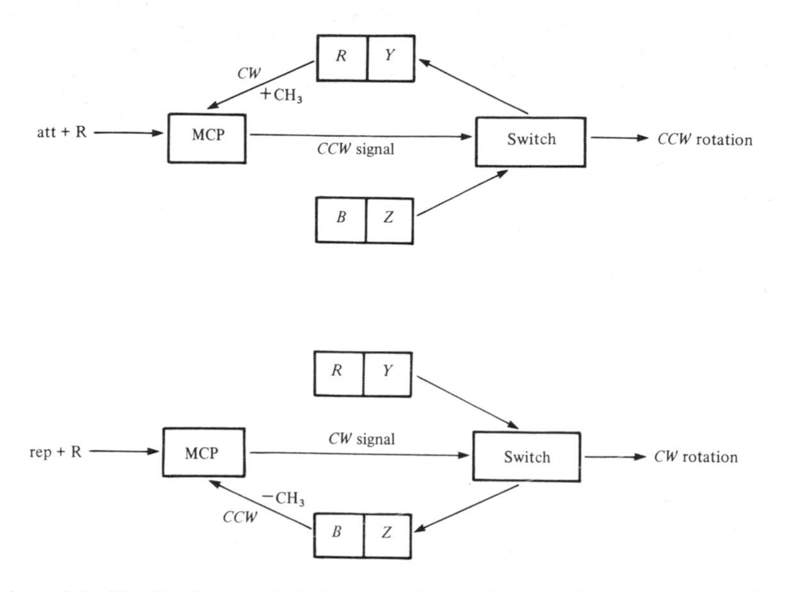

Fig. 6. A model of feedback control of the adaptation machinery. The enzymatic activity of the methyltransferase (*cheR* product) and methylesterase (*cheB* product) may be modulated simply by controlling access to the MCP substrates by means of reversible binding to the flagellar switch.

CONCLUSION

Genetic studies of chemotaxis and motility in *E. coli* and *S. typhimurium* have revealed a large number of functions involved in the excitation and adaptation phases of the chemotactic response. It seems unlikely that additional gene products will be found to have a direct role in these events, although functions essential to cell viability could also be involved. As discussed here, genetic studies of various chemotaxis mutants have contributed to understanding the overall strategy of information processing by the chemotaxis machinery. Identification of the excitatory signal that controls flagellar rotation remains a major unsolved problem in this field. However, biochemical and genetic studies of the adaptation process are rapidly converging on a consistent, detailed view of the role of protein methylation–demethylation events in sensory adaptation. It is not unreasonable to predict that the powerful combination of genetic and biochemical tools that have been brought to bear on this system will soon lead to a molecular description of the entire stimulus transduction machinery in bacteria.

REFERENCES

ADLER, J. (1969). Chemoreceptors in bacteria. *Science*, **166**, 1588–97.

ADLER, J. (1975). Chemotaxis in bacteria. *Annual Review of Biochemistry*, **44**, 341–56.

ADLER, J. & EPSTEIN, W. (1974). Phosphotransferase-system enzymes as chemoreceptors for certain sugars in *Escherichia coli* chemotaxis. *Proceedings of the National Academy of Sciences, USA*, **72**, 2895–99.

ADLER, J. & TSO, W.-W. (1974). 'Decision'-making in bacteria: chemotactic response of *Escherichia coli* to conflicting stimuli. *Science*, **184**, 1292–4.

AKSAMIT, R. & KOSHLAND, D. E. JR. (1974). Identification of the ribose binding protein as the receptor for ribose chemotaxis in *Salmonella typhimurium*. *Biochemistry*, **13**, 4473–8.

ARMSTRONG, J. B. (1972a). Chemotaxis and methionine metabolism in *Escherichia coli*. *Canadian Journal of Microbiology*, **18**, 591–4.

ARMSTRONG, J. B. (1972b). An S-adenosylmethionine requirement for chemotaxis in *Escherichia coli*. *Canadian Journal of Microbiology*, **18**, 1695–1701.

ARMSTRONG, J. B. & ADLER, J. (1969a). Complementation of nonchemotactic mutants of *Escherichia coli*. *Genetics*, **62**, 61–6.

ARMSTRONG, J. B. & ADLER, J. (1969b). Location of genes for motility and chemotaxis on the *Escherichia coli* genetic map. *Journal of Bacteriology*, **97**, 156–61.

ARMSTRONG, J. B., ADLER, J. & DAHL, M. M. (1967). Nonchemotactic mutants of *Escherichia coli*. *Journal of Bacteriology*, **93**, 390–8.

ASWAD, D. & KOSHLAND, D. E. JR. (1974). Role of methionine in bacterial chemotaxis. *Journal of Bacteriology*, **118**, 640–5.

ASWAD, D. & KOSHLAND, D. E. JR. (1975a). Isolation, characterization and complementation of *Salmonella typhimurium* chemotaxis mutants. *Journal of Molecular Biology*, **97**, 225–35.

ASWAD, D. & KOSHLAND, D. E. JR. (1975b). Evidence for an S-adenosylmethionine requirement in the chemotactic behavior of *Salmonella typhimurium*. *Journal of Molecular Biology*, **97**, 207–23.

BERG, H. C. (1974). Dynamic properties of bacterial flagellar motors. *Nature, London*, **249**, 77–9.

BERG, H. C. (1975). Chemotaxis in bacteria. *Annual Reviews of Biophysics and Bioengineering*, **4**, 119–36.

BERG, H. C. & ANDERSON, R. A. (1973). Bacteria swim by rotating their flagellar filaments. *Nature, London*, **245**, 380–2.

BERG, H. C. & BROWN, D. A. (1972). Chemotaxis in *Escherichia coli* analyzed by three-dimensional tracking. *Nature, London*, **239**, 500–4.

BERG, H. C. & TEDESCO, P. M. (1975). Transient response to chemotactic stimuli in *Escherichi coli*. *Proceedings of the National Academy of Sciences, USA*, **72**, 3235–9.

BROWN, D. A. & BERG, H. C. (1974). Temporal stimulation of chemotaxis in *Escherichia coli*. *Proceedings of the National Academy of Sciences, USA*, **71**, 1388–92.

CHELSKY, D. & DAHLQUIST, F. W. (1980). Structural studies of methyl-accepting chemotaxis proteins of *Escherichia coli*: Evidence for multiple methylation sites. *Proceedings of the National Academy of Sciences, USA*, **77**, 2434–8.

CLARKE, S. & KOSHLAND, D. E. JR. (1979). Membrane receptors for aspartate and serine in bacterial chemotaxis. *Journal of Biological Chemistry*, **254**, 9695–702.

COLLINS, A. L. & STOCKER, B. A. D. (1976). *Salmonella typhimurium* mutants generally defective in chemotaxis. *Journal of Bacteriology*, **128**, 754–65.

DeFranco, A. L. & Koshland, D. E. Jr. (1980). Multiple methylation in the processing of sensory signals during bacterial chemotaxis. *Proceedings of the National Academy of Sciences, USA,* **77,** 2429–33

DeFranco, A. L., Parkinson, J. S. & Koshland, D. E. Jr. (1979). Functional homology of chemotaxis genes in *Escherichia coli* and *Salmonella typhimurium. Journal of Bacteriology*, **139**, 107–14.

DePamphilis, M. L. & Adler, J. (1971). Attachment of flagellar basal bodies to the cell envelope: specific attachment to the outer, lipopolysaccharide membrane and the cytoplasmic membrane. *Journal of Bacteriology*, **105**, 396–407.

Engström, P. & Hazelbauer, G. L. (1980). Multiple methylation of methyl-accepting chemotaxis proteins during adaptation of *E. coli* to chemical stimuli. *Cell*, **20**, 165–71.

Goy, M. F. & Springer, M. S. (1978). In search of the linkage between receptor and response: The role of a protein methylation reaction in bacterial chemotaxis. In *Taxis and Behavior, Receptors and Recognition*, ser. B. vol. 5, ed. G. L. Hazelbauer, pp. 1–34. London: Chapman and Hall.

Goy, M. F., Springer, M. S. & Adler, J. (1977). Sensory transduction in *Escherichia coli*: Role of a protein methylation reaction in sensory adaptation. *Proceedings of the National Academy of Sciences, USA*, **74**, 4964–8.

Goy, M. F., Springer, M. S. & Adler, J. (1978). Failure of sensory adaptation in bacterial mutants that are defective in a protein methylation reaction. *Cell*, **15**, 1231–40.

Hayashi, H., Koiwai, O. & Kozuka, M. (1979). Studies on bacterial chemotaxis. II. Effect of *cheB* and *cheZ* mutations on the methylation of methyl-accepting chemotaxis protein of *Escherichia coli. Journal of Biochemistry, Tokyo*, **35**, 1213–23.

Hazelbauer, G. L. (1975). The maltose chemoreceptor of *Escherichia coli. Journal of Bacteriology*, **122**, 206–14.

Hazelbauer, G. L. & Adler, J. (1971). Role of the galactose binding protein in chemotaxis of *Escherichia coli* toward galactose. *Nature New Biology*, **230**, 101–4.

Hazelbauer, G. L., Engström, P. & Harayama, S. (1980). Methyl-accepting chemotaxis protein III is the product of the transducer gene *trg. Journal of Bacteriology* (in press).

Hazelbauer, G. L. & Harayama, S. (1979). Mutants in transmission of chemotactic signals from two independent receptors of *Escherichia coli. Cell,* **16,** 617–25.

Hazelbauer, G. L., Mesibov, R. E. & Adler, J. (1969). *Escherichia coli* mutants defective in chemotaxis toward specific chemicals. *Proceedings of the National Academy of Sciences, USA*, **64**, 1300–7.

Hazelbauler, G. L. & Parkinson, J. S. (1977). Bacterial Chemotaxis. In *Receptors and Recognition: Microbial Interactions*, ser. B, vol. 3, ed. J. Reissig, pp. 59–98. London: Chapman and Hall.

Iino, T. (1977). Genetics of structure and function of bacterial flagella. *Annual Review of Genetics*, **11**, 161–82.

Khan, S. & Macnab, R. M. (1980a). Proton chemical potential, proton electrical potential, and bacterial motility. *Journal of Molecular Biology,* **138,** 599–614.

Khan, S. & Macnab, R. M. (1980b). The steady-state counterclockwise/clockwise ratio of bacterial flagellar motors is regulated by proton motive force. *Journal of Molecular Biology*, **138,** 563–97.

Kleene, S. J., Toews, M. L. & Adler, J. (1977). Isolation of glutamic acid methyl ester from an *Escherichia coli* membrane protein involved in chemotaxis. *Journal of Biological Chemistry*, **252**, 3214–18.

Koiwai, O., Minoshima, S. & Hayashi, H. (1980). Studies on bacterial chemotaxis. V. Possible involvement of four species of the methyl-accepting chemotaxis protein in chemotaxis of *Escherichia coli. Journal of Biochemistry, Tokyo*, (in press).

Komeda, Y., Shimada, K. & Iino, T. (1977). Isolation of specialized lambda transducing bacteriophages for flagellar genes (*fla*) of *Escherichia coli* K-12. *Journal of Virology*, **22**, 654–61.

Kondoh, H. (1977). Isolation and characterization of nondefective transducing lambda bacteriophages carrying *fla* genes of *Escherichia coli* K12. *Journal of Bacteriology*, **130**, 736–45.

Kondoh, H., Ball, C. B. & Adler, J. (1979). Identification of a methyl-accepting chemotaxis protein for the ribose and galactose chemoreceptors of *Escherichia coli. Proceedings of the National Academy of Sciences, USA*, **76**, 260–4.

Kort, E. N., Goy, M. F., Larsen, S. H. & Adler, J. (1975). Methylation of a membrane protein involved in bacterial chemotaxis. *Proceedings of the National Academy of Sciences, USA*, **72**, 3939–43.

Koshland, D. E. Jr. (1978). Bacterial chemotaxis. In *The Bacteria*, vol. 7, ed. J. R. Sokatch & L. N. Ornston. New York: Academic Press.

Larsen, S. H., Adler, J., Gargus, J. J. & Hogg, R. W. (1974a). Chemomechanical coupling without ATP: the source of energy for motility and chemotaxis in bacteria. *Proceedings of the National Academy of Sciences, USA,* **71,** 1239–43.

Larsen, S. H., Reader, R. W., Kort, E. N., Tso, W.-W. & Adler, J. (1974b). Change in direction of flagellar rotation is the basis of the chemotactic response in *Escherichia coli. Nature, London*, **249**, 74–7.

Lengeler, J. (1975). Nature and properties of hexitol transport systems in *Escherichia coli. Journal of Bacteriology*, **124**, 39–47.

Macnab, R. M. (1978). Bacterial motility and chemotaxis: the molecular biology of a behavioural system. *Chemical Rubber Company Critical Reviews of Biochemistry*, **5**, 291–341.

Macnab, R. W. & Koshland, D. E. Jr. (1972). The gradient-sensing mechanism in bacterial chemotaxis. *Proceedings of the National Academy of Sciences, USA*, **69**, 2509–12.

Macnab, R. M. & Ornston, M. K. (1977). Normal-to-curly flagellar transitions and their role in bacterial tumbling. Stabilization of an alternative quaternary structure by mechanical force. *Journal of Molecular Biology*, **112**, 1–30.

Matsumura, P., Silverman, M. & Simon, M. (1977). Synthesis of *mot* and *che* gene products of *Escherichia coli* programmed by hybrid colE1 plasmids in minicells. *Journal of Bacteriology*, **132**, 996–1002.

Ordal, G. W. & Adler, J. (1974). Properties of mutants in galactose taxis and transport. *Journal of Bacteriology*, **117**, 517–26.

Parkinson, J. S. (1974). Data processing by the chemotaxis machinery of *Escherichia coli. Nature, London*, **252**, 317–19.

Parkinson, J. S. (1976). *cheA*, *cheB* and *cheC* genes of *Escherichia coli* and their role in chemotaxis. *Journal of Bacteriology*, **126**, 758–70.

Parkinson, J. S. (1977). Behavioral genetics of bacteria. *Annual Review of Genetics*, **11**, 397–414.

Parkinson, J. S. (1978). Complementation analysis and deletion mapping of *Escherichia coli* mutants defective in chemotaxis. *Journal of Bacteriology*, **135**, 45–53.

Parkinson, J. S. (1980). Novel mutations affecting a signaling component for chemotaxis of *Escherichia coli. Journal of Bacteriology*, **142**, 953–61.

Parkinson, J. S. & Huskey, R. J. (1971). Deletion mutants of bacteriophage lambda. I. Isolation and initial characterization. *Journal of Molecular Biology*, **56**, 369–84.

Parkinson, J. S. & Parker, S. R. (1979). Interaction of the *cheC* and *cheZ* gene products is required for chemotactic behavior in *Escherichia coli. Proceedings of the National Academy of Sciences, USA*, **76**, 2390–4.

Parkinson, J. S. & Revello, P. T. (1978). Sensory adaptation mutants of *E. coli. Cell*, **15**, 1221–30.

Reader, R. W., Tso, W.-W., Springer, M. S., Goy, M. F. & Adler, J. (1979). Pleiotropic aspartate taxis and serine taxis mutants of *Escherichia coli. Journal of General Microbiology*, **111**, 363–74.

Ridgway, H. F., Silverman, M. & Simon, M. (1977). Localization of proteins controlling motility and chemotaxis in *Escherichia coli. Journal of Bacteriology*, **132**, 657–65.

Rubik, B. A. & Koshland, D. E. Jr. (1978). Potentiation, desensitization, and inversion of response in bacterial sensing of chemical stimuli. *Proceedings of the National Academy of Sciences, USA*, **75**, 2820–4.

Silverman, M. & Simon, M. (1973). Genetic analysis of bacteriophage Mu-induced flagellar mutants in *Escherichia coli. Journal of Bacteriology*, **116**, 114–22.

Silverman, M. & Simon, M. (1974). Flagellar rotation and the mechanism of bacterial motility. *Nature, London*, **249**, 73–4.

Silverman, M. & Simon, M. (1976). Operon controlling motility and chemotaxis in *E. coli. Nature, London*, **264**, 577–9.

Silverman, M. & Simon, M. I. (1977a). Bacterial flagella. *Annual Review of Microbiology*, **31**, 397–419.

Silverman, M. & Simon, M. (1977b). Identification of polypeptides necessary for chemotaxis in *Escherichia coli. Journal of Bacteriology*, **130**, 1317–25.

Silverman, M. & Simon, M. (1977c). Chemotaxis in *Escherichia coli*: Methylation of *che* gene products. *Proceedings of the National Academy of Sciences, USA*, **74**, 3317–21.

Smith, R. A. & Parkinson, J. S. (1980). Evidence for overlapping genes at the *cheA* locus of *Escherichia coli. Proceedings of the National Academy of Sciences, USA*, (in press).

Springer, M., Goy, M. F. & Adler, J. (1977a). Sensory transduction in *Escherichia coli*: A requirement for methionine in sensory adaptation. *Proceedings of the National Academy of Sciences, USA*, **74**, 183–7.

Springer, M. S., Goy, M. F. & Adler, J. (1977b). Sensory transduction in *Escherichia coli*: Two complementary pathways of information processing that involve methylated proteins. *Proceedings of the National Academy of Sciences, USA,* **74,** 3312–16.

Springer, M. S., Goy, M. F. & Adler, J. (1979). Protein methylation in behavioral control mechanisms and in signal transduction. *Nature, London,* **280,** 279–84.

Springer, M. S., Kort, E. N., Larsen, S. H., Ordal, G. O., Reader, R. W. & Adler, J. (1975). Role of methionine in bacterial chemotaxis: Requirement for tumbling and involvement in information processing. *Proceedings of the National Academy of Sciences, USA*, **72**, 4640–4.

Springer, W. R. & Koshland, D. E. Jr. (1977). Identification of a protein methyltransferase as the *cheR* gene product in the bacterial sensing system. *Proceedings of the National Academy of Sciences, USA*, **74**, 533–7.

Spudich, J. L. & Koshland, D. E. Jr. (1975). Quantitation of the sensory response in bacterial chemotaxis. *Proceedings of the National Academy of Sciences, USA*, **72**, 710–13.

Stock, J. B. & Koshland, D. E. Jr. (1978). A protein methylesterase involved in bacterial sensing. *Proceedings of the National Academy of Sciences, USA*, **75**, 3659–63.

STRANGE, P. G. & KOSHLAND, D. E. JR. (1976). Receptor interactions in a signalling system: competition between ribose receptor and galactose receptor in the chemotaxis response. *Proceedings of the National Academy of Sciences, USA*, **73**, 762–6.

TOEWS, M. L. & ADLER, J. (1979). Methanol formation *in vivo* from methylated chemotaxis proteins in *Escherichia coli. Journal of Biological Chemistry*, **254**, 1761–4.

TSANG, N., MACNAB, R. & KOSHLAND, D. E. JR. (1973). Common mechanism for repellents and attractants in bacterial chemotaxis. *Science*, **181**, 60–3.

TSO, W.-W. & ADLER, J. (1974). Negative chemotaxis in *Escherichia coli. Journal of Bacteriology*, **118**, 560–76.

VAN DER WERF, P. & KOSHLAND, D. E. JR. (1977). Identification of a γ-glutamyl methyl ester in a bacterial membrane protein involved in chemotaxis. *Journal of Biological Chemistry*, **252**, 2793–5.

WARRICK, H. M., TAYLOR, B. L. & KOSHLAND, D. E. JR. (1977). The chemotactic mechanism of *Salmonella typhimurium*: preliminary mapping and characterization of mutants. *Journal of Bacteriology*, **130**, 233–31.

GENETIC ANALYSIS OF THE CELL CYCLE

PAUL NURSE

School of Biological Sciences, The University of Sussex,
Brighton, Sussex BN1 9QG, UK

INTRODUCTION

The cell cycle is the period between the birth of a cell and the time when the cell itself divides into two daughters. It is made up of a series of events, the most conspicuous of which are those concerned with the replication and partition of the hereditary material between the two daughters, namely DNA replication, chromosome segregation and cell division. For the cell to successfully complete division, these events must occur with the correct cell cycle timing, and certain events such as wall division must occur in the correct position within the cell. In addition the rate of division must be co-ordinated with cell growth, and there must be controls which ensure that a cell only becomes committed to the mitotic cycle when conditions are appropriate to vegetative growth. How the cell achieves this temporal and spatial organisation is a major unsolved problem of cell biology.

In the present chapter I want to discuss these cell cycle controls in the light of genetic analysis. Both prokaryotic and eukaryotic microorganisms will be considered, since the fundamental problems of cell cycle organisation are similar for both types of organism. Most attention will be given to the yeasts *Saccharomyces cerevisiae* and *Schizosaccharomyces pombe*, and to *Escherichia coli*. In the first part of the chapter a brief survey will be made of the various types of cell cycle mutants and of their use in cell cycle analysis. This is followed by a large speculative section dealing with the basis of temporal organisation of the cell cycle. Finally there are short sections dealing with commitment and termination and problems of spatial organisation. Some topics are considered rather briefly and for a fuller discussion of *E. coli* the reader is referred to Slater & Schaechter (1974). Weschsler (1978) and Donachie (1981); for discussion of *S. cerevisiae* to Hartwell (1974, 1978), Simchen (1978) and Nurse (1981); and for discussion of *S. pombe* to Nurse & Fantes (1981).

CELL CYCLE MUTANTS

Cell cycle mutants have been isolated in a variety of microorganisms, including *E. coli*, *Bacillus subtilis*, *S. cerevisiae*, *S. pombe*, *Aspergillus nidulans*, *Ustilago maydis*, *Chlamydomonas reinhardii*, *Paramecium tetraaurelia* and *Tetrahymena pyriformis* (for lists of references see Slater & Schaechter, 1974; Hartwell, 1978; Morris, 1980). Most of these mutants are temperature-sensitive lethals which are unable to complete the cell cycle at the restrictive temperature. They are likely to be mis-sense mutants, coding for polypeptides with altered activities at the restrictive temperature as a consequence of an amino acid substitution. Nonsense mutants which do not complete the cell cycle can also be isolated using temperature-sensitive nonsense suppressors (Lutkenhaus & Donachie, 1979; Donachie *et al.*, 1979). At the restrictive temperature the nonsense suppressor does not function and the cell cycle nonsense mutant is expressed. Non-lethal cell cycle mutants which can complete the cell cycle but are altered in some way in the process have also been found. For example minicell mutants of *E. coli* undergo cell division at very variable cell sizes (Adler, Fisher, Cohen & Hardigree, 1967), and *wee* mutants of *S. pombe* are altered in both cell size at division and in the relative duration of the G1 and G2 phases (Nurse, 1975; Nurse & Thuriaux, 1977; Thuriaux, Nurse & Carter, 1978).

Cell cycle mutants can be used to identify those gene products required for normal traverse of the cell cycle. This approach has been particularly useful for the investigation of DNA replication in *E. coli* where over 200 mutants have been isolated, which define 10 or 11 *dna* genes (Sevastopoulos, Wehr & Glaser, 1977). It is possible that in this instance all the gene functions concerned have been identified, but this is unlikely to be the case for any other system. Some cell cycle gene functions have been characterised biochemically. The genes involved in DNA replication are the most amenable to biochemical characterisation since reasonable guesses can be made about which biochemical activities may be involved. Several of the *E. coli dna* gene functions have been characterised (Wechsler, 1978), as have DNA ligase and thymidylate synthetase in yeasts (Nasmyth, 1977; Johnston & Nasmyth, 1978; Game, 1976). The advent of 2D-gel electrophoresis should help in the identification of further cell cycle gene products. This technique allows many polypeptides to be rapidly screened for slight changes in structure, and has been used with great success for the identification of the genes coding for α and β tubulin in *A. nidulans* (Morris, 1980). This approach is even more powerful when used in conjunction with nonsense mutants producing dramatically altered polypeptides, and with gene cloning techniques (Lutkenhaus & Donachie, 1979; Donachie *et al.*, 1979). However one difficulty with this approach is that it only physically identifies the gene product and does not necessarily identify its biochemical function.

Apart from identifying genes and the gene products required for cell cycle progress, the mutants can be used in a number of other ways for analysing the cell cycle. Investigation of the aberrant behaviour of a mutant cell cycle can give some insight into the processes which control the normal cell cycle. The mutants can be used for identifying cell cycle events or 'landmarks' (Hartwell, 1978) which might otherwise have gone undetected (Byers & Goetsch, 1976). They also provide a convenient method for producing large numbers of cells blocked at specific points in the cell cycle (Reid & Hartwell, 1977). Examples of these various uses will be seen throughout the rest of this chapter.

TEMPORAL ORGANISATION

One of the more remarkable features of cells is their ability to organise all the biochemical and cytological events which make up the cell cycle into a strictly controlled order. How this temporal order may be achieved is considered in this section.

Dependent sequences of events

Many cell cycle mutants which specifically block an event early in the cell cycle also block other events which take place later. This phenomenon has been investigated most fully in *S. cerevisiae* (Hartwell, Cullotti, Pringle & Reid, 1974), where it was found that mutants blocked in nuclear division also became blocked in further cytokinesis and DNA replication, and mutants blocked in DNA replication became blocked in further nuclear division and cytokinesis. In contrast, mutants blocked in cytokinesis continued to undergo DNA replication and nuclear division. These observations have led to a model of the *S. cerevisiae* cell cycle incorporating two dependent sequences of events (Hartwell *et al.*, 1974). The events of

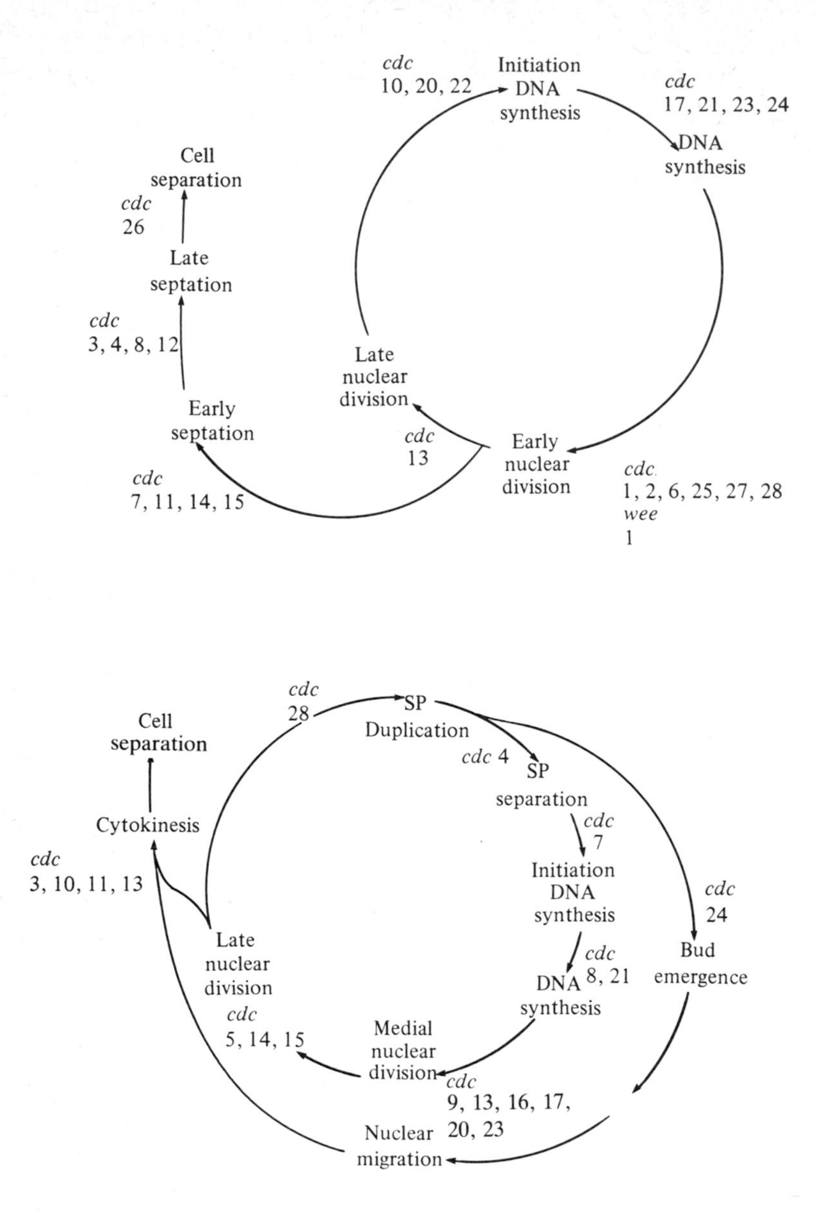

Fig. 1. Dependent sequences of cell cycle gene-controlled events in *S. cerevisiae* (upper panel) and *S. pombe* (lower panel). (Redrawn from Hartwell (1978) with permission.)

DNA replication–nuclear division make up one sequence and a second one links bud emergence and nuclear migration with cytokinesis (Fig. 1). A similar set of dependent sequences has been proposed for *S. pombe* (Nurse, Thuriaux & Nasmyth, 1976 and

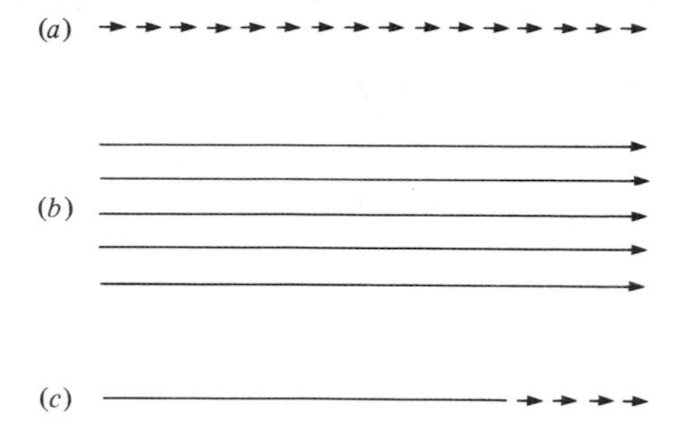

Fig. 2. Dependent sequences of events. The lengths of the arrows represent the time it takes for each event to be completed. (*a*) Linear sequence of events all equally rate controlling. (*b*) Parallel sequence of events all equally rate controlling. (*c*) Linear sequence of events one being a major rate controlling step.

Fig. 1), and various dependencies between cell cycle gene-controlled events have been established for other organisms such as *T. pyriformis* (Frankel, Jenkins & DeBault, 1976) and *A. nidulans* (Orr & Rosenberger, 1976).

The simplest way to view the dependent sequence is to assume that each event is dependent upon the completion of a previous event and that the cell cycle is made up of a large number of these events occurring one after each other (Hartwell, 1974; Hartwell *et al.*, 1974). Such a view of the basis of temporal order in the cell cycle is schematised in Fig. 2a. In the original conception of the model the completions of the various gene-controlled functions were distributed throughout the cell cycle (Culotti & Hartwell, 1971). This can be tested by determining the times at which the cell cycle genes complete their functions by shifting temperature-sensitive mutants to the restrictive temperature and measuring how many cells go on to complete division. Only those cells that have completed the temperature-sensitive function will be able to divide allowing the cell cycle timing at which this occurs to be calculated. This timing has been called the execution point (Hartwell, 1974), block point (Howell & Naliboff, 1973), and transition point (Nurse *et al.*, 1976). The timing of this point is open to ambiguous interpretation. For example, if the gene product takes some time to decay at the restrictive temperature, or if it is thermosensitive for synthesis rather than for function, the execution point will appear earlier in the cell cycle than it actually is. Alternatively, if the gene product is only partially active at the permissive temperature, the execution

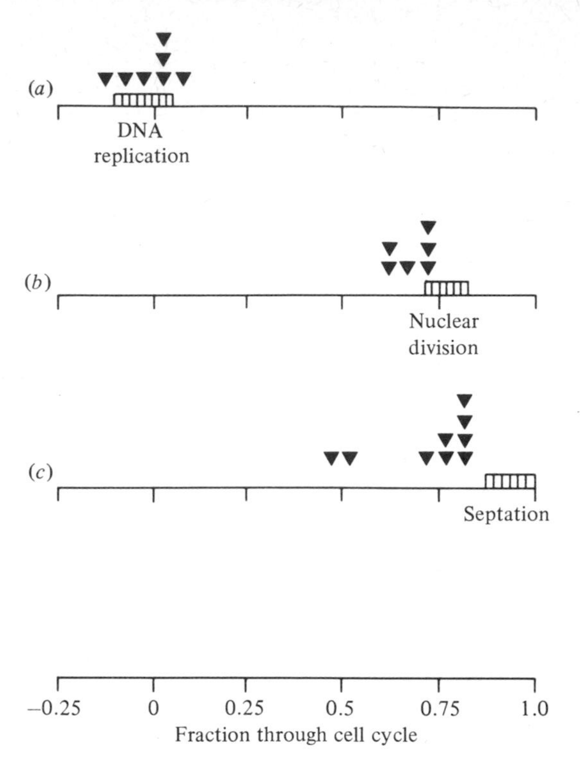

Fig. 3. Execution points of mutants in *cdc* genes of *S. pombe.* (*a*) DNA replication genes. (*b*) Nuclear division genes. (*c*) Septation genes. Hatched boxes mark timing of DNA replication, nuclear division and septation, and closed triangles mark the execution points.

point will appear later than it really is. Despite these difficulties some overall pattern of execution points has emerged. A few execution points significantly precede the event in which they are involved, but most coincide closely with it. This applies to many of the *S. cerevisiae cdc* (cell division cycle) genes (Hartwell, 1978), to 7 of the 9 mutants of *A. nidulans* (Orr & Rosenberger, 1976), to all the 3 penetrant mutants of *T. pyriformis* (J. Frankel, quoted in Hartwell, 1978), and to 19 of the 22 *cdc* genes of *S. pombe* (Nurse *et al.*, 1976; Fantes, 1979; K. Nasmyth & P. Nurse, unpublished). The execution points for *S. pombe* are plotted in Fig. 3 where it can be seen that no cell cycle gene completes its function more than 0.4 of a cell cycle before the event in which it is involved, and only 3 are more than 0.15 of a cell cycle before it. This distribution of execution points argues against the simple model shown in Fig. 2*a* which would predict a more even distribution.

The uneven distribution can be interpreted in two ways. It is possible that the cell cycle genes act in parallel over a large part of the cell cycle and only complete their functions close to the events in which they are involved (Fig. 2*b*). Alternatively certain cell cycle events could act as major rate-controlling steps for progress through the cell cycle. Whilst these rate-controlling steps are in progress no other cell cycle gene functions can take place. When the rate-controlling steps are completed many of the other dependent events will rapidly take place accounting for the clustering of the execution points (Fig. 2*c*). Support for the latter view is provided by *wee* mutants of *S. pombe* (Nurse, 1975; Thuriaux *et al.*, 1978). Temperature sensitive *wee* mutants are advanced prematurely into cell division on shift to high temperature showing that the *wee* gene products act in a major rate-controlling step for completion of the cell cycle. Nutritional shifts can phenocopy these mutants in *S. pombe* (Fantes & Nurse, 1977) and there is some evidence for similar nutritional effects on the rate of bud emergence in *S. cerevisiae* (Lorincz & Carter, 1979; Johnston, Ehrhardt, Lorincz & Carter, 1979). Further evidence in favour of some cell cycle events being rate controlling for cell division are considered in the next section.

Implicit in the above analysis is that most of the gene-controlled functions are not rate controlling for cell cycle progress. An analogy can be made to a metabolic pathway where most enzymes with the exception of the pacemaker are not rate limiting (Nurse & Fantes, 1980). The non-rate controlling gene products are likely to be present in excess. Support for this supposition is provided by various *cdc* mutants of *S. cerevisiae* which undergo several cell cycles before arrest (Hartwell, Mortimer, Culotti & Culotti, 1973). It has been argued that these mutants are thermosensitive for synthesis of the gene product and that previously synthesised material remains functional at the restrictive temperature (Hartwell, 1978). If this is the case then the amount of the gene product must be present in excess to enable the cell to complete several more cycles before it is sufficiently diluted to become limiting. The temperature-sensitive nonsense suppressor system developed for *E. coli* is a better means for investigating this problem since only the synthesis of the gene product is switched off at the restrictive temperature and consequently gene products present in rate-limiting amounts will show first cycle arrest (Donachie, 1980). So far nonsense mutants showing both first cycle and multi-cycle arrest have been identified.

The picture of temporal order in the cell cycle provided by these mutant studies is best interpreted in terms of a dependent sequence of events as schematised in Fig. 2*c*. Certain cell cycle events take a long time to complete and so are major rate-controlling steps for cell cycle progress. Most other events are completed rapidly and are not rate limiting, but they cannot take place until some earlier event has been completed. This results in bursts of developmental activity during the cell cycle when the major rate-controlling events are completed. The nature of these rate-controlling steps is considered in the next section.

Major rate-controlling events

Growth in cell mass

For most if not all microbial cells the rate of cell division is set by the rate of increase in cell mass. This follows from the facts that in steady-state cultures the rate of cell division follows that of cell mass and also that in a particular steady-state culture, cell mass at division remains constant. Therefore the need to attain a critical cell mass before certain cell cycle events can take place is likely to act as the major rate-controlling step of the cell cycle.

In *E. coli* it is the initiation of DNA replication which appears to require growth to a particular cell mass. At various growth rates the ratio of cell mass to the number of chromosomes undergoing initiation was always constant (Donachie, 1968). In addition, if DNA replication was blocked and cell mass allowed to increase, on release of the block many new chromosome initiations took place (Donachie, 1969). Both sets of observations may be explained if cells have to attain a critical mass before initiating a new round of DNA replication.

In *S. cerevisiae* bud emergence takes place when cells attain a certain size (Johnston, Pringle & Hartwell, 1977). Abnormally small cells produced by nutritional deprivation took much longer to initiate a bud on re-inoculation in fresh media than did larger cells. Three gene-controlled steps specific to G1, *cdc* 4, *cdc* 7 and *cdc* 28, were not completed until the cells had grown to a critical size. The *cdc* 28 gene function acts before the other two genes (Hereford & Hartwell, 1974) and is also required for bud emergence. These data suggest that attainment of a critical cell mass is required before the *cdc* 28 event is completed and that bud emergence together with the other G1 events then follow afterwards (Hartwell, 1974). This hypothesis explains the behaviour of cells cultured at slow growth rates since the small daughter cells take much longer before they complete G1 and undergo bud emergence than do larger cells (Hartwell & Unger, 1977; Jagadish & Carter, 1977). It also explains why the *cdc* 28 execution point can shift from 0.12 of a cell cycle in fast-growing cells to 0.72 of a cell cycle in slow-growing cells (Jagadish & Carter, 1977).

Cell size is also important for controlling the rate of cell division in *S. pombe* (Fantes, 1977). Cells larger than average at birth have shortened subsequent cell cycles during which they grow less than average and consequently divide at a size closer to the population mean. Conversely, cells smaller than average at birth have lengthened subsequent cell cycles during which they grow more than average and consequently also divide at a size closer to the population mean. These and other observations (Fantes & Nurse, 1978) can be best explained if cells monitor their mass and undergo mitosis and cell division when the mass attains a critical value. The *wee* mutants mentioned previously undergo cell division at a reduced cell size and act in a major rate-controlling step determining when mitosis takes place (Nurse, 1975; Thuriaux *et al.*, 1978). Two genes, *wee* 1 and *cdc* 2, are defined by the *wee* mutants, and it is possible that these two genes are involved in the control which monitors cell mass and determines when mitosis takes place (P. Nurse & P. Thuriaux, unpublished observations).

A second size control in the cell cycle has been revealed by the *wee* mutants to act on DNA replication (Nurse, 1975; Nurse & Thuriaux, 1978; Nasmyth, Nurse & Fraser, 1979). The *wee* mutants have larger G1 periods because S phase is delayed by the requirement that a cell must attain a minimum critical mass before DNA replication can take place (Fig. 4). Wild-type cells begin their cell cycles above this size and so S phase takes place rapidly after mitosis. *wee* mutant cells begin their cell cycle beneath this size and so have to grow for a longer period before S phase can take place. The timing of DNA replication and the cell size at which it occurs in a range of mutants dividing at different sizes and in small cells generated by nitrogen starvation, sporulation of growth in a chemostat (Nurse & Thuriaux, 1977; Nasmyth, 1979) can all be explained in terms of this second control.

Unlike *E. coli* (Donachie, 1968), there is good evidence for the yeasts that nutritional level or cell growth rate directly interact with the size controls. When growing slowly, *S. cerevisiae* cells undergo

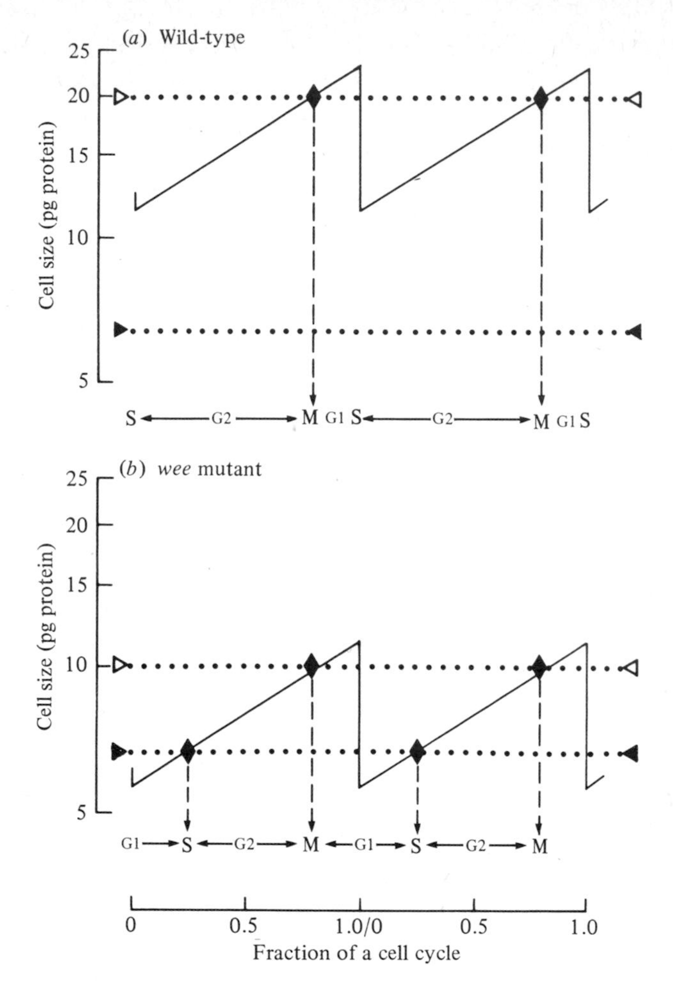

Fig. 4. Model of cell size controls acting over DNA replication and nuclear division in *S. pombe*. (*a*) Wild-type. (*b*) *wee* mutant. Figures show increase in cell mass during two consecutive cell cycles. Nuclear division is initiated when the cell attains the mass shown by the open triangles. This is reduced in the *wee* mutant compared with wild-type. In wild-type, S-phase follows rapidly on after mitosis whilst in the *wee* mutant the cell has to grow for a quarter of a cell cycle to attain the minimum mass required for DNA replication shown by the closed triangles. (Redrawn from Nurse & Thuriaux (1977) with permission.)

bud emergence of a reduced size (Lorencz & Carter, 1979; Johnston *et al.*, 1979), and *S. pombe* cells undergo both mitosis (Fantes & Nurse, 1977) and DNA replication at reduced sizes (Nasmyth, 1979). This so called nutritional or growth rate modulation of the size control acting at mitosis is lost in *wee* mutant cells. The modulation may exert its effect through the translational apparatus since mutants which alter nonsense suppression in *S. pombe* also alter the size control at mitosis (Munz, P. Nurse & P. Thuriaux, unpublished results). Perhaps the mechanism monitoring mass shares common components with the translational apparatus.

'Timers'

Certain cell cycle events which take a long time to be completed can be considered as major rate-controlling steps and are not affected by the mass of the cell. The most obvious example of such a 'timer' is the time required for the replication of the chromosome in *E. coli*, called the C period. Over a wide range of growth rates this timer is constant (Cooper & Helmstetter, 1968), and is only extended at very slow growth rates (Kubitschek & Newman, 1978). In cells growing with faster doubling times than C, new rounds of replication are initiated before the original one is completed which ensures that the doubling time for DNA matches that of mass. Another example of a 'timer' is the period between S phase and cell division, which is relatively constant over a range of growth rates in *S. cerevisiae* (Hartwell & Unger, 1977; Jagadish & Carter, 1977). A similar constant period is seen in slow-growing *S. pombe* cells (Nasmyth, 1979) and possibly in more rapidly growing *wee* mutant cells (Fantes & Nurse, 1978).

Dependency

If temporal order within the cell cycle is based on dependent sequences of events then the nature of the dependency relationship between events is important for cell cycle control. Methods used for sequencing the order of function of events during phage morphogenesis (Jarvick & Botstein, 1973) have been applied to various cell cycle events in *S. cerevisiae* (Hereford & Hartwell, 1974). These studies have been successful in establishing that the order of function of the genes involved in G1 is *cdc* 28–*cdc* 4–*cdc* 7 and that the mating hormone functions interdependently with *cdc* 28. A possible basis for the dependency in function between the three genes is their role in the duplication of the spindle plaque (Byers & Goetsch, 1975). This structure first undergoes duplication and then separation during G1, and mutants in *cdc* 28 are blocked before duplication, in *cdc* 4 after duplication and before separation, and in *cdc* 7 after separation. If the three gene products were involved in different stages of this process then the action of the later functions would be dependent upon the earlier ones. The later-acting gene products could not be expressed until the earlier ones were complete. The *cdc* 28 gene function is also required for bud emergence and the basis of this dependency relationship may also involve the spindle plaque. Extranuclear microtubules have been found emanating from the duplicated plaque in the direction of the emerging bud (Byers & Goetsch, 1975). Thus the *cdc* 28 effect on bud emergence may be the result of the block in spindle plaque duplication.

The examples of dependency just quoted together with others concerned with the relationships between DNA replication and mitosis (Hartwell, 1974; Nurse *et al.*, 1976) are cases where the events are closely related functionally. However, other events are not so closely related and yet still show dependency relationships. One example is the dependency of septation in *E. coli* upon the completion of DNA replication. This dependency rule is broken in *dna*A and *dna*C mutants blocked in the initiation of DNA replication which continue to undergo cell division and produce anucleate cells (Hirota *et al.*, 1968). If the *dna*A mutants are shifted to the restrictive temperature in the presence of nalidixic acid which stops any further rounds of DNA replication from being completed, then no anucleate cells are produced. This can be explained by a model in which an inhibitor of septation is produced after initiation of DNA replication which is only removed after termination of replication (Donachie, 1980). In the *dnaA* mutant which blocks initiation, all chromosomes complete replication, and the inhibition is removed. Since there are no more initiations, no more inhibitor is made and cell division can take place producing anucleate cells. If termination is prevented by nalidixic acid then septation is also blocked. The production and removal of the septation inhibitor is therefore the basis of the dependency of septation upon DNA replication.

Another example of dependency between septation and DNA in *E. coli* is found with the induction of the SOS functions when DNA is damaged by agents such as UV (Witkin, 1976). These functions include lambda prophage induction, error-prone DNA repair and, more relevant to the present discussion, a block in septation leading to filament formation. The block ensures that cell division will not take place whilst damaged DNA is still being repaired and therefore septation is dependent upon the removal of DNA damage. Induction of SOS functions involves two genes *recA* and *lexA* and the participation of DNA breakdown products (Gudas & Pardee, 1975:

Oishi & Smith, 1978). It was originally postulated that the *recA* gene product, protein X, was synthesised as a consequence of DNA damage and acted as an inhibitor of septation (Satta & Pardee, 1978). This was based on the observation that temperature-sensitive alleles of *lexA* were constitutive for protein X and formed filaments. More recent results suggest that protein X cannot act as the inhibitor, since *rnmB* mutants closely linked to *recA* are also constitutive for protein X and yet do not filament (Volkert, Spencer & Clark, 1979). Therefore it is possible that the inhibition of septation is a direct effect of the *lexA* gene product rather than an indirect one mediated through protein X. This dependency relationship could be related to the one previously discussed between the completion of DNA replication and septation. Whilst DNA replication is in progress there is a block on septation and a similar block can also be induced by the SOS functions. This possibility gains support from the observation that the *dnaA* mutants, which block the initiation of DNA replication but still allow septation, are altered in the SOS functions and do not make protein X (Satta & Pardee, 1978). Although the details have yet to be finalised, this is clearly a promising system for working out the nature of a dependency relationship between cell cycle events.

Quantal events

A dependency relation between cell cycle events will ensure that they occur in the correct sequence but not that each event only takes place once per cell cycle. That there must be a mechanism ensuring such quantal behaviour is illustrated by the behaviour of a mutant of *cdc* 16 in *S. pombe* (Minet, Nurse, Thuriaux & Mitchison, 1979). This mutant is blocked at a late state in septation. A septum is formed but is defective in some way so the cells cannot separate. The cell continues accumulating septa in the centre of the cell although only a single septum is made at a time. Cells shifted to the restrictive temperature when they are making a septum proceed to cell division, but one of the daughter cells produces one further septum at the pole of the cell that was the original site of septation in the mother cell. The formation of this septum is unusual since it occurs before nuclear division and forms an anucleate compartment. A model explaining these data proposes that the septal manufacturing apparatus activated after mitosis can only support the synthesis of one septum at a time. At a late stage in septation a signal destroys this apparatus, but this signal is not generated in the mutant of *cdc* 16. Consequently the apparatus is reused and another septum is initiated, accounting for the pattern of accumulation of septa in the middle of the cell. The single septum produced at the pole of the cell is a consequence of the inhibitory signal being delayed sufficiently for one further septum to be initiated which is formed in the daughter.

There is a similar quantal mechanism in *E. coli* ensuring a single septum is made each cell cycle. Minicell-producing strains behave as if division sites are not inactivated once they have been formed (Adler *et al.*, 1967; Teather, Collins & Donachie, 1974). Consequently septa are produced at cell poles which were once division sites producing anucleate minicells. Investigation of the pattern of septation has shown that the normal number of septa are produced per unit increase in cell mass, suggesting that a quantum of division factor is being produced per cell cycle which is distributed at random between the new and old division sites. The quantum of division factor ensures only one septum is made at a time and if it is consumed during the septation process then no more septa will be made that cell cycle.

In *S. cerevisiae* only one bud undergoes emergence per cell cycle. In mutants of *cdc* 4, which are blocked in DNA replication, the cell continues to undergo bud emergence but only one at a time (Hartwell, 1971). The association between the duplicated spindle plaque and bud emergence and the fact that mutants of *cdc* 4 are blocked with a duplicated spindle plaque suggests that this mutant is stuck in a state which is permanently 'competent' for bud emergence. In a normal cell cycle the duplicated plaques separate and this may destroy the 'competence' of the cell to produce any further buds and hence only one bud is formed per cell cycle.

In all three examples above there is some sort of quantal mechanism, a single septal manufacturing apparatus in *S. pombe*, a quantum of division factor in *E. coli*, and a single duplicated spindle plaque in *S. cerevisiae*. In addition there is some process to inactivate the quantal mechanism once it has performed its function, a signal destroying the apparatus in *S. pombe*, consumption of the division factor during septation in *E. coli* and separation of the duplicated spindle plaque in *S. cerevisiae*. These types of mechanism are sufficient to ensure that the cell cycle event will occur only once per cell cycle.

Mechanisms involved in controlling events

The genetic analysis of some cell cycle controlling events has progressed sufficiently for some comments to be made on the possible mechanisms involved. There is good evidence that *dnaA* is concerned with the rate-controlling step for initiation of DNA replication in *E. coli* since a revertant of a temperature-sensitive *dnaA* mutant called *dnaA cos* results in over-initiation of DNA replication (Kellenberger-Gujer, Podhajska & Caro, 1978). *dnaA cos* is likely to be an intragenic suppressor of *dnaA* and is thought to cause over-production of the rate-controlling *dnaA* gene product. One possible model is that the *dnaA* gene product may regulate its own synthesis by feedback inhibition. Further evidence in favour of this model is provided by Hansen & Rasmussen (1977) and Zahn & Messer (1979). It has been argued that the *dnaA* gene product cannot be involved in the control of initiation since it does not determine the rate at which each chromosome is initiated but only the number of replicating chromosomes per cell (Fralick, 1978). If cell mass is important in the control then another interpretation is possible. The *dnaA* mutants may be altered in cell mass at initiation, which over a certain range will not alter the rate at which each chromosome is initiated, but will reduce the number of replicating chromosomes per cell.

Considerable evidence has accumulated which suggests that DNA replication in *E. coli* requires a replication complex (Lark, 1979). This evidence is based on observations that various mutant gene products concerned with replication behave in a manner suggestive of interaction between them. For example the presence of certain *dnaC* alleles modify the activity of an external suppressor of *dnaB*, suggesting direct interaction between the *dnaB* and *dnaC* gene products (Schuster *et al.*, 1977). Similar types of observations have implicated various gene products as possible components in the complex, including the products of *dnaA*, *dnaB*, *dnaC*, *dnaT* and even *rpoB* which codes for a sub-unit of RNA polymerase (Kellenberger-Gujer & Podhajska, 1978; Bagdasarian, Izakowska & Bagdasarian, 1977; K. G. Lark unpublished observations quoted in Lark, 1979). The latter is of interest given the role that origin RNA may play in initiation (Lark, 1972). More direct evidence in favour of a replication complex has been obtained using *in vitro* systems where the *dnaB* and *dnaC* proteins form a stable complex before replication begins (Wickner & Hurwitz, 1975; Weiner, McMacken &

Kornberg, 1976). It has also been proposed that certain components in the complex are usually responsible for its inactivation at the end of each round of replication. *Sdv*[c] mutants are defective in inactivation (Kogoma, 1978) whilst in *dnaT* mutants the complex can never be stabilised against inactivation (Lark, 1979).

The involvement of a complex in a cell cycle event has interesting regulatory implications. If its assembly was of a co-operative nature then it could only be assembled when its constituent components were present in sufficient concentration and, after assembly, the concentration of these components would be reduced preventing assembly of any further complexes (Lark, 1979). If an inactivation mechanism was built into the complex which destroyed the structure once it had completed its function, then the event is likely to occur only once per cell cycle. It is possible that other cell cycle events may involve complexes or structures and there is already some evidence that the *cdc* 28 gene product of *S. cerevisiae* is a component of a nucleus-restricted structure (Dutcher & Hartwell, 1978).

It has been argued in this chapter that attainment of a particular cell mass is important for regulating cell cycle progress and therefore mechanisms for monitoring cell mass may be important as cell cycle controls. For this reason I want to consider plasmid copy number, because monitoring of cell mass may be involved in its control (Pritchard, 1978). Mutants of the *E. coli* R1 plasmid have been isolated which show increased copy number and therefore must be altered in some gene function which is normally rate controlling (Uhlin, Gustafsson, Mulin & Nordström, 1978). One of these mutants is suppressible by an amber nonsense suppressor (Gustafsson & Nordström, 1978), suggesting that the mutant gene usually codes for an inhibitor or negative element in the control. Negative elements have also been proposed in the control of Col E1 replication. Studies of the behaviour of a hybrid Col E1 and pSC 101 plasmid are suggestive of a negative element (Cabello, Timmis & Cohen, 1976), but an interpretation in terms of a positive element is also possible. More telling evidence in favour of a negative element is the observation that a mutant with a high plasmid copy number is recessive to wild-type (Shepard, Gelfand & Polisky, 1979). The gene altered in this mutant has been cloned and is localised near the origin of replication. Another argument against a positive element has been provided by Kahn & Helinski (1978), who showed that plasmid-coded protein synthesis was not required for Col E1 DNA replication. Overall, these experiments favour a negative element as a possible component in the mass monitoring mechanism.

Some attempt has been made to do a similar analysis of the *wee* 1 and *cdc* 2 genes concerned with the control of mitosis in *S. pombe* (P. Nurse & P. Thuriaux, unpublished observations). The frequency of mutation and the presence of a nonsense mutant at the *wee* 1 locus suggest that *wee* 1 codes for an inhibitor whilst the occurence of frequent temperature-sensitive lethals at the *cdc* 2 locus suggests that *cdc* 2 codes for some sort of positive element. As these studies and those concerned with plasmid copy number control become more defined a more complete attempt at building a model for a mechanism monitoring mass may be possible.

COMMITMENT AND TERMINATION

Commitment

The controls involved in commitment to the mitotic cell cycle have been most thoroughly studied in *S. cerevisiae*. Since they have been reviewed extensively in the past (Hartwell, 1974; Hartwell, 1978; Simchen, 1978; Nurse, 1981), I shall consider them only briefly here. Commitment to the mitotic cell cycle as against the developmental alternative of conjugation was investigated by Reid & Hartwell (1977). Using various cell cycle mutants, cells were blocked at different stages of the cycle and their ability to undergo conjugation was tested. Most cell cycle mutants were unable to conjugate, and so it was argued that they were blocked sufficiently late in the mitotic cycle to have become fully committed to that cycle, and consequently were unable to undergo conjugation. Only cells blocked at the function controlled by *cdc* 28 could still conjugate. Since this was the earliest known gene-controlled event in the cell cycle it was called ‘start’, and it was proposed that cells became committed to the cell cycle when they had passed ‘start’ (Hartwell, 1974). Cells become arrested at ‘start’ if they are treated with the mating hormone α-factor (Bücking-Throm, Duntze, Hartwell & Manney, 1973), or are starved of nutrients (Hartwell & Unger, 1977), and as discussed earlier cells need to attain a critical mass before the ‘start’ function can be completed (Johnston, Pringle & Hartwell, 1977). These and other observations led to the picture of 'start' as a critical point early in the cycle where the cell monitors the presence or absence of conditions favouring conjugation and sporulation, its nutritional status and its mass and if these conditions are all satisfactory then the cell passes 'start' and becomes committed to the cycle (Hartwell, 1974).

The concept of 'start' has been very useful for understanding commitment to the cell cycle, although more recent experiments suggest that it may have to undergo some modification. For example, there is some evidence that cells arrested by α-factor, or nutritional deprivation or the *cdc* 28 mutant differ in their terminal phenotypes (Byers & Goetsch, 1975; Pinon & Pratt, 1979). This suggests that 'start' may be more complex than originally conceived with the possibility that a cell can shift from one state to another within a 'start' area (Nurse, 1981). In addition, cells blocked at *cdc* 4, which is later in the cell cycle than *cdc* 28, can still undergo sporulation, suggesting that cells which have completed the *cdc* 28 function are not necessarily committed to the mitotic cell cycle as compared with all other alternative developmental pathways (Hirschenberg & Simchen, 1977).

Further support for the 'start' concept has come from *S. pombe* where preliminary experiments indicate that cells blocked using *cdc* 10 mutants still conjugate well, whereas other cell cycle mutants do not (P. Nurse & Bissett, unpublished observations). The *cdc* 10 gene function is required for DNA replication (Nurse *et al.*, 1976) and the completion of the function requires the attainment of a critical cell mass (Nasmyth, 1979). Therefore *cdc* 10 may define a 'start' gene analogous to *cdc* 28 in *S. cerevisiae*. The fundamental control mechanism at 'start' is not known, but one exciting possibility is that it may involve some sort of gene rearrangement. Such rearrangement has been shown to determine mating type in *S. cerevisiae* (Hicks, Strathern & Klar, 1979), and if this turns out to be a widespread mechanism for defining cellular development fates then it may also be important in committing cells to the mitotic cycle.

Termination

When microbial cells become deprived of nutrients they terminate progress through the cell cycle. Usually they accumulate at the beginning of the cycle by completing cell cycles in progress and not starting new ones. This is brought about in *S. cerevisiae* as a result of a fairly constant time period between 'start' and cell division and the fact that the cell has to attain a certain mass before passing 'start' (Hartwell & Unger, 1977; Jagadish & Carter, 1977). When growth is slowed down by nutrient deprivation, cells will grow little during the constant time period and will produce small daughter buds which are too small to pass 'start'. As a consequence cells accumulate at the beginning of the cycle. A similar mechanism can also operate in bacterial cells which also have a constant time period between the initiation of DNA replication and cell division and a constant cell mass at initiation of DNA replication (Donachie, Jones & Teather, 1973). In *S. pombe* the mechanism is different. In poor nutritional conditions the cell mass required for mitosis is reduced so that cell division takes place at a smaller size (Fantes & Nurse, 1977). These smaller daughter cells eventually become too small to initiate DNA replication which also requires growth to a critical cell mass (Nurse & Thuriaux, 1977) and as a consequence cells accumulate at the beginning of the cell cycle.

SPATIAL ORGANISATION

The problem of spatial organisation of cell cycle events has received much less attention than temporal organisation. In bacterial cells the position of the septum is located normally in the middle of the dividing cell. In minicell-forming mutants this positioning is disturbed (Adler *et al.*, 1967). This is because old division sites are not turned off and so septation can take place at the ends of cells which were once division sites in the middle of the cell. Therefore, these mutants are not particularly informative about the mechanism determining the position of the septum in the cell. More informative are mutants of *B. subtilis* which become blocked in DNA replication and continue to undergo septation (McGuiness & Wake, 1979). If DNA replication is blocked early, then septation occurs asymmetrically, but if more than 70% of the chromosome is replicated septation occurs centrally and cuts through the nucleoid. An analogous situation has also been reported for *E. coli*. A mutant defective in DNA gyrase which prevents nucleoid segregation results in an asymmetrically placed septum (Orr, Fairweather, Holland & Pritchard, 1979). Other non-allelic mutants called *ran* (*dom*) which are defective in nucleoid segregation also produce asymmetrically or randomly located septa (unpublished results quoted in Donachie, 1980). These experiments suggest a coupling between nucleoid behaviour and the position of the septum.

A coupling between nuclear behaviour and the position of septation or bud emergence has also been observed in the yeasts. In *S. pombe* a mutant of *cdc* 13 blocked late in mitosis undergoes septation and the septum cuts the nucleus in two (P. Nurse, unpublished observations). In *S. cerevisiae* the position of a new bud is related to the nucleus by microtubules emanating from the duplicated spindle plaque (Byers & Goetsch, 1975). The relationship between the nucleus and budding may be similar to that between a dividing nucleus and septation. This is because mitosis in *S. cerevisiae* can be considered to be initiated in G1 when a short spindle is formed, and bud emergence can be considered to be the initiation of cell division or at least to define the position of cell division.

These experiments suggest that either the nuclear position determines the site of cell division or that some common mechanism determines the position of both the nucleus and the site of cell division. At present, little can be said about the mechanisms determining position within the cell, but the involvement of microtubules mentioned above suggests that cytoskeletal elements may play a role.

CONCLUSIONS

The major conclusions and speculations of the chapter can be summarised as follows:

1. The cell cycle is made up of sequences of dependent events. Temporal order is maintained because later events cannot take place until earlier ones are completed.

2. The dependency of one event on another can either be due to a close functional relationship between the events (e.g. DNA replication and mitosis), or because there is a control system linking the two events (e.g. SOS function linking septation and DNA metabolism).

3. Quantal mechanisms exist which ensure that events only take place once every cell cycle.

4. Structures or complexes, the assembly of which is co-operative in nature, may be involved in cell cycle events. Such systems ensure that the periodicity of the event can be well controlled.

5. Certain events are rate controlling for progress through the cell cycle and occupy large periods of the cycle. Other events take

place rapidly over a limited period resulting in bursts of developmental activity at particular stages of the cell cycle such as DNA replication and mitosis.

6. Rate-controlling events often require that the cell attains a critical mass before they can be completed. Such a requirement ensures that the rate of cell division is matched to the rate of increase in mass.

7. Cells become committed to the cell cycle sometime during G1 if conditions do not favour alternative developmental pathways and if the cell has attained a critical mass. Mechanisms exist which result in cells terminating growth before this point of commitment if conditions become unfavourable for vegetative growth.

8. The position of the nucleus and the site of cell division are well controlled and may be determined by cytoskeletal elements within the cell.

I would like to thank Willie Donachie for his advice during the preparation of this manuscript, and Peter Fantes and Murdoch Mitchison for stimulating discussions on cell cycle problems.

REFERENCES

ADLER, H. I., FISHER, W. D., COHEN, A. & HARDIGREE, A. A. (1967). Miniature *Escherichia coli* cells deficient in DNA. *Proceedings of the National Academy of Sciences, USA*, **57**, 321–6.

BAGDASARIAN, M. M., IZAKOWSKA, M. & BAGDASARIAN, M. (1977). Suppression of the *dnaA* phenotype by mutations in the *rpoB* cistron of ribonucleic acid polymerase in *Salmonella typhimurium* and *Escherichia coli*. *Journal of Bacteriology*, **130**, 577–82.

BÜCKING-THROM, E., DUNTZE, W., HARTWELL, L. H. & MANNEY, T. R. (1973). Reversible arrest of haploid yeast cells at the initiation of DNA synthesis by a diffusible sex factor. *Experimental Cell Research*, **76**, 99–110.

BYERS, B. & GOETSCH, L. (1975). Behaviour of spindles and spindle plaques in the cell cycle and conjugation of *Saccharomyces cerevisiae*. *Journal of Bacteriology*, **124**, 511–23.

BYERS, B. & GOETSCH, L. (1976). Loss of the filamentous ring in cytokinesis defective mutants of the budding yeast. *The Journal of Cell Biology*, **70** (2, Part 2., 35a (Abstract).

CABELLO, F., TIMMIS, K. & COHEN, S. (1976). Replication control in a composite plasmid constructed by *in vitro* linkage of two distinct replicons. *Nature, London*, **259**, 285–90.

COOPER, S. & HELMSTETTER, C. E. (1968). Chromosome replication and the division cycle of *Escherichia coli* B/r. *Journal of Molecular Biology*, **31**, 519–40.

CULOTTI, J. & HARTWELL, L. H. (1971). Genetic control of the cell division cycle in yeast. III. Seven genes controlling nuclear division. *Experimental Cell Research*, **67**, 389–401.

DONACHIE, W. D. (1968). Relationship between cell size and time of initiation of DNA replication. *Nature, London*, **219**, 1077–9.
DONACHIE, W. D. (1969). Control of cell division in *Escherichia coli*: Experiments with thymine starvation. *Journal of Bacteriology*, **100**, 260–8.
DONACHIE, W. D. (1981). The cell cycle of *Escherichia coli*. In *The Cell Cycle*, ed. P. C. L. John. Cambridge University Press. (In press.)
DONACHIE, W. D., BEGG, K. J., LUTKENHAUS, J. F., SALMOND, G. P. C., MARTINEZ-SALAS, E. & VICENTE, M. (1979). Role of the *ftsA* gene product in control of *Escherichia coli* cell division. *Journal of Bacteriology*, **140**, 388–94.
DONACHIE, W. D., JONES, N. C. & TEATHER, R. (1973). The bacterial cell cycle. In *Microbial Differentiation*, ed. J. M. Ashworth & J. E. Smith, *Symposium of the Society for General Microbiology, 23,* pp. 9–44. Cambridge University Press.
DUTCHER, S. & HARTWELL, L. H. (1978). The involvement of *cdc* gene products in conjugation. In *Abstracts of the 9th International Conference on Yeast Genetics and Molecular Biology*, p. 70. University of Rochester, Rochester, New York, USA.
FANTES, P. (1977). Control of cell size and cycle time in *Schizosaccharomyces pombe. Journal of Cell Science*, **24**, 51–67.
FANTES, P. (1979). Epistatic gene interactions in the control of division in fission yeast. *Nature, London*, **279**, 428–30.
FANTES, P. & NURSE, P. (1977). Control of cell size at division in fission yeast by a growth modulated size control over nuclear division. *Experimental Cell Research*, **107**, 377–86.
FANTES, P. & NURSE, P. (1978). Control of the timing of cell division in fission yeast. *Experimental Cell Research*, **115**, 317–29.
FRALICK, J. A. (1978). Studies on the regulation of initiation of chromosome replication in *Escherichia coli. Journal of Molecular Biology*, **122**, 271–86.
FRANKEL, J., JENKINS, L. M. & DEBAULT, L. E. (1976). Causal relations among cell cycle processes in *Tetrahymena pyriformis*: an analysis employing temperature-sensitive mutants. *Journal of Cell Biology*, **71**, 242–60.
GAME, J. (1976). Yeast cell-cycle mutant *cdc* 21 is a temperature-sensitive thymidylate auxotroph. *Molecular and General Genetics*, **146**, 313–15.
GUDAS, L. J. & PARDEE, A. B. (1975). Model for regulation of *Escherichia coli* DNA repair functions. *Proceedings of the National Academy of Sciences, USA*, **72**, 2330–4.
GUSTAFSSON, P. & NORDSTRÖM, K. (1978). Temperature-dependent and amber copy mutants of plasmid R1 *drd-19* in *Escherichia coli. Plasmid*, **1**, 134–44.
HANSEN, P. G. & RASMUSSEN, K. V. (1977). Regulation of the *dnaA* product in *E. coli. Molecular and General Genetics*, **155**, 219–25.
HARTWELL, L. H. (1971). Genetic control of the cell division cycle in yeast. II. Genes controlling DNA replication and its initiation. *Journal of Molecular Biology*, **59**, 183–94.
HARTWELL, L. H. (1974). *Saccharomyces cerevisiae* cell cycle. *Bacteriological Reviews*, **38**, 164–98.
HARTWELL, L. H. (1978). Cell division from a genetic perspective. *Journal of Cell Biology*, **77**, 627–37.
HARTWELL, L. H., CULLOTTI, J., PRINGLE, J. R. & REID, B. J. (1974). Genetic control of the cell division cycle in yeast: a model. *Science*, **183**, 46–51.
HARTWELL, L. H., MORTIMER, R. K., CULOTTI, J. & CULOTTI, M. (1973). Genetic control of the cell division cycle in yeast. V. Genetic analysis of *cdc* mutants. *Genetics*, **74**, 267–86.

Hartwell, L. H. & Unger, M. W. (1977). Unequal division in *Saccharomyces cerevisiae* and its implications for the control of cell division. *Journal of Cell Biology*, **75**, 422–35.

Hereford, L. M. & Hartwell, L. H. (1974). Sequential gene function in the initiation of *Saccharomyces cerevisiae* DNA synthesis. *Journal of Molecular Biology*, **84**, 445–61.

Hicks, J., Strathern, J. N. & Klar, J. S. (1979). Transposable mating type genes in *Saccharomyces cerevisiae. Nature, London*, **282**, 478–83.

Hirota, Y., Jacob, F., Ryter, A., Buttin, G. & Nakai, T. (1968). On the process of cellular division in *Escherichia coli. Journal of Molecular Biology*, **35**, 175–92.

Hirschenberg, J. & Simchen, G. (1977). Commitment to the mitotic cell cycle in yeast in relation to meiosis. *Experimental Cell Research*, **105**, 245–52.

Howell, S. H. & Naliboff, J. A. (1973). Conditional mutants in *Chlamydomonas reinhardi* blocked in the vegetable cell cycle. I. An analysis of cell cycle block points. *Journal of Cell Biology*, **57**, 760–72.

Jagadish, M. N. & Carter, B. L. A. (1977). Genetic control of cell division in yeast cultured at different growth rates. *Nature, London*, **269**, 145–7.

Jarvick, J. & Botstein, D. (1973). A genetic method for determining the order of events in a biological pathway. *Proceedings of the National Academy of Sciences, USA*, **70**, 2046–50.

Johnston, G. E., Ehrhardt, C. W., Lorincz, A. & Carter, B. L. A. (1979). Regulation of cell size in the yeast *Saccharomyces cerevisiae. Journal of Bacteriology*, **137**, 1–5.

Johnston, G. C., Pringle, J. R. & Hartwell, L. H. (1979). Coordination of growth with cell division in the yeast *Saccharomyces cerevisiae. Experimental Cell Research*, **105**, 79–98.

Johnston, L. & Nasmyth, K. (1978). *Saccharomyces cerevisiae* cell cycle mutant *cdc* 9 is defective in DNA ligase. *Nature, London*, **274**, 891–3.

Kahn, M. & Helinski, D. R. (1978). Construction of a novel plasmid-phage hybrid: use of the hybrid to demonstrate Col E1 DNA replication *in vivo* in the absence of a Col E1-specified protein. *Proceedings of the National Academy of Sciences, USA*, **75**, 2200–4.

Kellenberger-Gujer, G. & Podhajska, A. J. (1978). Interactions between the plasmid λdv and *Escherichia coli dnaA* mutants. *Molecular and General Genetics*, **162**, 17–22.

Kellenberger-Gujer, G., Podhajska, A, J. & Caro, L. (1978). A cold sensitive *dnaA* mutant of *E. coli* which overinitiates chromosome replication at low temperature. *Molecular and General Genetics*, **162**, 9–16.

Kogoma, T. (1978). A novel *Escherichia coli* mutant capable of DNA replication in the absence of protein synthesis. *Journal of Molecular Biology*, **121**, 55–69.

Kubitschek, H. E. & Newman, C. N. (1978). Chromosome replication during the division cycle in slowly growing steady state cultures of three *Escherichia coli* B/r strains. *Journal of Bacteriology*, **136**, 179–90.

Lark, K. G. (1972). Evidence for the direct involvement of RNA in the initiation of DNA replication in *Escherichia coli* $15T^{-}$. *Journal of Molecular Biology*, **64**, 47–60.

Lark, K. G. (1979). Some aspects of the regulation of DNA replication in *Escherichia coli.* In *Biological Regulation and Development*, vol. 1, ed. R. F. Goldberger, *Gene Expression*, pp. 201–17. Plenum Press, New York.

Lorencz, A. & Carter, B. L. A. (1979). Control of cell size at bud initiation in *Saccharomyces cerevisiae. Journal of General Microbiology*, **113**, 287–95.

Lutkenhaus, J. F. & Donachie, W. D. (1979). Identification of the *ftsA* gene product. *Journal of Bacteriology*, **137**, 1074–88.

McGuinness, T. & Wake, R. (1979). Division septation in the absence of chromosome termination in *Bacillus subtilis*. *Journal of Molecular Biology*, **134**, 251–64.

Morris, N. R. (1980). Chromosome structure and the molecular biology of mitosis in eukaryotic micro-organisms. In *The Eukaryotic Microbial Cell*, ed. G. W. Gooday, D. Lloyd & A. P. J. Trinci, *Symposium of the Society for General Microbiology 30*, pp. 41–76. Cambridge University Press.

Minet, M., Nurse, P., Thuriaux, P. & Mitchison, J. M. (1979). Uncontrolled septation in a cell division cycle mutant of the fission yeast *Schizosaccharomyces pombe. Journal of Bacteriology*, **137**, 440–6.

Nasmyth, K. (1977). Temperature-sensitive lethal mutants in the structural gene for DNA ligase in the yeast *Schizosaccharomyces pombe. Cell*, **12**, 1109–20.

Nasmyth, K. (1979). A control acting over the initiation of DNA replication in the yeast *Schizosaccharomyces pombe. Journal of Cell Science*, **36**, 155–68.

Nasmyth, K., Nurse, P. & Fraser, R. (1979). The effect of cell mass on the cell cycle timing and duration of S-phase in fission yeast. *Journal of Cell Science*, **39**, 215–33.

Nurse, P. (1975). Genetic control of cell size at cell division in yeast. *Nature, London*, **256**, 547–55.

Nurse, P. (1981). Genetic analysis of cell cycle control – a reappraisal of 'start'. In *The Fungal Nucleus*, ed. K. Gull. Cambridge University Press. (In press.)

Nurse, P. & Fantes, P. (1981). Cell cycle controls in fission yeast – a genetic analysis. In *The Cell Cycle*, ed. P. C. L. John. Cambridge University Press. (In press.)

Nurse, P. & Thuriaux, P. (1977). Controls over the timing of DNA replication during the cell cycle of fission yeast. *Experimental Cell Research,* **107,** 365–75.

Nurse, P., Thuriaux, P. & Nasmyth, K. (1976). Genetic control of the cell division cycle in the fission yeast *Schizosaccharomyces pombe. Molecular and General Genetics*, **146**, 167–78.

Oishi, M. & Smith, C. L. (1978). Inactivation of phage repressor in a permeable cell system: role of *recBC* DNase in induction. *Proceedings of the National Academy of Sciences, USA*, **75**, 3569–73.

Orr, E., Fairweather, N. F., Holland, B. & Pritchard, R. H. (1979). Isolation and characterisation of a strain carrying a conditional lethal mutation in the *Cou* gene of *Escherichia coli* K12. *Molecular and General Genetics*, **117**, 103–12.

Orr, E. & Rosenberger, R. F. (1976). Determination of the execution points of mutations in the nuclear replication cycle of *Aspergillus nidulans. Journal of Bacteriology*, **126**, 903–6.

Peterson, E. L. & Berger, J. D. (1976). Mutational blockage of DNA synthesis in *Paramecium tetraaurelia. Canadian Journal of Zoology*, **54**, 2089–97.

Pinon, R. & Pratt, D. (1979). Folded chromosomes of mating-factor arrested yeast cells: comparison with Go arrest. *Chromosoma,* **73,** 117–29.

Pritchard, R. H. (1978). Control of DNA replication in bacteria. In *DNA Synthesis Present and Future*, ed. I. Molineux & M. Kohiyama, pp. 1–26. New York and London: Plenum Press.

Reid, B. J. & Hartwell, L. H. (1977). Regulation of mating in the cell cycle of *S. cerevisiae. Journal of Cell Biology*, **75**, 355–65.

Satta, G. & Pardee, A. B. (1978). Inhibition of *Escherichia coli* division by protein X. *Journal of Bacteriology*, **133**, 1492–500.

Schuster, H., Schlicht, M., Lanka, E., Mikolajczyk, M. & Edelbluth, C. (1977). DNA synthesis in an *Escherichia coli dnaB dnaC* mutant. *Molecular and General Genetics*, **151**, 11–16.

SEVASTOPOULOS, C. G., WEHR, C. T. & GLASER, D. A. (1977). Large-scale automated isolation of *Escherichia coli* mutants with thermo-sensitive DNA replication. *Proceedings of the National Academy of Sciences, USA*, **74**, 3485–9.

SHEPARD, H. M., GELFAND, D. H. & POLISKY, B. (1979). Analysis of a recessive plasmid copy number mutant: evidence for negative control of Col E1 replication. *Cell*, **18**, 267–75.

SIMCHEN, G. (1978). Cell cycle mutants. *Annual Review of Genetics*, **12**, 161–91.

SLATER, M. & SCHAECHTER, M. (1974). Control of cell division in bacteria. *Bacteriological Reviews*, **38**, 199–221.

TEATHER, R. M., COLLINS, J. F. & DONACHIE, W. D. (1974). Quantal behaviour of a division factor involved in the initiation of cell division at potential division sites in *E. coli*. *Journal of Bacteriology*, **118**, 407–13.

THURIAUX, P., NURSE, P. & CARTER, B. (1978). Mutants altered in the control co-ordinating cell division with cell growth in the fission yeast *Schizosaccharomyces pombe*. *Molecular and General Genetics*, **161**, 215–20.

UHLIN, B. E., GUSTAFSSON, P., MOLIN, S. & NORDSTRÖM, K. (1978). Copy number control mutants of the R plasmid R1 in *Escherichia coli*. In *DNA Synthesis Present and Future*, ed. I. Molineaux & M. Kohiyama, pp. 773–86. New York and London: Plenum Press.

UNGER, M. W. & HARTWELL, L. H. (1976). Control of cell division in *Saccharomyces cerevisiae* by methionyl-tRNA. *Proceedings of the National Academy of Sciences, USA*, **73**, 1664–8.

UNRAU, P. & HOLLIDAY, R. (1970). A search for temperature-sensitive mutants of *Ustilago maydis* blocked in DNA synthesis. *Genetical Research*, **15**, 157–69.

VOLKERT, M. R., SPENCER, D. F. & CLARKE, A. J. (1979). Indirect and intragenic suppression of the *lexA102* mutation in *E. coli* B/r. *Molecular and General Genetics*, **177**, 129–37.

WECHSLER, J. A. (1978). The genetics of *E. coli* DNA replication. In *DNA Synthesis Present and Future*, ed. I. Molineux & M. Kohiyama, pp. 49–70. New York and London: Plenum Press.

WEINER, J. H., MCMACKEN, R. & KORNBERG, A. (1976). Isolation of an intermediate which precedes *dnaG* RNA polymerase participation in enzymatic replication of bacteriophage φX174 DNA. *Proceedings of the National Academy of Sciences, USA*, **73**, 752–6.

WICKNER, S. & HURWITZ, J. (1975). Interaction of *Escherichia coli dnaB* and *dnaC* (*D*) gene products *in vitro*. *Proceedings of the National Academy of Sciences, USA*, **72**, 921–5.

WITKIN, E. M. (1976). Ultraviolet mutagenesis and inducible DNA repair in *Escherichia coli*. *Bacteriological Reviews*, **40**, 869–907.

ZAHN, G. & MESSER, W. (1979). Control of the initiation of DNA replication in *Escherichia coli*. 2. Function of the *dnaA* product. *Molecular and General Genetics*, **168**, 197–209.

GENETIC ANALYSIS OF THE DIFFERENTIATING BACTERIUM: *CAULOBACTER CRESCENTUS*

LUCILLE SHAPIRO, PERRY NISEN AND BERT ELY*

Albert Einstein College of Medicine, Bronx, New York, USA
**University of South Carolina, Columbia, South Carolina, USA*

INTRODUCTION

Morphogenesis in unicellular prokaryotes, whether a normal function of the cell cycle or induced by external conditions, reflects regulation of the timing of gene expression and the localization of gene products within the cell structure, as well as control at the level of gene organization. The bacterium *Caulobacter crescentus* is an excellent system for studying these types of cell regulation because its cell cycle is characterized by a series of spacially-localized and temporally-defined morphological and biochemical transitions (Fig. 1). Specifically: (a) defined sets of structural gene products are synthesized during limited time periods in the cell cycle, independent of environmental fluctuations, (b) some of these gene products are sent through the cell membrane where they are assembled into structures such as flagella and stalks, and (c) the special orientation of such structures on the cell surface is strictly controlled. Many additional metabolic functions, such as DNA replication (Degnan & Newton, 1972), phospholipid synthesis (Mansour, Henry & Shapiro, 1980) and the synthesis of both cytoplasmic and membrane proteins (Shapiro & Maizel, 1973; Cheung & Newton, 1977; Iba, Fukuda & Okada, 1978; Lagenaur & Agabian, 1978; Agabian, Evinger & Parker, 1979; Evinger & Agabian, 1979) are similarly differentially expressed during the cell cycle.

It has become increasingly apparent that there are no 'simple' model differentiation systems and that understanding the regulatory network, at a molecular level, of any system which displays selective gene expression is a difficult task. Furthermore, answers to questions of how genetic information is somehow translated into spacial or positional information will take innovative approaches. The use of classical and molecular genetics, together with biochemical

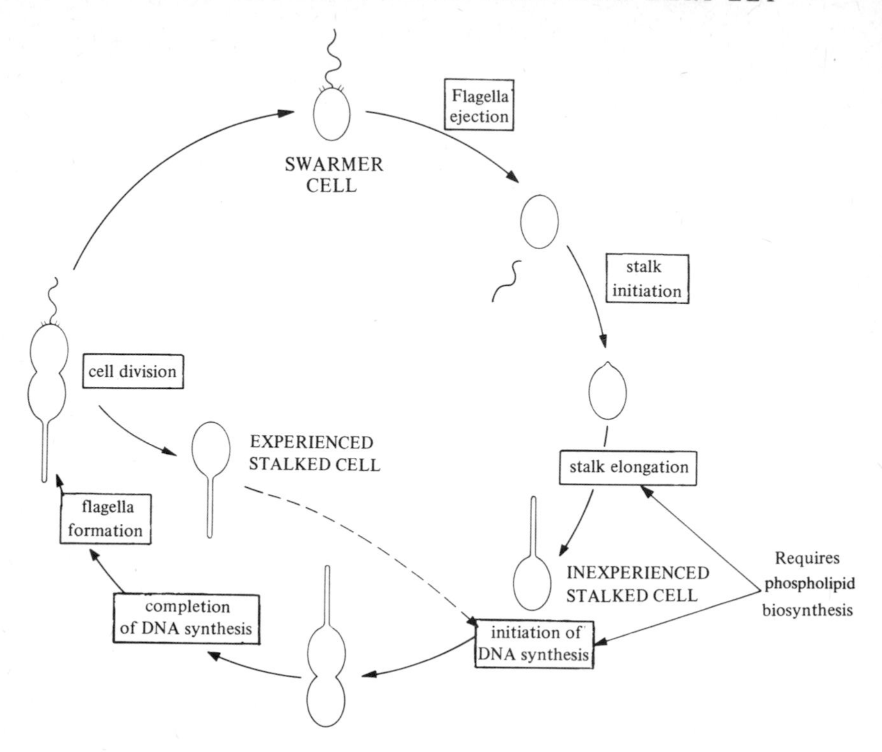

Fig. 1. Diagram of the *Caulobacter crescentus* cell cycle.

analysis, are essential if these regulatory phenomena are to be understood. The discovery of transducing phage and conjugative plasmids capable of mediating genetic exchange in *Caulobacter* coupled with the availability of recombinant DNA technology presented us with the capability to decipher at least some modes of differential gene expression.

CAULOBACTER CRESCENTUS LIFE CYCLE

Two different *C. crescentus* cell types, swarmer cells and stalked cells, share a common origin but follow distinct developmental programmes. Cell polarity is maintained by the presence of organelles at specific sites on the cell surface. An elongated stalked cell divides asymmetrically to produce two different cell types, a non-motile stalked cell and a motile swarmer cell with a single flagellum, pili and phage receptor sites at one pole (Fig. 1). As the cell cycle proceeds the stalked cell develops a flagellum, pili and phage receptor sites at the pole opposite the stalk. Asymmetric cell division again yields a new swarmer cell and the original stalked cell. Thus, the stalked cell acts like a stem cell generating new swarmer cells. The swarmer cells follow a different developmental sequence: after a constant fraction of the cell cycle, the swarmer cell loses its flagellum, pili and phage receptor sites and synthesizes a stalk at the same pole. This differentiation produces a mature stalked cell which is capable of cell division and carries out the same developmental programme as the stalked cell produced by cell division.

Concurrent with the distinct morphogenetic pattern observed during each cell cycle, synthesis of DNA, RNA and specific classes of proteins has been demonstrated. Either upon chromosome completion or at the time of cell division an event occurs that delays chromosome replication in the progeny swarmer, but not in the progeny stalked cell (Degnan & Newton, 1972; Iba, Fukuda & Okada, 1978). It has been demonstrated that *de novo* RNA synthesis is required at defined periods in the cell cycle for motility, stalk formation, cell division and DNA synthesis to occur (Newton, 1972). Differential expression of certain proteins during the cell cycle, such as flagellin (Shapiro & Maizel, 1973; Lagenaur & Agabian, 1978; Osley, Sheffery & Newton, 1977), cytoplasmic proteins (Agabian *et al.*, 1979), and membrane-associated proteins (Evinger & Agabian, 1977, 1979; Agabian *et al.*, 1979), have been observed. A number of proteins are found to be synthesized during specific stages of the cell cycle, although most are synthesized continuously (Cheung & Newton, 1977; Agabian *et al.*, 1979). Pulse chase experiments have shown that some newly synthesized proteins are found exclusively in stalked cells immediately after cell division, while others are found exclusively in swarmer cells (Agabian *et al.*, 1979). When *Caulobacter* spheroplasts are fractionated to separate outer membranes, inner membranes and soluble constituents (Agabian *et al.*, 1979; Agabian & Unger, 1978), newly synthesized proteins are found in each of the fractions (Evinger & Agabian, 1979). Furthermore, pulse chase experiments showed that some of these proteins move from one fraction to another (Evinger & Agabian, 1979). Taken together, these results indicate that differentiation in *Caulobacter* involves both differential gene expression and the maintenance and creation of unique cellular regions where morphogenesis can occur.

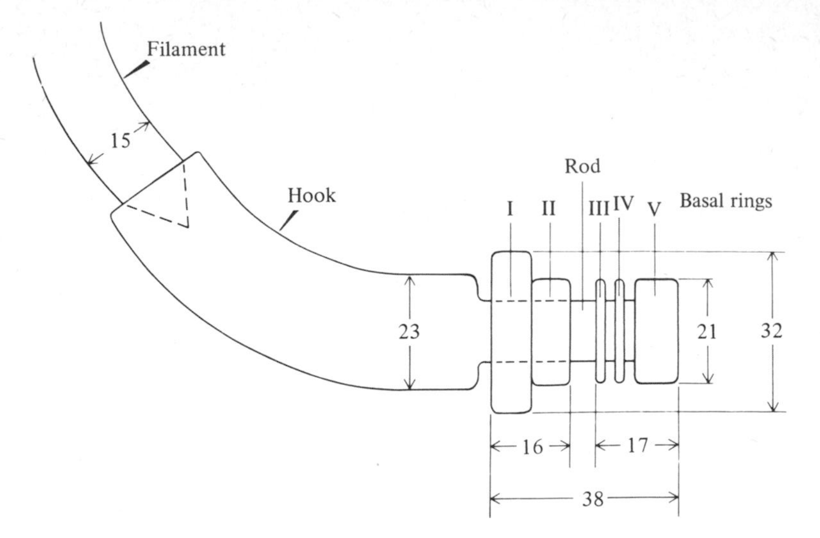

Fig. 2. Diagram of *C. crescentus* flagella basal complex (dimensions in nanometres) (Johnson *et al.*, 1979).

THE FLAGELLUM: A DIFFERENTIALLY EXPRESSED SURFACE STRUCTURE

Caulobacter crescentus contains a single polar flagellum for approximately one-third of the cell cycle. The formation, maintenance and release of this flagellum is co-ordinated with the cell cycle. The major 25 K filament protein is synthesized at the time of flagellar assembly (Shapiro & Maizel, 1973; Lagenaur & Agabian, 1978; Osley *et al.*, 1977). Furthermore, it has been shown that flagellin gene expression is coupled to the completion of chromosome replication (Osley *et al.*, 1977). During the transition of the motile swarmer cell to the sessile stalked cell the flagellum is released into the medium and a stalked structure is then assembled at the site previously occupied by the flagellum. Flagella purified from the culture medium have both a hook and a rod structure at one end of the filament, but do not have the basal ring complex found in intact flagella isolated from both Gram-positive and Gram-negative bacteria (Shapiro & Maizel, 1973). The basal complex of the *C. crescentus* flagellum has been defined by analysis of intact structures artificially released from cell membrane preparations of wild-type cells and of incomplete structures obtained from membranes of non-flagellate mutants (Johnson, Walsh, Ely & Shapiro, 1979). The basal complex consists of five rings mounted on a rod (Fig. 2). Two rings are in the hook-proximal upper set and three rings (two narrow and one wide) are in the lower set.

Flagella components

Although the major component of the flagella filament is a 25 K protein (Fla A), a 27 K protein (Fla B) is also released from the cell at the time when the flagella is shed into the medium and the two proteins copurify (Lagenaur & Agabian, 1976, 1978; Marino, Ammer & Shapiro, 1976; Osley *et al.*, 1977; Fukuda, Koyasu & Okada, 1978). There is evidence that Fla B is a minor component of the flagella filament (Lagenaur & Agabian, 1978; Fukuda *et al.*, 1978). The two flagellins have been separated by DEAE chromatography (Fukuda *et al.*, 1978) and purified to homogeneity using an affinity column containing anti-flagellin antibody (Weissborn & Shapiro, unpublished). The two purified flagellins have been studied by one- and two-dimensional electrophoresis, amino acid analysis (Lagenaur & Agabian, 1976, 1978; Fukuda *et al.*, 1978; Osley *et al.*, 1977; Weissborn & Shapiro, unpublished), tryptic peptide map analysis and immunoprecipitation. Amino acid analysis revealed that the two flagellins have nearly identical compositions. Peptide maps showed that the two fingerprints are very similar, except that there are more peptides in Fla B and a few peptides with differing mobility in Fla A. These results coupled with those obtained by two-dimensional electrophoresis suggest that modification of specific amino acids may have occurred. Immunoelectrophoresis shows complete cross-reaction between the two flagellins. There is no cross-reaction between antibody raised against the hook protein and either of the flagellins. When whole cell extracts are treated with antibody to Fla A, several proteins are precipitated, including Fla A, Fla B, hook protein and a 29 000 MW protein. Immunoprecipitation analysis of several motility-minus mutants of *Caulobacter* (Johnson & Ely, 1979) have shown that the 29 000 MW protein cross-reacts with the flagellins. The isolated hook structure was purified and was found to be composed of 70 000 MW protein component (Lagenaur, De Martini & Agabian, 1978; Johnson *et al.*, 1979; Sheffery & Newton, 1979). When anti-hook antibody is used to precipitate proteins from a wild-type cell extract, in addition to the 70 000 MW hook protein, Fla A and Fla B, but not the 29 000 MW protein, are precipitated (Contreras *et al.*, 1979), whereas in an assembly-minus mutant only hook protein is precipitated. It therefore appears that either hook or flagellin antibody can precipitate the *assembled* structure as well as the specific monomer against which the antibody is directed. The 29 000 MW protein is therefore not assembled. Since it is precipitated by the anti-Fla antibody, however, it may be a precursor to the flagellins.

Genetic analysis of motility mutants

Transductional analysis of 69 non-motile mutants has resulted in the identification of 30 genes involved in motility (Johnson & Ely, 1979). Since 15 of these genes were identified by a single mutant, it is likely that many more genes remain to be detected. Mutants defective in 27 of the 30 genes did not make a flagellum (*fla*$^-$) and synthesized flagellin monomers at a reduced rate. Mutants representing four of these genes were able to assemble a short stub of a flagellar filament. Mutants defective in any of the remaining three genes were non-motile (*mot*$^-$) even though they synthesized a flagellum with a normal morphology. Genetic analysis by φCr30-mediated transduction revealed 27 linkage groups for the *fla* and stub-forming mutations, and three linkage groups for the *mot* mutations. Mutants representing nine of the 27 *fla*$^-$ genes lacked the flagellar hook as well as the flagellar filament (Johnson *et al.*, 1979). An additional mutant produced straight elongated hooks even though the hook protein was apparently normal. In another study, Sheffery & Newton (1979) identified a mutant which produced polyhooks which appeared to be a polymer of morphologically normal hook structures.

GENETIC ANALYSIS OF *CAULOBACTER CRESCENTUS*

A combination of genetic, biochemical and physiological studies are being used to analyse the regulation of *Caulobacter* cell cycle events. The advent of molecular genetics, which permits the isolation and manipulation of individual genes and nucleotide sequences, has made available new modes of genetic dissection not readily available in many bacterial systems. However, the power of recombinant DNA technology is limited without the use of classical genetic techniques. Consequently, methods have been developed for mutant isolation, strain manipulations, the complementation of mutants, and the construction of a genetic map.

Mutant isolation

A wide variety of *Caulobacter* mutants have been isolated. Auxotrophs were isolated initially by direct screening of survivors after irradiation with ultraviolet light (Jollick & Schervish, 1972) or nitrosoguanidine mutagenesis (Newton & Allebach, 1975). More recently, auxotrophs have been isolated from cultures which were not subjected to mutagenesis (Johnson & Ely, 1977). To compensate for the low frequency of mutations in unmutagenized cultures, Johnson and Ely used an enrichment procedure to increase the probability of finding mutants. After enrichment for non-growing cells with either fosfomycin or *D*-cycloserine (*C. crescentus* has penicillinase activity), spontaneously occurring auxotrophs were present at a frequency of approximately 5% (Johnson & Ely, 1977). Using these techniques, over 250 spontaneously occurring auxotrophs have been obtained from the *C. crescentus* wild-type CB15. These mutants have been characterized with regard to their nutritional requirements and a number of them have been subjected to reversion analysis to identify the kinds of mutations present. Reversion analyses of these mutants indicated the presence of frameshifts, base substitutions and deletions. In addition, unlike the mutants from earlier studies, all of the spontaneously derived mutants remained sensitive to the generalized transducing phage φCr30.

Groups of mutants with similar nutritional requirements have been subjected to more detailed analyses, including the fulfilment of nutritional requirements with biosynthetic intermediates, cross-feeding experiments and transductional mapping. From these data a picture of the genetics and physiology of *C. crescentus* is gradually emerging. For example, *Caulobacter* assimilates ammonia from the medium using a combination of the glutamate synthase and glutamine synthetase enzymes (Ely, Amarasinghe & Bender, 1978). Mutants in glutamate synthetase (*glt*$^-$) cause a requirement for glutamate and map at two locations on the *C. crescentus* chromosome. Data from experiments with various amino acid auxotrophs suggest that the amino acid biosynthetic pathways of *Caulobacter* are similar to those of *E. coli* but that the genes for the biosynthetic pathways are not clustered in operons as they are in the enteric bacteria.

In addition to auxotrophs, other types of mutants have been isolated by screening survivors of mutagenesis or spontaneous mutants for particular structural defects. Mutants affecting stalk elongation were identified microscopically after treatment with ethylmethane sulfonate (EMS) (Schmidt, 1968, 1969). Mutants with an additional division plane at or near the base of the stalk were obtained from both EMS-mutagenized and unmutagenized cultures after serial enrichment for bacteria with altered sedimentation properties (Poindexter, 1978). Treatment of *Caulobacter crescentus* with nitrosoguanidine or irradiation with ultraviolet light has been used to obtain temperature-sensitive DNA replication and division mutants (Osley & Newton, 1977), conditional non-motile mutants (Marino *et al.*, 1976; Sheffery & Newton, 1979) and pleiotropic surface structure mutants (Kurn, Ammer & Shapiro, 1974; Fukuda, Miyakawa, Iida and Okada, 1976). Spontaneously occurring motility mutants have been obtained by serial enrichment for non-motile cells (Johnson & Ely, 1979) and pleiotropic surface structure mutants have been obtained from unmutagenized cultures by selection for resistance to swarmer-specific bacteriophage (Lagenaur, Farmer & Agabian, 1977; Johnson & Ely, unpublished).

Bacteriophages

Bacteriophages capable of propagating on *Caulobacter crescentus* have been isolated in a number of laboratories (Schmidt & Stanier, 1966; Driggers & Schmidt, 1970; Bendis & Shapiro, 1970; Agabian-Keshishian & Shapiro, 1970; West, Lagenaur & Agabian, 1976; Fukuda *et al.*, 1976; Miyakawa, Fukuda & Okada, 1976; Johnson, Wood & Ely, 1977). All of the RNA phages that have been isolated resemble the *E. coli* RNA phages (Shapiro & Bendis, 1975) and attach to the vegetative pili found at the flagellar pole of *C. crescentus* swarmer cells (Schmidt, 1966; Bendis & Shapiro, 1970; Miyakawa *et al.*, 1976; Johnson *et al.*, 1977). Most of the *Caulobacter* DNA phages are also specific for the flagellar pole of swarmer cells (Schmidt & Stanier, 1966; Johnson *et al.*, 1977). In one study 28 of 36 independent isolates of DNA phages had a similar morphology and were swarmer-specific (Johnson *et al.*, 1977). The remaining eight phages were capable of attaching at random locations on all *Caulobacter* cell types. Another phage of this type was isolated by West *et al.* (1976) and appears to be similar to coliphage T7. Twenty of the 28 swarmer-specific DNA phages were capable of forming a stable lysogenic relationship with their host. These temperate bacteriophages are a potential source of specialized transducing phages.

Two bacteriophages, φCr30 and φCr35, have been isolated which are capable of mediating generalized transduction (Johnson *et al.*, 1977; Ely & Johnson, 1977). The two phages are quite similar in most respects and have similar but non-identical fragment patterns when examined by agarose gel electrophoresis after cleavage with restriction endonuclease *Hin*dIII (Ely & Johnson, 1977). One of these phages, φCr30, was shown to have the ability to transfer every marker tested at frequencies ranging from 10^{-4} to 10^{-7} transductants per plaque-forming unit depending on the recipient marker chosen (Ely & Johnson, 1977). Transductional analyses of *C. crescentus* mutants with φCr30 have resulted in the identification of nearly 100 genes (Johnson & Ely, 1977, 1979). The φCr30 chromosome is quite large and has a molecular weight of approximately 130×10^6 (Schoenlein & Ely, unpublished). Thus, φCr30 may be capable of transferring up to 4% of the *C. crescentus* chromosome. Unlike most other transducing phages, φCr30 is virulent and does not form any sort of lysogenic relationship with its host (Ely & Johnson, 1977).

Plasmids

Plasmids have not been detected in wild-type strains of *Caulobacter crescentus* (Wood, Rake & Shapiro, 1976; Schoenlein & Ely, unpublished; Nisen, unpublished). However, recent studies have demonstrated the presence of large plasmids in wild-type strains of several other species of *Caulobacter*, including *C. leidyi*, *C. fusiformis* and *C. henricii* (Schoenlein & Ely, unpublished). Attempts to transfer these plasmids to *C. crescentus* have been unsuccessful. In contrast, P-type and N-type drug resistance factors can be transferred to *C. crescentus* from *E. coli* and *P. aeruginosa* (Alexander & Jollick, 1977; Ely, 1979). These plasmids are stably maintained and can be transferred serially. The antibiotic resistance genes present on the plasmids are expressed in *Caulobacter* and provide the basis for selection of plasmid transfer (Table 1). Other plasmids with a more limited host range, such as F and R100, are not transferred to *C. crescentus*. Transfer frequencies for the P-type plasmids approach 100% under ideal conditions in crosses using *E. coli* donors. Transfer frequencies are 3- to 10-fold lower when *C. crescentus* strains are used as donors (Ely, 1979).

Table 1. *Transfer of plasmids from* Escherichia coli *to* Caulobacter crescentus

Plasmid	Plasmid type	Transfer	Properties of transferred plasmids		
			Kan^R	Tet^R	Chromosome mobilization
RP4	P	+	+	+	+
RK2	P	+	+	+	+
R68.45	P	+	+	+	+
R702	P	+	+	n.d.[b]	n.d.
R46	N	+	n.a.[a]	+	+
R100	FII	−	n.a.	n.a.	n.a.
F	FI	−	n.a.	n.a.	n.a.

[a] n.a. = not applicable.
[b] n.d. = not determined.

C. crescentus cells containing the P-type plasmid RP4 synthesize peritrichous pili which are shorter and thicker than the vegetative pili normally found at the *Caulobacter* cell pole. The RP4-coded pili are present throughout the cell cycle, appear identical to those produced by RP4-infected *E. coli*, function in plasmid transfer and serve as attachment sites for RP4-specific phages (Table 2). Thus, the factors causing a polar location and temporal expression of the vegetative pili do not affect the synthesis and location of the sex pili. RP4-specific phages PRD1, PRR1 and φGU5 cause lysis of *C. crescentus* strains harbouring RP4 even though they are incapable of propagation on these hosts. Consequently, they can be used to isolate plasmid-free clones from strains harbouring RP4.

RP4 and several other P-type drug resistance factors have been shown to mobilize the *C. crescentus* chromosome at frequencies of 10^{-6} to 10^{-8} in crosses between *C. crescentus* auxotrophs (Ely, 1979). Similar results have been obtained in a variety of organisms

Table 2. *Expression of RP4 genes*

Host	Amp^R	Kan^R	Tet^R	Phage sensitivity[a]	Phage production[b]	RP4 pili
E. coli	+	+	+	+	+	+
C. crescentus	n.d.[c]	+	+	+	−	+

[a] Sensitivity to RP4-specific phages PRD1, PRB1, φGU5 in complex media.
[b] Phage production after infection with PRD1, PRR1, φGU5 in complex media.
[c] n.d. = not determined, since wild-type *C. crescentus* is Amp^R.

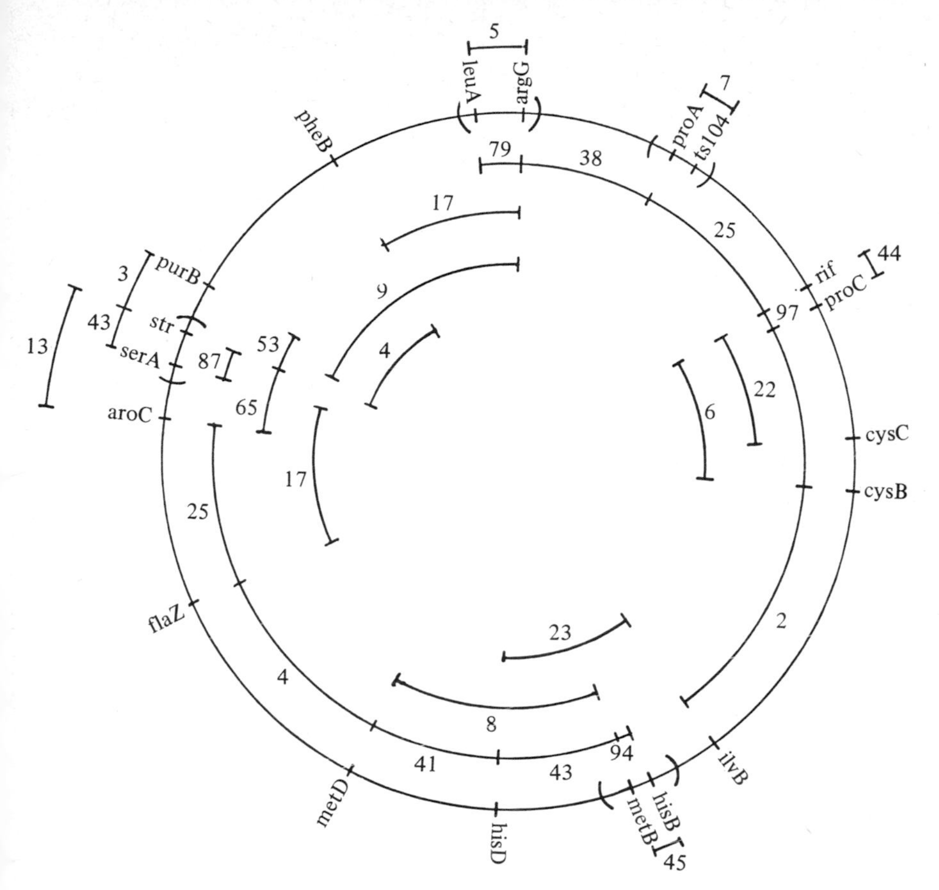

Fig. 3. Genetic map of *Caulobacter crescentus*. The numbers in the interior of the circle represent percentage cotransfer of the designated markers by conjugation, while the numbers lying outside the circle represent percentage cotransfer of the designated markers by transduction.

(Holloway, 1979). Chromosomal transfer also occurs between *C. crescentus* and *E. coli* (Ely, 1979). Further studies with RP4 suggest that chromosome mobilization occurs from random points of origin and that donor fragments can replace nearly 20% of the recipient chromosome (Barrett *et al.*, 1980). Three- four- and five-factor crosses mediated by RP4 have been combined with transductional analyses to provide data for construction of a preliminary genetic map for *C. crescentus* (Fig. 3; Barrett *et al.*, 1980). In addition to chromosome mobilization, chromosomal fragments on plasmids can be exchanged between *E. coli* and *C. crescentus* to provide a system for interspecies complementation (Ely, 1979). RP4 plasmids containing either the *E. coli trp* or the *K. aerogenes his* operons can be transferred to *C. crescentus* and will complement tryptophan- and histidine-requiring auxotrophs, respectively (Barrett & Ely, unpublished). Also, RP4 can be used to mobilize colE1 plasmids containing fragments of the *E. coli* genome. These plasmids are transferred to *C. crescentus* at a frequency of 10^{-6} (Ely, 1979) and will complement the appropriate *C. crescentus* auxotrophs (Ely, 1978). The colE1 plasmids are stably maintained and can be serially transferred.

Transposons such as Tn7 and Tn5 can be transferred to *Caulobacter* using plasmid vectors (Ely, 1979). After transfer they express their drug resistance genes and can be observed to transpose at frequencies similar to those observed in *E. coli*. Recent studies indicate that Tn5 is an effective mutagen when introduced to *Caulobacter* on a suicide plasmid which contains both Tn5 and a defective Mu phage (Beringer, Hoggan & Johnson, 1978). The presence of Mu prevents the plasmid from becoming established when transferred to a new host (van Vliet, Silva, van Montagu & Schell, 1978). As a consequence, the frequency of transposition is approximately 100 times greater than the frequency of establishment of the plasmid. In *Caulobacter*, transposition of this plasmid occurs at a frequency of 10^{-6} and a variety of mutants have been detected among the survivors. Auxotrophs and motility mutants occur at a frequency of 2% and 1%, respectively.

MUTANT ANALYSIS AND PERTURBATION OF THE CELL CYCLE

To understand the relationship among major cell cycle events and the specific expression of a small subset of genes coding for surface structure proteins, various biochemical parameters of cell growth and differentiation have been established and mutants in known biochemical pathways have been selected. The goal of this type of mutant analysis is to perturb specific cellular functions, such as DNA replication or membrane biosynthesis, and to then observe the effects of a specific mutation on the progression of the cell cycle. Most studies of this type have been complemented by an analysis of the effects of specific inhibitors and antibiotics which mimic the defined mutations.

Cell cycle mutants

Temperature-sensitive cell cycle mutants have been isolated after ultraviolet irradiation of *C. crescentus* (Osley & Newton, 1977). These mutants were found to cause blocks in the initiation of DNA replication, DNA chain elongation, the initiation of cell division, or cell separation. Temperature-shift experiments were used to define the nature of the lesion and to perform an order of function analysis (Osley & Newton, 1978, 1980). Nine mutants were subjected to these analyses and the time of action of the temperature-sensitive product was determined. These results, along with a determination of the effects of penicillin and hydroxyurea, have resulted in the identification of five discrete steps in the DNA replication and the cell division pathways. As a result of these studies Osley & Newton (1980) have proposed that the two pathways are independent but that initiation of the cell division pathway requires some event during DNA chain elongation. In addition, cell separation requires the completion of DNA replication. In another study, flagellin synthesis was shown to be inhibited in DNA chain elongation mutants at the restrictive temperature (Osley *et al.*, 1977). Therefore, the completion of DNA replication may be required to initiate flagellum formation.

Mutants in membrane biosynthesis

The acid phospholipids, phosphatidylglycerol and cardiolipin, comprise approximately 87% of the total phospholipid of *Caulobacter crescentus*. Neither phosphatidylethanolamine nor its precursor, phosphatidylserine, are present at detectable levels (Contreras *et al.*, 1978). Since glycerol was shown to be an essential precursor to all *C. crescentus* phospholipids, mutants dependent on exogenous glycerol for growth were sought both to analyse phospholipid biosynthesis and to study the effect of the cessation of phospholipid biosynthesis on the progression of the cell cycle.

A glycerol auxotroph has been isolated in a strain of *C. crescentus* unable to use glycerol as a carbon source. The defect is due to a single mutation in *L*-glycerol-3-phosphate, NAD(P) oxidoreductase, commonly referred to as glycerol-3-phosphate dehydrogenase (Contreras *et al.*, 1979). Growth of this mutant in the presence of glucose as a carbon source, but in the absence of added glycerol, caused the immediate cessation of net phospholipid biosynthesis.

The rate of DNA synthesis decreased concomitantly with the sharply decreased rate of phospholipid synthesis. In fact, it was determined that the cessation of net phospholipid biosynthesis caused the termination of DNA elongation. Both RNA and protein synthesis, however, continued for just under a generation when the rate of synthesis decreased concomitantly with cell death. Using immunoprecipitin assays, with antibody directed against either cell cycle proteins (i.e., hook, Fla A or Fla B) or 'housekeeping proteins' such as RNA polymerase and outer and inner membrane proteins, it was shown that a specific group of surface structure proteins, including flagellin (the expression of which is normally regulated within the cell cycle) stopped early after the cessation of phospholipid synthesis, whereas the synthesis of other specific intracellular and membrane proteins continued for at least a generation (Contreras *et al.*, 1980). Parallel results were obtained when phospholipid synthesis was inhibited with the antibiotic cerulenin.

These experiments suggest a tight coupling between membrane biogenesis, DNA replication and the expression of specific cell cycle-regulated proteins. Furthermore, these observations corroborate those of Osley *et al.* (1977), who have shown that temperature-sensitive DNA elongation mutants of *Caulobacter* are unable to synthesize flagellin at the restrictive temperature. It appears as if a given gene sequence must be replicated just prior to the synthesis of the cell cycle-mediated gene product. It will be of major interest to determine how this coupling is related to site-specific biogenesis of the polar organelles in *Caulobacter*. Perhaps the replicating chromosome, in contact with the membrane at the cell pole, signals the start of flagellin synthesis at the time of flagellin gene replication, and the localized product is then available at the correct site on the membrane to be assembled into a flagellar structure. The key signal on the replicating chromosome could involve a localized rearrangement of nucleotide sequences, much like the inversion shown to occur at the promoter site of the *Salmonella* flagellin H2 gene (Silverman, Zieg, Hilman and Simon, 1979).

MOLECULAR GENETIC APPROACHES AND THE ORGANIZATION OF THE GENOME

Recombinant DNA methodology in concert with the classical genetic approaches described above provides the means to study the molecular mechanisms which regulate the temporally and spacially restricted expression of specific subsets of genes during the *Caulobacter* cell cycle.

Identification and isolation of specific genetic sequences

Restriction endonuclease-generated fragments of *C. crescentus* chromosomal DNA have been cloned into several bacteriophage λ and plasmid vectors. Banks of cloned DNA fragments from separated populations of stalked and swarmer cells have also been prepared (Shapiro & Skalka, unpublished). One goal of using recombinant DNA technology in this system is to isolate differentially transcribed genes which code for surface structure proteins. However, genes within recombinant DNA clones are most easily identified by assay for specific functions. The requirement for functional expression in the case of *Caulobacter* genes being expressed in an *E. coli* host can limit identification of clones encoding proteins which must be processed or assembled in special ways, such as foreign flagellin or pilin. A direct method has been devised for detecting cloned gene translation which is independent of the expression of protein functions. This method is based on specific antigen–antibody complex formation within a vector phage plaque or surrounding a vector-containing bacterial colony (Skalka & Shapiro, 1976; Anderson, Shapiro & Skalka, 1979). In the former case the antigen–antibody precipitin is easily identified within a 'cloudy' plaque and in the latter case the antibody complexes with antigen released from the plasmid-bearing colony by thermo-induction of a λ prophage carried by the host or by treatment with lysozyme and detergent. In each case the relevant antibody is simply included in the plate agar. The techniques proved to be simple and an extremely sensitive means of measuring specific protein synthesis directed by a gene in a reconstructed chromosome. Several clones of *Caulobacter* DNA in a λWES vector have been selected which produce antigens that cross-react with antibody directed against the entire flagella structure (Weissborn, Shaw & Shapiro, unpublished).

Inverted repeat sequences in the Caulobacter *genome*

It is becoming increasingly apparent that a specific configuration of nucleotides, e.g. inverted repeat sequences, play a significant role in the regulation of gene expression in both prokaryotes and eukary-

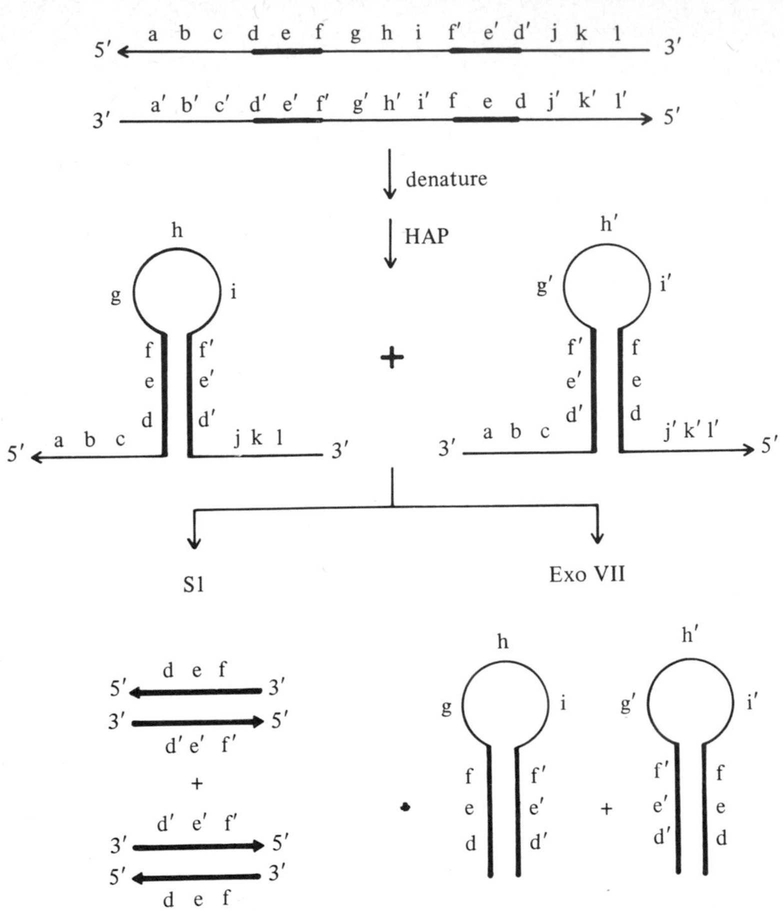

Fig. 4. Effects of various treatments on *Caulobacter* chromosomal DNA. Letters correspond to nucleotide bases. Letters with a prime (e.g. a′) correspond to the complement of the letter without the prime (e.g. a). The thickened lines depict an inverted repeat. When sheared chromosomal DNA (top line) is heat-denatured, quick-cooled and run through an hydroxylapatite (HAP) column, the hairpin loop structures with non-self-complementary tails (seen in the middle of the figure) are bound to and can be purified from the column. Treatment of these structures with S1 nuclease removes all single-stranded DNA and results in the double-stranded DNA seen at the bottom left. Treatment of the DNA which bound to the HAP column with exonuclease VII results in the hairpin structure seen at the bottom right of the figure.

otes. These sequences, which have their complement in inverted order nearby on the same strand (Fig. 4), can form intramolecular duplex (hairpin) structures. The inverted repeat sequences comprise 1 to 10% of the chromosomal DNA in all organisms studied. Furthermore, small inverted repeat sequences (less than 50 base pairs) have been identified at most regulatory regions of the genome and at the termini of translocatable DNA elements (Cohen, 1976; Kleckner, 1977). There are two cases in prokaryotes where inversion of DNA sequences serves as an on–off switch for gene expression: phase variation in *Salmonella* (Zieg, Silverman, Hilman & Simon, 1977; Silverman *et al.*, 1979), and bacteriophage Mu infectivity (Kamp *et al.*, 1978). In both cases it has now been demonstrated that the inverting region is bordered by inverted repeat sequences (Zieg & Simon, personal communication). From studies of the role played by inverted repeat sequences in the translocation of transposable elements and insertion sequence elements (Cohen, 1976; Kleckner, 1977) as well as their presence at the termini of inverting sequences known to control specific gene expression, one is led to speculate that the organization of a given DNA sequence within an inverted repeat may confer upon that sequence the capacity to rearrange. Since at least 3% of the *Caulobacter crescentus* genome is comprised of inverted repeat DNA sequences (Wood *et al.*, 1976; Nisen *et al.*, 1979) (a considerably higher proportion than most bacterial genomes studied), an analysis of the role of these sequences in the regulation of the *Caulobacter* cell cycle has been initiated (Nisen *et al.*, 1979). The complexity of the *C. crescentus* genome, $C_0t\frac{1}{2} = 17.8$, as determined by DNA:DNA solution hybridization, is in close agreement with the expected complexity of a non-repetitive DNA which has a molecular weight similar to that of the *E. coli* chromosome. A significant repetitive component (greater than 5%) was not observed.

The *Caulobacter crescentus* inverted repeat sequences occur in two size classes; one 100–600 base pairs long and the other 1 500–3 000 base pairs long. These sequences reassociate with concentration-independent first-order, zero-time kinetics. To analyse the complexity of this fraction, reassociation of the denatured S1-resistant DNA (which can no longer snap back because hairpin loops were cut by S1) was monitored: no significant repetitive component was observed. Therefore, the bulk of the inverted repeat DNA of *C. crescentus* appears to be comprised of unique DNA sequences (Nisen *et al.*, 1979; Nisen & Medford, unpublished).

Each of the different cell types of the *Caulobacter* cycle can be isolated from synchronous cell cultures (Shapiro, 1976). Chromosomal DNA was isolated from different cell types, digested with *Bam*H1 restriction endonuclease, and analysed by agarose gel electrophoresis. The pattern of restriction fragments appeared identical for each of the different DNA preparations. Stalked and swarmer cell-derived DNA similarly yielded indistinguishable patterns when cleaved with the restriction endonucleases *Eco*RI or *Hin*dIII and, as expected, these patterns differed from those obtained with *Bam*H1 (Nisen *et al.*, 1979).

To identify the location of inverted repeat DNA in the stalked and swarmer cell chromosome, whole cell populations and individual inverted repeat sequences which were cloned into bacteriophage λWES were hybridized to restriction fragments of the genomic DNA (Nisen *et al.*, 1979). Hybridization of unfractionated inverted repeat DNA to Southern blots of stalked and swarmer cell DNA treated with either *Bam*H1 or *Eco*RI showed that, in general, the inverted repeat DNA was homologous to a limited number of similar sites in the two genomes. However, some differences were consistently noted. In the high molecular weight region of the gel containing *Eco*RI digested DNA, regions of homology present in the swarmer genome were absent or displaced in the stalked genome. Hybridization of unfractionated inverted repeat (IR) DNA to *Bam*H1 digested genomic DNA revealed an additional band in swarmer cell-derived DNA which was not observed in stalked cell-derived DNA. To localize these differences ^{32}P-labeled λWES·IR DNA clones were individually hybridized to portions of the same Southern blot of *Bam*H1 digests of stalked and swarmer cell-derived DNA. It was found that with two clones tested, IR1 and IR2, both hybridized to the same four regions in stalked and swarmer cell-derived DNA; however, IR1 also hybridized to two additional bands in the swarmer cell-derived DNA. This experiment was repeated with different stalked and swarmer cell DNA preparations that had been treated with *Bam*H1; IR1 DNA consistently hybridized to the same two additional regions in the swarmer cell DNA. When ^{32}P-labelled λWES·IR1 DNA was hybridized to a Southern blot of *Eco*RI-generated fragments of stalked and swarmer cell DNA, IR1 DNA also hybridized to additional regions of the swarmer cell DNA (Nisen *et al.*, 1979).

The observation that IR1 and IR2 DNA hybridized to the same region of the *C. crescentus* chromosome, but are not cross-homologous, suggests that the inverted repeat DNA elements are clustered in the genome. This is consistent with the observations of Wood *et al.* (1976), who concluded from hydroxylapatite binding studies that the organization of inverted repeat DNA in the genome was non-random and presumably clustered.

The additional regions of homology between IR1 DNA and the swarmer cell chromosome may have resulted from a sequence rearrangement of a part of the IR1 DNA sequence in the genome. Although the nature of this rearrangement cannot be ascertained from these experiments, a likely possibility is that the IR-DNA sequence may have inverted or translocated. The observation that two new bands are present and none have disappeared may mean that the sequence duplicated before translocation, or that the segment that moved was originally located on a fragment to which another region of the IR1 DNA insert was still able to hybridize. Electron microscope studies and digestion of heated, quick-cooled DNA with exonuclease VII indicate that there are non-self-complementary DNA sequences between IR-DNA segments. It remains to be determined whether or not these sequences can rearrange as well. The fact that a different pattern of IR1 DNA hybridization is consistently obtained with genomic DNA isolated from two distinctly different cell types suggests that rearrangement of the inverted repeat DNA sequences may be involved in the cell differentiation process.

CONCLUSIONS

A major task in the study of the *Caulobacter* cell cycle is to understand the mechanisms which regulate both the temporal expression of specific structural proteins and the assembly of those proteins at specific sites on the cell. At the molecular level, elucidation of the strategies that cells employ to control this type of gene expression remains elusive. In this review we have described both classical and molecular genetic approaches that are being employed in the study of the dimorphic bacterium *Caulobacter crescentus*. Differentiation in this organism represents a phenomenon in which a multitude of genes and gene products must be co-ordinately regulated and expressed. Because *Caulobacter* is a unicellular prokaryote which expresses a complex regulatory pattern, this organism is being studied intensively in several laboratories. Many fundamental parameters of *Caulobacter* growth and differentiation are being approached. Numerous biochemical and developmental mutants have been isolated and characterized, a genetic map is under construction, membrane protein and lipid biosynthesis, carbohydrate utilization and DNA synthesis are being analysed and the formation of the stalk and flagella is being investigated.

A parallel and perhaps more difficult task is to understand how these specific observations interrelate with one another in the overall regulation of differentiation within the organism. It has been demonstrated that DNA replication is required for flagellin synthesis and for the initiation of the cell division pathway (Degnan & Newton, 1972; Newton, Osley & Terrana, 1975; Osley *et al.*, 1977). In addition, membrane phospholipid synthesis in *C. crescentus* has been shown to be related to the expression of specific cell cycle events (Contreras *et al.*, 1980). DNA replication is inhibited if phospholipid synthesis is terminated either by glycerol starvation of a glycerol auxotroph or by treatment with cerulenin. Termination of phospholipid synthesis by either method results in the inhibition of stalk elongation, flagellum biogenesis and cell division. The inability to form a stalk appears to be directly due to the cessation of phospholipid synthesis, whereas the inhibition of flagella formation and cell division is likely a result of the secondary effect on DNA replication. Two cell cycle events, the ejection of the flagellum and stalk initiation, appear to be independent of phospholipid synthesis and DNA replication (Fig. 1).

Genome organization can function as an additional level of cell regulation. In this context analysis of genome fluidity is being approached in *Caulobacter*. As described earlier, experiments have been done which suggest that certain inverted repeat DNA sequences rearrange during the cell cycle (Nisen *et al.*, 1979). Since inverted repeat DNA is interspersed throughout much of the *Caulobacter* genome, it is conceivable that the inverted repeat rearrangements can function in the modulation of numerous genes in a highly organized, concerted fashion.

Ultimately, to compose a picture of the regulatory phenomena which govern morphogenesis and differential gene expression in *Caulobacter* data will have to be derived from different conceptual approaches and then integrated to form a network which results in a controlled cell cycle. Fundamental to these studies is the development and application of both classical and molecular genetics.

REFERENCES

Agabian, N., Evinger, M. & Parker, G. (1979). Generation of asymmetry during development: Segregation of type-specific proteins in *Caulobacter*. *Journal of Cell Biology*, **81**, 123–36.

AGABIAN-KESHISHIAN, N. & SHAPIRO, L. (1970). Stalked bacteria: Properties of DNA bacteriophage φCbK. *Journal of Virology*, **5**, 795–800.

AGABIAN, N. & UNGER, B. (1978). *Caulobacter crescentus* cell envelope: Effect of growth conditions on murein and outer membrane protein composition. *Journal of Bacteriology*, **133**, 987–94.

ALEXANDER, J. L. & JOLLICK, J. D. (1977). Transfer and expression of *Pseudomonas* plasmid RP1 in *Caulobacter*. *Journal of General Microbiology*, **93**, 325–31.

ANDERSON, D., SHAPIRO, L. & SKALKA, A. (1980). *In situ* immunoassays for translation products. *Methods in Enzymology* (in press).

BARRETT, J. T., RHODES, C. S., FERBER, D. M., KUHL, S. A. & ELY, B. (1980). Genetic map of *Caulobacter crescentus*. *Genetic Maps*, **1**, 124–5.

BENDIS, I. K. & SHAPIRO, L. (1970). Properties of *Caulobacter* RNA bacteriophage φCb5. *Journal of Virology*, **6**, 847–54.

BERINGER, J. E., HOGGAN, S. A. & JOHNSON, A. W. B. (1978). Linkage mapping in *Rhizobium leguminosarum* by means of R plasmid-mediated recombination. *Journal of General Microbiology*, **104**, 201–7.

CHEUNG, K. K. & NEWTON, A. (1977). Patterns of protein synthesis during development in *Caulobacter crescentus*. *Developmental Biology*, **56**, 184–92.

COHEN, S. N. (1976). Transposable genetic elements and plasmid evolution. *Nature, London*, **263**, 731–8.

CONTRERAS, I., BENDER, R. A., MANSOUR, J., HENRY, S. & SHAPIRO, L. (1979). *Caulobacter crescentus* mutant defective in membrane phospholipid synthesis. *Journal of Bacteriology*, **140**, 612–19.

CONTRERAS, I., SHAPIRO, L. HENRY, S. (1978). Membrane phospholipid composition of *Caulobacter crescentus*. *Journal of Bacteriology*, **135**, 1130–6.

CONTRERAS, I., WEISSBORN, A., AMEMIYA, K., MANSOUR, J., HENRY, S., BENDER, R. & SHAPIRO, L. (1980). The effect of termination of membrane phospholipid synthesis on cell cycle-dependent events in *Caulobacter*. *Journal of Molecular Biology*, **138**, 401–9.

DEGNAN, S. T. & NEWTON, A. (1972). Chromosome replication during development in *Caulobacter crescentus*. *Journal of Molecular Biology*, **64**, 671–80.

DRIGGERS, L. J. & SCHMIDT, J. M. (1970). Induction of defective and temperate bacteriophages in *Caulobacter*. *Journal of General Virology*, **6**, 421–7.

ELY, B. (1978). Transfer of drug resistance factors to *Caulobacter crescentus*. *Abs., Annual Meeting of the American Society for Microbiology*, p. 109.

ELY, G. (1979). Transfer of drug resistance factors to the dimorphic bacterium *Caulobacter crescentus*. *Genetics*, **91**, 371–9.

ELY, G., AMARASINGHE, A. B. C. & BENDER, R. A. (1978). Ammonia assimilation and glutamate formation in *Caulobacter crescentus*. *Journal of Bacteriology*, **133**, 225–31.

ELY, B. & JOHNSON, R. C. (1977). Generalized transduction in *Caulobacter crescentus*. *Genetics*, **87**, 391–9.

EVINGER, M. & AGABIAN, N. (1977). Envelope-associated nucleoid from *Caulobacter crescentus* stalked and swarmer cells. *Journal of Bacteriology*, **132**, 294–301.

EVINGER, M. & AGABIAN, N. (1979). *Caulobacter crescentus* nucleoid: Analysis of sedimentation behavior and protein composition during the cell cycle. *Proceedings of the National Academy of Sciences, USA*, **76**, 175–8.

FUKUDA, A., KOYASU, S. & OKADA, Y. (1978). Characterization of two flagella-related proteins from *Caulobacter crescentus*. *FEBS Letters*, **95**, 70–5.

FUKUDA, A., MIYAKAWA, K., IIDA, H. & OKADA, Y. (1976). Regulation of polar surface structures in *Caulobacter crescentus*: Pleiotropic mutations affect the coordinate morphogenesis of flagella, pili and phage receptors. *Molecular and General Genetics*, **149**, 167–73.

HOLLOWAY, B. W. (1979). Plasmids that mobilize bacterial chromosomes. *Plasmid*, **2**, 1–36.

IBA, H., FUKUDA, A. & OKADA, Y. (1978). Rate of major protein synthesis during the cell cycle of *Caulobacter crescentus*. *Journal of Bacteriology*, **135**, 647–55.

JOHNSON, R. C. & ELY, B. (1977). Isolation of spontaneously-derived mutants of *Caulobacter crescentus*. *Genetics*, **86**, 25–32.

JOHNSON, R. C. & ELY, B. (1979). Analysis of non-motile mutants of the dimorphic bacterium *Caulobacter crescentus*. *Journal of Bacteriology*, **137**, 627–34.

JOHNSON, R. C., WALSH, J. P., ELY, B. & SHAPIRO, L. (1979). Flagella hook and basal complex of *Caulobacter crescentus*. *Journal of Bacteriology*, **138**, 984–9.

JOHNSON, R. C., WOOD, N. B. & ELY, B. (1977). Isolation and characterization of bacteriophage for *Caulobacter crescentus*. *Journal of General Virology*, **37**, 323–35.

JOLLICK, J. D. & SCHERVISH, E. M. (1972). Genetic recombination in *Caulobacter*. *Journal of General Microbiology*, **73**, 403–7.

KAMP, K., KAHMANN, R., ZIPSER, D., BROOKER, T. & CHOW, L. (1978). Inversion of the G DNA segment of phage Mu controls phage infectivity. *Nature, London*, **271**, 577–80.

KLECKNER, N. (1977). Translocatable elements in prokaryotes. *Cell*, **11**, 11–23.

KURN, N., AMMER, S. & SHAPIRO, L. (1974). A pleiotropic mutation affecting expression of polar development events in *Caulobacter crescentus*. *Proceedings of the National Academy of Sciences, USA*, **71**, 3157–61.

LAGENAUR, C. & AGABIAN, N. (1976). Physical characterization of *Caulobacter crescentus* flagella. *Journal of Bacteriology*, **128**, 435–44.

LAGENAUR, C. & AGABIAN, N. (1978). *Caulobacter* flagella organelle: Synthesis, compartmentation and assembly. *Journal of Bacteriology*, **135**, 1062–9.

LAGENAUR, C., DE MARTINI, M. & AGABIAN, N. (1978). Isolation and characterization of *Caulobacter crescentus* flagella hooks. *Journal of Bacteriology*, **136**, 795–6.

LAGENAUR, C., FARMER, S. & AGABIAN, N. (1977). Adsorption properties of stage-specific *Caulobacter* phage φCbK. *Virology*, **77**, 401–7.

MANSOUR, J., HENRY, S. & SHAPIRO, L. (1980). Differential membrane phospholipid synthesis during the cell cycle of *Caulobacter crescentus*. *Journal of Bacteriology*, **141**, 262–9.

MARINO, W., AMMER, S. & SHAPIRO, L. (1976). Conditional surface structure mutants of *Caulobacter crescentus*: Temperature-sensitive flagella formation due to an altered flagellin monomer. *Journal of Molecular Biology*, **107**, 115–30.

MIYAKAWA, K., FUKUDA, A. & OKADA, Y. (1976). Isolation and characterization of RNA phages for *Caulobacter crescentus*. *Virology*, **73**, 442–53.

NEWTON, A. (1972). Role of transcription in the temporal control of development in *Caulobacter crescentus*. *Proceedings of the National Academy of Sciences, USA*, **69**, 447–51.

NEWTON, A. & ALLEBACH, E., (1975). Gene transfer in *Caulobacter crescentus*: Polarized inheritance of genetic markers. *Genetics*, **80**, 1–11.

NEWTON, A., OSLEY, M. A. & TERRANA, B. (1975). *Caulobacter crescentus*: A model for the temporal and spacial control of development. In *Microbiology-1975*, ed. D. Schlesinger, pp. 442–52. Washington, DC: American Society for Microbiology.

NISEN, P., MEDFORD, R., MANSOUR, J., PURUCKER, M., SKALKA, A. & SHAPIRO, L. (1979). Cell cycle-associated rearrangement of inverted repeat DNA sequences. *Proceedings of the National Academy of Sciences, USA*, **76**, 6240–4.

OSLEY, M. A. & NEWTON, A. (1977). Mutational analysis of developmental control in *Caulobacter crescentus. Proceedings of the National Academy of Sciences, USA*, **74**, 124–8.
OSLEY, M. A. & NEWTON, A. (1978). Regulation of cell cycle events in asymmetrically dividing cells: Functions required for DNA initiation and chain elongation in *Caulobacter crescentus. Journal of Bacteriology*, **135**, 10–17.
OSLEY, M. A. & NEWTON, A. (1980). Temporal control of the cell cycle in *Caulobacter crescentus*: Roles of DNA chain elongation and completion. *Journal of Molecular Biology*, **138**, 109–28.
OSLEY, M. A., SHEFFERY, M. & NEWTON, A. (1977). Regulation of flagellin synthesis in the cell cycle of *Caulobacter*: Dependence on DNA replication. *Cell*, **12**, 393–400.
POINDEXTER, J. S. (1978). Selection for nonbuoyant morphological mutants of *Caulobacter crescentus. Journal of Bacteriology*, **135**, 1141–5.
SCHMIDT, J. M. (1966). Observations on the adsorption of *Caulobacter* bacteriophages containing RNA. *Journal of General Microbiology*, **45**, 347–53.
SCHMIDT, J. M. (1968). Stalk elongation in mutants of *Caulobacter crescentus. Journal of General Microbiology*, **53**, 291–8.
SCHMIDT, J. M. (1969). *Caulobacter crescentus* mutants with short stalks. *Journal of Bacteriology*, **98**, 816–17.
SCHMIDT, J. M. & STAINIER, R. Y. (1966). The development of cellular stalks in bacteria. *Journal of Cell Biology*, **28**, 423–36.
SHAPIRO, L. (1976). Differentiation in the *Caulobacter* cell cycle. *Annual Reviews of Microbiology*, **30**, 377–407.
SHAPIRO, L. & BENDIS, I. (1975). RNA phages of bacteria other than *E. coli*. In *RNA phage*, ed. N. Zinder, p. 397. *Cold Spring Harbor Symposium on Quantitative Biology*.
SHAPIRO, L. & MAIZEL, J. V. (1973). Synthesis and structure of *Caulobacter crescentus* flagella. *Journal of Bacteriology*, **113**, 478–85.
SHEFFERY, M. & NEWTON, A. (1977). Reconstitution and purification of flagellar filaments from *Caulobacter crescentus. Journal of Bacteriology*, **132**, 1027–30.
SHEFFERY, M. & NEWTON, A. (1979). Purification and characterization of a polyhook protein from *Caulobacter crescentus. Journal of Bacteriology*, **138**, 575–83.
SILVERMAN, M., ZIEG, J., HILMAN, M. & SIMON, M. (1979). Phase variation in *Salmonella*: Genetic analysis of a recombinational switch. *Proceedings of the National Academy of Sciences, USA*, **76**, 391–5.
SKALKA, A. & SHAPIRO, L. (1976). *In situ* immunoassays for gene translation products in phage plaques and bacterial colonies. *Gene*, **1**, 65–79.
VAN VLIET, F., SILVA, B., VAN MONTAGU, M. & SCHELL, J. (1978). Transfer of RP4::Mu plasmids to *Agrobacterium tumefaciens. Plasmid*, **1**, 446–51.
WEST, D., LAGENAUR, C. & AGABIAN, N. (1976). Isolation and characterization of *Caulobacter crescentus* bacteriophage φCd1. *Journal of Virology*, **17**, 568–75.
WOOD, N., RAKE, A. & SHAPIRO, L. (1976). Structure of *Caulobacter* DNA. *Journal of Bacteriology*, **126**, 1305–15.
ZIEG, J., SILVERMAN, M., HILMAN, M. & SIMON, M. (1977). Recombinational switch for gene expression. *Science*, **196**, 170–2.

GENETICS OF BACTERIAL VIRULENCE

WERNER K. MAAS

Department of Microbiology,
New York University Medical Center,
New York, NY 10016, USA

INTRODUCTION

Virulence is an attribute of certain bacteria that results in damage to the host they inhabit. If the damage is severe, the host will die. In general, this is detrimental for the bacteria, since it destroys the source of their livelihood. It is therefore not surprising that most species of bacteria inhabiting a living host, such as the bacteria in the human intestine, are harmless or even beneficial; that is, they are commensals or symbionts. Pathogenic species are the exceptions.

Although pathogenic species are relatively rare, they have persisted tenaciously during the course of human evolution and they are of course of great medical importance as causative agents of infectious diseases. Moreover, as the conditions of human existence have changed, so have the bacterial species that produce disease. Even the spectacular advances of the 20th Century in sanitary conditions and the development of antimicrobial agents have hardly diminished the importance of bacteria as infectious agents. In the light of the past, the existence of pathogenic bacteria appears to be a *sine qua non* of human existence.

It is against this background that molecular genetics has now entered the field of microbial pathogenicity. The concepts and methodologies that have been developed during the past 20 years in the areas of gene structure and gene expression are now being applied to the genetic basis of pathogenicity. This chapter describes some of the advances that have been made in this field and attempts to point out possible areas for subsequent research. It is the hope of this research that understanding of pathogenic mechanisms at the molecular level will lead to new control measures for infectious diseases and that insight gained through understanding of genetic mechanisms will enable us to cope more effectively with pathogenic bacteria than has been possible with the more empirical methods of the past.

The salient properties of bacteria that enable them to be virulent can be classified into categories in terms of the ends they achieve in the host. For the present chapter these properties will be grouped under the following headings: (1) adhesion to host cells, (2) damage to host cells, (3) invasion of host cells and tissues, (4) resistance to host defences. It is realized that these headings do not include all traits associated with virulence and that they are not completely mutually exclusive. In spite of these limitations, these headings have been adopted with the expectation that they will permit a clear organization of the material to be presented. It must be kept in mind that for a given pathogen, virulence is an expression of the sum total of its genetic traits.

GENETIC APPROACHES TO VIRULENCE

In general, the kind of information that can be obtained from genetic studies involves several aspects of virulence. Thus studies on the movement and location of genes will be useful for understanding population changes and the epidemiology of bacterial infections. Studies on gene expression will give information about the biosynthesis of substances and structures associated with virulence. Finally, studies of mutants with altered virulence properties may throw light on the mode of action of virulence factors and their interactions with host cells. The plan for the following sections of this chapter is to discuss for each of the four categories of virulence traits listed in the previous paragraph the different genetic approaches that have been used to elucidate different aspects of bacterial virulence. Because of the large scope of the subject matter the treatment will not be comprehensive. Examples will be chosen for each of the four categories, the choice of these examples being determined largely by the author's interest in enteric infections, especially those by *Escherichia coli*, as well as by the recent demonstration of plasmids as carriers of genes for virulence factors.

Adhesion to host cells

For bacteria to produce disease within a given site of the host, such as the urogenital tract or the respiratory tract or the alimentary tract, they have first to colonize this site. Colonization usually involves adhesion of the bacteria to surface cells of the region, followed by growth and proliferation. Without adhesion there is a good chance for the bacteria to be washed out by the movement of the fluids that bathe these sites.

Adhesion has been studied in many species of bacteria and has usually been found to be specific both for the surface components of the bacteria and the receptor components of the host cells. Specificity of bacterial components exists for the site of the host that is affected as well as for the species of the host. Familiar examples of site specificity are the attachment of group A streptococci to the pharyngeal epithelium, of gonococci to urethral and cervical epithelium and of enteropathogenic *E. coli* to intestinal epithelium. In the case of *E. coli*, bacterial adhesion is largely due to surface pili or pili-like structures classified among the K antigens. In regard to specificity for host species it has been shown that one of these pili-like K antigens, K88, occurs mainly in strains infecting pigs, whereas another, K99, occurs mainly in calves and lambs, although it is occasionally found in strains isolated from pigs. In strains isolated from humans, two different pili-like adherence antigens have been found, CFA/I and CFA/II (CFA = colonization factor antigen). For a detailed review of bacterial adhesion factors the reader is referred to two recent books (Ellwood, Melling & Rutter, 1979; Inouye, 1979).

In regard to receptor sites of the host, specificity is also present, although in most cases the chemical nature of the host receptors has not been elucidated. Specificity can be demonstrated by genetic evidence: for example, it has been shown that K88-positive *E. coli* do not attach to the intestinal brush-borders of certain pigs, the animals in which adhesion does not take place being resistant to infection by these strains (Rutter, Burrows, Sellwood & Gibbons, 1975). Susceptibility is due to the presence of an intestinal receptor and is inherited in a simple Mendelian fashion, with adhesion being dominant over non-adhesion (Gibbons, Sellwood, Burrows & Hunter, 1978).

Genes for several bacterial adhesion factors, such as K88, K99 and CFA/I, have been found to be located in plasmids. This is perhaps not surprising, since adhesion factors are not essential for survival of the bacteria and plasmids are dispensable genetic elements. On the other hand, not all adherence factors are plasmid-determined. The genes for type I pili are on the chromosome (Brinton, Gemski, Falkow & Baron, 1961) and these pili have been shown to have a role in adhesion (Duguid, Anderson & Campbell, 1966).

Location on a plasmid has facilitated studies on the molecular genetics of several adhesion factors, especially K88. So far these studies have concentrated more on the structure and phenotypic properties of the plasmids, such as conjugal transfer, presence of other genes, plasmid dissociation, etc., than on the nature and expression of the adhesion factor genes themselves. For example, the gene for K88 has been found on large plasmids, 75 or 135 kilobases (kb) in size and these plasmids often carry genes for raffinose fermentation (Smith & Parsell, 1975; Shipley, Dallas, Dougan & Falkow, 1979). For some of these plasmids conjugal transfer has been demonstrated and there is evidence that the K88 plasmid forms a recombinant with a transfer plasmid (Shipley, Gyles & Falkow, 1978). The gene for K99 has been found in one study on a large (78 kb) conjugative plasmid (So, Boyer, Betlach & Falkow, 1976). This plasmid carried genes for resistance to tetracycline and streptomycin. The gene for CFA/I has been shown in several studies to be located on large (about 90 kb) non-conjugative plasmids, in linkage with a gene for heat stable enterotoxin (ST) (Smith *et al.*, 1979). We have found such a plasmid in strains isolated in Brazil and in one strain a second transfer plasmid was present, which brought about efficient transfer of the CFA/I-ST plasmid to other strains (Reis *et al.*, 1980).

Cloning of HindIII restriction enzyme fragments containing genes for K88 has been carried out with two plasmids. Three serological variants of K88 have been described, K88ab, K88ac and K88ad; in one case the cloned fragment determined K88ab (Mooi, de Graaf & von Emden, 1979) and in the other K88ac (Shipley, Dallas, Dougan & Falkow, 1979). The cloning vehicle in both cases was the multicopy plasmid pBR322. Some studies have been carried out on the expression of the K88 genes in these cloned derivatives. For the K88ab plasmid, the amount of K88 antigen produced depended on the orientation of the inserted fragment in pBR322. In one direction, there was a four-fold increase over that produced by the parental plasmid; in the other direction, there was a considerable decrease (Mooi *et al.*, 1979). For the K88ac cloned plasmid, expression was studied in minicells and a 23 500 dalton polypeptide was identified by immunoprecipitation with antiK88ac serum and *S. aureus* protein A (Shipley *et al.*, 1979). This molecular weight is in good agreement with the reported size of purified K88ac antigen (Mooi & de Graaf, 1979).

So far no studies have been reported on the use of mutants to investigate the biosynthesis and structure–activity relationships of K88 and other adhesion factors, but it seems likely that with the knowledge gained about the structure of K88 plasmids and the location of the K88 genes on these plasmids, such studies will soon be forthcoming. Other questions that will be of interest in the future include the possible location of adhesion factor genes on transposons and the possibility of phase variation for these genes, resulting in turning on and turning off of gene expression. Such a mechanism is well known for flagellar genes in *Salmonella* and has been shown for type I pili genes (Swaney *et al.*, 1977). Finally, one intriguing fact about adhesion factor plasmids is that in nature they are confined to a small fraction of the existing serotypes of *E. coli*. This is also true for enterotoxin plasmids, as will be discussed below. The reasons for the association of these virulence plasmids with only a few serotypes are not known, but elucidation of underlying causes should lead to increased understanding of the physiology of adhesion factor synthesis and the population dynamics of strains carrying adhesion factor plasmids.

Damage to host cells

Bacteria can cause damage to host cells either physically or chemically. Physical damage, although it undoubtedly occurs in certain infections as a result of continued replication of bacteria in a confined environment, has not been studied extensively. Damage due to substances produced by bacteria, on the other hand, has been the subject of many investigations. Such substances vary in their potencies from those that need a high concentration to produce inhibitory effects, such as weak organic acids, to those that act in very low concentrations. These latter substances usually attack specific targets, are proteins and are referred to as toxins. For many toxins, enzymatic activity has been demonstrated; for example, both diphtheria toxin and cholera toxin have NADase activity and bring about ADP ribosylation of target proteins. For diphtheria toxin, this results in inhibition of protein synthesis; for cholera toxin, in stimulation of adenylate cyclase.

Genes for toxin production may be located on plasmids, or on phages or on the chromosome. Examples of each type to be discussed in this section are listed in Table 1. It should be noted that the genes described in this table are not necessarily the structural genes for the toxins. They are recognized by the fact that mutations

Table 1. *Location of genes for toxin production*

Location	Bacteria	Toxin	Type of gene
Plasmid	*Escherichia coli*	LT	structural
Plasmid	*Escherichia coli*	ST	structural
Plasmid	*Staphylococcus aureus*	enterotoxin B	regulatory
Plasmid	*Staphylococcus aureus*	exfoliative toxin	structural (?)
Phage	*Corynebacterium diphtheriae*	diphtheria toxin	structural
Phage	*Clostridium botulinum*	botulinum toxin	?
Phage	*Clostridium novyi*	α-toxin	?
Chromosome	*Vibrio cholerae*	CT	regulatory
Chromosome	*Vibrio cholerae*	CT	structural (?)
Chromosome	*Staphylococcus aureus*	enterotoxin B	structural

occurring in them abolish toxin production. They may therefore be regulatory genes or may affect toxin synthesis in a secondary manner.

In the following section we shall deal mainly with toxin genes located on plasmids, since they have been the subject of recent studies at the molecular level. Work on phage genes and chromosomal genes will be summarized briefly.

Plasmid-located genes

These have been studied mainly in enteric bacteria, notably *E. coli*, and in *Staphylococcus aureus*. In *E. coli* much work has been done on the genetics of enterotoxin production. This toxin is able to cause diarrhoea in man and domestic animals. Actually, two types of enterotoxin are known, one heat labile (LT) and similar to cholera toxin (CT) in its structure and mode of action, the other heat stable (ST). Genes for both type of toxin are found in most cases to be located in plasmids. Some plasmids carry genes for LT only, others for ST only and yet others for both LT and ST.

The size of enterotoxin plasmids (Ent plasmids) varies from 30 to 150 kb, but most Ent plasmids are 80–100 kb in length and are conjugative. They usually do not carry other genes besides those for conjugal transfer, but recently several examples of chimeric Ent plasmids carrying drug resistance genes have been described (McConnell *et al.*, 1979; Gyles, Palchaudhuri & Maas, 1977). We mentioned above a plasmid with genes for ST and CFA/I.

In regard to the mobility of these genes, it is of interest that the ST gene is, at least in some cases, located on a transposon. In one case, So *et al.* found ST to be present on a short DNA segment bounded by inverted repeat segments identical with the insertion element IS1 (So, Heffron & McCarthy, 1979). Transposition of this stem-loop structure was demonstrated. In another LT-ST plasmid, we have also found the ST gene (but not the LT gene) to be on a segment bounded by inverted repeats, but we have not yet demonstrated transposition of this structure. The STs involved in these two cases differ from each other, both in their host specificity and chemical nature, and no homology between the two ST genes was found in hybridization experiments. The inverted repeat segments were also different. Further studies should reveal whether or not genes for ST are always located on transposons. In this connection, it should be noted that drug resistance genes are often located on transposons and that the chimeric Ent plasmids with drug resistance genes mentioned above may have arisen as a result of insertion of one or more transposons into an Ent plasmid.

In general, the findings mentioned so far indicate that toxin genes located on plasmids have the capacity to move not only from cell to cell, but also from one genetic structure to another within the same cell.

In regard to gene expression, the most extensive studies so far have been carried out with the gene for LT. The plasmids employed in these studies were two conjugative LT-ST plasmids, Ent P307 and pCG86. The latter carries also genes for drug resistance and its structure is shown in Fig. 1. As can be seen, there is considerable homology between the two plasmids, including the region containing the toxin genes. Details about the structure and properties of these plasmids have been published (Gyles *et al.*, 1977; Santos, Palchaurdhuri & Maas, 1975). A BamHI restriction enzyme fragment of length 8.7 kb obtained from Ent-P307 has been cloned and carries the gene for LT production (So, Dallas & Falkow, 1978).

The LT molecule consists of two sub-units, A and B, and in the complete molecule, one A sub-unit is surrounded by five B sub-units. The same arrangement is found in the CT molecule. The cistrons for both A and B were localized within a 1.8 kb segment (Dallas, Gill & Falkow, 1979) and a clone carrying only this segment was found to have all the genetic information for LT synthesis. Expression of this cloned fragment and deletions derived from it was studied in minicells. The results led to the conclusion that the cistrons for A (*elt*A) and B (*elt*B) are transcribed as a unit, with a promoter being located at the N-terminal end of A. DNA sequencing has been carried out with both *elt*A and *elt*B (Dallas & Falkow, 1980; Spicer

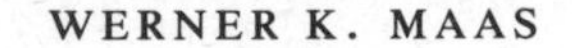

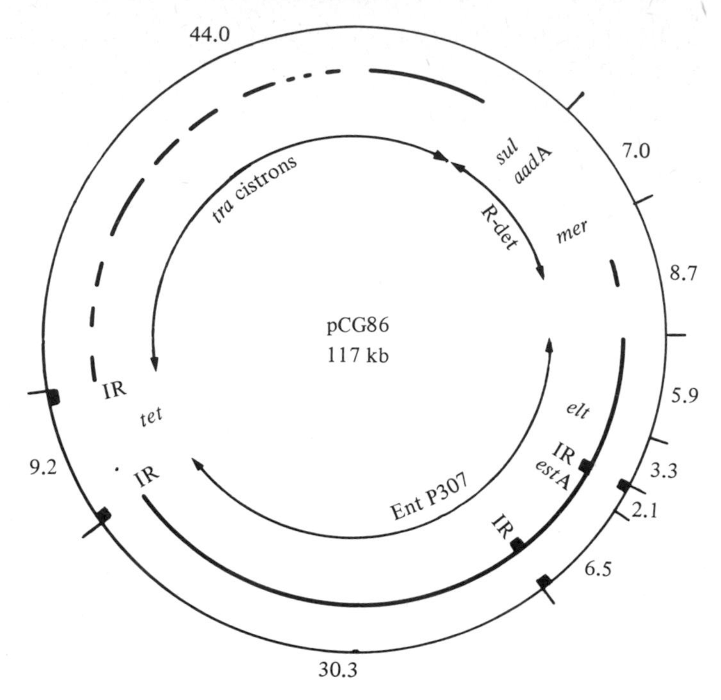

Fig. 1. Physical map of plasmid pCG86. Numbers in the outer circle are distances in kilobases. The middle circle represents regions of homology with plasmid Ent P307. Abbreviations: IR = inverted repeat; the phenotypic expressions of the genes are as follows: *sul* = resistance to sulfonamides; *aad*A = resistance to streptomycin and spectinomycin; *mer* = resistance to mercuric irons; *elt* = production of LT; *est*A = production of ST; *tet* = resistance to tetracycline; *tra* cistrons = cistrons for conjugal transfer.

et al., 1980). The two cistrons were found to be separated by a 200 base pair (bp) intercistronic region. The most striking results of the DNA sequence studies are that both *elt*A and *elt*B code for a signal peptide at the N-terminal end which is presumably required for passage through the inner membrane, that both are preceded by a ribosome binding site, and that the amino acid sequences deduced from the DNA sequence are very similar to the amino acid sequences of the A and B sub-units of CT.

The results obtained strongly suggest that *elt*A and *elt*B are transcribed into a single messenger RNA. The subsequent steps in LT synthesis, translation, passage through the cell membranes, processing of the A and B sub-units and assembly of the LT molecule, are so far obscure. Also, very little is known about the regulation of LT synthesis. It is expected that studies with mutants affecting LT production, to be mentioned next, will be helpful for elucidating these steps in toxin production.

Isolation of mutants affecting toxin production is laborious, since no direct selection procedures are available and assay methods for screening large numbers of isolates are inconvenient. The availability of pCG86 with its drug resistance genes has enabled us to use an enrichment procedure by co-mutagenesis with nitrosoguanidine (Silva, Maas & Gyles, 1978). In this procedure, the frequency of LT^- mutants among the mutagenized colonies is about 2% and we have characterized 60 mutants so far by testing for enzyme activity (for the A sub-unit) and by radioimmunassay (for either A or B). The mutants were found to be of different types, some with a defect in A, others in B. Complementation tests have been carried out with a cloned fragment carrying only *elt*A and it was found that most mutants defective in A could be complemented by the A sub-unit produced by the cloned fragment. However, a few mutants with defective A could not be complemented. Studies of such non-complementing mutants, now in progress, may throw light on the mechanism of LT synthesis and assembly.

Besides mutants defective in LT synthesis, mutants have also been isolated that overproduce LT(*htx* mutants). Such mutants can be scored on blood agar plates by using a passive radial immune haemolysis procedure developed by Bramucci & Holmes (1978). These authors have studied *htx* mutants isolated after nitrosoguanidine mutagenesis and found most of the mutations to be in the chromosome (Bramucci & Holmes, 1979). In some mutants, total LT production was increased. However, in others, total LT production was not increased, but more than the normal amount of toxin was released from the cells. It should be noted that in wild-type cells, most of the LT remains cell associated and is found largely in the outer membrane (Wensink *et al.*, 1978). The finding of different chromosomal mutants affecting LT synthesis point up the complicated nature of LT synthesis and its regulation.

We have obtained *htx* mutants for LT following insertions of a transposon either into the chromosome or into the plasmid in strains carrying pCG86 (Maas, 1979). The transposons used were Tn3 and Tn5. The insertions in the chromosome were presumably into genes affecting the control of LT synthesis and with a selective marker gene being present on these transposons (ampicillin resistance for Tn3, kanamycin resistance for Tn5), this will facilitate the mapping of chromosomal genes affecting LT synthesis. Insertions of transposons into the plasmid frequently gave rise to *htx* mutants, in contrast to the absence of nitrosoguanidine-induced plasmid *htx* mutants. The mechanism by which these insertions bring about hyperproduction of LT has remained obscure.

Studies similar to those described for LT have also been carried out with ST. Thus the gene for the ST previously mentioned to be located on an IS1-flanked DNA segment has been sequenced by So *et al.* and the amino acid sequence has been deduced (So & McCarthy, 1980). Here also a signal peptide was found at the N-terminal end. It has been reported that most of the ST produced is excreted into the culture medium, in accordance with the presence of a signal peptide. ST is a small protein consisting of 72 amino acids. It produces its effects on the host cells by stimulating guanylate cyclase (Field, Graf, Laird & Smith, 1978). As in the case of LT, little is known about the steps in the biosynthesis of ST. So far, no mutants have been isolated in the ST gene sequenced by So *et al.* A few mutants were found in the ST gene present on pCG86 (Silva *et al.*, 1978), which, as was pointed out, is different from the ST gene studied by So *et al.* The ST gene on pCG86, although mapped (Mazaitis, Maas & Maas, 1980), has not been sequenced. Thus studies on ST have been hampered somewhat by the delay in recognizing that there are several different STs. The main difficulty with ST, however, has been the assay, which has to be carried out either in suckling mice or in intestinal loops of pigs or rabbits. For genetic studies, in which many isolates have to be tested, this is laborious and time consuming (Gyles, 1979).

Another toxin of *E. coli* whose genetic determinants are located on a plasmid and whose genetics has been studied at the molecular level is α-haemolysin. In contrast to LT and ST, a definite role of α-haemolysin in the pathogenesis of infections has not been found. Nevertheless, this toxin represents an interesting model system for the genetic control of toxin production.

The toxin is a largely extracellular protein with a molecular weight of 58 000. A conjugative plasmid carrying genes for toxin production has been studied in detail (Noegel, Rdest, Springer & Goebel, 1979). The analysis involved the isolation and mapping of nitrosoguanidine-induced and transposon insertion toxin-negative mutants, the cloning of restriction enzyme fragments carrying toxin genes and complementation tests among mutant plasmids and cloned restriction enzyme fragments. It was found that three cistrons necessary for toxin production are clustered on a 4.8 kb segment of the plasmid. One cistron determines the structure of a 90 000 molecular weight cytoplasmic precursor of the toxin, a second cistron controls the conversion of the precursor to the toxin and the third cistron is involved in the release of the toxin from the cell. It can be seen that the genetic machinery on the plasmid for α-haemolysin is more complex than that for LT or ST in that there are separate plasmid genes for the processing of the toxin molecule and its transport to the outside, whereas for the enterotoxins the only genetic information on the plasmid appears to be for the structure of the toxin molecules.

In *S. aureus*, the genetics of two toxins has been investigated recently, enterotoxin B and exfoliative toxin. The former is responsible for food poisoning, the latter for a condition in children called the scalded skin syndrome. For both, genetic determinants for toxin production have been found on plasmids.

The genetics of enterotoxin B synthesis has been confusing for some time; the presence of a 1.3 kb plasmid seems to be required for toxin synthesis, yet in transduction experiments the gene for toxin synthesis appears to be on the chromosome. The confusion has been resolved, at least partially, by experiments of Khan & Novick (1980), who have determined the DNA sequence of the 1.3 kb plasmid and have studied its protein products. Two polypeptides are coded for by the plasmid, of molecular weights 20 000 and 12 000, and neither bears any resemblance to the toxin, which has a molecular weight of 29 000. It seems therefore that one or both polypeptides are required for toxin synthesis, but do not contribute to its structure. The structural gene for the toxin, which seems to be on the chromosome, exhibits unusual properties in transduction experiments and Novick has proposed on the basis of these peculiarities in gene transfer that the toxin gene is part of a 'hitchhiking transposon' (Phillips & Novick, 1979). Thus far, little is known about the biosynthesis of the toxin and its control, but with the recent clarification of the situation the way is now open for systematic exploration of this process.

Genetic studies on exfoliative toxin have also been a source of confusion, in this case because it was not recognized until recently that two different toxins, rather than a single moiety, are produced by some toxigenic strains. This is reminiscent of the situation with ST. A gene for one of these toxins is located on a 40 kb plasmid (Warren, Rogolsky, Wiley & Glasgow, 1975), but it is not known if this is the structural gene or a regulatory gene. The genetic determinants for the other toxin are most likely chromosomal, although there is a possibility that they may be located on a small cryptic plasmid. Studies on exfoliative toxin, like those with enterotoxin B, have now reached a stage where rapid progress may be expected.

Phage-located genes

The usual manner of demonstrating toxin genes on phages is to cure a lysogenic, toxigenic strain of its phage and show that loss of the phage results in loss of toxin production. Re-lysogenization with the phage then restores toxigenicity. The acquisition of a new phenotypic property such as toxigenicity as a result of lysogeny is called phage conversion. The best studied case of phage-mediated toxigenesis is diphtheria toxin production in *Corynebacterium diphtheriae* (Barksdale & Arden, 1974), but phage conversion resulting in toxigenicity has also been found with other toxins, notably botulinum toxin and other toxins of several species of *Clostridium* (Eklund, Povsky, Meyers & Pelroy, 1974; Eklund, Povsky, Peterson & Meyers, 1976).

Diphtheria toxin, like LT and CT, consists of two domains, one with enzymatic activity, the other involved in binding to host cell receptors. However, diphtheria toxin, in contrast to the other two toxins, consists of a single polypeptide chain. The structural gene for this polypeptide is located on a temperate phage and several mutants in this structural gene have been isolated and characterized. One advantage in these genetic studies was that the chemical nature of the toxin had been elucidated previously so that the mutations could be defined precisely in terms of the affected site on the molecule. These mutants were very helpful in elucidating the mode of action of the toxin, especially the different roles played by the two domains of the molecule. For a detailed account, the reader is referred to recent articles (Collier, 1979; Murphy, 1976).

Regulation of diphtheria toxin production has also been the subject of genetic studies. It has been known for a long time that toxin synthesis is repressible by iron. Presumably iron acts as a co-repressor, together with a repressor protein. Phage mutants have been isolated in which toxin is produced constitutively (Murphy, Michel & Teng, 1978). These are probably operator mutants, insensitive to the repressor. Chromosomal mutants have also been isolated which produce high levels of toxin in the presence of iron. These mutants appear to be defective in the transport of iron into the cell (Cryz & Holmes, 1979).

So far, no cloning and DNA sequencing studies on the phage

carrying the structural gene for diphtheria toxin have been published, but with the available methodologies this will undoubtedly occur soon. Such studies should help to clarify some points that have remained obscure, especially the nature of the proposed control region on the DNA preceding the structural gene.

Chromosomal genes

The best-studied example of chromosomally determined toxigenicity is CT in *Vibrio cholerae*. A variety of mutants have been isolated. Some produce no active toxin, but do produce either the B sub-unit (Honda & Finklestein, 1979) or high levels of A sub-unit in relation to B. Others produce reduced amount of normal toxin (Ltx) and yet others produce increased amounts of normal toxin (Htx) (Mekalanos & Murphy, 1980). Presumably the latter two categories are regulatory mutants. Two sites on the chromosome for Htx and Ltx mutants have been mapped. It has been suggested that one of these genes controls the production of a repressor (Mekalanos & Murphy, 1980). So far none of the mutants that appear to be in the structural gene (B only or high A:B ratio) have been mapped. However, it appears from studies on the presence or absence of plasmids in toxigenic strains as well as from hybridization studies with a probe containing the LT gene of *E. coli* that the structural gene for CT is also located on the chromosome. It has been shown that there is a great deal of homology (about 80% in the amino acid sequences) between LT and CT for both the A and the B sub-units (Dallas & Falkow, 1980; Spicer *et al.*, 1980). This strongly suggests a common origin for the LT and CT genes. The intriguing picture is emerging in which in *V. cholerae* the cistrons for the A and B sub-units are located on the chromosome, while in *E. coli* they are located on a plasmid. It will be of interest to see if in *V. cholerae* the cistrons for A and B are next to each other, as they are in *E. coli*, and whether or not they constitute a transcriptional unit.

Invasion of host cells and tissues

Some bacteria produce disease by penetrating into the cells of the host and multiplying there. In some cases, these bacteria pass through cells and enter body cavities and tissues not directly accessible from the outside, such as the blood stream. 'Invasion' although considered by some workers to include both penetration and multiplication in the host tissue, shall in the present chapter refer only to the ability of bacteria to penetrate into host cells.

The specific properties that enable bacteria to penetrate host cells are poorly understood. In one genus, *Shigella*, these properties have been investigated by genetic methods. These bacteria enter the epithelial cells and cells of the lamina propria of the intestine, multiply there and cause destruction of the surrounding tissue. The resulting disease is bacillary dysentery. To study the genetics of invasiveness, hybrids were obtained from matings of non-virulent *E. coli* strains with *Shigella flexneri* strains (Formal & Hornick, 1978). They were tested for invasiveness in experimental animals. It was found that replacement of certain segments of the *Shigella* chromosome by the corresponding segment of the *E. coli* chromosome resulted in the loss of virulence, while replacement of other segments did not. It was not possible to pin down any one gene as a major 'virulence gene' and the conclusion was reached that pathogenicity is determined by a multiplicity of genes. One factor that was found to be important was the specific constitution of the lipopolysaccharide (LPS, contains 0 antigen). It was shown that hybrids that inherited the ability to synthesize the *E. coli* 08 somatic antigen lost the ability to invade the intestinal mucosa, while those that expressed the 025 antigen did not.

So far, plasmids have not been implicated in invasiveness, although studies now in progress with *Shigella sonnei* suggest that a 210 kb plasmid may play a role (S. Formal, personal communication). In general not much work has been done on the genetics of invasiveness in other bacteria besides *Shigella* and this is a large subject for future explorations. For example there are a number of *E. coli* serotypes that produce a gastrointestinal disease similar to that of *Shigella* and with the available knowledge of *E. coli* genetics, the genetics of these strains is open to investigation.

Resistance to host defences

The classical example of bacterial resistance to host defences is the production of a polysaccharide capsule, which makes the bacteria resistant to phagocytosis. It was the study of genetics of pneumococcal capsules that launched the field of molecular genetics in the early 1940s. Since then, work has continued on the genetics of pneumococcal capsular polysaccharide synthesis (Bernheimer, Wermundsen & Austrian, 1968) as well as on the genetics of capsule formation in other species.

Recently other properties of bacteria have come to light that render them resistant to host defences. These involve resistance to the bactericidal action of serum, which is presumably due to the action of complement and antibodies. These properties are of interest, because their genetic determinants are located on plasmids.

One group of plasmids, which has been implicated in the virulence of *E. coli* strains producing extra-intestinal infections, is the Col V plasmids (Smith, 1974). From one such plasmid, cloned restriction enzyme fragments were obtained and by this means the region on the plasmid containing the gene(s) for serum resistance was localized. This region is different from, though nearby, the region containing the genes for colicin production and immunity to colicin V (Binns, Davies & Hardy, 1979).

Another Col V plasmid was found to enable *E. coli* strains that carry it to sequester iron more efficiently from the medium than strains not carrying the plasmid (Williams, 1979). Iron is an essential growth factor for the bacteria and since available iron is present in low concentrations in serum, the ability to utilize iron efficiently helps the bacteria to survive and grow in serum. It is therefore not surprising that the presence of this Col V plasmid enhances the virulence of the bacteria. The genes responsible for more effective iron utilization have not been mapped on the plasmid, but as in the case of serum resistance, this property appears to be independent of colicin V production.

Another plasmid whose presence confers resistance to the bactericidal action of serum is a well-studied resistance (R) plasmid, R6-5. Many cloned restriction enzyme fragments of this plasmid are available and many genes have been mapped on this plasmid. By studying serum resistance of strains carrying cloned fragments, it was possible to narrow down the resistance to one of the genes of the *tra* operon, which governs conjugal transfer. This gene, *traT*, is reponsible for the property of surface exclusion (Moll, Manning & Timmis, 1980). Its product, the *traT* protein, is an outer membrane protein, and it is presumably the presence of this protein that protects against killing in serum.

In addition to R6-5, other R plasmids structurally related to R6-5 have been shown to confer increased serum resistance on their hosts (Taylor & Hughes, 1978).

Plasmid genes have recently also been shown to play a role in the intracellular survival of pathogens. *Yersinia enterocolitica*, which can cause gastroenteritis, is an invasive organism. It requires calcium for growth and in the presence of low calcium it produces two surface proteins, V and W. These properties are plasmid mediated and are also essential for virulence (Portnoy, Moseley & Falkow, 1980). They are expressed at 37 °C, but not at 26 °C. With the availability of the methods of molecular genetics it should not be long before the role of the calcium requirement and the V and W proteins in intracellular survival will be clarified. A similar calcium requirement and formation of V and W proteins has been found in *Yersinia pestis*, the causative agent of bubonic plague (Brubaker, 1979).

CONCLUSIONS

From the examples discussed in this chapter it is evident that the approaches used by molecular genetics are beginning to make inroads into the elucidation of mechanisms underlying bacterial virulence. In these endeavours it has been helpful that genes controlling virulence are often located on extrachromosomal elements, since this has facilitated their cloning and sequencing, as well as studies on gene expression. Much remains to be learned about the properties of bacteria that are responsible for pathogenicity, but the tools are available and the directions to be followed are becoming apparent.

Besides its usefulness as a tool for the study of basic mechanisms of virulence, molecular genetics has also provided a means for attacking practical problems. For example, the development of vaccines will be aided by the availability of cloned fragments that carry genetic determinants of antigenicity. In the case of LT, the B sub-unit is largely responsible for antitoxic immunity and cloned fragments carrying only the B cistron are available. Another practical use of molecular genetics methods is in diagnostic bacteriology. Again, using the example of the cloned B sub-unit of LT, one can prepare monoclonal antibodies that are highly specific for B and can use them for the detection of LT-producing bacteria. Other applications will undoubtedly be found, especially through the fusion of genes to create hybrid proteins. It is gratifying to see that discoveries made in molecular genetics during the past 30 years are now finding useful applications in the management of bacterial infections.

REFERENCES

BARKSDALE, L. & ARDEN, S. B. (1974). Persisting bacteriophage infections, lysogeny, and phage conversions. *Annual Review of Microbiology*, **28**, 265–99.

BERNHEIMER, H. P., WERMUNDSEN, I. E. & AUSTRIAN, R. (1968). Mutation in *Pneumococcus* type III affecting multiple cistrons concerned with the synthesis of capsular polysaccharide. *Journal of Bacteriology*, **96**, 1099–102.

BINNS, M. M., DAVIES, D. L. & HARDY, K. G. (1979). Cloned fragments of the plasmid ColV, I-K94 specifying virulence and serum resistance. *Nature, London*, **279**, 778–81.

BRAMUCCI, M. G. & HOLMES, R. K. (1978). Radial passive immune hemolysis assay for detection of heat-labile enterotoxin produced by individual colonies of *Escherichia coli* or *Vibrio cholerae*. *Journal of Clinical Microbiology*, **8**, 252–5.

BRAMUCCI, M. G. & HOLMES, R. K. (1979). Chromosomal mutations in *Escherichia coli* resulting in increased extracellular heat-labile enterotoxin. In *Abstracts of the Annual Meeting of the American Society of Microbiology*, p. 16. Washington DC: American Society of Microbiology.

BRINTON, C. C., GEMSKI, P., FALKOW, S. & BARON, L. S. (1961). Location of the piliation factor on the chromosome of *Escherichia coli*. *Biochemical and Biophysical Research Communications*, **5**, 293.

BRUBAKER, R. R. (1979). Expression of virulence in *Yersiniae*. In *Microbiology-1979*, ed. D. Schlesinger, pp. 168–75. Washington, DC: American Society of Microbiology.

COLLIER, R. J. (1979). Genetic approaches to structure and activity in ADP-ribosylating exotoxins. In *Microbiology-1979*. Washington, DC: American Society of Microbiology.

CRYZ, S. J. & HOLMES, R. K. (1979). Defective transport of ferric iron in mutants of *Corynebacterium diphtheriae* C7 (β) that produce diphtherial toxin under high-iron conditions. In *Abstracts of the Annual Meeting of the American Society of Microbiology*, p. 16. Washington, DC: American Society of Microbiology.

DALLAS, W. S., GILL, D. M. & FALKOW, S. (1979). Cistrons encoding *Escherichia coli* heat-labile toxin. *Journal of Bacteriology*, **139**, 850–8.

DALLAS, W. S. & FALKOW, S. (1980). Amino acid sequence homology between cholera toxin and the *Escherichia coli* heat-labile toxin. *Nature*. (In press.)

DUGUID, J. P., ANDERSON, E. S. & CAMPBELL, I. (1966). Fimbriae and adhesive properties in salmonellae. *Journal of Pathology and Bacteriology*, **92**, 107.

EKLUND, M. W., POYSKY, F. T., MEYERS, J. A. & PELROY, G. A. (1974). Interspecies conversion of *Clostridium botulinum* Type C to *Clostridium novyi* Type A by bacteriophage. *Science*, **186**, 456–8.

EKLUND, M. W., POYSKY, F. T., PETERSON, M. E. & MEYERS, J. A. (1976). Relationship of bacteriophages to alpha toxin production in *Clostridium novyi* Types A and B. *Infection and Immunity*, **14**, 793–803.

ELLWOOD, D. C., MELLING, J. & RUTTER, I. (eds.) (1979). *Adhesion of Microorganisms to Surfaces*. London: Academic Press.

FIELD, M., GRAF, L. H., LAIRD, W. J. & SMITH, P. L. (1978). Heat-stable enterotoxin of *Escherichia Coli: In vitro* effects on guanylate cyclase activity, cyclic GMP concentration, and iron transport in small intestine. *Proceedings of the National Academy of Sciences, USA*, **75**, 2800–4.

FORMAL, S. B. & HORNICK, R. B. (1978). Invasive *Escherichia coli*. *Journal of Infectious Diseases*, **137**, 641–4.

GYLES, C. L., PALCHAUDHURI, S. & MAAS, W. K. (1977). Naturally occurring plasmid carrying genes for enterotoxin production and drug resistance. *Science*, **198**, 198–9.

Gyles, C. L. (1979). Limitations of the infant mouse test for *Escherichia coli* heat stable enterotoxin. *Canadian Journal of Comparative Medicine*, **43**, 371–9.

Gibbons, R. A., Sellwood, R., Burrows, M. R. & Hunter, P. A. (1978). Inheritance of resistance to neonatal *E. coli* diarrhoea in the pig: Examination of the genetic system. *Theoretical and Applied Genetics*, **51**, 65–70.

Heffron, R., Rubens, C. & Falkow, S. (1975). Translocation of a plasmid DNA sequence which mediates ampicillin resistance: Molecular nature and specificity of insertion. *Proceedings of the National Academy of Sciences, USA*, **72**, 3623–7.

Honda, T. & Finklestein, R. A. (1979). Selection and characteristics of a *Vibrio cholerae* mutant lacking the A (ADP-ribosylating) portion of the cholera enterotoxin. *Proceedings of the National Academy of Sciences, USA*, **76**, 2052–6.

Inouye, M. (1979). *Bacterial Outer Membranes. Biogenesis and Function*. New York: John Wiley and Sons.

Khan, S. & Novick, R. (1980). Plasmid control of enterotoxin B production in *Staphylococcus aureau. Nature, London*, (submitted).

Maas, W. K. (1979). *Abstracts of a Symposium on 'Plasmids of Medical, Environmental and Commercial Importance'*. Spitzingsee, W. Germany.

McConnell, M. M., Willshaw, G. A., Smith, H. R., Scotland, S. M. & Rowe, B. (1979). Transposition of ampicillin resistance to an enterotoxin plasmid in an *Escherichia coli* strain of human origin. *Journal of Bacteriology*, **139**, 346–55.

Mazaitis, A. J., Maas, R. & Maas, W. K. (1980). Structure of a naturally occurring plasmid with genes for enterotoxin production and drug resistance. *Journal of Bacteriology*. (In press.)

Mekalanos, J. J. & Murphy, J. R. (1980). Regulation of cholera toxin production in *Vibrio cholerae*: Genetic analysis of phenotypic instability in hypertoxinogenic mutants. *Journal of Bacteriology*, **141**, 570–6.

Moll, A., Manning, R. A. & Timmis, K. N. (1980). Plasmid-determined resistance to serum bactericidal activity: a major outer membrane protein, the *traT* gene product, is responsible for plasmid-specified serum resistance in *Escherichia coli. Infection and Immunity*, **28**, 359–67.

Mooi, F. R., de Graaf, F. K. & von Emden, J. D. A. (1979). Cloning, mapping, and expression of the genetic determinant that encodes for the K88 antigen. *Nucleic Acid Research*, **3**, 849–65.

Mooi, F. R. & de Graaf, F. K. (1979). Isolation and characterization of K88 antigens. *FEMS Letters*, **5**, 17–20.

Murphy, J. R. (1976). Structure activity relationships of diphtheria toxin. In *Mechanisms in Bacterial Toxinology*, ed. A. W. Bernheimer, pp. 31–51. New York: John Wiley and Sons.

Murphy, J. R., Michel, J. L. & Teng, M. (1978). Evidence that the regulation on diphtheria toxin production is directed at the level of transcription. *Journal of Bacteriology*, **135**, 511–16.

Noegel, A., Rdest, V., Springer, W. & Goebel, W. (1979). Plasmid cistrons controlling synthesis and excretion of the exotoxin α-haemolysin of *Escherichia coli. Molecular and General Genetics*, **175**, 343–50.

Phillips, S. & Novick, R. P. (1979). Tn554-a site-specific repressor-controlled transposon in *Staphylococcus aureus. Nature, London*, **278**, 476–8.

Portnoy, D. A., Moseley, S. L. & Falkow, S. (1980). Possible plasmid mediated determinants of pathogenesis in *Yersinia enterocolitica* (Ye). In *Abstracts of the Annual Meeting of the American Society of Microbiology*, p. 19. Washington, DC: American Society of Microbiology.

Reis, M. H. L., Alfonso, H. T., Trabulsi, L. R., Mazaitis, A. J., Maas, R. &

MAAS, W. K. (1980). Transfer of a CFA/I plasmid promoted by a conjugative plasmid in a strain of *Escherichia coli* of serotype 0128:H12. *Injection and Immunity*, **29,** 140–3.

RUTTER, J. H., BURROWS, M. R., SELLWOOD, R. & GIBBONS, R. A. (1975). A genetic basis for resistance to enteric disease caused by *E. coli. Nature, London*, **257**, 135–6.

SANTOS, D. S., PALCHURDHURI, S. & MAAS, W. K. (1975). Genetic and physical characteristics of an enterotoxin plasmid. *Journal of Bacteriology*, **124**, 1240–7.

SHIPLEY, P. L., GYLES, C. L. & FALKOW, S. (1978). Characterization of plasmids that encode for the K88 colonization antigen. *Infection and Immunity*, **20**, 559–66.

SHIPLEY, P. L., DALLAS, W. S., DOUGAN, G. & FALKOW, S. (1979). Expression of plasmid genes in pathogenic bacteria. In *Microbiology-1979*, ed. D. Schlesinger, p. 176. Washington, DC: American Society of Microbiology.

SILVA, M. L. M., MAAS, W. K. & GYLES, C. L. (1978). Isolation and characterization of enterotoxin-deficient mutants of *Escherichia coli. Proceedings of the National Academy of Sciences, USA*, **75**, 1384–8.

SMITH, H. W. (1974). A search for transmissible pathogenic characters in invasive strains of *Escherichia coli*: The discovery of a plasmid-controlled toxin and a plasmid-controlled lethal character closely associated, or identical, with colicine V. *Journal of General Microbiology*, **83**, 95–111.

SMITH, H. W. & PARSELL, Z. (1975). Transmissible substrate-utilizing ability in enterobacteria. *Journal of General Microbiology*, **87**, 129–40.

SMITH, H. R., CRAVIATO, A., WILLSHAW, G. A., MCCONNELL, M. M., SCOTLAND, S. M., GROSS, R. J. & ROWE, B. (1979). A plasmid coding for the production of colonization factor antigen I and heat-stable enterotoxin in strains of *Escherichia coli* of serogroup 078. *FEMS Letters*, **6**, 255–60.

SO, M., BOYER, H. W., BETLACH, M. & FALKOW, S. (1976). Molecular cloning of an *Escherichia coli* plasmid determinant that encodes for the production of heat-stable enterotoxin. *Journal of Bacteriology*, **128**, 463–72.

SO, M., DALLAS, W. S. & FALKOW, S. (1978). Characterization of an *Escherichia coli* plasmid encoding for synthesis of heat-labile toxin: Molecular cloning of the toxin determinant. *Infection and Immunity*, **21,** 405–11.

SO, M., HEFFRON, F. & MCCARTHY, B. J. (1979). The *E. coli* gene encoding heat-stable toxin is a bacterial transposon flanked by inverted repeats of IS1. *Nature, London*, **277**, 453–5.

SO, M. & MCCARTHY, B. J. (1980). Nucleotide sequence of the bacterial transposon Tn1681 encoding a heat stable (ST) toxin and its identification in enterotoxigenic *E. coli* strains. *Proceedings of the National Academy of Sciences, USA, 77*, 4011–15.

SPICER, E. K., KAVANAUGH, W. M., DALLAS, W. S., FALKOW, S., KONIGSBERG, W. H. & SCHAFER, D. E. (1980). Sequence homologies between A subunits of *E. Coli* and *V. Chloerae* enterotoxins. *Proceedings of the National Academy of Sciences, USA* (in press).

SWANEY, L. M., LILL, Y.-P., TO, C.-M., IPPEN-IHLER, K. & BRINTON, C. C. (1977). Isolation and characterization of *Escherichia coli* phase variants and mutants deficient in Type 1 pilus production. *Journal of Bacteriology*, **130**, 495.

TAYLOR, P. W. & HUGHES, C. (1978). Plasmid carriage and the serum sensitivity of Enterobacteria. *Infection and Immunity*, **22,** 10–17.

WARREN, R., ROGOLSKY, M., WILEY, B. B. & GLASGOW, L. A. (1975). Isolation of extrachromosomal DNA for exfoliative toxin from phage group 2 *Staphylococcus aureus. Journal of Bacteriology*, **122,** 99–105.

WENSINK, J., GANKEMA, H., JANSEN, W. H., GUINEE, P. A. M. & WITHOLT, B.

(1978). Isolation of the membranes of an enterotoxigenic strain of *Escherichia coli* and distribution of enterotoxin activity in different subcellular fractions. *Biochimica et Biophysica Acta*, **514**, 128–36.
WILLIAMS, P. H. (1979). Novel iron uptake system specified by Col V plasmids: an important component in the virulence of invasive strains of *Escherichia coli*. *Infection and Immunity*, **26**, 925–32.

ACKNOWLEDGEMENTS

The author thanks Dr Patricia Shipley for having made her excellent review article available prior to publication (Elwell, L. P. & Shipley, P. L. (1980): Plasmid-associated factors associated with virulence of bacteria to animals. *Annual Reviews of Microbiology*, **34**, 465–496). The author is the holder of Public Health Service Research Career Award GM-15129 from the National Institute of General Medical Science.

GENETICS OF PATHOGENIC FUNGI

P. R. DAY

Plant Breeding Institute, Maris Lane, Trumpington, Cambridge CB2 2LQ, UK

INTRODUCTION

In 1974 I wrote at some length on the subject of this paper. An adequate revision of those ideas would require far more time and space than is available to me now. I have attempted instead to trace some of the contributions genetics has made to plant pathology and to deal with several developments in the genetics of fungal pathogens since 1974.

Early genetic studies in fungi were principally concerned with the controls of sexual reproduction and mating type. Mating types were ready-made markers distinguishable by tester stocks. Although fungi are eukaryotes their cytology is difficult because of the small size of nuclei and chromosomes and because the spindle develops, and mitosis takes place, inside the unbroken nuclear envelope. The cytological evidence for meiosis, even in the Ascomycetes, now one of the best known families, was at first confusing. The place in the life cycle at which meiosis occurs in the Oomycetes was only firmly established 11 years ago by microspectrophotometric analysis of single Feulgen-stained nuclei which revealed that gamete nuclei contained half as much DNA as the nuclei of vegetative hyphae (Bryant & Howard, 1969).

The most significant development in fungal genetics was the discovery of auxotrophic mutants in *Neurospora crassa* in 1945. This had two important consequences. The first was the principle of investigating the steps in the regulation of biosynthetic pathways by inducing and selecting mutant blocks and analysing their effects by identifying intermediates that accumulate before the block, or intermediates which relieve the block and hence occur after it. The second was that these markers provided selection techniques that allowed the measurement of low recombination frequencies which, in turn, led to fine-structure mapping, to the discovery of gene conversion in tetrads, and to many of our current concepts of DNA repair and the mechanism of recombination in eukaryotes. The discovery of the parasexual cycle in *Aspergillus* pointed to the significance of mitotic recombination in the absence of normal sexual reproduction.

One would expect to find many examples where the application of this body of knowledge in fungal genetics to plant pathogenic fungi was revealing and illuminating for the special problems associated with pathogens. Indeed nothing that we know about plant pathogenic fungi suggests that they are any different from their genetically better known saprophytic cousins. However, they grow on living substrates, that are also subject to genetic controls, for all (obligate parasites) or part (facultative parasites) of their life cycles. Much of the rationale for studying the genetics of plant pathogens comes from the belief that this knowledge will allow the development of better methods of controlling plant disease. Two such methods are pre-eminent, namely breeding for resistance and the use of fungicides. However, biological controls can also be used and these introduce a third genetic component in addition to the host and parasite. The following discussion deals with each of these approaches.

BREEDING FOR RESISTANCE

Although differences in the degree to which different varieties of crops were damaged by a fungal pathogen were recorded by Theophrastus (372–287 BC) the modern concept of inherited resistance dates from Biffen's work in Cambridge. He showed that resistance to yellow rust (*Puccinia striiformis*) in wheat was determined by a recessive gene (Biffen, 1905, 1912). The discovery was soon followed by others that showed dominant single gene resistance to be far more common. The search for easily manipulated single dominant genes for disease resistance in all crops subject to improvement by plant breeders began. The early N. American wheat breeders believed that, with resistance, they would be able to conquer stem rust (*P. graminis tritici*) once and for all.

Physiologic races

Unfortunately the early promise of disease resistance did not hold. Resistant varieties selected variants from the pathogen population to which they were as susceptible, sometimes more so, as the original varieties which they replaced. The physiological races of plant disease fungi presented breeders with a bewildering complexity of forms that could only be identified by test inoculations on a set of differential host varieties. Resistance determined by single dominant genes seemed to be especially vulnerable to the appearance of new pathogen races. This led to the boom and bust cycles of a successful new stem rust-resistant wheat becoming grown on an increasingly large scale only for it eventually to succumb to a new rust race. Later called vertical resistance by Vanderplank (1963), it was effective against some races but not at all against others. Horizontal resistance on the other hand appeared to be effective wherever and for as long as it was exposed but it was more difficult to handle in a breeding programme. Horizontal resistance is usually not expressed by seedlings and is often under the control of many genes of individually small effect. A survey carried out by Sidhu (1975) showed that of 1042 papers published since 1905 describing the inheritance of disease resistance, some 927 (approximately 89%) described monogenic resistance. In retrospect the problem of races was a result of 'man-guided evolution' (Johnson, 1961). Sidhu's survey also showed that the earlier genetic studies of resistance in plants were generally only conclusive when genetically homogeneous pathogen cultures were used to identify host resistance genes in segregating populations.

The genetics of pathogenicity

Research on the origin of physiologic races showed that recombination and mutation were both important. For heteroecious rusts, which need two distinct hosts to complete their life cycles, attempts were made to eradicate the non-economic or alternate hosts. For wheat stem rust in North America these were the several species of *Berberis* on which sexual reproduction and overwintering occurred. Barberry eradication, although carried out efficiently, would have been very much more effective in limiting inoculum if spore clouds were not blown up by winds from the south where the alternate host was not needed for overwintering. Another example, white pine blister rust (*Cronartium ribicola*), seemed a more promising candidate for this kind of control since white pine is only infected by spores formed on the alternate host, the wild currant (*Ribes* spp.). In spite of large-scale efforts at *Ribes* eradication over many years, in the forests of the north-west United States enough *Ribes* persisted to ensure infection of white pine. Breeding white pines for blister rust resistance is now being actively pursued.

There are considerable technical difficulties in carrying out genetic studies with heteroecious rusts. In nature they are obligate parasites. Although several rusts have been successfully grown on synthetic media (Scott & Maclean, 1969) their growth rates are slow and only vegetative spores are formed. When inoculated plants are used in the greenhouse precautions must be taken to prevent airborne cross contamination by spores. In genetic studies involving crosses the generation times are 4–6 months or more long and complicated by the need to break the dormancy of diploid teliospores before they will germinate to undergo meiosis and release haploid products.

For autoecious rusts some of the problems are simplified by the fact that all five spore forms: uredospores (dikaryotic n + n), teliospores (diploid 2n), basidiospores (haploid), spermatia (haploid), and aeciospores (dikaryotic, n + n), are produced on the same host. The best known genetically of these is flax rust (*Melampsora lini*) which can be a severe problem on flax (*Linum usitatissimum*). Over the period 1942–71 Flor, working in Fargo, North Dakota, established the genetic basis of rust resistance in a number of flax varieties and also carried out studies of the inheritance of virulence in the rust. I use the term virulence to mean ability to produce disease on a host variety carrying one or more resistance genes. When resistance is effective against a given pathogen genotype the pathogen phenotype is avirulent. In this usage pathogenicity is a general property, describing the ability to produce disease on the host species, while virulence and avirulence refer to reactions on specific host genotypes. With obligate parasites like the rusts, or mildews, non-pathogenic forms cannot be studied, unless conditional mutants are available. In facultative parasites that can readily be cultured on synthetic media, non-pathogenic forms are common and have been studied in attempts to throw light on the nature of pathogenicity (Day, 1974).

Gene-for-gene hypothesis

Flor (1956) summarized his findings on flax rust in the gene-for-gene hypothesis. This recognized that to be effective a host gene for resistance must be matched with an allele for avirulence at a corresponding locus in the pathogen. Multiple allelism and independent action is common at loci governing resistance in the host and is therefore a handicap for plant breeders who wish to incorporate two or more resistance alleles at the same locus in a diploid pure line. The extent of multiple allelism at loci controlling virulence in the fungi is not known. The only examples I am aware of are in the flax rust pathogen (*M. lini*) (Lawrence, 1977) and *Erysiphe graminis hordei* (Moseman, 1963).

The nature of virulence

It is disappointing to note that relatively few advances have been made in our understanding of the molecular biology of the systems governed by the gene-for-gene interaction. Simple interpretations, such as that the resistant host fails to provide a nutrient needed by the avirulent pathogen or that it produces a toxic substance and the avirulent pathogen is sensitive, are clearly inadequate. Neither model will explain the interactions of 19 identified avirulence loci in *Venturia* (apple scab) with a test series of apple cultivars. Most wild type strains of *V. inaequalis* grow on a simple defined culture medium and no differentially toxic materials have been isolated from resistant apple leaf tissue prior to exposure to *Venturia.* A currently fashionable model suggests that host resistance is an induced response triggered by the interaction of the products of the alleles for resistance and avirulence. The product of the avirulence allele is considered to be an elicitor (Keen, 1981) and the product of the resistance gene a membrane binding site for the elicitor. Several candidate elicitor molecules have been advanced. These include high molecular weight glycoproteins recovered from the cell walls of the bacterium *Pseudomonas glycinea,* the cause of soybean bacterial blight (Bruegger & Keen, 1979), and glycopeptides secreted into the culture medium by the fungus *Phytophthora megasperma sojae* which causes a stem and root rot of soybean (Wade & Albersheim, 1979). The supporting evidence is that elicitors prepared from different races of these two pathogens duplicate them in specificity.

Assays for elicitor activity are not simple. One is to show that resistant tissue expected to show an incompatible response produces a phytoalexin when exposed to the elicitor. Phytoalexins are toxic polyphenolic compounds which accumulate in plant tissues within a short time of exposure to a variety of mildly injurious stimuli. The precise role of phytoalexins in bringing about resistant (incompatible) host responses is not clear since they may appear in comparable amounts in susceptible (compatible) as well as resistant responses (Ayers, Ebel, Valent & Albersheim, 1976). The assay used by

Wade & Albersheim (1979) was to test the extracellular glycoprotein elicitors to see if they protected soybean seedlings exposed to a virulent strain of *P. megasperma sojae.* Applied 90 minutes before, or with the mycelial inoculum, to the site of inoculation a preparation from an incompatible race protected resistant soybean seedlings from infection by a compatible race. The same preparation, however, failed to protect a susceptible cultivar. In terms of eliciting phytoalexin production the glycoproteins were about 1% as active as a non-specific glucan cell wall component of *P. megasperma sojae.* The glycoproteins were also non-specific in the low levels of phytoalexin induced. How they protect the host is still unknown.

Another point of view is that disease resistance results from a failure to bring about a compatible response. I have argued elsewhere (Day, 1976b) that susceptibility is an induced response. More recently Doke & Tomiyama (1980) have shown that the hypersensitive response of potato tuber protoplasts to cell wall components of *Phytophthora infestans* is to some extent specifically suppressed by water-soluble glucans extracted from compatible races of the pathogen. A hypersensitive response involves the death of host cells adjacent to the invading pathogen. Generally the toxic materials produced prevent further pathogen growth. It is characteristic of the expression of single gene race-specific resistance. Doke & Tomiyama's findings are supported by the observation that similar substances are released by germinating zoospores (Doke, Garas & Kuć, 1980) and suggest that physiologic races of potato blight differ in their ability to suppress the development of resistance or defence mechanisms. Most pathologists retain the habit of thinking that races differ in their ability to trigger defence mechanisms. Ouchi *et al.* (1979) studied the host range of the powdery mildews, *Erysiphe graminis tritici* (wheat), *E.g. hordei* (barley) and *Sphaerotheca fuliginea* (melon). They found that leaves of all three host species could be induced to support germination of conidia and the growth of secondary hyphae of the two mildews that do not attack them in nature if the leaves were preinoculated with conidia of a compatible race. Thus if barley leaves were preinoculated with *E.g. hordei,* and the superficial mycelium was removed with a cotton wool swab after 48 hours, they would support development of *S. fuliginea* conidia. These authors called this phenomenon induced accessibility. It no doubt also is a result of suppression of the defence mechanism like the potato example. Compatible host-parasite interactions are uncommon in nature and it would be surprising if the degree of mutual adjustment that has taken place in their evolution were not complex and likely to be disrupted in a variety of ways which the gene-for-gene hypothesis seems to account for. In some respects the gene-for-gene hypothesis has been misleading in that it tends to suggest to the unsuspecting that its underlying mechanism is simple. It clearly is not.

The early hope that the problems of host-specificity would yield to a rigorous analysis of a plant pathogen as though it were *Neurospora crassa* was not sustained. Although *Venturia inaequalis* was explored in this way in the 1950s the results were disappointing (some of the problems of such an approach are discussed by Day (1979)). The gene-for-gene hypothesis also only takes account of resistant or susceptible and virulent or avirulent reactions. There are of course intermediate reactions, often with important epidemiologic consequences (Scott *et al.*, 1980).

Putative genotypes – putting the hypothesis to work

Another implication of the gene-for-gene hypothesis is its predictive role in enabling the geneticist to assign putative genotypes to one of the components in a host–parasite interaction when there is some knowledge of the genetics of the other. An example of this is the pattern of reactions shown by several apple varieties inoculated with isolates of *Venturia inaequalis,* the cause of scab disease. Genetic studies with *Venturia* showed that one, or several, unlinked avirulence alleles were responsible for the flecking (resistant) reactions of various isolates on a given variety. A cross between an avirulent and a virulent isolate gives rise to a segregating haploid progeny which is tested on the varieties in question. The observation that a flecking allele responsible for avirulence to one variety is also responsible for avirulence to another is thus evidence that the two varieties have a common resistance gene (Day, 1974).

Even when no genetic studies have been carried out in either host or parasite the interactions of an array of races with an array of host varieties can be examined with the aid of a computer to establish best fit sets of genotypes of both components to account for most of the interactions (Loegering, McIntosh & Burton, 1971). Dinoor & Peleg (1972) have done this for natural populations of oats and the oat crown rust pathogen (*Puccinia coronata*) from Israel.

The order created from the confusion of physiologic races by application of the gene-for-gene hypothesis led also to various schemes for applying systematic race designations (Day, 1976a). The more flexible of these schemes assigned arbitrary numbers, or positions in a series, to host varieties without analysis of their genotypes. The object was simply to relate race names to the varieties they attacked in a systematic way. In spite of the order it was inevitable that the more resistant varieties, or genes for resistance, that the plant pathologist had to take into account, the more cumbersome the race designations became. By the early 1970s pathologists began to realise that it was more important to measure the frequencies of particular virulence genes in pathogen populations than to try to identify all the combinations of the virulence genes that could be recognized (Wolfe & Schwarzbach, 1975).

Population genetics of pathogens

The development of new races following the release of new resistant varieties generated three kinds of response among breeders and plant pathologists. The first was to continue breeding varieties based on single gene hypersensitive resistance that were likely to fail because of their eventual vulnerability to a new race. The second was to regard major gene, race-specific, resistance as something to be avoided – a time bomb that would sooner or later destroy what the breeder had built. This caused a shift to non-race-specific resistance. The third response was to ask if a strategy of relieving crop monoculture by growing a mixture of genotypes would delay the spread and build-up of the pathogen so that harvests could be gathered with minimal loss. The genetic vulnerability of modern crop cultivars to pests and diseases is a consequence of their uniformity (Horsfall, 1972; Day, 1978a). The idea of introducing variation to reduce the risk is appealing, but for many years seemed impractical because of demands for a uniform product. Jensen (1952) first suggested multilines in oats as a means of satisfying the need for agronomic uniformity and at the same time providing heterogeneity for disease resistance. The idea was pursued by Borlaug for wheat heterogeneous for resistance to stem rust (*P. graminis tritici*) and stripe rust (*P. striiformis*) and by breeders at Ames, Iowa for oats heterogeneous for resistance to crown rust (*P. coronata*) (Frey, Browning & Simons, 1977). To produce a multiline a number of different sources of resistance are separately crossed to a common parent selected for its agronomic qualities. By backcrossing and selection a series of isolines each homozygous for resistance are developed which can be combined to form a composite blend of as many as 16 different components. Although multilines are successful in reducing disease losses they take a long time to prepare. By the time they are ready they are likely to be outclassed in respects other than resistance by the varieties selected in the meantime. In current work on breeding multilines there is less emphasis on uniformity of the component isolines. Backcrossing is now limited to two or three generations with selection for the desired plant type among the components heterogeneous for resistance (Wolfe & Barrett, 1980).

An alternative strategy already applied to barley is to blend mixtures of existing varieties chosen because they differ considerably in their resistance genes, but less so in other characters. This approach has great promise for controlling powdery mildew. Trials of three component mixtures have shown yields in diseased fields that are 109% of the mean yields of monocultures of the components (Wolfe & Barrett, 1980). With little or no disease, there may be other benefits since mixtures have yielded 103% of the mean of their components under these conditions. This suggests that mixtures may be better buffered to withstand other stresses in the environment than monocultures. Perhaps heterogeneity for other characters such as drought tolerance, resistance to lodging, and temperature extremes may be as useful as heterogeneity for resistance.

The success of varietal mixtures and multilines in controlling plant disease depends on placing the pathogen in an evolutionary dilemma. Possession of unnecessary genes for virulence will reduce its fitness on any one component of a mixture but without additional genes for virulence it will be unable to attack other components of the mixture (Wolfe & Barrett, 1980). To prevent the selection of a super-race with increased fitness that can attack all components the make-up of the mixture must be changed frequently. Any single race adapted to one of the components will produce conidia that are wasted when they reach the surrounding plants that they are unable to attack. In this way the development of an epidemic may be delayed sufficiently to enable grain filling to be completed.

The design and use of mixtures requires careful monitoring of the virulence gene frequencies in the pathogen population. During the last five years surveys have shown that races which combine virulence genes that would enable them to attack all components of three and four variety mixtures are uncommon and appear to increase only slowly, if at all (Wolfe & Barrett, 1980). However, to minimise the risk, regular changes in the composition of the mixtures are advisable.

SOME GENETIC CONSEQUENCES OF FUNGICIDES

Fungi are genetically versatile and it would be surprising if resistance to fungicides was not sometimes a problem in controlling plant diseases. Before 1969 there were few examples of fungicide-resistant pathogens. However, by 1976 resistance to 40 fungicides among some 52 species of plant pathogenic fungi had been described from field or laboratory studies (Ogawa, Gilpatrick & Chiarappa, 1977). The mechanisms of resistance and their genetic controls were reviewed by Georgopoulos (1977). Prior to 1969 the fungicides in use had a broad spectrum of activity interfering with a number of metabolic pathways. Phytotoxicity was limited provided penetration of the plant cuticle and cell wall did not occur. As more highly specific, and less phytotoxic, compounds, including systemics, were discovered they were more vulnerable because relatively simple changes in the pathogens made them resistant.

It is difficult to generalize about the effects of genes for fungicide resistance on pathogen survival in the absence of the fungicide. In Greece even after three years in which benzimidazole fungicides had not been used as sprays to control sugar beet leaf spot, no decline in frequency of resistant spores of *Cercospora beticola* was reported (Dovas, Skylakakis & Georgopoulos, 1976). On the other hand avoiding the use of seed treatments with the systemic fungicide ethirimol on winter barley appears to have been successful in prolonging its use on spring barley to control powdery mildew (*Erysiphe graminis hordei*). Wolfe & Schwarzbach (1978) noted the shorter persistence of fungicidal activity compared with host plant resistance and pointed out that fungicide-resistant genotypes may only have a relatively short part of the crop growth period when they have a selective advantage. Furthermore, fungicides applied as seed dressings will normally be active when the pathogen population density is low and will thus exert a greater selection pressure than foliar sprays applied when the population density is high.

Dekker (1977) pointed to a number of examples of fungicides for which resistant mutants can be recovered in the laboratory but have not appeared in the field. Most of these affect either the synthesis or function of sterols in the fungal cell (triforine, fenarimol, imazalil, triadimefon and the antibiotics nystatin, lucensomycin and pimaricin are examples). In several cases laboratory studies with the resistant mutants showed that they were less vigorous, less aggressive, and produced fewer spores, than sensitive strains. However, laboratory selection experiments involve small, finite populations and there is no opportunity for compensating selection for fitness. In comparison with large-scale field observations over several years, such studies are of doubtful significance.

The use of fungicide-treated seeds (Wolfe & Barrett, 1980) of a mildew-susceptible barley variety, as one member of a three-component mixture to control mildew, could exploit the possible lowered fitness of resistant forms on untreated seedlings. The strategy also adds to the scope for designing mixtures by relieving the limitations set by the availability of suitable mildew-resistant varieties. These must not only be heterogeneous for resistance but also compatible in terms of harvest date and end use.

In Western Europe the use of fungicide to protect cereal crops is now so widespread that the question arises of why we should continue to breed for resistance to diseases that can be chemically controlled. There are several reasons for continuing. If breeders relaxed selection for resistance, more frequent applications of fungicide would be required to maintain present levels of protection. The increased expense would be accompanied by a greater risk of spray damage to both the crop and the environment. It is not always possible, because of weather conditions or the pressure of other work, for farmers to spray when they should; resistance affords protection when they do not or cannot. Seed costs are in any case a small percentage of farm production costs (Day, 1977) and high-yielding disease-resistant varieties cost the farmer no more than similar but susceptible ones. The use of disease resistance and fungicides together as an integrated control has the potential of being more effective than either used alone.

FUNGAL VIRUSES

Hypovirulence in Endothia

The discovery of fungal viruses as the cause of a disorder of cultivated mushrooms sparked interest in the possibility that fungal viruses might be used as biological controls of plant pathogens. Among several candidates the most successful so far discovered is the virus responsible for hypovirulence in the chestnut blight pathogen *Endothia parasitica*. Discovered by Biraghi (1953) in Italy, it was at first interpreted as an example of induced resistance in the chestnut. Trees of *Castanea sativa,* which several years before were noted as infected, had remained alive and were not diseased when healthy branches were inoculated with fungus isolated from the healing cankers. Grente (1965) showed that pathogenicity was attenuated and that this character was transmitted by hyphal anastomosis. He called it hypovirulence and began to use hypovirulent strains of *E. parasitica* to cure blight infections on orchard trees in Southern France. Tests of Grente's hypovirulent strains in the USA showed that they would cure French virulent strains inoculated on *Castanea dentata* and that local hypovirulent strains could be derived which would cure other vegetatively compatible strains of US origin (Anagnostakis, 1978).

Native hypovirulent strains of *E. parasitica* were subsequently found in States east of the Mississippi River. These and nearly all other hypovirulent strains of French and Italian origin contain double-stranded RNA (dsRNA) (Day & Dodds, 1979). The strains vary considerably in the degree to which pathogenicity is reduced (Elliston, 1978). There is also considerable variation in the complement and sizes of dsRNA extracted from mycelia as judged by electrophoresis in polyacrylamide gels. The stability and form of the dsRNA present was shown by Anagnostakis & Day (1979) to be dependent on the recipient strain. Hypovirulent strains tend to be unstable and may give rise to forms with apparently normal pathogenicity.

In Italy the blight epidemic, which was first noticed in 1938, appears now to be almost completely under control (Mittempergher, 1978). In large areas chestnut regeneration has occurred and a high proportion of the *E. parasitica* cankers are evidently hypovirulent. Representative isolations have confirmed the presence of dsRNA and inoculation tests showed them to be hypovirulent. The process of recovery has occurred without man's deliberate intervention. It is of considerable interest therefore to compare the Italian and American epidemics. The latter began at the turn of the century as a result of the accidental introduction of diseased nursery stock from the Far East to the city of New York. The disease spread very rapidly through the chestnuts which comprised 25% of the eastern hardwood forest that formed the climax vegetation over thousands of square miles of woodlands from the Alleghenies eastwards and from Georgia north to Vermont and New Hampshire. *Castanea dentata* and *C. sativa* are similarly susceptible to *E. parasitica* but *C. dentata* does not respond with a healing response to inoculation with a hypovirulent strain as rapidly as *C. sativa* (J. Grente, personal communication).

E. parasitica is an ascomycete which is normally heterothallic (Anagnostakis, 1979). In Italy perithecia are uncommon and the number of vegetative compatibility groups, which govern the outcome of hyphal anastamoses between different strains, is small (Anagnostakis, 1977 & personal communication). In North America perithecia are common. The rapid spread of the epidemic is considered to have occurred through aerial dispersal of ascospores. There are also many vegetative compatibility groups, some of which are common to the European population.

In Italy in the 1930s chestnut was prized as a food source and dense stands or groves were encouraged. In the US the trees of the native American chestnut, although never 'cultivated', also formed contiguous stands. However, as a result of the destruction by blight, at the present time these are reduced to scattered clumps of stems formed by sprouts from old stumps.

The discovery of hypovirulence near Genoa in 1950 occurred at a time when the chestnut population density, consisting mostly of coppiced sprouts, was still high. Although discovered in the US only recently, hypovirulent cankers of considerable age are present in trees beyond the extreme edge of the natural chestnut range in Michigan. In 1978 I saw a grove of some 20 trees, planted by settlers 126 years earlier, near the eastern shore of Lake Michigan. All of these trees had cankers that were probably at least 50 years old. The nearest other chestnuts, also planted, were some 25 or 30 miles away (L. Brewer, personal communication). It is likely that under these conditions the only *Endothia* strains that could survive would be hypovirulent. Virus-free virulent strains would have eliminated their hosts. Further east, in Connecticut, only virulent strains have been observed. The density of chestnut, even though a tiny fraction of what it was originally, is high enough that pathogen survival does not depend on attenuation of its virulence. In places it is sufficiently uncommon that trees up to 10 inches in diameter at breast height may develop without becoming infected. Why is hypovirulence not more common in the eastern US? Vegetative incompatibility is clearly a barrier to the spread of the virus. However, it is not complete since either the virus can sometimes override it or else some incompatible confrontations permit cytoplasmic continuity for long enough for virus transfer to occur (Anagnostakis & Day, 1979). Much more information is needed on the presence of vectors, on the nature of the host response and on the density of infections to answer this question (Day, 1978b).

In France the introduction of highly debilitated hypovirulent strains has been very successful for the treatment of individual cankers on orchard trees but there has been little natural spread. Again we have an example of an evolutionary dilemma since viral dsRNAs that produce strongly hypovirulent strains which do not sporulate will be unable to perpetuate themselves. Thus viral dsRNAs able to spread must not restrict sporulation and growth of *Endothia* to the extent that no living infected propagules (spores or mycelial fragments) are formed that can be carried by insect and animal vectors (birds or mice) to uninfected virulent cankers. The variation in dsRNA genomes that occurs in some *Endothia* strains may well be important in generating a range of forms to ensure a balance between the virus and its host ensuring the survival of both. In Italy this has worked to the advantage of the chestnut. In the US progress in this direction has been much slower and whether such a balance can be achieved there by deliberate intervention remains to be seen.

Ustilago maydis

The killer system in *U. maydis* is determined by several dsRNA viruses but has no obvious application for the biological control of corn smut. It has rather attracted attention because of the opportunities it affords for exploring the molecular biology of these viruses in a plant pathogen unusually amenable to genetic analysis. The killer phenotype is associated with the presence of dsRNA viruses with spherical particles 41 nm in diameter (see review by Day & Dodds, 1979). Three different killer specificities, P1, P4 and P6, are each associated with characteristic dsRNA species with molecular weights ranging from 50 000 to 2.9 million daltons. Killer activity is the result of the secretion of a polypeptide of about 10 000 daltons which is toxic to sensitive strains. Purified killer protein was shown to possess nuclease activity against single-stranded RNA and single and double-stranded DNA (Levine, Koltin & Kandel, 1979). Non-killer derivatives from killer cells may retain dsRNA that determines resistance to the killer polypeptide. Resistance is also determined by nuclear genes that are recessive in diploids. Analysis of the segmented genomes of killers indicates that a 0.7×10^6 daltons segment is responsible for synthesis of the killer protein and that a segment of 0.05×10^6 daltons, common to all killer strains, is necessary for killer expression (Koltin, Levine & Peery, 1980). The analysis was carried out using non-killer mutants isolated following treatment with either N-methyl-N′-nitro-N-nitrosoguanadine, or UV, or non-killer progeny from crosses between killer and sensitive strains.

In fact dsRNA viruses are very common in *U. maydis.* Day (1981) examined a sample of 49 smut galls from Connecticut and found none without dsRNA. A total of nine different segmented genome classes were found. Inoculation experiments were carried out to detect whether the dsRNA genomes influenced pathogenicity on field grown maize plants. No significant interactions were found.

Stock cultures of meiotic products retained from teliospore material of 13 galls collected in Poland in 1970 (Silva, 1972) were tested and dsRNA was found in 36 out of 40 cultures. There were four classes of segmented genomes of which three were common to the Connecticut sample and one was unique.

CONCLUSION

Genetic studies of plant pathogens are generally directed towards finding better ways to control plant diseases. They reveal the strengths and weaknesses of the pathogen; what the breeder and pathologist must contend with and what they can exploit. The capacity of the pathogen for genetic change must be considered in choosing the best control strategy. It may invalidate single gene hypersensitive resistance to control rust in wheat monocultures. On the other hand, coupled with a knowledge of population dynamics, such resistance is seen to have great promise in barley mixtures to control mildew. Pathogen populations wax and wane. The epidemiological models that account for environmental effects are incomplete if they omit the genetic basis of plant and pathogen responses and their interactions. Wiser and more effective use of resistance, fungicides and biological controls cannot fail to follow as our knowledge of their genetics grows.

I wish to thank Martin Wolfe for his constructive comments.

REFERENCES

Anagnostakis, S. L. (1977). Vegetative incompatibility in *Endothia parasitica. Experimental Mycology*, **1**, 306–16.

Anagnostakis, S. L. (1978). The American Chestnut: new hope for a fallen giant. Bulletin 777, The Connecticut Agricultural Experiment Station, 9 pp.

Anagnostakis, S. L. (1979). Sexual reproduction of *Endothia parasitica in the laboratory. Mycologia*, **71**, 213–15.

Anagnostakis, S. L. & Day, P. R. (1979). Hypovirulence conversion in *Endothia parasitica. Phytopathology*, **69**, 1226–9.

Ayers, A. R., Ebel, J., Valent, B. & Albersheim, P. (1976). Host-pathogen interactions × fractionation and biological activity of an elicitor isolated from the mycelial walls of *Phytophthora megasperma sojae. Plant Physiology*, **57**, 760–5.

Biffen, R. H. (1905). Mendel's laws of inheritance and wheat breeding. *Journal of Agricultural Science*, **1**, 4–48.

Biffen, R. H. (1912). Studies in inheritance in disease resistance. II. *Journal of Agricultural Science*, **4**, 421–9.

Biraghi, A. (1953). Possible active resistance of *Endothia parasitica* in *Castanea sativa. Reports of Eleventh Congress of the International Union of Forest Research Organizations*, Rome, 643–5.

Bruegger, B. B. & Keen, N. T. (1979). Specific elicitors of glyceollin accumulation in the *Pseudomonas glycinea*-soybean host-parasite system. *Physiological Plant Pathology*, **15**, 43–51.

Bryant, T. R. & Howard, K. L. (1969). Meiosis in the Oomycetes I. A microspectrophotometric analysis of nuclear DNA in *Saprolegnia terrestris* Cookson. *American Journal of Botany*, **56**, 1075–83.

Day, P. R. (1974). *Genetics of Host-Parasite Interaction,* pp. 238. San Francisco: Freeman.

Day, P. R. (1976a). The taxonomy of physiologic races. In *Taxonomy of Fungi*, ed. C. V. Subramanian, pp. 164–9. University of Madras.

Day, P. R. (1976b). Gene functions in host-parasite systems. In *Specificity in Plant Diseases,* ed. R. K. S. Wood & A. Graniti, pp. 65–73. New York: Plenum Press.

Day, P. R. (1977). Plant genetics: increasing crop yield. *Science*, **197**, 1334–9.

Day, P. R. (1978a). The genetic basis of epidemics. In *Plant Pathology*, vol. 3, ed. J. G. Horsfall & E. Cowling, pp. 263–85. New York: Academic Press.

Day, P. R. (1978b). Epidemiology of hypovirulence. *Proceedings of the American Chestnut Symposium*, Morgantown, W. Va, 118–22.

Day, P. R. (1979). Modes of gene expression in disease reaction. In *Recognition and Specificity in Plant Host-Parasite Interactions*, ed. J. M. Daly & I Uritani, pp. 19–31. Tokyo: Japan Scientific Societies Press.

Day, P. R. (1981). Fungal virus population in corn smut from Connecticut. *Mycologia* (in preparation).

Day, P. R. & Dodds, J. A. (1979). Viruses of plant pathogenic fungi. In *Viruses and Plasmids in Fungi*, ed. P. A. Lemke, pp. 201–38. New York: Dekker.

Dekker, J. (1977). Tolerance and the mode of action of fungicides. *Proceedings 1977 British Crop Protection Conference – Pests and Diseases*, **3**, 689–97.

DINOOR, A. & PELEG, N. (1972). The identification of genes for resistance or virulence without genetic analyses, by the aid of the 'gene-for-gene' hypothesis. *Proceedings of the Cereal Rusts Conference*, Prague, Czechoslovakia, **2**, 115–19.

DOKE, N., GARAS, N. A. & KUĆ, U. (1980). Effect on host hypersensitivity of suppressors released during the germination of *Phytophthora infestans* cystospores. *Phytopathology*, **70**, 35–9.

DOKE, N. & TOMIYAMA, T. (1980). Suppression of the hypersensitive response of potato tuber protoplasts to hyphal wall components by water soluble glucans isolated from *Phytophthora infestans*. *Physiological Plant Pathology*, **16**, 177–86.

DOVAS, C., SKYLAKAKIS, G. & GEORGOPOULOS, S. K. (1976). The adaptability of the benomyl-resistant population of *Cercospora beticola* in Northern Greece. *Phytopathology*, **66**, 1452–6.

ELLISTON, J. E. (1978). Pathogenicity and sporulation of normal and diseased strains of *Endothia parasitica* in American Chestnut. *Proceedings of the American Chestnut Symposium*, Morgantown, W. Va, 95–100.

FLOR, H. H. (1956). The complementary genic systems in flax and flax rust. *Advances in Genetics*, **8**, 29–54.

FREY, K. J., BROWNING, J. A. & SIMONS, M. D. (1977). Management systems for host genes to control disease loss. *Annals of the New York Academy of Sciences*, **287**, 255–74.

GEORGOPOULOS, S. G. (1977). Development of fungal resistance to fungicides. In *Antifungal Compounds*, vol. 12, ed. M. R. Siegel & H. D. Sisler, pp. 439–95. New York: Dekker.

GRENTE, J. (1965). Les formes hypovirulentes d'*Endothia parasitica* et les espoirs de lutte contre le chancre du châtaignier. *Comptes Rendus Hebdomadaire des Séances de l'Academie d'Agriculture de France*, **51**, 1033–7.

HORSFALL, J. G. (1972). *Genetic Vulnerability of Major Crops*, pp. 307. Washington: National Academy of Sciences.

JENSEN, N. F. (1952). Intra-varietal diversification in oat breeding. *Agronomy Journal*, **44**, 30–4.

JOHNSON, T. (1961). Man-guided evolution in plant rusts. *Science*, **133**, 357–62.

KEEN, N. (1981). Recognition in gene-for-gene systems. *Advances in Plant Pathology*. London: Academic Press. (In press).

KOLTIN, Y., LEVINE, R. & PEERY, T. (1980). Assignment of functions to segment of the dsRNA genome of the *Ustilago* virus. *Molecular & General Genetics*, **178**, 173–8.

LAWRENCE, G. J. (1977). Genetics of pathogenicity in flax rust. Ph.D. Thesis, University of Adelaide.

LEVINE, R., KOLTIN, Y. KANDEL, J. (1979). Nuclease activity associated with the *Ustilago maydis* virus induced killer proteins. *Nucleic Acids Research*, **6**, 3717–31.

LOEGERING, W. Q. MCINTOSH, R. A. & BURTON, C. H. (1971). Computer analysis of disease data to derive hypothetical genotypes for reaction of host varieties to pathogens. *Canadian Journal of Genetics & Cytology*, **13**, 742–8.

MITTENPERGHER, L. (1978). The present status of chestnut blight in Italy. *Proceedings of the American Chestnut Symposium*, Morgantown, W. Va., 34–7.

MOSEMAN, J. G. (1963). Relationship of genes conditioning pathogenicity of *Erysiphe graminis* f.sp. *hordei* on barley. *Phytopathology*, **53**, 1326–30.

OGAWA, J. M., GILPATRICK, J. D. & CHIARAPPA, L. (1977). Review of plant pathogens resistant to fungicides and bactericides. *FAO Plant Protection Bulletin*, **25**, 97–111.

OUCHI, S., HIBINO, C., OKU, H., FUJIWARA, M. NAKABAYASHI, H. (1979). The induction of resistance or susceptibility. In *Recognition and Specificity in Plant*

Host-Parasite Interactions, ed. J. M. Daly & I. Uritani, pp. 49–65. Tokyo: Japan Scientific Societies Press.

Scott, K. J. & MacLean, D. J. (1969). Culturing of rust fungi. *Annual Review of Phytopathology,* **7**, 123–46.

Scott, P. R., Johnson, R., Wolfe, M. S., Lowe, H. J. B. & Bennett, F. G. A. (1980). Host-specificity in cereal parasites in relation to their control. *Applied Biology*, **5**, 349–93.

Sidhu, G. S. (1975). Gene-for-gene relationships in plant parasitic systems. *Science Progress, Oxford*, **62**, 467–85.

Silva, J. (1972) Alleles at the b incompatibility locus in Polish and North American populations of *Ustilago maydis* (DC) Corda. *Physiological Plant Pathology*, **2**, 333–7.

Vanderplank, J. E. (1963). *Plant Diseases: Epidemics and Control*, pp. 349. New York: Academic Press.

Wade, M. & Albersheim, P. (1979). Race-specific molecules that protect soybeans from *Phytophthora megasperma* var. *sojae. Proceedings of the National Academy of Sciences, USA*, **76**, 4433–7.

Wolfe, M. S. & Barrett, J. (1980). Can we lead the pathogen astray? *Plant Disease*, **64**, 148–55.

Wolfe, M. S. & Schwarzbach, E. (1975). The use of virulence analysis in cereal mildew. *Phytopathologische Zeitschrift*, **82**, 297–307.

Wolfe, M. S. & Schwarzbach, E. (1978). Patterns of race changes in powdery mildews. *Annual Review of Phytopathology*, **16**, 159–80.

GENETICS OF PHOTOSYNTHESIS AND THE CHLOROPLAST

SUE G. BARTLETT*, JOHN E. BOYNTON† AND NICHOLAS W. GILLHAM†

*The Rockefeller University, New York, NY, USA
†Departments of Botany and Zoology, Duke University, Durham, NC, USA

INTRODUCTION

Chloroplasts are the semiautonomous organelles responsible for photosynthesis in eukaryotic cells. They are bounded by a double membrane envelope and contain the photosynthetic lamellae (thylakoids) and a soluble phase (stroma). The double membrane envelope can be isolated free from other chloroplast components in certain higher plants (reviewed by Chua & Schmidt, 1979; Douce & Joyard, 1979). Although the inner and outer membranes of this envelope cannot yet be separated from one another, permeability studies with intact chloroplasts have shown that the inner membrane is the osmotic barrier. Specific translocator polypeptides mediate the transport of ions and metabolites through this membrane. Receptors in the envelope membranes are very likely involved in the transport of chloroplast proteins synthesized in the cytoplasm into the interior of the chloroplast. The envelope also plays a prominent role in the synthesis of galactolipids.

The thylakoid membrane system contains the pigments and proteins associated with the light reactions of photosynthesis (reviewed by Chua & Schmidt, 1979; Gillham, 1978; Kirk & Tilney-Bassett, 1978). Recent ultrastructural and biochemical studies have established that the inner envelope membrane and thylakoid membrane systems are distinct from one another (Douce & Joyard, 1979; Gunning & Steer, 1975; Heldt, 1976). The stroma includes a variety of soluble enzymes involved in the dark reactions of photosynthesis, the biosynthesis of chlorophyll and haeme, carotenoids, and certain amino acids and lipids (reviewed by Givan & Harwood, 1976; Kirk & Tilney-Bassett, 1978; Leech & Murphy, 1976).

The chloroplast stroma also contains chloroplast DNA and the entire apparatus for the transcription and translation of this DNA including polymerases, tRNAs, tRNA synthetases, ribosomes etc.

(reviewed by Bedbrook & Kolodner, 1979; Boynton, Gillham & Lambowitz, 1980; Edelman, 1980; Ellis, 1977; Gillham, 1978; Gillham & Boynton, 1980; Kirk & Tilney-Bassett, 1978). Chloroplast DNA in all algae and higher plants so far examined with the exception of the giant siphonaceous alga *Acetabularia* ranges in molecular weight from 60×10^6 to 126×10^6 d. and is circular in conformation. In *Acetabularia* the molecular weight of the chloroplast genome is comparable to that of *E. coli* (1.5×10^9 d.), but its conformation has not been established (Green, Muir & Padmanabhan, 1977; Padmanabhan & Green, 1978).

The chloroplast genome is small relative to the nuclear genome and so far appears only to specify chloroplast components. Thus the most economical approach to understanding how the genes specifying the multitude of chloroplast proteins are divided between the chloroplast and nuclear genomes is to establish which proteins are coded by the chloroplast genome. By elimination, the remaining proteins are assigned to the nucleus. This approach has proved highly successful in establishing which mitochondrial functions are governed by the mitochondrial genome (Borst & Grivell, 1978). In this article we describe briefly the three general methods that have been used to probe chloroplast genome function. Then we turn to the application of one of these methods, formal genetic analysis, to the chloroplast genome of a single model organism, the unicellular green alga *Chlamydomonas reinhardtii*. Not only is *C. reinhardtii* an appropriate subject for this symposium on microorganisms, but this alga is also the only plant in which the formal genetics of the chloroplast can presently be investigated. Finally, we consider the chloroplast components thought to be coded by the chloroplast genome in both *Chlamydomonas* and higher plants.

METHODS FOR STUDYING CHLOROPLAST GENOME FUNCTION

The genetic approach

Non-Mendelian (plastome) mutations affecting chloroplast structure and function were reported in 1909 by Correns in the four-o'clock (*Mirabilis*) and Baur in geranium (*Pelargonium*), shortly after the rediscovery of Mendel's principles of inheritance. Since that time a great many nuclear and plastome mutations affecting the normal biogenesis of the chloroplast have been isolated in higher plants (reviewed by Gillham, 1978; Gillham, Boynton & Chua, 1978; Hagemann & Börner, 1978; Kirk & Tilney-Bassett, 1978; Nasyrov & Šesták, 1975). Because the literature on plastome mutants in higher plants has been reviewed in detail in the foregoing references, we will make only a few general remarks here to serve as background against which the *Chlamydomonas* system can be contrasted.

First, in higher plants the pattern of inheritance of mutations affecting chloroplast structure and function has been used traditionally to distinguish plastome from nuclear mutations. Thus, nuclear mutations exhibit biparental inheritance without somatic segregation while plastome mutations are either maternally inherited or are inherited biparentally and show somatic segregation.

While rigorous proof that plastome mutations in higher plants actually reside in chloroplast DNA is still lacking, three lines of correlative evidence support this contention. (1) Light and electron microscopic evidence strongly suggests that plastids show the same pattern of inheritance as plastome mutations whether that pattern is biparental or maternal (reviewed by Hagemann, 1976, 1979; Sears, 1980; Tilney-Bassett, 1975; Tilney-Bassett & Abdel-Wahab, 1979). (2) In those species of plants where biparental inheritance of plastids occurs, somatic segregation of plastome mutants can be correlated with the presence of mixed cells containing normal and mutant plastids (reviewed by Gillham, 1978; Kirk & Tilney-Bassett, 1978). (3) In maize, where plastome mutations exhibit maternal inheritance, Pring & Levings (1978) have shown that the same is true of chloroplast DNA restriction pattern differences. In *Nicotiana*, Frankel, Scowcroft & Whitfeld (1979) analysed male sterile backcross derivatives from interspecific hybrids. In four out of six cases, these male sterile derivatives had the chloroplast DNA restriction patterns of the maternal parent. In two cases the chloroplast DNA of the male sterile derivative was distinct from either parent. These two exceptions to maternal inheritance could have arisen by nucleotide substitution in the chloroplast DNA of a maternal parent during the many generations of backcrossing.

The foregoing results with *Nicotiana* are particularly important because Wildman and his colleagues have used *Nicotiana* hybrids to provide the most rigorous genetic evidence so far available from higher plants that certain chloroplast components are coded by chloroplast DNA whereas others are coded by nuclear genes (reviewed by Chen, Johal & Wildman, 1977; Gillham, 1978;

Gillham *et al.*, 1978; Kung, 1976; Wildman *et al.*, 1975). In these experiments interspecific crosses were made between *Nicotiana* species differing in specific chloroplast constituents, most notably the large and small subunits of the CO_2-fixing enzyme ribulose-1,5-bisphosphate carboxylase (RuBPCase), and the pattern of inheritance was determined in reciprocal F_1 hybrids. The assumption was that biparental inheritance indicated that the component was coded by the nuclear genome while maternal inheritance meant that the structural gene resided in the chloroplast. Using this approach Wildman and his colleagues provided the first convincing genetic evidence that the large subunit of RuBPCase was coded by chloroplast DNA while the small subunit was the product of a nuclear gene.

Second, if formal genetic analysis of the chloroplast genome is to proceed further in higher plants, the first step must be to develop a system in which recombination of chloroplast genes can be detected. The most likely candidate would be a plant such as *Nicotiana* in which protoplast fusion, tissue culture and whole plant regeneration are all possible. Protoplast fusion obviates the problem of maternal inheritance as shown already by Chen, Johal & Wildman (1977). These authors followed the inheritance of the large and small subunits of RuBPCase in interspecific hybrids of *N. glauca* and *N. langsdorfii* where both subunits of the two parents can be distinguished electrophoretically. As expected, small subunits from both parents were present in the hybrids. However, only the large subunit from one or the other parent was found in all but one parasexual hybrid, suggesting that expression of the large subunit gene was random in the hybrids. The one exception which expressed both large subunit genes may have been a recombinant or contained both species of chloroplast DNA. Assuming that chloroplast fusion occurs, tissue culture provides the opportunity to screen large populations of cells for rare recombinants. Mutants resistant to different antibiotics known to block chloroplast protein synthesis should prove ideal for this. Maliga and his colleagues have already obtained one such mutation to streptomycin resistance (Maliga, Sz-Breznovits & Marton, 1973; Maliga, Sz-Breznovits, Joo & Marton, 1975).

Third, non-photosynthetic mutants of higher plants require special conditions for their propagation, e.g. growth on enriched media containing fixed carbon sources or grafting on wildtype plants. Recessive nuclear mutations completely blocking photosynthesis behave as seedling lethals in the homozygous condition. Plastome mutations which segregate somatically can easily be obtained as variegated plants. Although non-photosynthetic mutants among algae such as *Chlamydomonas, Chlorella*, *Euglena* and *Scenedesmus* are readily grown on a source of fixed carbon such as acetate, genetic analysis is possible only in *Chlamydomonas.*

Products of chloroplast translation

This approach to understanding chloroplast genome function relies on the assumption that chloroplast translation products are coded by chloroplast DNA. Chloroplast translation products have been assayed *in vivo* using inhibitors of cytoplasmic or chloroplast protein synthesis and *in vitro* in isolated plastids. In the *in vivo* method (reviewed by Gillham, 1978; Gillham *et al.*, 1978), the plant under study is pulse labelled in the presence of each inhibitor separately. Those polypeptides synthesized in the presence of the inhibitor of cytoplasmic protein synthesis (e.g. anisomycin or cycloheximide) must be made on chloroplast ribosomes whereas those formed in the presence of an inhibitor of chloroplast protein synthesis (e.g. chloramphenicol, erythromycin, lincomycin, spectinomycin) must be made on cytoplasmic ribosomes. Ideally, those polypeptides affected by the cytoplasmic inhibitor should not be affected by the inhibitor of chloroplast protein synthesis and vice versa. Although this result is obtained for many polypeptides, occasional polypeptides are found whose synthesis appears to be sensitive to both classes of inhibitors. The latter polypeptides are usually elements of multicomponent systems (e.g. chloroplast ribosomes, thylakoid membranes) and their apparent sensitivity to both classes of inhibitors can be explained by the fact that they require a partner protein for integration (Chua & Gillham, 1977). Thus, a polypeptide made on chloroplast ribosomes may require the presence of a second polypeptide synthesized on cytoplasmic ribosomes for its assembly. Alternatively, antibodies for a given polypeptide can be used to detect synthesis of this polypeptide directly, circumventing the assembly problem.

The *in vivo* method is particularly well adapted for use in unicellular plants such as *Chlamydomonas* and *Euglena* where homogenous populations of cells can be labelled in the presence of inhibitors for precisely controlled times. However, the method has also been applied successfully in regreening and excised shoot systems in higher plants (reviewed by Gillham *et al.*, 1978).

In contrast, the *in vitro* method is particularly well adapted to higher plant systems such as pea and spinach (reviewed by Ellis, 1976, 1977; Ellis & Barraclough, 1978; Ellis, Highfield & Silverthorne, 1977). where excellent preparations of intact chloroplasts are readily made. Ellis and his colleagues have perfected this system for the study of chloroplast translation products in pea. In this system light energy or ATP is required to drive the synthesis of polypeptides from labelled amino acids. Chloroplast preparations from *Euglena* have been used by Price, Ortiz & Gaynor (1978) to investigate chloroplast translation products in this flagellate. Unfortunately, the *in vitro* method cannot be used so far in *Chlamydomonas* where the preparation of intact chloroplasts has proved an elusive goal. Ideally, the *in vivo* and *in vitro* methods should give essentially concordant results. So far, this seems to be the case, as we discuss later in this review.

Transcripts of chloroplast DNA

Recently, chloroplast DNA transcripts with known functions have been localized to specific regions of the restriction map of the chloroplast genome. The method has been used to map the chloroplast rRNA genes in several higher plants, *Chlamydomonas* and *Euglena* (reviewed by Boynton *et al.*, 1980). In addition, tRNA genes have been localized in spinach (Driesel *et al.*, 1979), 4S (presumably tRNA) genes in *Chlamydomonas* (Malnoë & Rochaix, 1978) and messenger RNAs for several chloroplast proteins (Bogorad, Link, McIntosh & Jolly, 1979; Hallick *et al.*, 1979; Malnoë, Rochaix, Chua & Spahr, 1979).

CHLOROPLAST GENETICS OF *CHLAMYDOMONAS*

The chloroplast genome and its transmission

Chlamydomonas reinhardtii is a biflagellate green alga containing a single, large cup-shaped chloroplast and several mitochondria (reviewed by Gillham, 1978). Both organelles possess their own unique genetic information and protein synthesizing systems distinct from their counterparts in the nucleus and cytoplasm (reviewed by

Boynton *et al.*, 1980; Gillham *et al.*, 1979). Four species of DNA have been identified in this alga: α, nuclear DNA (1.723 g/cc); β, chloroplast DNA (1.696 g/cc); γ, cytoplasmic ribosomal DNA (1.715 g/cc); δ, mitochondrial DNA (1.706 g/cc). In vegetative cells chloroplast DNA constitutes about 14% of the cellular DNA, but contains only about 0.2% of the total information (reviewed by Gillham, 1978). This DNA is organized in 62 μm circles (1.34×10^8 d.) present in 70–80 copies per cell (Behn & Herrmann, 1977). Restriction mapping of this genome yields a molecular weight of 1.26×10^8 d. and reveals that each molecule contains inverted repeats of 1.3×10^7 d. located approximately opposite one another (Rochaix, 1978). Kinetic complexity estimates are somewhat higher (2×10^8 d.), probably because the molecular weights of the standards were overestimated (Gillham, 1978).

The sexual cycle in *C. reinhardtii* is induced by depriving vegetative cells of nitrogen (Sager & Granick, 1954; Sears, Boynton & Gillham, 1980). When gametes of opposite mating type (designated mt^+ and mt^-) are mixed, they quickly fuse to form diploid zygotes. Shortly after zygote formation, not only nuclei but chloroplasts fuse (Cavalier-Smith, 1970). Although the chloroplasts of the gametes contain equal amounts of chloroplast DNA (Sears *et al.*, 1980), most zygotes transmit only chloroplast genes and genomes from the mt^+ (maternal) parent to the meiotic progeny upon germination (Gillham, 1978; Grant, Gillham & Boynton, 1980; Sager, 1972). In contrast, nuclear genes segregate 2:2 and putative mitochondrial genes are transmitted in a biparental, but non-Mendelian fashion (Alexander, Gillham & Boynton, 1974; Wiseman, Gillham & Boynton, 1977).

In every cross a few zygotes transmit chloroplast genes from both parents (biparental zygotes) or rarely only from the mt^- parent (paternal zygotes). The frequency of biparental zygotes can be increased greatly by ultraviolet (UV) irradiation of mt^+ gametes prior to mating (Sager & Ramanis, 1967; Gillham, Boynton & Lee, 1974). Segregation and recombination of chloroplast genes are studied among the progeny of biparental zygotes.

Isolation of chloroplast mutants and their genetic analysis

While molecular mapping techniques have been applied to study chloroplast genome function in higher plants, *C. reinhardtii* is unique in that molecular methods can be combined with the classical methods of transmission genetics in probing the chloroplast genome. In this regard *C. reinhardtii* now has the same advantages for the study of chloroplast genome function that baker's yeast, *Saccharomyces cerevisiae*, has for investigating the mitochondrial genome. First, methods have been developed for the selective isolation and identification of mutations in the chloroplast genome. Second, these mutations can be mapped and allele tested. Third, mutations can be assigned to complementation groups. Fourth, the genetic and physical maps can be aligned.

Until recently most chloroplast mutants isolated in *C. reinhardtii* were resistant to or dependent upon antibacterial antibiotics such as erythromycin and streptomycin (Gillham, 1978; Gillham *et al.*, 1978; Sager, 1972). These mutations confer antibiotic resistance on chloroplast ribosomes *in vitro* (reviewed by Boynton *et al.*, 1980). Since antibiotic resistant and dependent mutants represent a limited class with which to probe the chloroplast genome, development of new methods for the isolation of chloroplast mutations with different phenotypes became imperative. As discussed by Gillham *et al.* (1979), non-photosynthetic (acetate-requiring) mutations were the logical mutations for which to search. To facilitate isolation of these mutants, three conditions had to be met: (1) selective killing of photosynthetically competent (wild type) cells; (2) elimination of the high background of nuclear non-photosynthetic mutations; (3) reduction in the redundancy of the chloroplast genome to increase the likelihood of expression of chloroplast mutations. The frequency of photosynthetically competent cells can be decreased by exposing mutagenized cells to arsenate (Harris, Boynton & Gillham, 1974; Togasaki & Hudock, 1972) or metronidazole (Schmidt, Matlin & Chua, 1977). Growth of *C. reinhardtii* in media containing the thymidine analogue 5-fluorodeoxyuridine (FdUrd) leads to a selective reduction in the amount of chloroplast DNA (Wurtz, Boynton & Gillham, 1977) and an increase in the frequency of chloroplast mutations with no concomitant increase in the frequency of nuclear mutations (Gillham *et al.*, 1979; Wurtz *et al.*, 1979). This reduction in ploidy presumably facilitates expression of chloroplast mutations and the frequency of these mutations is also very likely raised because of mutation induction by thymine deprivation. The FdUrd method yielded for the first time numerous chloroplast mutations with specific defects in photosynthesis (Shepherd, Boynton & Gillham, 1979) and has been exploited by other laboratories to obtain additional non-photosynthetic chloroplast mutations

(Bennoun, Masson, Piccioni & Chua, 1978; Spreitzer & Mets, 1980).

Subjecting FdUrd treated cells to arsenate (Shepherd *et al.*, 1979) or metronidazole (Schmidt *et al.*, 1977) selection in the light to kill photosynthetically competent cells has yielded chloroplast mutations with defects in chloroplast ribosome assembly, photosystems I and II and the CF_1-ATPase. However, neither method has yielded mutations specifically affecting RuBPCase, the large subunit of which is coded by chloroplast DNA (Malnoë *et al.*, 1979). Recently, Spreitzer & Mets (1980) isolated for the first time a mutation affecting the large subunit. They obtained this mutant by growing cells in the dark in the presence of FdUrd, mutagenizing the cells with ethyl methane sulfonate and selecting colonies which grew on acetate in the dark, but were killed in the light with CO_2 as the sole carbon source. In addition, light-sensitive nuclear and chloroplast mutations with defects in either photosystem I or II were obtained using this protocol. Spreitzer and Mets theorized that previous failures to obtain mutants with clear defects in RuBPCase resulted from the photosensitivity of cells unable to fix CO_2. This hypothesis is reasonable since both photorespiration and photosynthesis are catalysed by the same active site on RuBPCase (Andrews, Lorimer & Tolbert, 1973; Lorimer, Andrews & Tolbert, 1973; Tolbert & Ryan, 1976). Evidence is mounting that photorespiration is necessary for draining excess photochemical energy when conditions for the Calvin cycle are suboptimal (Powles & Osmond, 1979). Thus, mutations leading to impaired CO_2 fixing ability often may be lethal when selection is carried out in the light. However, this should not mean that mutations blocking the function of other Calvin cycle enzymes should necessarily be lethal. For example the nuclear mutant F60 which lacks phosphoribulokinase activity grows in the light (Moll & Levine, 1970).

Analysis of recombination and segregation of chloroplast genes among the progeny of biparental zygotes has not only made possible the construction of chloroplast gene maps, but has led to the conundrum that these genes behave as if they were present in far fewer copies than the 70 to 80 DNA molecules known to be present in each chloroplast. Since models explaining chloroplast gene behaviour and methods for mapping chloroplast genes have been extensively criticized and reviewed elsewhere (Adams, Van Winkle-Swift, Gillham & Boynton, 1976; Birky, 1978; Forster *et al.*, 1980; Harris *et al.*, 1977; Sager, 1972, 1977; Sager & Ramanis, 1976a, b;

Singer, Sager & Ramanis, 1976; VanWinkle-Swift, 1980), we will simply summarize the results briefly here.

Sager and colleagues have proposed a diploid model for the chloroplast genome in which chloroplast genes segregate equationally to account for the results thay have obtained by pedigree analysis. Gillham, Boynton & Lee (1974) proposed a multiple copy model for the chloroplast genome in which most copies were normally transmitted by the maternal parent and in which chloroplast gene segregation was also equational. Their results were obtained with zygote clones. Neither model really accounts satisfactorily for the behaviour of chloroplast genes. The diploid model is not in accord with the finding that chloroplast DNA is present in multiple copies in the chloroplast of *C. reinhardtii*. The multiple copy model does not adequately explain the rapid segregation rates of chloroplast genes in pedigrees (Forster *et al.*, 1980). Recently, VanWinkle-Swift (1980) has proposed a model in which the many copies of chloroplast DNA are grouped into a small number of discrete nucleoids which are distributed to daughter chloroplasts non-randomly. Thus, nucleoids on opposite sides of the plane of chloroplast division will be directed to opposite daughter cells. This model is consistent with a large number of physical copies, but a low number of segregating units. The recent cytological observations of Coleman (1978, 1979) are in accord with such a model. Using a DNA-specific fluorochrome (4′6-diamidino-2-phenylindole, DAPI), Coleman has shown that the chloroplasts of *Chlamydomonas* and other algae possess small numbers of DNA-containing nucleoids.

Chloroplast genes have been mapped to a single linkage group using several different methods for analysis. Sager (Sager, 1977; Sager & Ramanis, 1976b; Singer *et al.*, 1976) has mapped chloroplast genes on the basis of the frequency with which they segregate together in parental combination (cosegregation analysis) in pedigrees and using the polarity of the segregation rate of different pairs of chloroplast alleles in liquid culture. Sager has been able to reconcile the two linear maps obtained using these methods by drawing a single, circular map.

Allele testing and mapping of chloroplast genes by conventional recombination analysis in the progeny of biparental zygote clones (Bartlett, Harris, Grabowy, Gillham & Boynton, 1979; Conde *et al.*, 1975; Gillham *et al.*, 1979; Harris *et al.*, 1977) yields a linear map which presently includes nine loci. Several of the same antibiotic resistance mutants have been analysed by both groups of workers and Sager's circular map can be reconciled with the linear recombination map if the circle is broken at a specific point (Gillham, 1978). Whether the existing genetic map is indeed circular and the markers studied are distributed around the entire circumference of the circular chloroplast DNA molecule or are tightly linked in a single linear array on a small segment of the circle will be apparent once the genetic and physical maps have been aligned.

Grouping mutations into recombinationally distinct loci is not sufficient proof that they affect different functional genes. While nuclear mutations are easily tested for complementation in vegetative diploids of *C. reinhardtii* (Gillham, 1978), chloroplast mutations segregate somatically and cannot be tested for complementation in dividing cells. This problem has recently been circumvented by Bennoun, Masson & Delosme (1980) by assaying complementation of chloroplast genes in terms of restoration of photosynthetic function at the level of fluorescence kinetics in young, non-dividing zygotes. Not only did chloroplast mutants defective in photosystems I and II complement, but complementation was also observed for recombinationally distinct mutants deficient in the CF_1 ATPase. Togasaki and Hudock (personal communication) have also used light-induced oxygen evolution to measure complementation of nuclear mutants in young, non-dividing zygotes and this method can very likely be applied to chloroplast mutants as well. Recombination of chloroplast genes in these young zygotes is a highly unlikely explanation for the restoration of photosynthetic function, because the majority of these zygotes ultimately transmit only maternal chloroplast genes.

The availability of methods for selective isolation of chloroplast mutants, for mapping these mutants, and for assigning them to functional groups, means that all of the methods of classical transmission genetics can now be brought to bear in dissecting the chloroplast genome of *C. reinhardtii*. What remains is to align the genetic and physical maps of the chloroplast genome. There are several ways in which this can be done. Malnoë *et al.* (1979) have localized the gene coding for the large subunit of RuBPCase on the physical map of the chloroplast genome and Spreitzer & Mets (1980) have succeeded in isolating a chloroplast mutant which alters the large subunit. By assuming this mutation maps in the structural gene coding for the large subunit and by establishing the genetic map position of this mutant with regard to the chloroplast linkage group, the mutant fixes one point on the genetic and physical maps. To fix a second point on these two maps, either mutations must be isolated in other chloroplast genes that have been mapped physically or else physical markers must be obtained in the chloroplast genome and these mapped genetically.

In order to obtain the requisite physical markers Gillham *et al.* (1979) screened over 30 different chloroplast mutants with the restriction enzymes *Eco*RI and *Hae*III and reported that two non-photosynthetic mutants contained deletions in chloroplast DNA. One of the mutants, *ac-u-g-2-3*, had symmetrically located deletions in the inverted repeat regions of about 100 base pairs in length (Grant *et al.*, 1980). However, since a revertant of *ac-u-g-2-3* has been isolated and because genetic recombinants have been obtained with the mutant phenotype which lack both deletions, we know that the *ac-u-g-2-3* mutation and the deletions are distinct. Grant *et al.* (1980) have used the stock containing the *ac-u-g-2-3* mutant and the two deletions to demonstrate that the maternal pattern of inheritance of chloroplast genetic markers is reflected precisely by the inheritance of the deletions. Additional deletion mutations have now been isolated and their analysis, presently underway, should prove invaluable for the precise orientation of the genetic and physical maps of the chloroplast genome.

GENETIC CONTROL OF CHLOROPLAST STRUCTURE AND FUNCTION

Many nuclear and chloroplast (plastome) mutations blocking the normal biogenesis or function of the chloroplast have been isolated and characterized in both algae and higher plants (reviewed by Kirk & Tilney-Bassett, 1978; Gillham *et al.*, 1978; Gillham, 1978; Nasyrov & Šesták, 1975; Hagemann & Börner, 1978; Akoyunoglou & Argyroudi-Akoyunoglou, 1978). The pigment deficiency in many of these mutants is but one of several pleiotropic consequences of the failure to make specific gene products necessary for normal biogenesis of the chloroplast and usually does not result from a block in the biosynthetic pathway for chlorophyll *per se*. The primary genetic defect or gene product involved has been identified in only a very few of the many mutations known in higher plants and algae. In certain cases, the genetic defect has been localized to a specific metabolic pathway, e.g. chlorophyll biosynthesis (Gough,

1972, Von Wettstein *et al.*, 1971; Wang, 1978), to a specific portion of the photosynthetic electron-transport chain (Levine, 1968, 1969, 1971; Bishop 1971; Bishop & Jones, 1978), or to a specific organelle such as the chloroplast ribosome (Boynton *et al.*, 1980).

The genetic control of specific chloroplast proteins, as deduced using the three approaches outlined at the beginning of this article, was thoroughly reviewed by Gillham *et al.* (1978). Therefore only the major conclusions will be summarized here as background for the salient new findings with special emphasis on those from *Chlamydomonas*.

Chloroplast protein synthesizing systems

Chloroplast protein synthesis bears many similarities to prokaryotic protein synthesis. Chloroplasts contain *N*-formylmethionyl tRNA and active transformylases (reviewed by Gillham, 1978; Stutz & Boschetti, 1976). Chloroplasts also contain unique elongation factors similar to those in prokaryotes and these have been characterized in detail in *Chlorella* and spinach (Tiboni, Pasquale & Ciferri, 1978). Elongation factors $EF\text{-}G_{chl}$ and $EF\text{-}T_{u\,chl}$ have been purified from spinach chloroplasts and both factors are labelled in isolated chloroplasts, indicating that they are products of chloroplast protein synthesis.

(1) *Ribosomes*

The structure and function of chloroplast (70S) ribosomes have been reviewed recently by Boynton *et al.* (1980). Chloroplast ribosomes are resistant to inhibitors of eukaryotic protein synthesis such as cycloheximide and anisomycin, but sensitive to virtually all inhibitors of prokaryotic protein synthesis including chloramphenicol, the aminoglycosides (kanamycin, neamine, neomycin, spectinomycin, streptomycin), the macrolides (carbomycin, clindamycin, erythromycin) and lincomycin. All of these antibiotics are effective inhibitors of protein synthesis by chloroplast ribosomes of *Chlamydomonas in vitro* with the exception of chloramphenicol. Many of these antibiotics, including chloramphenicol, also inhibit chloroplast protein synthesis *in vivo* in *Chlamydomonas* (Conde *et al.*, 1975; Chua & Gillham, 1977).

In *Chlamydomonas*, chloroplast ribosomes account for about 30% of the total cellular ribosomes whereas the 80S cytoplasmic ribosomes constitute most of the remainder and mitochondrial ribosomes are negligible in quantity (reviewed by Boynton *et al.*, 1980). The small (41S) subunit of the *Chlamydomonas* chloroplast ribisome contains 16S rRNA and 22–34 proteins whereas the large (54S) subunit has 23S, 7S, 5S and 3S rRNA as well as 26–34 proteins. The structure of chloroplast ribosomes from higher plants is similar except that the large subunit contains only 23S, 5S and 4.5S species of rRNA. The chloroplast rRNA genes have been mapped on the chloroplast genome in *Chlamydomonas* (Rochaix & Malnoë, 1978); *Euglena* (Gray & Hallick, 1978, 1979; Hallick *et al.*, 1979; Jenni & Stutz, 1978; Rawson *et al.*, 1978); maize (Bedbrook, Kolodner & Bogorad, 1977; Bogorad *et al.*, 1979) and spinach (Crouse *et al.*, 1978; Whitfeld, Herrmann & Bottomley, 1978). In *Chlamydomonas,* maize and spinach there are two sets of rRNA genes, one located in each of the two inverted repeat regions which range in size from 1.25×10^7 to 1.62×10^7 d. and are widely separated from one another in the chloroplast genome (Bedbrook & Kolodner, 1979). In *Euglena* there are three sets of rRNA genes closely linked in straight tandem repeats. Inverted repeats appear to be absent from the chloroplast DNA of pea (Kolodner & Tewari, 1979) but the rRNA genes have not been mapped. In *Chlamydomonas* each inverted repeat of the chloroplast DNA codes for (5′)-5S-23S-3S-7S-spacer-16S-(3′). The 23S gene contains a 940 base pair intervening sequence and the spacer contains at least one 4S (tRNA) gene.

Only one detailed study of the sites of synthesis of individual chloroplast ribosomal proteins has so far been published. In *Euglena* Freysinnet (1978) has used inhibitors of chloroplast and cytoplasmic protein synthesis *in vivo* to show that at least nine chloroplast ribosomal proteins are made within the chloroplast while at least 12 are made on cytoplasmic ribosomes. Many chloroplast and nuclear mutants affecting chloroplast ribosome structure and function have been isolated in *Chlamydomonas* (reviewed by Boynton *et al.*, 1980). These mutations either confer resistance to antibiotics which block protein synthesis on chloroplast ribosomes or cause defects in chloroplast ribosome assembly. Subunit exchange experiments employing synthetic polynucleotide messages and labelled amino acids show that the antibiotic resistant and dependent mutations have the same subunit specificity as similar mutations in *E. coli* (Bartlett *et al.*, 1979). So far the structural gene for only one chloroplast ribosomal protein has been identified unequivocally. In *Chlamydomonas*, Davidson, Hanson & Bogorad (1974) concluded that a nuclear gene codes for protein LC6, since erythromycin resistant mutations in this gene either delete a portion of this protein or alter its electrophoretic mobility. In several other cases chloroplast gene mutations in *Chlamydomonas* resistant to specific antibiotics have been reported to alter particular proteins of a given chloroplast ribosomal subunit, although each of these cases needs more rigorous confirmation (Boynton *et al.*, 1980). Bourque, Horn & Capel (1977) have reported that the chloroplast ribosome small subunits from the maternally inherited streptomycin resistant mutant of *Nicotiana* isolated by Maliga *et al.* (1973, 1975) bind about one-tenth as much labelled streptomycin as wildtype subunits and show differences in the electrophoretic mobility of at least two ribosomal proteins.

(2) *tRNAs and amino acyl tRNA synthetases*
In spinach most of the 35 tRNAs found within the chloroplast have been mapped on the chloroplast genome (Driesel *et al.*, 1979). In *Chlamydomonas* between 12 and 14 out of 33 *Eco*RI chloroplast DNA restriction fragments have been found to hybridize 4S RNAs (Malnoë & Rochaix, 1978). This is a minimal estimate of the number of chloroplast tRNA genes since any one fragment could contain more than one tRNA gene. Saturation hybridization experiments using 4S RNA in *Euglena* indicate that the chloroplast genome of this flagellate contains 25 tRNA genes (McCrea & Hershberger, 1976; Schwartzbach, Hecker & Barnett, 1976). Similar experiments with maize indicate the presence of 20–26 tRNA genes (Haff & Bogorad, 1976). In *Euglena* at least several aminoacyl tRNA synthetases seem to be coded by nuclear genes (reviewed by Gillham *et al.*, 1978). So far, no evidence has been published indicating that any of the tRNA synthetases are coded by chloroplast genes.

Stroma proteins

(1) *Ribulose-1,5-bisphosphate carboxylase/oxygenase (RuBPCase)*
This enzyme performs two catalytic reactions in the chloroplast, one of which is the first step in the Calvin cycle and the other of which is involved in photorespiration (properties reviewed by Jensen & Bahr, 1976, 1977; Tolbert & Ryan, 1976).

The RuBPCase holoenzyme is an extremely large soluble protein (about 550 000 d. in higher plants and green algae). The enzyme is oligomeric, consisting of eight identical large subunits (55 000 d.)

and eight identical small subunits (14 000 d.). The small subunit is synthesized on cytoplasmic ribosomes (Gooding, Roy & Jagendorf, 1976; Roy, Patterson & Jagendorf, 1976; Gray & Kekwick, 1974). Synthesis of the large subunit in isolated chloroplasts (Blair & Ellis, 1973) as well as in an *in vitro* bacterial translation system primed with chloroplast RNA (Hartley, Wheeler & Ellis, 1975) suggested that this subunit is synthesized on 70S ribosomes.

As discussed earlier, the best evidence that the cytoplasmic and chloroplast sites of synthesis reflected the coding sites of these two subunits was derived from studies of Wildman's group using interspecific hybrids of *Nicotiana* (Chen *et al.*, 1976; Wildman *et al.*, 1975; Kung, 1976). They found that, after carboxymethylation, the large and small subunits of some species yielded different polypeptide patterns on isoelectric focusing gels. The patterns obtained from large and small subunits of F_1 hybrids of reciprocal crosses of several species indicated that the small subunit is coded in the nucleus while the large subunit is coded in the chloroplast. The gene for the small subunit has now been cloned from nuclear DNA of pea (Bedbrook *et al.*, 1979b). Recent studies have defined precisely the region of chloroplast DNA which hybridizes with mRNA for the large subunit in *Chlamydomonas* (Malnoë *et al.*, 1979), maize (Coen, Bedbrook, Bogorad & Rich, 1977; Bedbrook *et al.*, 1979a; Bogorad *et al.*, 1979) and *Euglena* (Hallick *et al.*, 1979). In both maize and *Chlamydomonas* specific cloned fragments of chloroplast DNA have been shown to synthesize a protein that can be immunoprecipitated by antibody directed against authentic large subunit and has the same peptide composition as the large subunit. In *Euglena* the major translation product of chloroplast RNA in the wheat germ system is the large subunit polypeptide (Sagher, Grosfeld & Edelman, 1976). Hallick *et al.* (1979) have shown that production of this polypeptide in the reticulocyte lysate system is completely blocked by hybrid arrest with a cloned chloroplast DNA fragment established to contain the large subunit gene by hybridization to a homologous fragment from *Chlamydomonas*.

As discussed earlier, mutations clearly affecting the structure of RuBPCase had not been isolated until Spreitzer & Mets (1980) described a chloroplast mutation of *C. reinhardtii* in which the enzyme is non-functional and contains a large subunit with altered electrophoretic mobility in isoelectric focusing gels. This mutation is lethal in the light with CO_2 as carbon source for reasons considered earlier. Several other mutations affecting the activity of RuBPCase have been described in *Chlamydomonas* (Nelson & Surzycki, 1976a, b), bacteria (Anderson, 1979), and tobacco (King & Marsho, 1976). None of these mutations has been shown clearly to alter the structure of either subunit of the RuBPCase enzyme however.

When the small subunit of RuBPCase is synthesized *in vitro* using poly (A) RNA from *Chlamydomonas* (Dobberstein, Blobel & Chua, 1977) as well as higher plants (Highfield & Ellis, 1978; Chua & Schmidt, 1979), this polypeptide is made as a 20 000 d. precursor. In higher plants, the post-translational import of this precursor by intact chloroplasts, its processing to mature small subunit and its assembly to form the holoenzyme have been demonstrated (Chua & Schmidt, 1978; Smith & Ellis, 1979). The amino terminal extension of the small subunit precursor, which presumably functions in transport of the polypeptide across the chloroplast envelope, has been sequenced in *Chlamydomonas* (Schmidt *et al.*, 1979). Since the small subunit is a nuclear gene product synthesized in the cytoplasm and the large subunit is a chloroplast gene product made in the chloroplast stroma, RuBPCase is an excellent model for studying how these two genomes are co-regulated during chloroplast biogenesis. Results of studies of the regulation of RuBPCase synthesis at the translational level in *Chlamydomonas* as well as higher plants (Iwanij, Chua & Siekevitz, 1975; Givan, 1979; Feierabend & Schrader-Reichardt, 1976; Feierabend, 1976; Feierabend & Mikus, 1977; Rademacher & Feierabend, 1976; Cashmore, 1976; Ellis, 1977; Barraclough & Ellis, 1979) suggest that different modes of regulation may occur in *Chlamydomonas* and higher plants. Now that both the large (Bogorad *et al.*, 1979; Malnoë *et al.*, 1979) and small subunit (Bedbrook *et al.*, 1979b) genes have been cloned, probes exist with which to study the regulation of both subunits at the transcriptional level. In fact, such probes have been used already to show that the large subunit gene is present in chloroplast DNA of both mesophyll and bundle sheath cells of maize, but it is transcribed only in the bundle sheath cells where the RuBPCase enzyme is known to be sequestered (Link, Coen & Bogorad, 1978). Studies of the relative rates of transcription of both subunit genes combined with further studies of the transport and processing mechanism of the small subunit may provide a clearer picture of how the synthesis of the RuBPCase protein is regulated.

(2) *Other soluble polypeptides*
Results of early experiments (Armstrong, Surzycki, Moll & Levine, 1971) involving the *in vivo* effects of protein synthesis inhibitors must be interpreted with caution because of long incubation times, high concentrations of inhibitors and/or secondary effects of the inhibitors used (reviewed by Gillham, 1978). Nevertheless, the results of Armstrong *et al.*, indicating that in *Chlamydomonas* ferredoxin and ferredoxin NADP reductase are products of cytoplasmic protein synthesis, have been borne out. Studies with *Euglena* and higher plants (Kirk & Tilney-Bassett, 1978) as well as *Chlamydomonas* (Huisman, Touw, Liebregts & Bernards, 1979) indicate that cycloheximide, but not inhibitors of chloroplast protein synthesis, blocks, the production of ferredoxin. Amino acid analysis of ferredoxin from progeny of crosses between two *Nicotiana* species, one containing methionine in this protein and the other lacking methionine, indicated that the protein is coded in the nucleus (Kwanyuen & Wildman, 1975). Ferredoxin (Bartlett, Grossman & Chua, 1979) and ferredoxin NADP reductase (S. G. Bartlett, unpublished) are translated *in vitro* from poly (A) RNA and imported into intact chloroplasts isolated from spinach. Results of inhibitor (Armstrong *et al.*, 1971) as well as genetic (Moll & Levine, 1970) studies in *Chlamydomonas* indicate that phosphoribulokinase is coded in the nucleus and synthesized on cytoplasmic ribosomes.

Since the stromal proteins discussed above make up only a small proportion of the polypeptides visualized on SDS gels from the chloroplast stroma (Grossman, Bartlett, Schmidt & Chua, 1980), the possibility exists that other soluble polypeptides will be found which are products of the chloroplast transcription and translation machinery. Thus, Ellis & Barraclough (1978) and Ellis, Highfield & Silverthorne (1977) have reported detection of 80 radioactive spots on two-dimensional gels of stroma following labelling of isolated pea chloroplasts. However, these results must be interpreted with caution since only the large subunit of RuBPCase clearly co-migrates with an authentic stroma polypeptide.

The chloroplast envelope

Most of the more than 25 polypeptides so far identified in the chloroplast envelope of higher plants have been characterized only according to their mobility in gels (Pineau & Douce, 1974; Douce & Joyard, 1979). A notable exception is the 29 000 d. polypeptide which clearly plays a role in phosphate translocation into the chloroplast (Flügge & Heldt, 1976, 1977). Two or three envelope polypeptides are thought to be synthesized inside isolated chloroplasts of spinach and pea (Joy & Ellis, 1975; Morgenthaler & Mendiola-Morgenthaler, 1976) and *in vivo* labelling patterns obtained using pea seedlings incubated with cycloheximide support this conclusion. Although a complementary labelling experiment in the presence of chloramphenicol was not performed, the remainder of the envelope polypeptides were assumed by these workers to be synthesized on cytoplasmic ribosomes. However, the possibility that envelopes obtained in these studies were contaminated with stroma and thylakoid components cannot be overlooked (Douce & Joyard, 1979). Furthermore, since protein synthesis is optimal only in chloroplasts isolated from very young seedlings and declines precipitously in chloroplasts isolated from older plants (Silverthorne & Ellis, 1980), the possibility remains open that additional envelope polypeptides are synthesized in isolated organelles in quantities too small to detect. Thus the relative contribution of the nuclear and chloroplast gene products to the polypeptide composition of the chloroplast envelope remains unclear.

Both the large and small subunits of RuBPCase are observed in gel profiles of highly purified envelopes (Pineau, Ledoight, Maillefer & Lefort-Tran, 1979). In SDS gels of high resolution the large subunit associated with the envelope sometimes appears to have slightly slower mobility than the large subunit in the stroma (Douce & Joyard, 1979). While the physiological significance of these observations is not yet known, they may relate to the assembly of the RuBPCase holoenzyme at the chloroplast envelope. The ability to reconstitute *in vitro* systems for synthesis and transport of many cytoplasmically-synthesized polypeptides into intact chloroplasts, combined with a relatively simple method for isolation of chloroplast envelopes (Douce & Joyard, 1979), will facilitate the search for possible receptor proteins in the envelope. In addition these systems may clarify the role played by the envelope in assembly of the RuBPCase holoenzyme.

Thylakoid polypeptides

Early studies of thylakoid membrane biogenesis centred on the isolation and characterization of mutants deficient in activities of photosynthetic electron transport (Levine, 1968, 1969, 1971) and on the structural rearrangements of etioplast prolamellar bodies to form the photosynthetic lamellar system during greening (Kirk & Tilney-Bassett, 1978). The *in vivo* approach using specific translational inhibitors in the re-greening system provided the first information on the sites of synthesis of specific thylakoid polypeptides (reviewed by Gillham *et al.*, 1978). More recent studies have involved the search for mutations which result in specific membrane alterations and, as methods of electrophoretic analysis have improved, sites of synthesis experiments have identified an increasing number of polypeptides of chloroplast or cytoplasmic origin. Chua & Gillham (1977) carried out *in vivo* experiments using translational inhibitors on wildtype cultures of *Chlamydomonas* grown in the light. They found that nine out of 33 major thylakoid membrane polypeptides resolved by SDS-gradient gel electrophoresis were synthesized on chloroplast ribosomes. Unfortunately, only a few of these polypeptides have been identified in terms of function. The discussion which follows will centre on the genetic control of those polypeptides whose functions have been identified or at least assigned to specific complexes associated with the thylakoid membranes.

(1) *Photosystem I (PSI)*

Both Mendelian and chloroplast mutants deficient in polypeptide 2, the chlorophyll-binding polypeptide of chlorophyll protein complex I (CPI), have been described in *Chlamydomonas* (Bennoun & Chua, 1976; Bennoun, Girard & Chua, 1977; Bennoun *et al.*, 1978; Girard, Schmidt, Chua & Bennoun, 1980; Shepherd *et al.*, 1979). Spectral analysis of all mutants examined so far reveals that they also lack the long wavelength forms of chlorophyll *a* which are associated with the reaction centre of PSI. The chloroplast mutations map at a minimum of three recombinationally distinct loci while the nuclear mutations map in 12 to 14 different genes. Chua & Gillham (1977) found that polypeptide 2 behaved as if its synthesis was sensitive to inhibitors both of cytoplasmic and of chloroplast protein synthesis. However, if the cells were incubated first in the presence of chloramphenicol to block chloroplast protein synthesis and then labelled in the presence of the cytoplasmic protein synthesis inhibitor anisomycin, incorporation of labelled polypeptide 2 into thylakoid membranes occurred. These results were interpreted to mean that polypeptide 2 was synthesized in the chloroplast, but required the synthesis of a partner protein in the cytoplasm for its integration into the thylakoid membranes. Recently, Zielinski & Price (1980) have demonstrated that the chlorophyll binding protein of CPI in spinach is synthesized in isolated plastids, which lends support to the interpretation of Chua and Gillham for their results with *Chlamydomonas*.

Since polypeptide 2 apparently is synthesized inside the chloroplast, the Mendelian mutants described above, which lack an assortment of thylakoid polypeptides, must be involved in the assembly of the PSI complex or its insertion into the thylakoid membrane. When PSI particles are isolated from wildtype cells using the method developed by Mullet, Burke & Arntzen (1980) for higher plants, the array of missing or deficient polypeptides in the several nuclear mutants analysed can be identified as part of this assembly (Girard *et al.*, 1980). In contrast, the chloroplast mutations seem more specific and appear to result in the absence of polypeptide 2 alone (Shepherd *et al.*, 1979), but these mutations have not been analysed in the same detail as the Mendelian mutants. The results of Girard *et al.* (1980) suggests that the PSI particle may be assembled entirely before insertion into the membrane.

(2) *Photosystem II (PSII)*

Chua & Bennoun (1975) isolated two Mendelian mutants in which Q, the primary electron acceptor of PSII, is either missing or inactive. SDS-gradient gel analysis revealed that polypeptide 6 was missing in the non-conditional mutant F34 and greatly reduced at the restrictive temperature in the conditional mutant T4. Polypeptide 5 was present in reduced amounts in F34 and in T4 at the restrictive temperature. When F34 was suppressed by another nuclear mutation, polypeptide 6 and PSII activity were recovered in practically a stoichiometric ratio, indicating that polypeptide 6 plays a pivotal role in PSII function. Since polypeptide 6 is synthesized on chloroplast ribosomes (Chua & Gillham, 1977), these Mendelian mutations most probably affect the function or assembly of the entire PSII complex. Bennoun *et al.* (1978) reported the isolation of a chloroplast mutant, FUD7, which had no PSII reactions centres and a membrane profile similar to that of F34. Spreitzer & Mets (1980) have also reported isolating chloroplast mutants affecting PSII, but the chloroplast mutants deficient in PSII activity reported by Shepherd *et al.* (1979) are pleiotropic because of a deficiency in chloroplast ribosomes, and the effect on PSII is a secondary consequence of a deficiency in chloroplast protein synthesis.

The mutants affecting PSII have not been characterized nearly as well as those affecting PSI. Since chlorophyll protein complex II(CPII) can be separated electrophoretically from other thylakoid polypeptides (Schmidt *et al.*, 1980), some mutants with altered PSII activity are amenable to analysis similar to that done by Girard *et al.* (1980) on their collection of PSI mutants. Such studies would be an important beginning for understanding the inheritance of the polypeptides in this particle.

In addition to CPI and CPII, Delepelaire & Chua (1979) have reported that three other rather labile chlorophyll protein complexes (CPIII, CPIV and CPV) can be isolated from *Chlamydomonas.* Delepelaire & Chua report that the apoproteins of CPIII and IV are polypeptides 5 and 6 respectively and that indirect evidence (which includes the behaviour of the F34 mutant) indicates that the two complexes are related to the PSII reaction centre. Both polypeptides are products of chloroplast protein synthesis in *Chlamydomonas* (Chua & Gillham, 1977) and apparently in spinach as well (Zielinski & Price, 1980). CPV has not been characterized.

(3) *Chloroplast coupling factor* (CF_1)

The CF_1 complex of chloroplasts consists of five polypeptides designated α, β, γ, δ and ε, present in a ratio of 2:2:2:1:1 (Nelson, 1976). These polypeptides are peripheral components which, together with integral membrane components designated subunits II and III, make up the ATPase complex in the thylakoids of higher plant chloroplasts (Nelson, 1976). In *Chlamydomonas* thylakoids an additional complex with ATPase activity, but apparently not involved in photophosphorylation, has been described (Grossman & Togasaki, in preparation; Piccioni, Bennoun & Chua, 1980). This complex accounts for about 90% of the ATPase activity of thylakoid preparations and contains polypeptides which are different from those of the CF_1 ATPase. Grossman and Togasaki suggest that the Mendelian mutant, F54, described by Sato, Levine & Neumann (1971), contains this activity rather than CF_1-ATPase activity. Clearly, these two complexes must be distinguished before assuming that photophosphorylation deficient mutants contain altered CF_1-ATPase.

Chua & Gillham (1977) showed that polypeptides 4.1 and 4.2, which are the α and β subunits of CF_1 (Bennoun *et al.*, 1978), are made on chloroplast ribosomes while the γ subunit (polypeptide 8.1; Bennoun *et al.*, 1978; Piccioni *et al.*, 1980) is made on cytoplasmic ribosomes. Piccioni *et al.* have also identified a low molecular weight polypeptide as ε and the δ subunit is presumably present in *Chlamydomonas* CF_1, but is lost during isolation of the complex.

Mutations in three recombinationally distinct chloroplast loci in *Chlamydomonas* are deficient in the α and β polypeptides, but whether other CF_1 polypeptides are also missing in these mutants has not been established (Shepherd *et al.*, 1979). Bennoun *et al.* (1978) reported the isolation of nine additional chloroplast mutants deficient in ATPase activity and analysed four of these on SDS-gradient gels. This analysis revealed that the α, β and γ subunits were all missing. So far, the allelic relationships of the mutants studied by Bennoun *et al.* to those of Shepherd *et al.* have not been established. The Mendelian mutant, F54, isolated by Sato *et al.* (1971) also appears to lack all the CF_1 subunits (Bennoun & Chua, 1976). Therefore, all CF_1-ATPase mutants isolated to date in *C. reinhardtii* are pleiotropic and result in the loss of most, if not all, of the CF_1-ATPase subunits.

Evidence that the α, β and ε subunits are synthesized inside the chloroplast has been obtained using isolated chloroplasts of pea and spinach (Ellis, 1977; Mendiola-Morgenthaler, Morgenthaler & Price, 1976). Recently, Nelson, Nelson & Schatz (1980) presented evidence that γ is also synthesized by isolated chloroplasts of spinach while δ is synthesized by poly (A) RNA as a higher molecular weight precursor. If γ is indeed polypeptide 8.1, then these results differ from those of Chua & Gillham (1977) in *Chlamydomonas*, since their results indicated that this polypeptide was synthesized in the cytoplasm. They also are inconsistent with the results of Ellis (1977) and Mendiola-Morgenthaler *et al.* (1976), who did not observe synthesis of the γ subunit in isolated chloroplasts from pea and spinach. Finally, Bouthyette & Jagendorf (1978) reported in pea that the synthesis of the α, β and ε subunits of CF_1 is inhibited by chloramphenicol while the synthesis of γ and δ is cycloheximide sensitive. The conflict between Nelson *et al.* (1980) and other investigators concerning the site of synthesis of the γ subunit remains to be resolved.

Subunit II of the membrane factor probably also is synthesized in the cytoplasm, while the DCCD-binding proteolipid (subunit III) reportedly is a product of chloroplast protein synthesis (Ellis, Smith & Barraclough, 1980; Nelson *et al.*, 1980). Nelson *et al.* further reported that the authentic ATPase complex is assembled in isolated chloroplasts in the presence of cycloheximide and suggested that a pool of δ and subunit II exists inside chloroplasts. The possibility that the appearance of label in the complex reflected subunit exchange rather than *de novo* assembly cannot be ruled out. However, the data indicate that, as in the case of RuBPCase, cytoplasmically synthesized components may regulate assembly of multimeric complexes in chloroplasts. Further, like the PSI particle, the entire CF_1 complex may be assembled before association with the thylakoids.

(4) *Other thylakoid polypeptides*

In the studies of Armstrong *et al.* (1971) increases in the amounts of cytochromes 552 (f) and 563, as measured spectrophotometrically, were inhibited by both cycloheximide and chloramphenicol in *Chlamydomonas* as well as by the chloroplast transcriptional inhibitor rifampicin. The conflicting results probably derive from the method of analysis. So far there is no evidence that the enzymes of the haeme (Borst & Grivell, 1978; Gillham, 1978) or chlorophyll (Gillham *et al.*, 1978; Wang, 1978) biosynthetic pathways are organelle gene products. These enzymes presumably are coded by nuclear genes whose messages are translated in the cytoplasm. Hence, the failure to synthesize haeme groups could result when the cells are treated with inhibitors of cytoplasmic protein synthesis and would have affected cytochrome synthesis, but the inhibition observed with chloramphenicol would be explained most easily if the apoproteins of these cytochromes were chloroplast translation products. More recent experiments with isolated higher plant chloroplasts are in accord with this hypothesis. Doherty & Gray (1979) demonstrated the synthesis of cytochrome f in intact chloroplasts isolated from pea. Likewise *in vitro* synthesis of cytochrome b_{559} and assembly with its lipid moiety has been accomplished in intact chloroplasts isolated from spinach (Zielinski & Price, 1980).

Several other polypeptides of unknown function are also chloroplast gene products. A rapidly labelled thylakoid component whose synthesis is induced by light and called peak D has been identified in isolated chloroplasts from pea (Eaglesham & Ellis, 1974). A similar protein which is under photocontrol has been identified in maize (Bedbrook *et al.*, 1978) and *Spirodela* (Edelman & Reisfeld, 1979). In both maize (Grebanier, Coen, Rich & Bogorad, 1978) and *Spirodela* this protein is known to be made as a larger precursor of molecular weight 33 000–34 500 d. Zielinski & Price (1980) have identified a similar polypeptide in isolated spinach chloroplasts and Chua & Gillham (1977) reported two rapidly turning over components in *Chlamydomonas*. D-1 has a molecular weight of about 40 000 d. and D-2 a molecular weight of about 31 000 d. Light induction of these polypeptides remains to be studied in *Chlamydomonas* and spinach.

Bogorad and his colleagues (Bogorad *et al.*, 1979) have also identified another chloroplast DNA transcript in maize. The so-called 2.2 kb gene is transcribed from the same chloroplast DNA fragment (Bam 9) as the large subunit of RuBPCase. However, only part of the 2.2 kb gene is included in this fragment. These two genes are transcribed from complementary strands in opposite directions (Link & Bogorad, 1980).

This work was supported by NIH grant GM-19427 to J.E.B. and N.W.G. and by NIH postdoctoral fellowship GM-06678 to S.G.B. We wish to thank our colleagues for allowing us to cite certain unpublished results.

REFERENCES

Adams, G. M. W., Van Winkle-Swift, K. P., Gillham, N. W. & Boynton, J. E. (1976). Plastid inheritance in *Chlamydomonas reinhardtii*. In *The Genetics of Algae*, ed. R. A. Lewin, pp. 69–118. Oxford: Blackwell Scientific.

Akoyunoglou, G. & Argyroudi-Akoyunoglou, J. H. (eds.) (1978). *Chloroplast Development*. Amsterdam: Elsevier/North-Holland Biomedical Press.

Alexander, N. J., Gillham, N. W. & Boynton, J. E. (1974). The mitochondrial genome of *Chlamydomonas*: Induction of minute colony mutations by acriflavin and their inheritance. *Molecular and General Genetics*, **130**, 275–90.

Anderson, K. (1979). Mutations altering the catalytic activity of a plant-type ribulose biphosphate carboxylase/oxygenase in *Alcaligenes eutrophus*. *Biochimica et Biophysica Acta*, **585**, 1–11.

Andrews, T. J., Lorimer, G. H. & Tolbert, N. E. (1973). Ribulose diphosphate oxygenase. I. Synthesis of phosphoglycolate by fraction I protein of leaves. *Biochemistry*, **12**, 11–18.

Armstrong, J. J., Surzycki, S. J., Moll, B. & Levine, R. P. (1971). Genetic transcription and translation specifying chloroplast components in *Chlamydomonas reinhardtii*. *Biochemistry*, **10**, 692–701.

Barraclough, R. & Ellis, R. J. (1979). The biosynthesis of ribulose bisphosphate carboxylase. *European Journal of Biochemistry*, **94**, 165–77.

Bartlett, S. G., Grossman, A. R. & Chua, N.-H. (1979). *In vitro* synthesis of ferredoxin and its transport into intact spinach chloroplasts. *Journal of Cell Biology*, **83**, 369a.

Bartlett, S. G., Harris, E. H., Grabowy, C. T., Gillham, N. W. & Boynton, J. E. (1979). Ribosomal subunits affected by antibiotic resistance mutations at seven chloroplast loci in *Chlamydomonas reinhardtii*. *Molecular and General Genetics*, **176**, 199–208.

BEDBROOK, J. R., COEN, D. M., BEATON, A. R., BOGORAD, L. & RICH, A. (1979a). Location of the single gene for the large subunit of ribulose bisphosphate carboxylase on the maize chloroplast chromosome. *Journal of Biological Chemistry*, **254**, 905–10.

BEDBROOK, J., GERLACH, W., SMITH, S., JONES, J. & FLAVELL, R. (1979b). Chromosome and gene structure in plants: a picture deduced from analysis of molecular clones of plant DNA. In *Genome Organization and Expression in Plants*, ed. C. J. Leaver, pp. 49–62. New York: Plenum Press.

BEDBROOK, J. R. & KOLODNER, R. (1979). The structure of chloroplast DNA. *Annual Review of Plant Physiology*, **30**, 593–620.

BEDBROOK, J. R., KOLODNER, R. & BOGORAD, L. (1977). *Zea mays* chloroplast ribosomal RNA genes are part of a 22 000 base pair inverted repeat. *Cell*, **11**, 739–49.

BEDBROOK, J. R., LINK, G., COEN, D. M., BOGORAD, L. & RICH, A. (1978). Maize plastid gene expressed during photoregulated development. *Proceedings of the National Academy of Sciences, USA*, **75**, 3060–4.

BEHN, W. & HERRMANN, R. (1977). Circular molecules of β satellite DNA of *Chlamydomonas reinhardtii. Molecular and General Genetics*, **157**, 25–30.

BENNOUN, P. & CHUA, N.-H. (1976). Methods for the detection and characterization of photosynthetic mutants in *Chlamydomonas reinhardi.* In *Genetics and Biogenesis of Chloroplasts and Mitochondria*, ed. T. Bücher, W. Neupert, W. Sebald & S. Werner, pp. 33–9. Amsterdam: Elsevier/North-Holland Biomedical Press.

BENNOUN, P., GIRARD, J. & CHUA, N.-H. (1977). A uniparental mutant of *Chlamydomonas reinhardtii* deficient in the chlorophyll-protein complex CP I. *Molecular and General Genetics*, **153**, 343–8.

BENNOUN, P., MASSON, A. & DELOSME, M. (1980). A method for complementation analysis of nuclear and chloroplast mutants of photosynthesis in *Chlamydomonas. Genetics*, **95**, 39–47.

BENNOUN, P., MASSON, A., PICCIONI, R. & CHUA, N.-H. (1978). Uniparental mutants of *Chlamydomonas reinhardi* defective in photosynthesis. In *Chloroplast Development*, ed. G. Akoyunoglou & J. H. Argyroudi-Akoyunoglou, pp. 721–6. Amsterdam: Elsevier/North-Holland Biomedical Press.

BIRKY, C. W. (1978). Transmission genetics of mitochondria and chloroplasts. *Annual Review of Genetics*, **12**, 471–512.

BISHOP, N. I. (1971). Photosynthesis: the electron transport system of green plants. *Annual Review of Biochemistry*, **40**, 197–226.

BISHOP, N. I. & JONES, L. (1978). Alternate states of the photochemical reducing power generated in photosynthesis: hydrogen production and nitrogen fixation. *Current Topics in Bioenergetics*, **8**, 3–31.

BLAIR, G. E. & ELLIS, R. J. (1973). Protein synthesis in chloroplasts I. Light-driven synthesis of the large subunit of fraction I protein by isolated pea chloroplasts. *Biochimica et Biophysica Acta*, **319**, 223–34.

BOGORAD, L., LINK, G., MCINTOSH, L. & JOLLY, S. O. (1979). Genes on the maize chloroplast chromosome. In *Extrachromosomal DNA*, ed. D. J. Cummings, P. Borst, I. B. Dawid, S. M. Weissman & C. F. Fox, pp. 113–26. New York: Academic Press.

BORST, P. & GRIVELL, L. A. (1978). The mitochondrial genome of yeast. *Cell*, **15**, 705–23.

BOURQUE, D. P., HORN, N. A. & CAPEL, M. S. (1977). Altered chloroplast ribosomes of a streptomycin resistant mutant of *Nicotiana tabacum. Plant Physiology*, **59**, S-110.

BOUTHYETTE, P.-Y. & JAGENDORF, A. T. (1978). The site of synthesis of pea chloroplast coupling factor 1. *Plant Cell Biochemistry and Physiology*, **19**, 1169–74.

Boynton, J. E., Gillham, N. W. & Lambowitz, A. M. (1980). Biogenesis of chloroplast and mitochondrial ribosomes. In *Ribosomes: Structure, Function & Genetics*, ed. G. Chambliss, G. R. Craven, J. Davies, K. Davis, L. Kahan & M. Nomura, pp. 903–50. Baltimore: University Park Press.

Cashmore, A. T. (1976). Protein synthesis in plant leaf tissue. *Journal of Biological Chemistry*, **251**, 2848–53.

Cavalier-Smith, T. (1970). Electron microscopic evidence for chloroplast fusion in zygotes of *Chlamydomonas reinhardi*. *Nature, London*, **228**, 333–5.

Chen, K., Johal, S. & Wildman, S. G. (1976). Role of chloroplast and nuclear DNA during evolution of fraction I protein. In *Genetics and Biogenesis of Chloroplasts and Mitochondria*, ed. T. Bücher, W. Neupert, W. Sebald & S. Werner, pp. 3–11. Amsterdam: Elsevier/North-Holland Biomedical Press.

Chen, K., Johal, S. & Wildman, S. G. (1977). Phenotypic markers for chloroplast DNA genes in higher plants and their use in biochemical genetics. In *Nucleic Acids and Protein Synthesis In Plants*, ed. J. H. Weil & L. Bogorad, pp. 183–94. New York: Plenum Press.

Chua, N.-H. & Bennoun, P. (1975). Thylakoid membrane polypeptides of *Chlamydomonas reinhardtii*. Wild-type and mutant strains deficient in photosystem II reaction center. *Proceedings of the National Academy of Sciences, USA*, **72**, 2175–9.

Chua, N.-H. & Gillham, N. W. (1977). The sites of synthesis of the principal thylakoid membrane polypeptides in *Chlamydomonas reinhardtii*. *Journal of Cell Biology*, **74**, 441–52.

Chua, N.-H. & Schmidt, G. W. (1978). Post-translational transport into intact chloroplasts of a precursor to the small subunit of ribulose-1,5-bisphosphate carboxylase. *Proceedings of the National Academy of Sciences, USA*, **75**, 6110–14.

Chua, N.-H. & Schmidt, G. W. (1979). Transport of proteins into mitochondria and chloroplasts. *Journal of Cell Biology*, **81**, 461–83.

Coen, D. M., Bedbrook, J. R., Bogorad, L. & Rich, A. (1977). Maize chloroplast DNA fragment encoding the large subunit of ribulose bisphosphate carboxylase. *Proceedings of the National Academy of Sciences, USA*, **74**, 5487–91.

Coleman, A. W. (1978). Visualization of chloroplast DNA with two fluorochromes. *Experimental Cell Research*, **114**, 95–100.

Coleman, A. W. (1979). Use of the fluorochrome 4′6-diamidino-2-phenylindole in genetic and developmental studies of chloroplast DNA. *Journal of Cell Biology*, **82**, 299–305.

Conde, M. F., Boynton, J. E., Gillham, N. W., Harris, E. H., Tingle, C. L. & Wang, W. L. (1975). Chloroplast genes in *Chlamydomonas* affecting organelle ribosomes. Genetic and biochemical analysis of antibiotic resistant mutants at several gene loci. *Molecular and General Genetics*, **140**, 183–220.

Crouse, E. J., Schmitt, J. M., Bohnert, H.-J., Gordon, K., Driesel, A. J. & Herrmann, R. G. (1978). Intramolecular compositional heterogeneity of *Spinacia* and *Euglena* chloroplast DNAs. In *Chloroplast Development*, ed. G. Akoyunoglou & J. H. Argyroudi-Akoyunoglou, pp. 565–72. Amsterdam: Elsevier/North-Holland Biomedical Press.

Davidson, J. N., Hanson, M. R. & Bogorad, L. (1974). An altered chloroplast ribosomal protein in *ery-M1* mutants of *Chlamydomonas reinhardtii*. *Molecular and General Genetics*, **132**, 119–29.

Delepelaire, P. & Chua, N.-H. (1979). Lithium dodecyl sulfate/polyacrylamide gel electrophoresis of thylakoid membranes at 4°C: Characterizations of two additional chlorophyll a-protein complexes. *Proceedings of the National Academy of Sciences, USA*, **76**, 111–15.

DOBBERSTEIN, B., BLOBEL, G. & CHUA, N.-H. (1977). In vitro synthesis and processing of a putative precursor for the small subunit of ribulose-1,5-bisphosphate carboxylase of *Chlamydomonas reinhardtii. Proceedings of the National Academy of Sciences, USA*, **74**, 1082–5.

DOHERTY, A. & GRAY, J. C. (1979). Synthesis of cytochrome f by isolated pea chloroplasts. *European Journal of Biochemistry*, **98**, 87–92.

DOUCE, R. & JOYARD, J. (1979). Structure and function of the plastid envelope. *Advances in Botanical Research*, **7**, 1–116.

DRIESEL, A. J., CROUSE, E. J., GORDON, K., BOHNERT, H.-J., HERRMANN, R. G., STEINMETZ, A., MUBUMBILA, M., KELLER, M., BURKHARD, G. & WEIL, J. H. (1979). Fractionation and identification of spinach chloroplast transfer RNAs and mapping of their genes on the restriction map of chloroplast DNA. *Gene*, **6**, 285–306.

EAGLESHAM, A. R. J. & ELLIS, R. J. (1974). Protein synthesis in chloroplasts. II. Light driven synthesis of membrane proteins by isolated pea chloroplasts. *Biochimica et Biophysica Acta*, **335**, 396–407.

EDELMAN, M. (1980). Nucleic acids of chloroplasts and mitochondria. In *The Biochemistry of Plants*, vol. 6, *Proteins and Nucleic Acids*, ed. A. Marcus. New York: Academic Press. (In press.)

EDELMAN, M. & REISFELD, A. (1979). Synthesis, processing and functional probing of P-32000, the major membrane protein translated within the chloroplast. In *Genome Organization and Expression in Plants*, ed. C. Leaver, pp. 353–62. New York: Plenum Press.

ELLIS, R. J. (1976). Protein and nucleic acid synthesis by chloroplasts. In *The Intact Chloroplast*, ed. J. Barber, pp. 335–64. Amsterdam: Elsevier/North-Holland Biomedical Press.

ELLIS, R. J. (1977). Protein synthesis by isolated chloroplasts. *Biochimica et Biophysica Acta*, **463**, 185–215.

ELLIS, R. J. & BARRACLOUGH, R. (1978). Synthesis and transport of chloroplast proteins inside and outside the cell. In *Chloroplast Development*, ed. G. Akoyunoglou & J. H. Argyroudi-Akoyunoglou, pp. 185–94. Amsterdam: Elsevier/North-Holland Biomedical Press.

ELLIS, R. J., HIGHFIELD, P. E. & SILVERTHORNE, J. (1977). The synthesis of chloroplast proteins by subcellular systems. *Proceedings of the Fourth International Congress on Photosynthesis*, ed. D. O. Hall, J. Coombs & T. W. Goodwin, pp. 497–506. London: The Biochemical Society.

ELLIS, R. J., SMITH, S. M. & BARRACLOUGH, R. (1980). Nuclear-plastid interactions in plastid protein synthesis. In *The Plant Genome, Proceedings of the Fourth John Innes Symposium*, ed. D. R. Davies & D. A. Hopwood, pp. 147–60. Norwich: The John Innes Charity.

FEIERABEND, J. (1976). Temperature-sensitivity of chloroplast ribosome formation in higher plants. In *Genetics and Biogenesis of Chloroplasts and Mitochondria*, ed. T. Bücher, W. Neupert, W. Sebald & S. Werner, pp. 99–102. Amsterdam: Elsevier/North-Holland Biomedical Press.

FEIERABEND, J. & MIKUS, M. (1977). Occurrence of a high temperature sensitivity of chloroplast ribosome formation in several higher plants. *Plant Physiology*, **59**, 863–7.

FEIERABEND, J. & SCHRADER-REICHARDT, U. (1976). Biochemical differentiation of plastids and other organelles in rye leaves with high-temperature-induced deficiency of plastid ribosomes. *Planta*, **129**, 133–45.

FLÜGGE, U.-I. & HELDT, H. W. (1976). Identification of a protein involved in phosphate transport of chloroplasts. *FEBS Letters*, **68**, 259–62.

Flügge, U.-I. & Heldt, H. W. (1977). Specific labelling of a protein involved in phosphate transport of chloroplasts by pyridoxal-5′-phosphate. *FEBS Letters*, **82**, 29–33.

Forster, J. L., Grabowy, C. T., Harris, E. H., Boynton, J. E. & Gillham, N. W. (1980). Behaviour of chloroplast genes during the early zygotic divisions of *Chlamydomonas reinhardtii. Current Genetics*, **1**, 137–53.

Frankel, R., Scowcroft, W. R. & Whitfield, P. R. (1979). Chloroplast DNA variation in isonuclear male-sterile lines of *Nicotiana. Molecular and General Genetics*, **169**, 129–35.

Freyssinet, G. (1978). Determination of the site of synthesis of some *Euglena* cytoplasmic and chloroplast ribosomal proteins. *Plant Science Letters*, **5**, 305–11.

Gillham, N. W. (1978). *Organelle Heredity*. New York: Raven Press.

Gillham, N. W., Boynton, J. E., Grant, D. M., Shepherd, H. S. & Wurtz, E. A. (1979). Genetic analysis of chloroplast DNA function in *Chlamydomonas*. In *Extrachromosomal DNA*, ed. D. J. Cummings, P. Borst, I. B. Dawid, S. M. Weissman & C. F. Fox, pp. 75–96. New York: Academic Press.

Gillham, N. W., Boynton, J. E. & Lee, R. W. (1974). Segregation and recombination of non-Mendelian genes in *Chlamydomonas. Genetics*, **78**, 439–57.

Gillham, N. W., Boynton, J. E. & Chua, N.-H. (1978). Genetic control of chloroplast proteins. *Current Topics in Bioenergetics*, **8**, 211–60.

Gillham, N. W. & Boynton, J. E. (1980). Evolution of organelle genomes and protein synthesizing systems. *Annals of the New York Academy of Sciences*, (in press).

Girard, J., Schmidt, G. W., Chua, N.-H. & Bennoun, P. (1980). Studies on mutants deficient in the photosystem I reaction centers in *Chlamydomonas reinhardtii. Current Genetics*, (in press).

Givan, A. (1979). Ribulose bisphosphate carboxylase from a mutant of *Chlamydomonas reinhardtii* deficient in chloroplast ribosomes. *Planta*, **144**, 271–6.

Givan, C. V. & Harwood, J. L. (1976). Biosynthesis of small molecules in chloroplasts of higher plants. *Biological Reviews of the Cambridge Philosophical Society*, **51**, 365–406.

Gooding, L. R., Roy, H. & Jagendorf, A. T. (1976). Identification of the small subunit of ribulose-1,5-bisphosphate carboxylase as a product of wheat leaf cytoplasmic ribosomes. *Archives of Biochemistry and Biophysics*, **172**, 64–73.

Gough, S. (1972). Defective synthesis of porphyrins in barley plastids caused by mutation in nuclear genes. *Biochimica et Biophysica Acta*, **286**, 36–54.

Grant, D. M., Gillham, N. W. & Boynton, J. E. (1980). Inheritance of chloroplast DNA in *Chlamydomonas reinhardtii. Proceedings of the National Academy of Sciences, USA*, (in press).

Gray, P. W. & Hallick, R. B. (1978). Physical mapping of the *Euglena gracilis* chloroplast DNA and ribosomal RNA gene region. *Biochemistry*, **17**, 284–9.

Gray, P. W. & Hallick, R. B. (1979). Isolation of *Euglena gracilis* chloroplast 5S ribosomal RNA and mapping the 5S gene on chloroplast DNA. *Biochemistry*, **18**, 1820–5.

Gray, J. C. & Kekwick, R. G. O. (1974). The synthesis of the small subunit of ribulose-1,5-bisphosphate carboxylase in the French bean *Phaseolus vulgaris. European Journal of Biochemistry*, **44**, 491–500.

Grebanier, A. E., Coen, D. M., Rich, A. & Bogorad, L. (1978). Membrane proteins synthesized but not processed by isolated maize chloroplasts. *Journal of Cell Biology*, **78**, 734–46.

Green, B. R., Muir, B. L. & Padmanabhan, U. (1977). The *Acetabularia* chloroplast genome: small circles and large kinetic complexity. In *Progress in Acetabularia Research*, ed. C. F. L. Woodcock, pp. 107–22. New York: Academic Press.

GROSSMAN, A. R., BARTLETT, S. G., SCHMIDT, G. W. & CHUA, N.-H. (1980). Post-translational uptake of cytoplasmically synthesized proteins by intact chloroplasts *in vitro*. *Annals of the New York Academy of Sciences*, **343,** 266–74.

GUNNING, B. E. S. & STEER, M. W. (1975). *Ultrastructure and Biology of Plant Cells*. London: Edward Arnold.

HAFF, L. A. & BOGORAD, L. (1976). Hybridization of maize chloroplast DNA with transfer ribonucleic acids. *Biochemistry*, **15**, 4105–9.

HAGEMANN, R. (1976). Plastid distribution and plastid competition in higher plants and the induction of plastom mutations by nitrosourea-compounds. In *Genetics and Biogenesis of Chloroplasts and Mitochondria*, ed. T. Bücher, W. Neupert, W. Sebald & S. Werner, pp. 331–8. Amsterdam: Elsevier/North-Holland Biomedical Press.

HAGEMANN, R. (1979). Genetics and molecular biology of plastids of higher plants. *Stadler Symposium*, **11**, 91–116.

HAGEMANN, R. & BÖRNER, T. (1978). Plastid ribosome-deficient mutants of higher plants as a tool in studying chloroplast biogenesis. In *Chloroplast Development*, ed. G. Akoyunoglou & J. H. Argyroudi-Akoyunoglou, pp. 709–20. Amsterdam: Elsevier/North-Holland Biomedical Press.

HALLICK, R. B., RUSHLOW, K. E., OROZCO, E. M., JR., STIEGLER, G. L. & GRAY, P. W. (1979). Chloroplast DNA *Euglena gracilis*: Gene mapping and selective *in vitro* transcription of the ribosomal RNA. In *Extrachromosomal DNA*, ed. D. J. Cummings, P. Borst, I. B. Dawid, S. M. Weissman & C. F. Fox, pp. 127–41. New York: Academic Press.

HARRIS, E. H., BOYNTON, J. E. & GILLHAM, N. W. (1974). Chloroplast ribosome biogenesis in *Chlamydomonas*. Selection and characterization of mutants blocked in ribosome formation. *Journal of Cell Biology*, **63**, 160–79.

HARRIS, E. H., BOYNTON, J. E., GILLHAM, N. W., TINGLE, C. L. & FOX, S. B. (1977). Mapping of chloroplast genes involved in chloroplast ribosome biogenesis in *Chlamydomonas reinhardtii*. *Molecular and General Genetics*, **155**, 249–65.

HARTLEY, M. R., WHEELER, A. & ELLIS, R. J. (1975). Protein synthesis in chloroplasts V. Translation of messenger RNA for the large subunit of Fraction I protein in a heterologous cell-free system. *Journal of Molecular Biology*, **91**, 67–77.

HELDT, H. W. (1976). Metabolite transport in intact spinach chloroplasts. In *The Intact Chloroplast*, ed. J. Barber, pp. 215–34. Amsterdam: Elsevier/North-Holland Biomedical Press.

HIGHFIELD, P. E. & ELLIS, R. J. (1978). Synthesis and transport of the small subunit of chloroplast ribulose bisphosphate carboxylase. *Nature, London,* **271** 420–4.

HUISMAN, J. G , TOUW, I., LIEBREGTS, P. & BERNARDS, A. (1979). Biosynthesis of ferredoxin in *Chlamydomonas reinhardii*. *Planta,* **145**, 351–6.

IWANIJ, V., CHUA, N.-H. & SIEKEVITZ, P. (1975). Synthesis and turnover of ribulose bisphosphate carboxylase and its subunits during the cell cycle of *Chlamydomonas reinhardtii*. *Journal of Cell Biology*, **64**, 572–85.

JENNI, B. & STUTZ, E. (1978). Physical mapping of the ribosomal DNA region of *Euglena gracilis* chloroplast DNA. *European Journal of Biochemistry*, **88**, 127–34.

JENSEN, R. G. & BAHR, J. T. (1976). Regulation of CO_2 incorporation via the pentose phosphate pathway. In *CO_2 Metabolism and Plant Productivity*, ed. R. H. Burris & C. C. Black, pp. 3–18. Baltimore: University Park Press.

JENSEN, R. G. & BAHR, J. T. (1977). Ribulose 1,5-bisphosphate carboxylase-oxygenase. *Annual Review of Plant Physiology*, **28**, 379–400.

JOY, K. W. & ELLIS, R. J. (1975). Protein synthesis in chloroplasts. IV. Polypeptides of the chloroplast envelope. *Biochimica et Biophysica Acta*, **378**, 143–51.

KIRK, J. T. O. & TILNEY-BASSETT, R. A. E. (1978). *The Plastids*. Amsterdam: Elsevier/North-Holland Biomedical Press.

KOLODNER, R. & TEWARI, K. (1979). Inverted repeats in chloroplast DNA from higher plants. *Proceedings of the National Academy of Sciences, USA*, **76**, 41–5.

KUNG, S.-D. (1976). Tobacco fraction I protein: A unique genetic marker. *Science*, **191**, 429–34.

KUNG, S.-D. & MARSHO, T. V. (1976). Regulation of RuDP carboxylase/oxygenase activity and its relationship to plant photorespiration. *Nature, London*, **259**, 325–6.

KWANYUEN, P. & WILDMAN, S. G. (1975). Nuclear DNA codes for the primary structure of plant ferridoxin. *Biochimica et Biophysica Acta*, **405**, 167–74.

LEECH, R. M. & MURPHY, D. J. (1976). The cooperative function of chloroplasts in the biosynthesis of small molecules. In *The Intact Chloroplast*, ed. J. Barber, pp. 365–401. Amsterdam: Elsevier/North-Holland Biomedical Press.

LEVINE, R. P. (1968). Genetic dissection of photosynthesis. *Science*, **162**, 768–71.

LEVINE, R. P. (1969). The analysis of photosynthesis using mutant strains of algae and higher plants. *Annual Review of Plant Physiology*, **20**, 523–40.

LEVINE, R. P. (1971). Preparation and properties of mutant strains of *Chlamydomonas reinhardi*. In *Methods in Enzymology*, vol. 23, ed. A. San Pietro, pp. 119–29. New York: Academic Press.

LINK, G. & BOGORAD, L. (1980). Sizes, locations, and directions of transcription of two genes on a cloned maize chloroplast DNA sequence. *Proceedings of the National Academy of Sciences, USA*, **77**, 1832–6.

LINK, G., COEN, D. M. & BOGORAD, L. (1978). Differential expression of the gene for the large subunit of ribulose bisphosphate carboxylase. *Cell*, **15**, 725–31.

LORIMER, G. H., ANDREWS, T. J. & TOLBERT, N. E. (1973). Ribulose diphosphate oxygenase. II. Further proof of reaction products and mechanism of action. *Biochemistry*, **12**, 18–23.

MCCREA, J. M. & HERSHBERGER, C. L. (1976). Chloroplast DNA codes for transfer RNA. *Nucleic Acids Research*, **3**, 2005–18.

MALIGA, P., SZ-BREZNOVITS, A. & MARTON, L. (1973). Streptomycin-resistant plants from callus culture of haploid tobacco. *Nature, New Biology, London*, **244**, 29–30.

MALIGA, P., SZ-BREZNOVITS, A., JOO, F. & MARTON, L. (1975). Non-Mendelian streptomycin-resistant tobacco mutant with altered chloroplasts and mitochondria. *Nature, London*, **255**, 401–2.

MALNOË, P. & ROCHAIX, J.-D. (1978). Localization of 4S RNA genes on the chloroplast genome of *Chlamydomonas reinhardii*. *Molecular and General Genetics*, **166**, 269–75.

MALNOË, P., ROCHAIX, J.-D., CHUA, N.-H. & SPAHR, P. F. (1979). Characterization of the gene and messenger RNA of the large subunit of ribulose 1,5-diphosphate carboxylase in *Chlamydomonas reinhardii*. *Journal of Molecular Biology*, **133**, 417–34.

MENDIOLA-MORGENTHALER, L. R., MORGENTHALER, J. J. & PRICE, C. A. (1976). Synthesis of coupling factor CF_1 protein by isolated spinach chloroplasts. *FEBS Letters*, **62**, 96–100.

MOLL, B. & LEVINE, R. P. (1970). Characterization of a photosynthetic mutant strain of *Chlamydomonas reinhardi* deficient in phosphoribulokinase activity. *Plant Physiology*, **46**, 576–80.

MORGENTHALER, J. J. & MENDIOLA-MORGENTHALER, L. (1976). Synthesis of soluble, thylakoid and envelope membrane proteins by spinach chloroplasts purified from gradients. *Archives of Biochemistry and Biophysics*, **172**, 51–8.

MULLET, J. E., BURKE, J. J. & ARNTZEN, C. J. (1980). Chlorophyll proteins of photosystem I. *Plant Physiology*, **65**, 814–22.

NASYROV, Y. S. & ŠESTÁK, Z. (EDS.) (1975). *Genetic Aspects of Photosynthesis*. The Hague: Junk.

NELSON, N. (1976). Structure and function of chloroplast ATPase. *Biochimica et Biophysica Acta*, **456**, 314–38.

NELSON, N., NELSON, H. & SCHATZ, G. (1980). Biosynthesis and assembly of the proton-translocating adenosine triphosphatase complex from chloroplasts. *Proceedings of the National Academy of Sciences, USA*, **77**, 1361–4.

NELSON, P. E. & SURZYCKI, S. J. (1976a). A mutant strain of *Chlamydomonas reinhardi* exhibiting altered ribulose bisphosphate carboxylase. *European Journal of Biochemistry*, **61**, 465–74.

NELSON, P. E. & SURZYCKI, S. J. (1976b). Characterization of the oxygenase activity in a mutant of *Chlamydomonas reinhardi* exhibiting an altered ribulose bisphosphate carboxylase. *European Journal of Biochemistry*, **61**, 475–80.

PADMANABHAN, U. & GREEN, B. R. (1978). The kinetic complexity of *Acetabularia* chloroplast DNA. *Biochimica et Biophysica Acta*, **521**, 67–73.

PICCIONI, R. G., BENNOUN, P. & CHUA, N.-H. (1980). A nuclear mutant of *Chlamydomonas reinhardtii* defective in photosynthetic phosphorylation. Characterization of the algal coupling factor ATPase. *European Journal of Biochemistry*, (submitted).

PINEAU, B. & DOUCE, R. (1974). Analyse électrophoretique des protéines de l'enveloppe des chloroplastes d'épinard. *FEBS Letters*, **47**, 255–9.

PINEAU, B., LEDOIGHT, G., MAILLEFER, C. & LEFORT-TRAN, M. (1979). Presence de sous-unités de la RubPcase dans les enveloppes des chloroplastes d'épinard. *Plant Science Letters*, **15**, 331–43.

POWLES, S. B. & OSMOND, C. B. (1979). Photoinhibition of intact attached leaves of C_3 plants illuminated in the absence of both carbon dioxide and of photorespiration. *Plant Physiology*, **64**, 982–8.

PRICE, C. A., ORTIZ, W. & GAYNOR, J. J. (1978). Regulation of protein synthesis in isolated chloroplasts of *Euglena gracilis*. In *Chloroplast Development*, ed. G. Akoyunoglou & J. H. Argyroudi-Akoyunoglou, pp. 257–66. Amsterdam: Elsevier/North-Holland Biomedical Press.

PRING, D. R. & LEVINGS, C. S. (1978). Heterogeneity of maize cytoplasmic genome among male-sterile cytoplasms. *Genetics*, **89**, 121–36.

RADEMACHER, E. & FEIERABEND, J. (1976). Formation of chloroplast pigments and sterols in rye leaves deficient in plastid ribosomes. *Planta*, **129**, 147–53.

RAWSON, J. R. Y., KUSHNER, S. R., VAPNEK, D., ALTON, N. K. & BOERMA, C. L. (1978). Chloroplast ribosomal RNA genes in *Euglena gracilis* exist as three clustered tandem repeats. *Gene*, **3**, 191–209.

ROCHAIX, J.-D. (1978). Restriction endonuclease map of the chloroplast DNA of *Chlamydomonas reinhardii*. *Journal of Molecular Biology*, **126**, 597–617.

ROCHAIX, J.-D. & MALNOË, P. (1978). Anatomy of the chloroplast ribosomal DNA of *Chlamydomonas reinhardii*. *Cell*, **15**, 661–70.

ROY, H., PATTERSON, R. & JAGENDORF, A. T. (1976). Identification of the small subunit of ribulose-1,5-bisphosphate carboxylase as a product of wheat leaf cytoplasmic ribosomes. *Archives of Biochemistry and Biophysics*, **172**, 64–73.

SAGER, R. (1972). *Cytoplasmic Genes and Organelles*. New York: Academic Press.

SAGER, R. (1977). Genetic analysis of chloroplast DNA of *Chlamydomonas*. *Advances in Genetics*, **19**, 287–340.

SAGER, R. & GRANICK, S. (1954). Nutritional control of sexuality in *Chlamydomonas reinhardi. Journal of General Physiology*, **37**, 729–42.

SAGER, R. & RAMANIS, Z. (1967). Biparental inheritance of non-chromosomal genes induced by ultraviolet irradiation. *Proceedings of the National Academy of Sciences, USA*, **58**, 931–7.

SAGER, R. & RAMANIS, Z. (1976a). Chloroplast genetics of *Chlamydomonas*. I. Allelic segregation ratios. *Genetics*, **83**, 303–21.

SAGER, R. & RAMANIS, Z. (1976b). Chloroplast genetics of *Chlamydomonas*. II. Mapping by cosegregation frequency analysis. *Genetics*, **83**, 323–40.

SAGHER, D., GROSFELD, H. & EDELMAN, M. (1976). Large subunit ribulose-bisphosphate carboxylase messenger RNA from *Euglena* chloroplasts. *Proceedings of the National Academy of Sciences, USA*, **73**, 722–6.

SATO, V. L., LEVINE, R. P. & NEUMANN, J. (1971). Photosynthetic phosphorylation in *Chlamydomonas reinhardi*. Effects of a mutation altering the ATP synthesizing enzyme. *Biochimica et Biophysica Acta*, **253**, 437–48.

SCHMIDT, G. W., BARTLETT, S., GROSSMAN, A. R., CASHMORE, A. R. & CHUA, N.-H. (1980). *In vitro* synthesis, transport, and assembly of the constituent polypeptides of the light-harvesting chlorophyll a/b protein complex. In *Genome Organization and Expression in Plants*, ed. C. J. Leaver, pp. 337–51. New York: Plenum Press.

SCHMIDT, G. W., DEVILLERS-THIERY, A., DESRUISSEAUX, H., BLOBEL, G. & CHUA, N.-H. (1979). NH_2-terminal amino acid sequences of precursor and mature forms of the ribulose-1,5-bisphosphate carboxylase small subunit from *Chlamydomonas reinhardtii. Journal of Biological Chemistry*, **83**, 615–22.

SCHMIDT, G. W., MATLIN, K. S. & CHUA, N.-H. (1977). A rapid procedure for selective enrichment of photosynthetic electron transport mutants. *Proceedings of the National Academy of Sciences, USA*, **74**, 610–14.

SCHWARTZBACH, S. D., HECKER, L. I. & BARNETT, W. E. (1976). Transcriptional origin of *Euglena* chloroplast tRNAs. *Proceedings of the National Academy of Sciences, USA*, **73**, 1984–8.

SEARS, B. B. (1980). Elimination of plastids during spermatogenesis and fertilization in the plant kingdom. *Plasmid*, (in press).

SEARS, B. B., BOYNTON, J. E. & GILLHAM, N. W. (1980). The effect of gametogenesis regimes on the chloroplast genetic system of *Chlamydomonas reinhardtii. Genetics*, (in press).

SHEPHERD, H. S., BOYNTON, J. E. & GILLHAM, N. W. (1979). Mutations in nine chloroplast loci of *Chlamydomonas* affecting different photosynthetic functions. *Proceedings of the National Academy of Sciences, USA*, **76**, 1353–7.

SILVERTHORNE, J. & ELLIS, R. J. (1980). Protein synthesis in chloroplasts. VIII. Differential synthesis of chloroplast proteins during spinach leaf development. *Biochimica et Biophysica Acta*, **607**, 319–30.

SINGER, B., SAGER, R. & RAMANIS, Z. (1976). Chloroplast genetics of *Chlamydomonas*. III. Closing the circle. *Genetics*, **83**, 341–54.

SMITH, S. M. & ELLIS, R. J. (1979). Processing of small subunit precursor of ribulose bisphosphate carboxylase and its assembly into whole enzyme are stromal events. *Nature, London*, **278**, 662–4.

SPREITZER, R. J. & METS, L. J. (1980). Non-Mendelian mutation affecting ribulose-1,5-bisphosphate carboxylase structure and activity. *Nature, London*, **285**, 114–15.

STUTZ, E. & BOSCHETTI, A. (1976). Chloroplast ribosomes. In *Handbook of Genetics*, vol. 5, ed. R. C. King, pp. 425–50. New York: Plenum Press.

TIBONI, O., DIPASQUALE, G. & CIFERRI, O. (1978). Purification, characterization and site of synthesis of chloroplast elongation factors. In *Chloroplast Develop-*

ment, ed. G. Akoyuonoglou & J. H. Argyroudi-Akoyuonoglou, pp. 675–8. Amsterdam: Elsevier/North-Holland Biomedical Press.

TILNEY-BASSETT, R. A. E. (1975). Genetics of variegated plants. In *Genetics and Biogenesis of Mitochondria and Chloroplasts*, ed. C. W. Birky, P. S. Perlman & T. W. Byers, pp. 268–308. Columbus: Ohio State University Press.

TILNEY-BASSETT, R. A. E. & ABDEL-WAHAB, O. A. L. (1979). Maternal effects and plastid inheritance. In *Maternal Effects in Development*, ed. D. R. Newth & M. Balls, pp. 29–45, *British Society for Developmental Biology, Symposium 4.* Cambridge University Press.

TOGASAKI, R. K. & HUDOCK, M. O. (1972). Effects of inorganic arsenate on the growth of *Chlamydomonas reinhardi. Plant Physiology*, **49**, 525-S.

TOLBERT, N. E. & RYAN, F. J. (1976). Glycolate biosynthesis and metabolism during photorespiration. In *CO_2 Metabolism and Plant Productivity*, ed. R. H. Burris & C. C. Black, pp. 141–59. Baltimore: University Park Press.

VANWINKLE-SWIFT, K. P. (1980). A model for the rapid vegetative segregation of multiple chloroplast genomes in *Chlamydomonas*: assumptions and predictions of the model. *Current Genetics*, **1,** 113–25.

WANG, W.-Y. (1978). Genetic control of chlorophyll biosynthesis. *International Review of Cytology, Supplement 8*, 335–54.

VON WETTSTEIN, D., HENNINGSEN, K. W., BOYNTON, J. E., KANNANGARA, G. C. & NIELSEN, O. F. (1971). The genetic control of chloroplast development in barley. In *Autonomy and Biogenesis of Mitochondria and Chloroplasts*, ed. N. K. Boardman, A. W. Linnane & R. M. Smillie, pp. 205–23. Amsterdam: North-Holland.

WHITFELD, P. R., HERRMANN, R. G. & BOTTOMLEY, W. (1978). Mapping of the ribosomal RNA genes on spinach chloroplast DNA. *Nucleic Acids Research*, **5**, 1741–51.

WILDMAN, S. G., CHEN, K., GRAY, J. C., KUNG, S.-D., KWANYUEN, P. & SAKANO, K. (1975). Evolution of ferridoxin and fraction I protein in *Nicotiana*. In *Genetics and Biogenesis of Chloroplasts and Mitochondria*, ed. C. W. Birky, P. S. Perlman & T. J. Byers, pp. 309–29. Columbus: Ohio State University Press.

WISEMAN, A., GILLHAM, N. W. & BOYNTON, J. E. (1977). The mitochondrial genome of *Chlamydomonas.* II. Genetic analysis of non-Mendelian obligate photoautotrophic mutants. *Molecular and General Genetics*, **150**, 109–18.

WURTZ, E. A., BOYNTON, J. E. & GILLHAM, N. W. (1977). Perturbation of chloroplast DNA amounts and chloroplast gene transmission in *Chlamydomonas reinhardtii* by 5-fluorodeoxyuridine. *Proceedings of the National Academy of Sciences, USA*, **74**, 4552–6.

WURTZ, E. A., SEARS, B. B., RABERT, D. K., SHEPHERD, H. S., GILLHAM, N. W. & BOYNTON, J. E. (1979). A specific increase in chloroplast gene mutations following growth of *Chlamydomonas* in 5-fluorodeoxyuridine. *Molecular and General Genetics*, **170**, 235–42.

ZIELINSKI, R. E. & PRICE, C. A. (1980). Synthesis of thylakoid membrane proteins by chloroplasts isolated from spinach: cytochrome b_{559} and P700-chlorophyll *a* protein. *Journal of Cell Biology*, **85**, 435–45.

INDEX